M. K. Rai, PhD

Handbook of Microbial Biofertilizers

Pre-publication
REVIEWS,
COMMENTARIES,
EVALUATIONS . . .

"This handbook presents in-depth and innovative research on microbial biofertilizers as an alternative to chemical fertilizers and pesticides. It is a self-contained book on almost all microbial biofertilizer agents ranging from bacterial and cyanobacterial to fungal species. *Handbook of Microbial Biofertilizers* is indeed a forward-looking publication which encompasses almost all emerging trends in production and usage of biological fertilizers, their taxonomy, environmental improvement by them, inoculum production, and IPNM. The book also provides access to numerous case studies with specific regional examples where this technology has been successfully applied. It is indispensable for biotechnologists, microbiologists, botanists, farmers, industrialists dealing with production of microbial fertilizers, environmentalists, and others, including students who are interested in the subject."

Manju Bansal, PhD
Reader, Department of Botany,
Sri Aurobindo College,
University of Delhi, India

"With twenty chapters covering topics as diverse as mycorrhiza helper bacteria, cyanobacteria, and mycorrhizal fungi, this book provides a comprehensive account of the current use and potential of bacteria and fungi as biofertilizers. Some chapters deal with the use of nitrogen-fixing organisms and mycorrhizal fungi in sustainable agriculture, while others provide a more practical guide to using biofertilizers. There is even a chapter on the physiological and molecular aspects of osmotic stress alleviation by arbuscular mycorrhizal fungi. This book will be of interest to those researching biofertilizers as well as to those wishing to use biofertilizers in practical agriculture and horticulture. The chapters dealing with biofertilizers' use in specific cropping systems such as rice production will be of particular value to those with more practical interests."

Professor Dale Walters, PhD, DSc
Crop & Soil Research Group,
Scottish Agricultural College,
Edinburgh, United Kingdom

NOTES FOR PROFESSIONAL LIBRARIANS AND LIBRARY USERS

This is an original book title published by Food Products Press®, an imprint of The Haworth Press, Inc. Unless otherwise noted in specific chapters with attribution, materials in this book have not been previously published elsewhere in any format or language.

CONSERVATION AND PRESERVATION NOTES

All books published by The Haworth Press, Inc., and its imprints are printed on certified pH neutral, acid-free book grade paper. This paper meets the minimum requirements of American National Standard for Information Sciences-Permanence of Paper for Printed Material, ANSI Z39.48-1984.

Handbook of Microbial Biofertilizers

FOOD PRODUCTS PRESS®
Crop Science
Amarjit S. Basra, PhD
Editor in Chief

Plant-Derived Antimycotics: Current Trends and Future Prospects edited by Mahendra Rai and Donatella Mares

Concise Encyclopedia of Temperate Tree Fruit edited by Tara Auxt Baugher and Suman Singha

Landscape Agroecology by Paul A. Wojtkowski

Concise Encyclopedia of Plant Pathology by P. Vidhyasekaran

Molecular Genetics and Breeding of Forest Trees edited by Sandeep Kumar and Matthias Fladung

Testing of Genetically Modified Organisms in Foods edited by Farid E. Ahmed

Fungal Disease Resistance in Plants: Biochemistry, Molecular Biology, and Genetic Engineering edited by Zamir K. Punja

Plant Functional Genomics edited by Dario Leister

Immunology in Plant Health and Its Impact on Food Safety by P. Narayanasamy

Abiotic Stresses: Plant Resistance Through Breeding and Molecular Approaches edited by M. Ashraf and P. J. C. Harris

Teaching in the Sciences: Learner-Centered Approaches edited by Catherine McLoughlin and Acram Taji

Handbook of Industrial Crops edited by V. L. Chopra and K. V. Peter

Durum Wheat Breeding: Current Approaches and Future Strategies edited by Conxita Royo, Miloudi M. Nachit, Natale Di Fonzo, José Luis Araus, Wolfgang H. Pfeiffer, and Gustavo A. Slafer

Handbook of Statistics for Teaching and Research in Plant and Crop Science by Usha Rani Palaniswamy and Kodiveri Muniyappa Palaniswamy

Handbook of Microbial Fertilizers edited by M. K. Rai

Eating and Healing: Traditional Food As Medicine edited by Andrea Pieroni and Lisa Leimar Price

Handbook of Plant Virology edited by Jawaid A. Khan and Jeanne Dijkstra

Physiology of Crop Production by N. K. Fageria, V. C. Baligar, and R. B. Clark

Plant Conservation Genetics edited by Robert J. Henry

Introduction to Fruit Crops by Mark Rieger

Sourcebook for Intergenerational Therapeutic Horticulture: Bringing Elders and Children Together by Jean M. Larson and Mary Hockenberry Meyer

Agriculture Sustainability: Principles, Processes, and Prospects by Saroja Raman

Introduction to Agroecology: Principles and Practice by Paul A. Wojtkowski

Handbook of Molecular Technologies in Crop Disease Management by P. Vidhyasekaran

Handbook of Precision Agriculture: Principles and Applications edited by Ancha Srinivasan

Dictionary of Plant Tissue Culture by Alan C. Cassells and Peter B. Gahan

Handbook of Potato Production, Improvement, and Postharvest Management edited by Jai Gopal and S. M. Paul Khurana

Handbook of Microbial Biofertilizers

M. K. Rai, PhD
Editor

Food Products Press®
An Imprint of The Haworth Press, Inc.
New York • London • Oxford

For more information on this book or to order, visit
http://www.haworthpress.com/store/product.asp?sku=5428

or call 1-800-HAWORTH (800-429-6784) in the United States and Canada
or (607) 722-5857 outside the United States and Canada

or contact orders@HaworthPress.com

Published by

Food Products Press®, an imprint of The Haworth Press, Inc., 10 Alice Street, Binghamton, NY 13904-1580.

Cover photo taken from *Prospects and Limitations* by Sarika Shende et al.

Cover design by Lora Wiggins.

Library of Congress Cataloging-in-Publication Data

Handbook of microbial biofertilizers / M. K. Rai, editor.
p. cm.
Includes bibliographical references and index.
ISBN 13: 978-1-56022-269-9 (hc. : alk. paper)
ISBN 10: 1-56022-269-7 (hc. : alk. paper)
ISBN 13: 978-1-56022-270-5 (pbk. : alk. paper)
ISBN 10: 1-56022-270-0 (pbk. : alk. paper)
1. Biofertilizers. I. Rai, M. K.

S654.5.H36 2005
631.8'6—dc22

2005004437

CONTENTS

About the Editor xv

Contributors xvii

Foreword xxi
Ian Robert Hall

Preface xxiii

Acknowledgments xxv

Chapter 1. Potential and Possible Uses of Bacterial and Fungal Biofertilizers 1
Francesco Gentili
Ari Jumpponen

Introduction 1
N_2-Fixing Bacteria As Microbial Biofertilizers 3
Symbiotic N_2-Fixing Bacteria 3
Cyanobacteria 6
N_2-Fixing Associated Bacteria 7
Free-Living N_2-Fixing Bacteria 8
Plant-Growth-Promoting Rhizobacteria 8
Phosphorus-Solubilizing Bacteria 9
Plant Hormone Production by Bacteria 9
Fungi and Their Potential As Biofertilizers 10
Use of Arbuscular Mycorrhizal Fungi As Biofertilizers 11
Use of Ectomycorrhizal Fungi As Biofertilizers 13
Other Root-Colonizing Fungi 14
Multiple Inoculations and Interactions Among Potential Biofertilizers 15
Future Perspectives and Conclusions 18

Chapter 2. Evaluation of the Functional Group of Microorganisms As Bioindicators on the Rhizosphere Microcosm **29**
Ulisses Brigatto Albino
Galdino Andrade

Introduction 29
The Functional Groups 31
Methods to Assess the Soil Functional Microorganism Groups 34
Protocols 35
Future Perspectives 46
Conclusions 46

Chapter 3. Tripartite Relationship of *Rhizobium,* AMF, and Host in Growth Promotion **51**
Nanjian Raman
Thangaswamy Selvaraj

Introduction 51
Rhizobium and AMF Interactions 52
Rhizobium, AMF, and Biochemical Activities 58
Enzyme Activities 60
Growth Hormones 62
Calcium and Phosphorus Requirements 63
Heavy Metals 64
Plant Uptake of Metals 68
Plant-Growth-Promoting Rhizobacteria 69
Coexistence of *Rhizobium* and AM Symbioses 69
Conclusions and Future Perspectives 71

Chapter 4. Biological Fertilizers for Sustainable Rice Production **89**
S. Kannaiyan
K. Kumar

Introduction 89
Azolla Biofertilizer for Rice Crop 90
Nitrogen-Fixing Blue Green Algae 92
Acid-Tolerant Cyanobacteria 94
N_2-Fixing Stem-Nodulating *Sesbania* Biofertilizer for Rice 96
Sesbania As Biofertilizer for Rice 100

Neem Cake and N Use Efficiency 100
Urea Super Granules and N Use Efficiency 103
Soil Enzyme Activity 104
Induction of Nodule-Like Structures in Rice Roots for N_2 Fixation 104
Azospirillum Biofertilizer for Rice 107
Phosphobacterial Biofertilizer 107
Future Perspectives 108
Conclusions 109

Chapter 5. Mycorrhiza Helper Bacteria: Their Ecological Impact in Mycorrhizal Symbiosis 117
Robin Duponnois

Introduction 117
The Mycorrhizosphere Effect 118
Evidence 123
Biological Properties 125
Mechanisms Involved in the MHB Effect 126
Application 128
Conclusions and Perspectives 129

Chapter 6. Plant-Growth-Promoting Rhizobacteria As Biofertilizers and Biopesticides 137
Manas Ranjan Banerjee
Laila Yesmin
Joseph Kevin Vessey

Introduction 137
Biofertilizers 139
Endophytic and Rhizospheric Relationships Between PGPR and Their Hosts 141
Mode of Action of PGPR As Biofertilizers 143
Biopesticides 154
Mode of Action of PGPR As Biopesticides 157
Viability and Shelf Life of PGPR 164
Benefits of PGPR 165
Limitations of PGPR 167
Future Perspectives 168
Conclusions 170

Chapter 7. Sustainable Agriculture and the Rhizobia/Legumes Symbiosis **183**
João Ruy Jardim Freire
Enilson Luiz Saccol de Sá

Introduction 183
Diversity of Rhizobia 184
The Selection of Rhizobia Strains 188
Desirable Characteristics 189
Production and Quality of the Inoculants 190
Techniques of Inoculation 190
Soil Limiting Factors 191
Some Management Practices 193
Future Perspectives 195
Conclusions 196

Chapter 8. Wild-Legume Rhizobia: Biodiversity and Potential As Biofertilizer **203**
Hamdi Hussein Zahran

Introduction 203
Taxonomy and Biodiversity of Wild-Legume Rhizobia 204
Symbiotic Nitrogen-Fixing Systems Established with Wild Rhizobia 206
Behavior of Wild-Legume Rhizobia Under Extreme Conditions 209
Future Perspectives 212
Conclusions 214

Chapter 9. Potential of Arbuscular Mycorrhizae in Organic Farming Systems **223**
Tanu
Anil Prakash
Alok Adholeya

Introduction 223
Mycorrhizae in Organic Farming 224
Factors Affecting Mycorrhizal Inoculum Potential 226
Selection of AM Fungi 230
Integration of AM Fungi and Composts for Disease Control 230
Organic Manure and AMF Inoculum Production 231
Application of AM Fungi 231

Future Perspectives 232
Conclusions 233

Chapter 10. Role of Mycorrhizae in Forestry **241**
Hanna Dahm

Introduction 241
Ecological Significance of Mycorrhizae 241
Types of Mycorrhizae on Forest Trees 243
Occurrence of Mycorrrhizal Fungi in Soils of Native Forests 248
Soil Type and Humus Properties 249
Ecological Specificity 250
Competition Among Ectomycorrhizal Fungi in Forest Soil 250
Favorable Effects of Mycorrhizae on Plant Growth 252
Factors Affecting Ectomycorrhizal Reduction 258
Receptiveness of Forest Soils to Ectomycorrhizal Association 261
Practical Aspects of Mycorrhizae: Perspectives 262
Conclusions 265

Chapter 11. Physiological and Molecular Aspects of Osmotic Stress Alleviation in Arbuscular Mycorrhizal Plants **271**
Juan Manuel Ruiz-Lozano

Introduction 271
Uptake and Transfer of Water Through the Fungal Hyphae to the Host Plant 274
Amelioration of Plant Gas Exchange 276
Changes in the Water Retention Properties of the Soil 277
Stimulation of Assimilative Activities Essential for Plant Growth 278
Osmotic Adjustment 278
Protection Against the Oxidative Damage Generated by Drought 280
Future Perspectives and Conclusions 283

Chapter 12. Arbuscular Mycorrhizal Inoculation in Nursery Practice **289**
Miguel Marin

Introduction 289

Types of Mycorrhizae 290
Mycorrhizae Effects on Plant Physiology 292
Interaction of Arbuscular Mycorrhizal Fungi with Soil Microbiota 295
Mycorrhizae As Substitute for Fertilizers 296
Mycorrhizal Symbiosis in Horticultural Systems 298
Endomycorrhizal Technology in Different Propagation Systems 300
Management of Arbuscular Mycorrhiza in Nursery Plant Production 302
Future Perspectives 309
Conclusions 310

Chapter 13. Interaction Between Arbuscular Mycorrhizal Fungi and Root Pathogens 325
Leonor Costa Maia
Norma Suely Sobral da Silveira
Uided Maaze Tiburcio Cavalcante

Introduction 325
Root Colonization 326
Changes Produced by Mycorrhization 326
Interaction: Arbuscular Mycorrhizal Fungi and Plant Pathogenic Fungi 329
Arbuscular Mycorrhizal Fungi × Bacteria 334
Arbuscular Mycorrhizal Fungi × Nematodes 336
Future Perspectives 340
Conclusions 341

Chapter 14. Production of Seedlings Inoculated with Arbuscular Mycorrhizal Fungi and Their Performance After Outplanting 353
Orivaldo José Saggin-Júnior
Eliane Maria Ribeiro da Silva

Introduction 353
Prerequisites for the Inoculation of Seedlings with AMF 355
Production of Inoculants for Seedlings 362
Production of Seedlings Inoculated with AMF 368
Future Perspectives of Mycorrhizal Seedling Production 382
Conclusions 383

Chapter 15. Status of Endomycorrhizal (AMF) Biofertilizer in the Global Market **395**
Polachi Raja

Introduction 395
World Fertilizer Consumption 396
Alternative Farming Systems 397
Sustainable Agriculture Systems 398
Effective Microorganisms 400
Evolution of AMF 401
Techniques of AMF Inoculum Propagation 404
AMF Benefits of Symbiosis 406
Impact of AMF in Agriculture 406
Availability and Cost of Mycorrhizal Inoculum 407
Producers and Formulators of AMF Inoculum 407
Future Perspectives 407
Conclusions 412

Chapter 16. Role of Cyanobacteria As Biofertilizers: Potentials and Limitations **417**
Pilar Irisarri

Introduction 417
Current Status of Cyanobacterial Use As Biofertilizers 418
Availability of Fixed Nitrogen to Rice Plants 419
Cultivation of Cyanobacteria for Biofertilizer 420
Inoculation and Crop Yield 421
Reasons for the Failure of Inoculation 422
Benefits of Cyanobacteria Beyond Nitrogen Input to the Agroecosystem 424
Is Inoculation with Cyanobacteria Worthwhile? 425
Integrated Soil Fertility Management 426
Future Perspectives 427
Conclusions 428

Chapter 17. Cyanobacterial Biofertilizers for Rice: Present Status and Future Prospects **433**
Siba Prasad Adhikary
Bagmi Pattanaik

Introduction 433
Cyanobacteria (Blue-Green Algae) 434

Factors Governing Distribution of Cyanobacteria During the Crop Growth Cycle 440
Seasonal Variation of Cyanobacterial Growth in Paddy Fields 441
Phosphorus 443
Cyanobacteria Biofertilizer for Increased Crop Yield and Maintenance of Soil Quality 445
Algalization Technology 446
Future Perspectives 449
Conclusions 450

Chapter 18. A Comparative Study on Nitrogen-Fixing Cyanobacteria in South American and European Rice Fields 459
Carlos Hugo Prosperi

Introduction 459
Measuring Nitrogen Fixation in the Field 462
Determination of Nitrogenase Activity in Vivo 463
Samples Collection and Identification 463
The Diversity of Cyanobacteria 464
Future Perspectives 471
Appendix: Identification Key for Most Common Cyanobacteria in Rice Fields of Argentina and Spain 472

Chapter 19. *Piriformospora indica* As a New and Emerging Mycofertilizer and Biotizer: Potentials and Prospects in Sustainable Agriculture 477
Sarika Shende *Mahendra Rai*
Kanchan Bhagwat *Ajit Varma*
Prasad Wadegaonkar

Introduction 477
The Morphology and Cultural Characteristics 478
Mycelial Penetration in Host 478
Ultrastructure of Hyphal Wall 478
Molecular Systemetics 479
Root Colonization Potential in Diverse Hosts 480
In Vitro Rapid Proliferation of Roots of *Withania somnifera* 480
Rhizoctonia-Like Growth Promotion of Orchids 480
Enhancement of Secondary Metabolites 481

Mechanisms Involved During Colonization 482
Mycorrhiza-Like Endophytes 483
Protoplast Fusion for Better Understanding of Interaction 483
Strategies for Mass Production of Inocula 485
Immobilization of Spores and Hyphae: A Step Forward for Inoculum Production 485
Application in Horticulture 486
Potential for Control of Phytopathogenic Rootborne Microbes 488
Conclusions and Future Directions 490

Chapter 20. Matsutake: A Natural Biofertilizer? 497
Wang Yun
Ian Robert Hall

Introduction 497
Matsutake in the Commercial World 499
Taxonomic History of Matsutake 502
Ecology 502
Morphology of Matsutake Infections 508
Cultivation of *Tricholoma matsutake* 519
Relationship Between Matsutake and Its Host Plants 526
Future Challenges 531
Conclusions 532

Index 543

ABOUT THE EDITOR

M. K. Rai, PhD, is Professor and Head of the Department of Biotechnology at Amravati University in Maharashtra, India. Dr. Rai has published 140 research papers, more than 100 articles in Indian and foreign journals, and six books. He has been teaching and conducting research for two decades. Dr. Rai is a member of several scientific societies and has been a National Scholar for five years. He has received several prestigious awards, including the Father T. A. Mathias Award (1989) from the All India Association for Christian Higher Education and the Medini Award (1999) from the Department of Environment and Forest (Government of India). He received a SERC Visiting Fellowship from the Department of Science and Technology (Government of India, 1996) and an INSA Visiting Fellowship from the Indian National Science Academy (1998), and is a TWAS-UNESCO Associate (Italy, 2002).

CONTRIBUTORS

Siba Prasad Adhikary, Department of Botany, Utkal University, Bhubaneswar, India.

Alok Adholeya, Centre for Mycorrhizal Research, Tata Energy Research Institute, New Delhi, India.

Ulisses Brigatto Albino, State University of Londrina, Biological Science Center, Department of Microbiology, Microbial Ecology Laboratory, Londrina, PR, Brazil.

Galdino Andrade, State University of Londrina, Biological Science Center, Department of Microbiology, Microbial Ecology Laboratory, Londrina, PR, Brazil.

Manas Ranjan Banerjee, Research and Development Division, Brett-Young Seeds Limited, Winnipeg, Manitoba, Canada.

Kanchan Bhagwat, Department of Biotechnology, Amravati University, Maharashtra, India.

Uided Maaze Tiburcio Cavalcante, Departamento de Biologia, Universidade Federal Rural de Pernambuco, Recife, Pernambuco, Brazil.

Hanna Dahm, Nicholas Copernicus University, Institute of General and Molecular Biology Department of Microbiology, Torun, Gagarina, Poland.

Robin Duponnois, Institut de Recherche pour le Développement (IRD), UR IBIS (Interactions Biologiques dans les Sols des Systèmes Anthropisés Tropicaux).

João Ruy Jardim Freire, Department of Soil Science—Universidade Federal do Rio Grande do Sul (UFRGS), Porto Alegre, RS, Brazil.

Francesco Gentili, Department of Agricultural Research for Northern Sweden; Swedish University of Agricultural Sciences, Umeå, Sweden.

Ian Robert Hall, New Zealand Institute for Crop and Food Research Limited, Invermay Agricultural Centre, Mosgiel, New Zealand.

Pilar Irisarri, Biochemistry Laboratory, Department of Plant Biology; Facultad de Agronomía, Universidade de la Républica, Montevideo, Uruguay.

Ari Jumpponen, Division of Biology, Kansas State University, Manhattan, Kansas.

S. Kannaiyan, Tamil Nadu Agricultural University, Coimbatore, India.

K. Kumar, Azolla Laboratory, Tamil Nadu Agricultural University, Coimbatore, India.

Leonor Costa Maia, Departamento de Micologia, Centro de Ciências Biológicas, Universidade Federal de Pernambuco, Recife, Pernambuco, Brazil.

Miguel Marin, Instituto Valenciano de Investigaciones Agrarias, Moncada, Valencia, Spain.

Bagmi Pattanaik, Department of Botany, Utkal University, Bhubaneswar, India.

Anil Prakash, Fungal Biotechnology Laboratory, Institute of Microbiology and Biotechnology, Faculty of Life Sciences, Barkatullah University, Bhopal, India.

Carlos Hugo Prosperi, National University of Cordoba (Argentina), Special Projects Area—School of Biology, Cordoba, Argentina.

Polachi Raja, Malaysian Agri Hi-Tech Sdn., Bhd., Block: Stroma, UKM-MTDC Smart Technology Center, 43600 Bangi, Selangor D.E. Malaysia.

Nanjian Raman, Centre for Advanced Studies in Botany, University of Madras, Guindy Campus, Chennai, India.

Eliane Maria Ribeiro da Silva, Embrapa Agrobiology, Seropédica, RJ, Brazil.

Juan Manuel Ruiz-Lozano, Departamento de Microbiología del Suelo y Sistemas Simbióticos. Estación Experimental del Zaidín, Grandada, Spain.

Enilson Luiz Saccol de Sá, Department of Soil Science—Universidade Federal do Rio Grande do Sul (UFRGS), Porto Alegre, RS, Brazil.

Orivaldo José Saggin-Júnior, Embrapa Agrobiology, Seropédica, RJ, Brazil.

Thangaswamy Selvaraj, Centre for Advanced Studies in Botany, University of Madras, Guindy Campus, Chennai, India.

Sarika Shende, Department of Biotechnology, Amravati University, Maharashtra, India.

Norma Suely Sobral da Silveira, Departamento de Biologia, Universidade Federal Rural de Pernambuco, Recife, Pernambuco, Brazil.

Tanu, Centre for Mycorrhizal Research, Tata Energy Research Institute, New Delhi, India.

Ajit Varma, School of Life Sciences, Jawaharlal Nehru University, New Delhi, India.

Joseph Kevin Vessey, Department of Plant Science, University of Manitoba, Winnipeg, Manitoba, Canada.

Prasad Wadegaonkar, Department of Biotechnology, Amravati University, Maharashtra, India.

Laila Yesmin, Research and Development Division, Brett-Young Seeds Limited, Winnipeg, Manitoba, Canada.

Wang Yun, New Zealand Institute for Crop and Food Research Limited, Invermay Agricultural Centre, Mosgiel, New Zealand.

Hamdi Hussein Zahran, Botany Department, Faculty of Science, Beni-Suef, Egypt.

Foreword

During the past 100 years chemical fertilizers, pesticides, and herbicides have rapidly become the cornerstone of highly productive forms of agriculture. However, accompanying their use, overuse, and abuse comes the risk of pollution, serious changes in ecological equilibria and poisonings. As a consequence, more sustainable alternatives are being sought by many to replace or complement the less natural methods. Biofertilizers play a major part in this strategy.

Although biofertilizers and other alternatives are considered with suspicion by many, growth-promoting rhizobacteria, arbuscular mycorrhizae, and ectomycorrhizal fungi are known to be essential symbionts without which the vast majority of plants could not survive in soils with "normal" levels of available phosphorus and nitrogen. Other microbial biofertilizers include phosphate-solubilizing bacteria and cyanobacteria. Additional benefits from biofertilizers include prevention of infection by phytopathogens, their effect as soil conditioners, and the ability to alleviate stress generated by heavy metals or alkaline soils.

This book contains 20 chapters dealing with arguably the most important of the biofertilizers—the arbuscular mycorrhizal fungi—and their roles in forestry, stress alleviation, applications in nursery practice, methods for inoculating plants with them, their performance in the field, and their potential future roles. The chapters dealing with the easily cultivated cyanobacteria are mainly concerned with the potential for their use in rice fields. Chapters are also included covering free and symbiotic bacteria and their effects on plant health, as well as an interesting chapter on wild rhizobia. A chapter is also devoted to *Piriformospora indica,* an exciting emerging biofertilizer. The chapter on matsutake challenges the idea that all ectomycorrhizal fungi are beneficial or whether in fact some have much more complicated trophic relationships.

This book provides a wealth of information for those interested in the well-understood biofertilizers as well as those biofertilizers that have potential for use in the future.

Ian Robert Hall
Programme Leader, Edible Fungi
Crop & Food Research
Invermay Agricultural Centre, Mosgiel, New Zealand

Preface

The excessive use of chemical fertilizers and pesticides has generated several environmental problems including the greenhouse effect, ozone layer depletion, and acidification of water. These problems can be tackled by use of biofertilizers and biopesticides, which are natural, beneficial, and ecologically and user-friendly. The biofertilizers provide nutrients to the plants, control soilborne diseases, and maintain soil structure. Microbial biofertilizers play a pivotal role in sustainable agriculture. Arbuscular mycorrhizal fungi (AMF) are important microbes of soil that form symbiotic associations with most of the terrestrial plants on the earth. These fungi are chiefly responsible for phosphorus (P) uptake. Early inoculation at the seedling stage has been proven beneficial. Even in tissue culture industries, early mycorrhization of micropropagated plantlets overcome "transplant shock" due to their sophisticated nature. However, AMF cannot be cultured axenically, which poses a problem in rapid and mass multiplication in vitro conditions.

Another group of microbes is nitrogen-fixing bacteria, which are potent microbial inoculates now universally used for growth promotion of plants. The use of *Rhizobium* cultures in legume crops is increasing day by day to sustain agricultural productivity. These inoculants are commonly available in the market. An additional advantage of these microbes is that their axenic culture can be prepared in the laboratory and stored for mass inoculation. Phosphate-solubilizing bacteria are also immensely important as they have been reported to increase uptake of P by converting insoluble forms to soluble ones. *Azotobacter* and *Azospirillum* are two other efficient bacteria. The response of these organisms in increasing crop yield has been commonly experienced. In addition to these microbes, blue-green algae also contribute to a great extent in nitrogen economy of sustainable agriculture. The tripartite relationship of legume-rhizobium-mycorrhizae is the most efficient combination for growth promotion and for gaining higher yield of crops, vegetables, and trees.

Blue-green algae are the great suppliers of nitrogen. Their importance for bumper production of crops cannot be overstated and has been realized the world over by farmers. Their multiplication is easy and they now can be obtained in packets in the open market.

As a matter of fact, application of microbial biofertilizers is a natural and effective way of increasing and maintaining the mineral economy of nature. However, they cannot be a treated as substitute for chemical fertilizers. Their use reduces the use of chemical fertilizers, which is the only alternative for sustainable agriculture.

There has been great demand for a book on microbial fertilizers, wherein different issues concerning microbial fertilizers can be taken up. This book will certainly provide useful information dealing with a diverse group of microbes, beneficial effects, and the bottlenecks in their implementation.

Biotechnologists, microbiologists, botanists, growers, industrialists dealing with production of microbial fertilizers, environmentalists, and, above all, the users should find this book immensely useful.

Acknowledgments

I am grateful to Dr. Sudhir Patil, honourable vice-chancellor, Amravati University Amravati, Maharastra, India, for constant encouragement.

My special thanks to Drs. Francesco Gentili, Robin Duponnois, Carlos Prosperi, Mitiko Habte, Manas Banerjee, and other scientists, who kindly reviewed the manuscript and offered critical suggestions.

Thanks are also due to my colleagues Dr. Nandkishore Chikhale, Dr. Prasad Wadegaonkar, Dr. Prashant Thakare, and Dr. Anita Ramteke for helpful discussion in some chapters.

I would like to thank my research scholars, particularly Sarika Shende, Anita Bhoge, Pritee Wagh, Sunita Bansod, Girish Tidke, Ravindra Ade, and Rajni Soundale, for their help.

Sincere thanks are due to Dr. Amarjit Basra and Bill Palmer of The Haworth Press for their encouragement and also for their cooperation in the publication of the book.

Finally, I would like to thank my children, Shivangi, Shivani, and Aditya, who stood by me during the editing of this book.

Chapter 1

Potential and Possible Uses of Bacterial and Fungal Biofertilizers

Francesco Gentili
Ari Jumpponen

INTRODUCTION

During the past four decades we have witnessed the doubling of the human population and a concurrent doubling of food production (Vance, 2001). Plant nutrition has played a key role in this dramatic increase in demand for and supply of food. Increases in crop production have been made possible through the use of commercial man-made fertilizers. The use of nitrogen (N) fertilizer has increased almost ninefold and phosphorus (P) more than fourfold (Vance, 2001). The tremendous increase of N and P fertilization, in addition to the introduction of highly productive and intensive agricultural systems, has allowed these developments to occur at relatively low costs (Schultz et al., 1995; Vance, 2001). The increasing use of fertilizers and highly productive systems have also created environmental problems such as deterioration of soil quality, surface water, and groundwater, as well as air pollution, reduced biodiversity, and suppressed ecosystem function (Schultz et al., 1995; Socolow, 1999; Vance, 2001).

Environmental pollution resulting from greater nutrient availability can be either direct or indirect. Directly, misuse and excessive or poorly managed use of fertilizers can result in leaching, volatilization, acidification, and denitrification. Indirectly, the production (use of fossil fuel in Haber-Bosch process) and transport (combustion of fossil fuel) of fertilizers result in airborne CO_2 and N pollution, which will be eventually deposited into terrestrial ecosystems. A more comprehensive and complete view of the N cycle and impacts of N deposition at the global level can be found in Socolow (1999).

Community waste and sewage sludge provide an inexpensive and attractive alternative. One problem with the use of these sources of plant nutrition is their high content of heavy metals, which may have adverse effects on crop growth, crop consumers, or microorganisms in soil or rhizosphere (Giller et al., 1998; Graham and Vance, 2000). Consequently, these sources of inexpensive plant nutrients may not be suitable for fertilization of crops intended for human consumption if the heavy metal content is not drastically reduced.

The most limiting nutrients for plant growth are N and P (Schachtman et al., 1998). Although soil may contain vast amounts of either nutrient, most is not readily available for plant use. Most of N is tied into the soil organic matter. Even after fertilization, plants have to compete with soil microbes for easily available soluble N. Problems with P are different. In acidic soils, even when added in substantial quantities as fertilizer, P precipitates with iron or aluminum, whereas in alkaline soils P precipitates as calcium phosphates (Hinsinger, 2001). Accordingly, P limitation may be a difficult problem to overcome through the addition of P-containing fertilizers. We refer the reader to Frossard et al. (1995) for a more comprehensive description of the reactions controlling P availability in soil. The extensive fertilization required to overcome N and P limitations may lead to (sub-)surface runoff, and the soluble nutrients can easily end up in surface-water bodies or groundwater. For example, the loss of P from agricultural systems is one of the main causes of eutrophication and hypoxia in lakes and estuaries in the developed world (Vance, 2001).

In conclusion, the recent great increases in crop yields and food production in developed countries have been achieved by intensive agricultural practices. These increases, however, have not come without tremendous environmental costs (Vance, 2001). In developing countries the problems are different. The lack of fertilizers and adequate agricultural practices do not allow intensive crop production, and a vast segment of the population remains undernourished. Clearly, there is an urgent need for sustainable agricultural practices on a global level. In the developed world a reduction of energy and environmental costs is necessary. In developing countries, efficient and sustainable practices are needed to allow cost-efficient production of adequate nutrition for the growing populations. To overcome the ecological problems resulting from the loss of plant nutrients and to increase crop yields in the absence of resources for obtaining costly fertilizers, microscopic organisms that allow more efficient nutrient use or increase nutrient availability can provide sustainable solutions for present and future agricultural practices. In this chapter we aim to provide a brief overview of potential use of various biological agents with shown or suggested crop-yield-improving properties. Further, we highlight a few well-studied systems and use those as examples for possible choices for practical large-scale applications.

N_2-FIXING BACTERIA AS MICROBIAL BIOFERTILIZERS

Atmospheric N_2 composes approximately 80 percent of the air we breathe. Although abundant and ubiquitous in the air, N is the most limiting nutrient to plant growth because the atmospheric N is not available for plant uptake. Some bacteria are capable of N_2 fixation from the atmospheric N pool. These bacteria form various associations with plants:

1. Many free-living N_2-fixing bacteria occur in soil.
2. Some have adapted to form symbioses; others have intimate endophytic associations with plants.
3. Others live in close association in the plant root zone (rhizosphere) without forming intimate endophytic symbioses.

The amount of N fixed by these different systems is considerable, although variation resulting from environmental conditions or different plant-microbe combinations is vast. The close proximity of these microorganisms to their host plants allows efficient plant use of fixed N and minimizes volatilization, leaching, and denitrification. In the following section, we discuss the different types of microorganisms capable of allowing plant access to the atmospheric N pool and consider their use as biofertilizers.

SYMBIOTIC N_2-FIXING BACTERIA

Rhizobia

The best known and most exploited symbiotic N_2-fixing bacteria are those belonging to the family Rhizobiaceae *(Rhizobia)* and include the following genera: *Rhizobium, Bradyrhizobium, Sinorhizobium, Azorhizobium, Mesorhizobium,* and *Allorhizobium* (Vance, 1998; Graham and Vance, 2000). More comprehensive information on rhizobial symbiosis can be found in Spaink et al. (1998). These bacteria infect legumes and have a global distribution ranging from high latitudes in Europe and North America to the equator, to tropics in Australia and South America. In equatorial and tropical areas, legumes are particularly important; they are utilized in sylvopastoral and agroforestry systems (Dommergues and Subba Rao, 2000). The N_2-fixing capability of rhizobia varies greatly (up to 450 Kg $N \cdot ha^{-1}$) among host plant species and bacterial strains (Stamford et al., 1997; see Table 9 in Unkovich et al., 1997; Unkovich and Pate, 2000). Therefore, selection of the best strains must take rhizobia-host compatibility into account for selection of biofertilizers. Additional key characteristics are essential for rhizobial strains that are considered for practical applications as biofertilizers. They must have a high N_2-fixation rate and be able to compete with the indigenous rhizobia to maximize infection of the target crops (Stephens and Rask, 2000). From a practical perspective, the inoculum must be easily produced and have a high survivorship in field applications once inoculated on target seeds (Date, 2000).

Legume inoculation is an old practice that has been carried out for more than a century in agricultural systems in the United States and the United Kingdom (Brockwell and Bottomley, 1995; Catroux et al., 2001). Inoculation is particularly important when local and resident soil rhizobial populations are either absent or known to be very low (Catroux et al., 2001). For example, acidic soils generally contain no or low population densities of the alfalfa rhizobial symbiont *Sinorhizobium meliloti,* whereas basic soils contain a low inoculum potential of *Bradyrhizobium* sp., a rhizobial symbiont of *Lupinus* spp. (Amager, 1980, quoted by Catroux et al., 2001). Under such conditions inoculation with compatible rhizobia is likely to prove highly advantageous.

Prior to initiating a large-scale inoculation program with rhizobia, it is essential to evaluate the need for inoculation and perform a cost-benefit analysis. If rhizobial population densities are high, the inoculation is likely to prove unnecessary and investment in inoculation practice might be wasted. Catroux et al. (2001) suggested that when rhizobial population density is lower than 100 rhizobia per gram of soil, inoculation is likely to be beneficial for crop productivity. In such low population densities inoculation would prove cost efficient regardless of the N_2-fixation efficiency of the indigenous rhizobia. Unfortunately, fast, simple, and inexpensive methods for rhizobial population quantification are not readily available. Field experiments are time-consuming and can take several months. Microbiological assays are more expedient and take a few weeks but require substantial expertise (Brockwell and Bottomley, 1995). We refer the reader to Brockwell et al. (1988) for a description of assays estimating the N_2-fixing capacity of resident soil rhizobia.

The rhizobial inoculum can be produced and applied in numerous ways. Inoculum can be prepared as powder, liquid, and granular formulations. Granular formulations are convenient as they allow control of placement and application rate (Stephens and Rask, 2000). An additional important feature of an inoculum is the selection of the carrier (e.g., peat, perlite, mineral soil, charcoal). Whichever the carrier, its sterilization is necessary to maximize inoculum survival and subsequent infection rate (Brockwell and Bottomley, 1995; Stephens and Rask, 2000; Catroux et al., 2001). We refer the reader to Bashan (1998), Stephens and Rask (2000), and Lupwayi et al.

(2000) who provide extensive recent reviews on inoculation practices and choices on carriers and formulations.

It is important to control the quality of the inoculum. Unfortunately, the inoculum quality is often questionable and possibly up to 90 percent of all available inocula have no practical effect on the productivity of legumes (Brockwell and Bottomley, 1995). There are several possible reasons for the lacking crop yield response to inoculation. These range from poor control of inoculum production (low density of infective propagules or inadequate inoculum storage conditions) to incompatible inoculum-crop species combinations or edaphic conditions, which may preclude successful inoculation. Both federally and internationally controlled and standardized regulations are needed to simplify and clarify the rules on the quality and trade of commercially produced inocula. Academic institutions and agricultural experimental stations would provide an already existing network that could conveniently facilitate services to evaluate available inocula and provide much-needed quality control. Currently, purchase of inoculum in many cases is investment in a commodity of poor or questionable value. More information on inoculum quality control is available in recent reviews by Bashan (1998), Stephens and Rask (2000), and Lupwayi et al. (2000).

Frankia

Frankia is the genus of N_2-fixing actinomycetes that are capable of infecting and nodulating a group of eight families of mainly woody plants (Benson and Silvester, 1993; Huss-Danell, 1997; Wall, 2000). These so-called actinorhizal plants are used in land reclamation, for timber and fuel wood production, in mixed plantations, for windbreaks, as well as for shelterbelts along deserts and coastlines (Schwencke and Carù, 2001). Actinorhizal *Hippophaë rhamnoides* is cultivated for its berries. *Frankia* N_2 fixation has been estimated to be similar to rhizobial symbioses (Torrey, 1978; Dawson, 1986; Dommergues, 1995). Despite the potential importance of *Frankia* symbiosis, only limited information is available for inoculation practices and their use. *Frankia* inoculation can be advantageous in arid environments, disturbed sites, and areas where native actinorhizal plants are absent (Schwencke and Carù, 2001). Key factors for strain selection are similar to those described for *Rhizobium* sp. In addition to the symbiotic properties of the strains, characteristics such as age of the inoculum culture, cellular concentration, and the method of preservation may greatly affect inoculum infectivity. Since a universal strain adapted to all different environments and host genotypes does not exist, the best plant-*Frankia* combination should be selected and customized for each target area and for each target species (Schwencke and Carù, 2001).

Inoculation and nodulation before seedling transplanting improves plant survival and performance (Prat, 1992). Consequently, inoculation in a nursery is likely to be most efficient (Sprent and Parsons, 2000). *Frankia* inoculum can be lyophilized, frozen in glycerol, or cultured on complex medium (Fontaine et al., 1986; Sougoufara et al., 1989). *Frankia* spores have also been successfully used for *Casuarina cunninghamiana* inoculation (Lalonde and Calvert, 1979; Burleigh and Torrey, 1990). Similar to rhizobial inocula, several different formulations and carriers can be utilized. This adds flexibility for selection of the most simple and applicable method of inoculation for each program.

We emphasize the need for developing techniques for nursery inoculation, which would allow the use of nodulated actinorhizal plants in land reclamation and revegetation applications. Studies from natural systems indicate clearly that the actinorhizal plants may facilitate and expedite early successional plant community development in nonvegetated terrain (Schwencke and Carù, 2001). More information on *Frankia* and actinorhizal plants can be found in Schwintzer and Tjepkema (1990).

CYANOBACTERIA

Cyanobacteria are ecologically important. For example, an aquatic cyanobacterium, *Trichodesmium,* contributes approximately 36 percent of global N_2 fixation (Gallon, 2001). Cyanobacterial N_2 fixation has been essential in the cultivation of rice. Until the end of the 1970s *Azolla-Anabaena* symbiosis was the major N source for the 6.5×10^6 ha of rice cultivation in China. Presently, population pressure and increased labor costs have decreased the reliance on cyanobacterial symbioses in Chinese rice cultivation (Graham and Vance, 2000). In Uruguay and many parts of Asia, the cyanobacteria (mainly *Nostoc* and *Anabaena*) still have a vital importance for rice-field fertility (Kundu and Ladha, 1995; Irisarri et al., 2001). Cyanobacteria and their N_2 fixation find further application in the remediation of arid soils. In the Sahelian soil in Niger, N_2-fixing cyanobacteria are present in the soil surface crust. These cyanobacteria have been shown to increase soil N content (Malam Issa et al., 2001), therefore bearing a great promise for reclamation of extreme, arid environments.

Despite the traditional importance of the N_2-fixing cyanobacteria in rice cultivation and their possible reclamation uses in arid environments (Malam Issa et al., 2001) or ecosystems prone to frequent and/or seasonal flooding (Bashan et al., 1998), the production and application of cyanobacteria is still fairly poorly developed (Hashem, 2001). However, cyanobacteria should be seriously considered as a biofertilizer supporting sustainable agricultural practices in various environments (Hashem, 2001).

N_2-FIXING ASSOCIATED BACTERIA

In addition to symbiotic bacteria infecting plant roots, numerous taxa of less intimately associated N_2-fixing bacteria can be considered for crop yield improvement. Examples of such bacteria include *Acetobacter diazotrophicus* and *Herbaspirillum* spp. associated with sugarcane, sorghum, and maize (Triplett, 1996; James et al., 1997; Boddey et al., 2000), *Azoarcus* spp. associated with kallar grass *(Leptochloa fusca)* (Malik et al., 1997), and *Alcaligenes, Azospirillum*, *Bacillus, Enterobacter, Herbaspirillum, Klebsiella, Pseudomonas,* and *Rhizobium* associated with rice and maize (James, 2000). The genus *Azospirillum* colonizes a great variety of annual and perennial plants, many of which have never been reported to be colonized by N_2-fixing bacteria. Accordingly, *Azospirillum* possesses a great potential as a general root colonizer, whose use is not limited by host specificity (Bashan and Holguin, 1997). Indeed, several studies indicate that *Azospirillum* can increase the growth of various crops. These include sunflower, carrot, oak, sugarbeet, tomato, eggplant, pepper, and cotton in addition to wheat and rice (Bashan et al., 1989; Bashan and Holguin, 1997). In two decades of field experiments, general consensus is that in 60 to 70 percent of the cases *Azospirillum* application results in a significant crop yield increase (Okon and Labandera-Gonzalez, 1994). The yield increases can be substantial, up to 30 percent, but generally range from 5 to 30 percent. These yield increases by *Azospirillum* are possibly a result of the production of growth-promoting substances rather than N_2 fixation (Okon, 1985).

The main problem that limits the use of *Azospirillum* on a large scale is the great uncertainty and unpredictability of the results. Regardless of these uncertainties *Azospirillum* bears great promise as a growth-promoting N_2-fixing biofertilizer. Its growth-promoting properties are fairly well documented, and its commercial production as well as field application are simple. Inoculum can be produced and applied as in peat formulation, production of which is inexpensive. The peat formulation can also be directly utilized in field research and agricultural applications. However, alternative carriers warrant and require further research to allow selection of a reliable and effective means for inoculum production and field application (Vande Broek et al., 2000).

To emphasize the potential significance of N_2 fixation by associated bacteria, we use *Acetobacter diazotrophicus* and sugarcane as an additional example. As with actinorhizal or rhizobial symbioses, the quantity of fixed N is substantial and therefore associated bacteria are attractive candidates for biofertilizers. *Acetobacter diazotrophicus* can fix up to 70 percent of the sugarcane N requirement (Boddey et al., 1995). This translates to an annual 150 kg N·ha^{-1} that is fixed from the atmosphere and made available to the crop plant (Boddey et al., 1995). These estimates and their generality require further work as it appears that the amount of N_2 fixed depends on the plant genotype and environmental or edaphic conditions.

FREE-LIVING N_2-FIXING BACTERIA

Many free-living bacteria also fix atmospheric N_2. Examples of such free-living bacteria include *Azotobacter, Beijerinckia,* and *Clostridium.* Furthermore, if environmental conditions allow, nodulating bacterial symbionts (e.g., *Frankia*) of plant roots can also fix N_2 when not in a symbiotic association (free-living culture) (Benson and Silvester, 1993) with their plant host. More interestingly, it has been found that *Frankia* can occur and possibly fixes atmospheric N_2 in the rhizosphere of nonhost plants. *Frankia* has been recorded in the rhizosphere of *Betula pendula* (Smolander and Sarsa, 1990) and in soil where actinorhizal plants were not present (Benson and Silvester, 1993). These observations strongly suggest *Frankia* N_2 fixation in the rhizosphere of nonactinorhizal plants. These results open exciting possibilities for utilization of *Frankia* in biofertilizer applications but should be confirmed and further evaluated.

Estimation of the N_2 fixation by free-living bacteria is difficult. In an alfalfa *(Medicago sativa)* stand, the contribution of free-living N_2-fixing bacteria was estimated to range from 3 to 10 kg N·ha^{-1} (Roper et al., 1995). In a greenhouse experiment using different types of bacterial inoculation methods (leaf spray, seed soaking, side dressing), *Beijerinckia mobilis* and *Clostridium* spp. stimulated growth in cucumber and barley plants (Polyanskaya et al., 2002). The mechanism of the growth stimulation, however, remained unclear. Although the N_2 fixation by the inoculated bacteria may have played a significant role, other mechanisms cannot be excluded. It is possible that the detected growth response was also due to the bacterial synthesis of secondary growth-promoting compounds, such as plant growth hormones (Polyanskaya et al., 2002).

PLANT-GROWTH-PROMOTING RHIZOBACTERIA

Various bacteria can promote plant growth (Bashan, 1998). Collectively such bacteria are called plant-growth-promoting rhizobacteria (PGPR).

PGPR is a generic acronym that indicates bacteria, in some often unknown way, can stimulate plant growth. These bacteria vary in their mechanism of plant growth promotion but generally influence growth via P solubilization, nutrient uptake enhancement, or plant growth hormone production (Bashan et al., 1990; Okon and Labandera-Gonzalez, 1994; De Freitas et al., 1997; Goldstein et al., 1999; Richardson, 2001). Bertrand et al. (2000) showed that a rhizobacterium belonging to the genus *Achromobacter* could enhance root hair number and length in oilseed rape *(Brassica napus). Achromobacter* increased NO_3 and K uptake and, consequently, shoot and root dry weights by 22 to 33 percent and 6 to 21 percent respectively (Bertrand et al., 2000).

PHOSPHORUS-SOLUBILIZING BACTERIA

We first discussed the importance of the bacterial N_2 fixation because N is generally the most limiting nutrient for plant growth. Phosphorus is the second most limiting plant nutrient after N (Schachtman et al., 1998). Total P content in soil is usually high, but most of this soil P pool is not in forms available for plant uptake. Bacteria that can mobilize P from unavailable soil pools and increase P availability to plants are of great importance. Most predominant phosphorus-solubilizing bacteria (PSB) belong to the genera *Bacillus* and *Pseudomonas* (Richardson, 2001).

Field experiments highlight the potential importance of PSB. Sundara et al. (2002) applied rock phosphate with a PSB (*Bacillus megaterium* var. *phosphaticum*) in lignite-based culture medium in a field experiment. They found that without P application PSB amendment could increase sugarcane yield by 12.6 percent. PSB and P fertilizer together reduced the P requirement by 25 percent. Furthermore, 50 percent of the costly superphosphate could be replaced with inexpensive rock phosphate. PSB also improved the sugar yield and juice quality (Sundara et al., 2002). In conclusion, PSB may be of greatest value in allowing use of cheaper P sources (e.g., rock phosphate instead of superphosphate).

PLANT HORMONE PRODUCTION BY BACTERIA

In addition to improving plant nutrition by fixing atmospheric N or solubilizing plant-unavailable pools of P, bacteria can influence plant growth by production of plant hormones. *Bacillus pumilus* and *B. licheniformis,* isolated from the rhizosphere of alder *(Alnus glutinosa),* can produce high amounts of physiologically active gibberellins (Gutierez-Mañero et al., 2001). *Paenibacillus polymyxa* (recently transferred from genus *Bacillus*) is able to produce cytokinins (Timmusk et al., 1999). According to Timmusk et al. (1999), *P. polymyxa* may possess a great variety of properties that are of interest in the development of biofertilizers. The reported properties include N_2 fixation, P solubilization, and production of antibiotics, chitinase, and other hydrolytic enzymes, as well as enhancement of soil porosity. Similarly, taxa in the genus *Azospirillum* can fix atmospheric N_2, enhance plant mineral uptake, and produce growth-promoting plant hormones (Bashan et al., 1990; Bashan and Holguin, 1997). *Azospirillum* and *P. polymyxa,* along with numerous other bacteria, are examples of bacteria, which may warrant further study due to their highly diverse, potentially beneficial effects on plant growth.

FUNGI AND THEIR POTENTIAL AS BIOFERTILIZERS

Mycorrhizal Fungi

Benefits of Mycorrhizal Symbiosis to Plant Growth

It is not our intention to provide an extensive review on the benefits of the mycorrhizal symbioses; we will briefly summarize the available literature with special emphasis on the crop yield improvement and potential applications that may prove useful in designing biofertilizers. For more in-depth review, we refer the reader to extensive published volumes on the importance and benefits of mycorrhizal symbiosis to plant growth and performance (Harley and Smith, 1983; Bolan, 1991; Newsham et al., 1995; Smith and Read, 1997).

Mycorrhizal fungi form mutualistic symbioses with a vast majority of land plants (Smith and Read, 1997) Possibly more than 80 percent of all land plants form mycorrhizal symbioses. The extent of mycorrhizal symbioses emphasizes the ancient evolutionary history and potential importance of fungal symbioses for plant production and physiology. The association between plants and their root-colonizing mycorrhizal fungi is a functional symbiosis in which the mycorrhizal fungus is obligately or facultatively dependent on host photosynthates and energy. The plant-acquired carbon is traded for various mycorrhizal benefits to the host plant. The fungal mycelium that extends from the root surfaces into the soil matrix captures nutrients from soil solution. The minuscule diameter of the fungal hyphae increases the surface area that the plants are able to utilize for their nutrient acquisition. Resulting from the more efficient nutrient uptake, plant growth is generally improved when mycorrhizal fungi colonize the root systems.

The increasing yield can easily be demonstrated in laboratory and greenhouse experiments, although host-fungus species combinations and environmental factors may cause variation in the host response to mycorrhizal inoculation (Johnson et al., 1997). Additional benefits from the mycorrhizal symbiosis include increased tolerance of heavy metal contamination or drought, as well as lesser susceptibility to root pathogens or herbivory. Mycorrhizal fungi may also improve soil quality by having a direct influence on soil aggregation (Rillig et al., 2002) and therefore aeration and water dynamics. An interesting potential application for mycorrhizal fungi is their ability to allow plant access to nutrient sources generally unavailable to the host plants. For example, crop plants may be able to use insoluble sources of P when inoculated with mycorrhizal fungi but not in the absence of inoculation (Smith and Read, 1997).

In summary, the various potential benefits of mycorrhizal symbiosis on plant performance and crop yield suggest that they have substantial applications in agriculture (Menge, 1983) and in land reclamation or vegetation restoration (Sylvia, 1990). We do, however, concur with the conclusions of Ryan and Graham (2002) that mycorrhizal inoculation practices are often not compatible with industrial-scale agriculture, but they may find applications in alternative agricultural and horticultural practices.

USE OF ARBUSCULAR MYCORRHIZAL FUNGI AS BIOFERTILIZERS

The unculturability and obligate biotrophy of arbuscular mycorrhizal (AM) fungi have precluded the development of large-scale inoculation programs (Wood and Cummings, 1992). The only feasible means for production of infective propagules is growing the inoculum in symbiosis with living host plants or in root organ cultures, in other words, never in the absence of living host tissue. Although such production systems provide an advantage by allowing a continuous monitoring of the infective capability of the inoculum, their major drawbacks include extensive production costs, slow turnover time, and difficulty excluding secondary root colonizers such as root pathogens.

AM inoculum can be applied as spores, fragments of roots colonized by AM fungi, or a combination of the two and incorporated soil mycelium. AM spores and hyphae can be isolated from the soil substrate and mixed with carrier substrate. Commonly used carriers include pumice or clay, sand, perlite, vermiculite, soilrite, and soil or glass pellets (Mallesha et al., 1992; Redecker et al., 1995; Gaur and Adholeya, 2000). AM taxa and strains may vary in their ability to colonize host plants depending on the source of inoculum (Biermann and Linderman, 1983; Klironomos and Hart, 2002). Spores may be the most reliable source of inoculum across various AM taxa, whereas fragments of colonized roots are effective for some taxa but not others (Biermann and Linderman, 1983; Klironomos and Hart, 2002). The entire substrate can also be used and homogenized into a crude soil carrier that includes plant roots and fungal spores as well as the soil mycelium. Alternative methods, including soil-free aeroponic (Sylvia and Jarstfer, 1992; Jarstfer and Sylvia, 1995), nutrient film (Elmes et al., 1983; Elmes and Mosse, 1984), and root organ culture (Mugnier and Mosse, 1987) systems, have been tried experimentally, but the extensive costs of these methods seem preclusive. Furthermore, large-scale production of inoculum using these alternative methods has been poorly developed.

Given the large costs and difficulty of producing inoculum, the need for AM inoculation should be carefully evaluated. The determining factors including expected crop response to AM inoculation, availability of soilborne inoculum, and alternative practices such as cropping system management for AM inoculum maintenance may need to be considered. Recent evidence also suggests that, contrary to general assumptions, AM fungi may have limited host specificity (Eom et al., 2000; Helgason et al., 2002). The possibility of host specificity patterns among the AM fungi underlines the importance of strain and taxon selection for each inoculation application. Although AM inoculum is commercially available, large-scale inoculation has not become practical or achievable in everyday agricultural practices. Accordingly, the AM inoculation has been limited to production of high value nursery stocks. In such nursery applications the inoculation is often highly advantageous, resulting in improved crop growth, more expedient development, and homogeneous end product. In the absence of practical applications for production of AM inoculum for agricultural practices, the importance of management for maintenance of soilborne fungi must be emphasized. The potential and relative benefits of different inoculation and land management practices have been reviewed and discussed elsewhere (Wood and Cummings, 1992; Dodd and Thomson, 1994). We summarize only a few general recommendations. Continuous plant cover by intercropping or sequential cropping systems may allow maintenance of adequate inoculum levels in soil (Tisdall and Adem, 1990). Similarly, minimizing disturbance is likely to support indigenous inoculum (Jasper et al., 1992; Miller and Jastrow, 1992). Extensive fertilization may also need to be avoided as P, in particular, often inhibits AM colonization (Allison and Goldberg, 2002). The last point underlines the importance of management practices for greater soil AM inoculum in agricultural systems, which may not allow economically viable application of commercial fertilizers. In intensive agricultural systems relying heavily on N and P fertilization, the management for AM inoculum is unlikely to prove cost-effective (Ryan et al., 2002).

One aspect of AM inoculation that has only recently received attention is the need for diverse populations and communities of fungi in soil. Van der Heijden, Klironomos, et al. (1998) concluded that a greater number of inoculated fungi significantly increased plant community yields. The impact of increasing mycorrhizal diversity can be attributed to functional complementarity: different fungi may perform different tasks in their natural environment (van der Heijden, Boller, 1998). Although only a limited body of literature exists in support of host specificity patterns among AM fungi, recent evidence suggests that different AM fungi vary in their effects in respect to their ability to increase host yield and/or nutrient use (Helgason et al., 2002). The mechanisms for these effects remain rather unclear to date, but it is possible that there is a need to test each crop species, possibly even variety, prior to recommendations on broad-scale inoculation programs. This way the most efficient host variety and fungal strain combinations can be identified for different environmental conditions.

USE OF ECTOMYCORRHIZAL FUNGI AS BIOFERTILIZERS

The importance of ectomycorrhizal (EM) symbiosis for tree growth and nutrient acquisition was already hypothesized by Frank (1885). Accordingly, there is a long history of technique development for incorporation of EM fungal inoculation into nursery or forestry plantation practices (White, 1941; Wilde, 1944; Mikola, 1969, 1970). General evidence suggests that tree plantations will fail unless endemic or inoculated EM fungi are available on site. Selection of the inoculated EM fungi has been largely based on tree crop enhancement and/or applicability for a large-scale inoculum practices (Smith and Read, 1997).

Although various alternatives have been used and surveyed, the most widespread inoculum programs have been developed for *Pisolithus tinctorius*. The general advantages for the use of *P. tinctorius* include its wide host range and extensive geographic distribution, as well as its occurrence on sites burdened by recent disturbance, drought, high temperatures, and/or chemical contaminants. *Pisolithus tinctorius* inoculum can be produced and applied as vegetative mycelium in a peat vermiculite carrier. The nutrient solution, which is necessary for the vegetative growth of *P. tinctorius* throughout the substrate, will also facilitate the competitive exclusion of other root-colonizing fungi (Smith and Read, 1997). Various alternative techniques and formulations for inoculation have been developed (Marx and Kenney, 1982; Marx et al., 1984, 1991). Although liquid or spore suspension techniques would avoid the problems resulting from bulky solid inoculum production and storage, they often suffer from delayed EM establishment (basidiospore inoculation) or mycelial fragmentation and shredding (axenic liquid culture techniques).

Inoculation programs with EM fungi have had some success. However, as with AM or bacterial inoculum applications, there seems to be no single fungal species or strain that could be universally applied across different sites and host species. When compared to local strains and species in the northwestern United States, the *P. tinctorius* strain that had proven extremely favorable for seedling growth and establishment elsewhere seemed less beneficial (Perry et al., 1987). In many cases, the strains that easily colonize seedlings in the nurseries and are easy to manipulate have only limited positive effects on the performance of the planted seedlings (Perry et al., 1987; le Tacon et al., 1992; Jackson et al., 1995). The limited success of the fungi, which have been selected for the inoculation programs, may be simply due to the ubiquitous presence of endemic mycorrhizal fungi in reforested sites and the competitive exclusion of the nursery-inoculated fungi in the field. Accordingly, the inoculation with EM fungi may be most important on sites with poor reforestation history or on plantations, which will be established on previously nonforested sites.

In the research focusing on the development of the forest nursery inoculation programs, one issue that has received relatively little attention is the impact that imported and possibly invasive EM fungi have on the endemic fungi and their community composition. The inoculated fungi may persist in the root systems for extended periods of time and outcompete less invasive endemic strains and species (de la Bastide et al., 1994). Although no direct evidence for such competitive exclusion currently exists, introductions may homogenize local fungal populations and communities. For example, *Eucalyptus* species imported and planted in the United States often support extensive colonization by a false truffle, *Hydnangium carneum*. Similarly, *Amanita phalloides,* which was likely imported from Europe with cork oaks, is spreading in the native oak stands in California. The question that remains is whether these successful invaders have had any negative impacts on the local EM communities.

OTHER ROOT-COLONIZING FUNGI

Root systems of all higher and lower plants support fungal communities, which are composed of fungi occupying the rhizosphere or rhizosplane environment and fungi forming intimate symbiotic associations with the plant roots colonizing them inter- and intracellularly. It is generally acknowledged that these root-associated fungal communities are diverse (Vandenkoornhuyse et al., 2002). However, the functions of these communities and of the individual species in these communities are poorly understood. Studies focusing on the fungi isolated from the root environment provide, not surprisingly, inconsistent results. Even studies focusing on a fairly limited number of anamorphic taxa or a limited number of fungal strains produce incongruent results (Jumpponen and Trappe, 1998a; Jumpponen, 2001). The bottom line is that these communities incorporate a vast diversity and contain species which may prove beneficial in biofertilizer applications. Based on the fairly limited number of existing studies, the benefits resulting from association with the root-colonizing miscellaneous fungi are similar to those reported for mycorrhizal symbioses. The root-associated fungi benefit plants by promoting plant growth and crop yield (Shivanna et al., 1994) and by reduction of root pathogen infection (Dewan and Sivasithamparam, 1989). Recent studies also indicate that these fungi may improve plant nutrient uptake (Jumpponen et al., 1998; Jumpponen and Trappe, 1998b) or allow plant access to otherwise unavailable nutrient sources (Barrow and Osuna, 2002). Several strains originating from crop roots have been shown to be able to promote plant growth and crop yields under field conditions, not only in controlled laboratory or greenhouse studies (Shivanna et al., 1994).

Very little effort has been made in the development of inoculum programs for these various potentially beneficial fungi. Although many of these fungi may be easily cultured and therefore also easily manipulated in inoculation applications, the problems of inoculation are similar to those described with root-associated bacteria and mycorrhizal fungi. The beneficial effects, growth promoting or otherwise, can be highly variable and depend on host genotype and selected fungal strain or species, and can change with plant development or environmental conditions. However, the ubiquitous presence and possible global distribution of many of these fungi suggest a great environmental tolerance. This, in addition to the suggested lacking host specificity (Jumpponen and Trappe, 1998a), makes these fungi a group of great potential interest.

MULTIPLE INOCULATIONS AND INTERACTIONS AMONG POTENTIAL BIOFERTILIZERS

We refer to the practice of inoculation and introduction of more than one fungus and/or bacterium into the target crops as *multiple inoculation.*

The rhizosphere presents a challenging environment to manipulate in agricultural practices. Plant roots and their carbohydrates control the bacterial and fungal populations and their dynamics in the soil matrix adjacent to the plant roots (Hadacek and Kraus, 2002). Such host-plant and root-microbe interactions may partly explain the species-specific responses to inoculation programs. The complexity of the rhizosphere environment is further emphasized by the various interactions among the different bacteria and fungi possibly competing for resources or facilitating presence and occurrence by modification of the rhizosphere environment. Although only poorly understood, various examples on inhibition of root pathogens by rhizosphere-associated bacteria and fungi presented elsewhere in this chapter are likely to be at least partly a result of competitive exclusion. Facilitation of other microscopic organisms in the rhizosphere environment is also possible. For example, bacteria have been reported to adhere superficially or intracellularly to fungal hyphae on the root surfaces and in soil or be generally associated with the rhizosphere of mycorrhizal plants (Bianciotto et al., 1996, 2000; Mogge et al., 2000; Poole et al., 2001; Minerdi et al., 2002). Although AM fungi are exclusively biotrophic, organic matter facilitates the growth of extramatrical mycelium—a response that has been suggested to be a result of bacterial activities in the organic matter (Green et al., 1999; Ravnskov et al., 1999). Some bacterial genera—for example *Paenibacillus* discussed earlier in this chapter—are able to stimulate mycorrhizal colonization and are often isolated from rhizosphere of mycorrhizal plants (Garbaye, 1994; Budi et al., 1999; Poole et al., 2001). These bacteria may also provide a crop yield or plant-growth-promoting effect. Ruiz-Lozano and Bonfante (2001) hypothesized that the bacterial association with the AM fungi may positively influence the host plant nutrient uptake and nutrient transport from the AM fungus to the plant.

Clearly, there is some evidence that bacteria may influence mycorrhizal fungi and fungal colonization of plant roots. Similarly, bacteria and their growth or activity are affected by fungi and their exudates in the rhizosphere (Christensen and Jakobsen, 1993; Olsson et al., 1996). Meyer and Linderman (1986) enumerated bacteria and surveyed their community structure in mycorrhizal and nonmycorrhizal plants. They observed no difference in the total numbers of culturable bacteria, but noted specific bacteria present more frequently in the mycorrhizal rhizospheres. Mansfeld-Giese et al. (2002), in contrast, reported large differences in the bacterial community structures between rhizosphere and nonrhizosphere soil, but found only few bacterial taxa whose occurrence differed between mycorrhizal and nonmycorrhizal plants. Most interestingly, they found *Paenibacillus* spp. almost exclusively associated with not only the rhizosphere of mycorrhizal plants but also the mycelium of the AM fungus *Glomus intraradices*. It remained unclear from that study whether the bacteria were living in the proximity of the mycelium, on the hyphal surface, or within the fungal hyphae. Whichever the case, the possibility of intimate association between fungi and helper bacteria exists (Garbaye, 1994; Minerdi et al., 2002). Further examples suggest that fungal symbionts of plant roots may stimulate bacterial activity. Coinoculation with AM fungi and *Frankia* stimulated N_2 fixation and AM development in *Hippophaë tibetana* (Tian et al., 2002). Alfalfa *(Medicago sativa)* N and P acquisition was stimulated by inoculation with *Rhizobium,* AM fungi, and PSB (Toro et al., 1998). From the examples presented here, one general conclusion can be drawn: Fungi and bacteria have various inhibitory and facilitative interactions, which may be of use in biofertilizer applications once the compatible combinations of fungi and bacteria are identified.

Compatibility and interactions among microscopic organisms are poorly understood. It is possible that different species of fungi have specific bacterial associates. Andrade et al. (1997) reported that bacterial populations in the rhizosphere environment were different depending on which species or strains of AM fungi were used as inoculum. Similar results are likely also in the EM systems as various bacteria have been reported to be associated with structures of various EM fungi (Poole et al., 2001). In addition, Danell et al. (1993) reported the association of *Pseudomonas* and ectomycorrhizal fungus *Cantharellus cibarius*.

Similar interactions also occur among bacteria. For example, we briefly list a few experimental systems that present evidence for facilitative interactions among bacteria. *Pseudomonas fluorescens* increased nodulation and nitrogenase activity of *Bradyrhizobium japonicum* in a soybean culture system (Chebotar et al., 2001). Zhang et al. (1996) found that certain PGPR belonging to the genera *Serratia* and *Aeromonas* could increase soybean nodulation and N_2 fixation of *Bradyrhizobium japonicum* at suboptimal root zone temperature. Bashan and Holguin (1997) have reviewed several examples of coinoculation with *Azospirillum* and *Rhizobium, Azotobacter, Arthrobacter, Enterobacter,* or *Klebsiella.* Based on a study on the coinoculation of the N_2-fixing *Phyllobacterium* sp. and the P-solubilizing *Bacillus licheniformis* in mangrove, Rojas et al. (2001) suggested that the interaction between different rhizosphere bacteria should be considered when evaluating the growth-promoting effects of these bacteria.

An interesting new application is the introduction of "artificial" associations. In addition to combining stimulatory bacteria and fungi, novel applications of microbes may allow extension of microbial symbioses beyond their usual host range. Koval'skaya et al. (2001) showed that treating rape *(Brassica napus* var. *napus)* with an abiogenic agent (auxinlike growth-promoting substance) or the bacteria *Micrococcus* sp. and *Rhodococcus* sp. induced formation of paranodules. The paranodules were then colonized by the N_2 fixing *Azotobacter nigricans* and *Bacillus* sp. As a result of the paranodulation and N_2 fixing, the rape N and protein content increased.

These examples strongly suggest the possibility of a complex web of interactions between root-associated fungi, rhizosphere bacteria. and host plants. The numerous benefits provided by various bacteria and fungi to plant growth and crop yield may open new avenues for developing biologically active fertilizers. Combinations of growth- and yield-promoting bacteria and fungi as well as combinations of organisms facilitating establishment or plant-derived benefit from inoculation deserve further study. For example, production of inoculum, which would integrate the growth-promoting and pathogen-inhibiting effects of *Paenibacillus* and AM fungi, seems like a reasonable first step in order to generate a product that may prove superior in agricultural systems. An add-on benefit from such inoculum would also be the facilitation of AM establishment and colonization by *Paenibacillus*. EM fungi and their mycorrhizal helper bacteria (Garbaye, 1994) are likely to provide new combination inocula with benefits similar to the AM fungal-bacterial systems.

FUTURE PERSPECTIVES AND CONCLUSIONS

In order to increase our understanding of the role of various root-associated organisms in plant growth and health as well as make use of their potential beneficial features as biofertilizers in plant production, more information is urgently needed on the interactions among plants and rhizosphere microorganisms. We have briefly presented a few examples of bacteria and fungi, which may bear a great promise as biofertilizers. We acknowledge that we selected simplified examples. However, studies utilizing simplified laboratory experiments are essential in uncoupling different factors and identifying the best possible candidates for biofertilizers. Such experiments and their results, however, are often difficult to extrapolate to field conditions. Furthermore, the rhizosphere presents additional challenges as an environment to be manipulated. The rhizosphere is a highly dynamic system with a vast number of fungi and bacteria interacting simultaneously; the difficulty of excluding endemic bacteria and fungi may preclude clear conclusions from inoculation experiments in the field. We understand the difficulty of conducting such experiments on a scale that would be meaningful in everyday agricultural practice. However, unless the positive effects can be repeatably shown in practical applications, the commercial viability of inoculation programs will be uncertain.

We emphasized earlier the importance of the evaluation of soil inoculum potential of N_2 fixers, PGPR, and mycorrhizal fungi. The inocula made commercially available should conform to minimal quality standards. Simple and clear governmental and international regulations are essential to guarantee efficiency of the inocula. Local or regional strains should preferably be selected and used for the target crops. We underlined in this chapter the difficulty of obtaining universal "wonder" strains that perform equally across different host taxa and environments.

Use of multiple inoculations can aim to stimulate N_2 fixation, P uptake, and mineral nutrition in general, but they can also help in controlling plant pathogens. Such applications would be welcomed as they allow reduction in chemical fertilizers and pesticides that are expensive and environmentally unsound. An interesting issue, in addition to the infection potential, is the persistence of the biofertilizer after inoculation. If inoculum potential can be built into agricultural soils, the interval between biofertilizer applications could be increased and costs lowered.

Various fungi and bacteria provide a battery of extracellular enzymes, which may be utilized for improved crop yields and reduced costs for inorganic fertilizers. We emphasize the need for field trials with multiple organism inoculations. These inoculum combinations may be of greatest value if various organisms with different proven or suspected benefits to the crop plants can be integrated. The integration of different microbial capabilities into combined biofertilizers with numerous potential yield-promoting effects is desirable. This is likely to be best achieved by approaching the application and research of biofertilizers at a scale that is relevant to agricultural practices. Finding avenues and funding for collaboration between research facilities and biotechnology industry seems like a reasonable first step. Producing inocula at large scales is essential for achievement of the research and practical application goals stated previously. Connection between research and industry allows not only the inoculum production for field trials but also testing of the industrial scale inoculum production for direct marketing.

Finally, we emphasize the importance of establishing federal and international guidelines for inoculum production and trade. To protect the end user of inoculum and to promote safe choice of commercial inocula collaboration among research facilities, federal agencies, farmers, and the inoculum producers are of pivotal importance. Such innovative collaborative approaches will allow a head start for commercial and economically viable production of biofertilizer inocula for marketing directly to primary target consumers.

REFERENCES

Allison, V.J. and Goldberg, D.E. (2002). Species-level versus community level patterns of mycorrhizal dependence on phophorus: An example of Simpson's paradox. *Functional Ecology* 16: 346-352.

Amarger, N. (1980). Aspect microbiologique de la culture des légumineuses. *Le Selectionneur Francais* 28: 61-66.

Andrade, G., Mihara, K.L., Linderman, R.G., and Bethlenfalvay, G.J. (1997). Bacteria from rhizosphere and hyphosphere soils of different arbuscular-mycorrhizal fungi. *Plant and Soil* 192: 71-79.

Barrow, J.R. and Osuna, P. (2002). Phosphorus solubilization and uptake by dark septate fungi in fourwing saltbrush *Atriplex canescens* (Pursh) Nutt. *Journal of Arid Environment* 51(3): 449-459.

Bashan, Y. (1998). Inoculants of plant growth-promoting bacteria for use in agriculture. *Biotechnology Advances* 16: 729-770.

Bashan, Y., Harrison, S.K., and Whitmoyer, R.E. (1990). Enhanced growth of wheat and soybean plant inoculated with *Azopirillum brasilense* is not necessary due to general enhancement of mineral uptake. *Appied Environmental Microbiology* 56: 769-775.

Bashan, Y. and Holguin, G. (1997). *Azospirillum*-plant relationships: Environmental and physiological advances (1990-1996). *Canadian Journal of Microbiology* 43: 103-121.

Bashan, Y., Puente, M.E., Myrold, D.D., and Toledo, G. (1998). In vitro transfer of fixed nitrogen from diazotrophic filamentous cyanobacteria to black mangrove seedlings. *FEMS Microbiology Ecology* 26: 165-170.

Bashan, Y., Ream, Y., Levanony, H., and Sade, A. (1989). Non-specific responses in plant growth, yield, and root colonization of noncereal crop plants to inoculation with *Azospirillum brasilense* Cd. *Canadian Journal of Botany* 67: 1317-1324.

Benson, D.R. and Silvester, W.B. (1993). Biology of *Frankia* strains, actinomycete symbionts of actinorhizal plants. *Microbiological Reviews* 57: 293-319.

Bertrand, H., Plassard, C., Pinochet, X., Touraine, B., Normand, P., and Cleyet-Marel, J.C. (2000). Stimulation of the ionic transport system in *Brassica napus* by a plant growth-promoting rhizobacterium (*Achromobacter* sp.). *Canadian Journal of Microbiology* 46: 229-236.

Bianciotto, V., Bandi, C., Minerdi, D., Sironi, M., Tichy, H.V., and Bonfante, P. (1996). An obligately endosymbiotic mycorrhizal fungus itself harbors obligately intracellular bacteria. *Applied and Environmental Microbiology* 62: 3005-3010.

Bianciotto, V., Lumini, E., Lanfranco, L., Minerdi, D., Bonfante, P., and Perotto, S. (2000). Detection and identification of bacterial endosymbionts in arbuscular mycorrhizal fungi belonging to Gigasporaceae. *Applied and Environmental Microbiology* 66: 4503-4509.

Biermann, B. and Linderman, R.G. (1983). Mycorrhizal roots, intraradical vesicles and extraradical vesicles as inoculum. *New Phytologist* 95: 97-105.

Boddey, R.M., Da Silva, L.G., Reis, V., Alves, B.J.R., and Urquiaga, S. (2000). Assessment of bacterial nitrogen fixation in grass species. In E.W. Triplett (ed.), *Prokaryotic nitrogen fixation: A model system for analysis of a biological process* (pp. 705-726). Wymondham, UK: Horizon Scientific Press.

Boddey, R.M., de Oliveira, O.C., Urquiaga, S., Reis, V.M., Olivares, F.L., Baldani, V.L.D., and Döbereiner, J. (1995). Biological nitrogen fixation associated with sugar cane and rice: Contributions and prospects for improvement. *Plant and Soil* 174: 195-209.

Bolan, N.S. (1991). A critical review on the role of mycorrhizal fungi in the uptake of phosphorus by plants. *Plant and Soil* 134: 189-207.

Brockwell, J. and Bottomley, P.J. (1995). Recent advances in inoculant technology and prospects for the future. *Soil Biology and Biochemistry* 27: 683-697.

Brockwell, J., Holliday, R.A., and Pilka, A. (1988). Evaluation of the symbiotic nitrogen-fixing potential of soils by direct microbiological means. *Plant and Soil* 108: 163-170.

Budi, S.W., van Tuinen, D., Martinotti, G., and Gianinazzi, S. (1999). Isolation from the *Sorghum bicolor* mycorhizosphere of a bacterium compatible with arbuscular mycorrhiza development and antagonistic towards soil borne fungal pathogens. *Applied and Environmental Microbiology* 65: 5148-5150.

Burleigh, S and Torrey, J.G. (1990). Effectiveness of different *Frankia* cell types as inocula for actinorhizal plant *Casuarina*. *Applied and Environmental Microbiology* 56: 2565-2567.

Catroux, G., Hartmann, A., and Revellin, C. (2001). Trends in rhizobial inoculant production and use. *Plant and Soil* 230: 21-30.

Chebotar, V.K., Constancio Asis, A. Jr., and Akao, S. (2001). Production of growth-promoting substances and high colonization ability of rhizobacteria enhance the nitrogen fixation of soybean when coinoculated with *Bradyrhizobium japonicum*. *Biology and Fertility Soils* 34: 427-432.

Christensen, H. and Jakobsen, I. (1993). Reduction of bacterial growth by a vesicular-arbuscular mycorrhizal fungus in the rhizosphere of cucumber (*Cucumis sativus* L.). *Biology and Fertility Soils* 15: 253-258.

Danell, E., Alström, S., and Ternström, A. (1993). *Pseudomonas fluorescens* in association with fruitbodies of ectomycorrhizal mushroom *Cantharellus cibarius*. *Mycological Research* 97: 1148-1152.

Date, R.A. (2000). Inoculated legumes in cropping systems of the tropics. *Field Crops Research* 65: 123-136.

Dawson, J.O. (1986). Actinorhizal plants: Their use in forestry amd agriculture. *Outlook on Agriculture* 15: 202-208.

De Freitas, J.R., Banerjee, M.R., and Germida, J.J. (1997). Phosphate-solubilizing rhizobacteria enhance the growth and yield but not phosphorus uptake of canola (*Brassica napus* L.). *Biology and Fertility Soils* 24: 358-364.

de la Bastide, P.Y., Kropp, B.R., and Piché, Y. (1994). Spatial and temporal persistence of discrete genotypes of the ectomycorrhizal fungus *Laccaria bicolor* (Maire) Orton. *New Phytologist* 127: 547-556.

Dewan, M.M. and Sivasithamparam, K. (1989). Growth promotion of rotation crop species by a sterile fungus from wheat and the effect of soil temperature and moisture on its suppression. *Mycological Research* 93: 156-160.

Dodd, J.C. and Thomson, B.D. (1994). The screening and selection of inoculant arbuscular-mycorrhizal and ectomycorrhizal fungi. *Plant and Soil* 159: 149-158.

Dommergues, Y.R. (1995). Nitrogen fixation by trees in relation to soil nitrogen economy. *Fertilizer Research* 42: 215-230.

Dommergues, Y.R. and Subba Rao, N.S. (2000). Introduction of N_2-fixing trees in non-N_2-fixing tropical plantations. In N.S. Subba Rao and Y.R. Dommergues (eds.), *Microbial interactions in agriculture and forestry,* Volume 2 (pp. 131-154). Enfield, NH: Science Publishers, Inc.

Elmes, R.P., Hepper, C.M., Hayman, D.S., and O'Shea, J. (1983). The use of vesicular-arbuscular mycorrhizal roots grown by the nutrient film technique as inoculum for field sites. *Annals of Applied Biology* 104: 437-441.

Elmes, R.P. and Mosse, B. (1984). Vesicular-arbuscular mycorrhizal inoculum production. II. Experiments with maize *(Zea mays)* and other hosts in nutrient flow culture. *Canadian Journal of Botany* 62: 1531-1536.

Eom, A.-H., Hartnett, D.C., Wilson, G.W.T. (2000). Host plant species effects on arbuscular mycorrhizal communities in tallgrass prairie. *Oecologia* 122: 435-444.

Fontaine, M.S., Young, P.H., and Torrey, J.G. (1986). Effects of long-term preservation of *Frankia* strains on infectivity, effectivity, and in vitro nitrogenase activity. *Applied and Environmental Microbiology* 51: 694-698.

Frank, A.B. (1885). Über die auf Wurzelsymbiose beruhende Ernährung gewisser Bäume durch unterirdische Pilze. *Berichte der Deutchen Botanischen Gesellschaft* 3: 128-145.

Frossard, E., Brossard, M., Hedley, M.J., and Metherell, A. (1995). Reactions controlling the cycling of P in the soils. In H. Tiessen (ed.), *Phosphorus in the global environment* (pp. 108-147). Chichester, UK: John Wiley and Sons Ltd.

Gallon, J.R. (2001). N_2 fixation in phototrophs: Adaptation to a specialized way of life. *Plant and Soil* 230: 39-48.

Garbaye, J. (1994). Helper bacteria: A new dimension to the mycorrhizal symbiosis. *New Phytologist* 128: 197-209.

Gaur, A. and Adholeya, A. (2000). Effects of the particle of soil-less substrates upon AM fungus inoculum production. *Mycorrhiza* 10: 43-48.

Giller, K.E., Witter, E., and Mcgrath, S.T. (1998). Toxicity of heavy metals to microorganisms and microbial processes in agricultural soils: A review. *Soil Biology and Biochemistry* 30: 1389-1414.

Goldstein, A.H., Braverman, K., and Osorio, N. (1999). Evidence for mutualism between a plant growing in a phosphate-limited desert environment and a mineral phosphate solubilizing (MPS) rhizobacterium. *FEMS Microbiology Ecology* 30: 295-300.

Graham, P.H. and Vance, C.P. (2000). Nitrogen fixation in perspective: An overview of research and extension needs. *Field Crops Research* 65: 93-106.

Green, H., Larsen, J., Olsson, P.A., and Jensen, D.F. (1999). Suppression of the biocontrol agent *Trichoderma harzianum* by mycelium of the arbuscular mycorrhizal fungus *Glomus intrradices* in root-free soil. *Applied and Environmental Microbiology* 65: 1428-1434.

Gutierez-Mañero, F.J., Ramos-Solano, B., Probanza, A., Mehouachi, J., Tadeo, F.R., and Talon, M. (2001). The plant-growth-promoting rhizobacteria *Bacillus pumilus* and *Bacillus licheniformis* produce high amounts of physiologically active gibberellins. *Physiologia Plantarum* 111: 206-211.

Hadacek, F. and Kraus, G.F. (2002). Plant root carbohydrate affects growth and behaviour of endophytic microfungi. *FEMS Microbiology Ecology* 41: 161-170.

Harley, J.L. and Smith, S.E. (1983). *Mycorrhizal symbiosis.* New York: Academic Press.

Hashem, M.A. (2001). Problems and prospects of cyanobacterial biofertilizer for rice cultivation. *Australian Journal of Plant Physiology* 28: 881-888.

Helgason, T., Merryweather, J.W., Denison, J., Wilson, P., Young, J.P.W., and Fitter, A.H. (2002). Selectivity and functional diversity in arbuscular mycorrhizas of co-occurring fungi and plants from a temperate deciduous woodland. *Journal of Ecology* 90: 371-384.

Hinsinger, P. (2001). Bioavailability of soil inorganic P in the rhizosphere as affected by root-induced chemical changes: A review. *Plant and Soil* 237: 173-195.

Huss-Danell, K. (1997). Actinorhizal symbioses and their N_2 fixation. *New Phytologist* 136: 375-405.

Irisarri, P., Gonnet, S., and Monza, J. (2001). Cyanobacteria in Uruguayan rice fields: Diversity, nitrogen fixing ability and tollerance to herbicides and combined nitrogen. *Journal of Biotechnology* 91: 95-103.

Jackson, R.M., Walker, C., Luff, S., and McEvoy, C. (1995). Inoculation of Sitka spuce and Douglas fir with ectomycorrhizal fungi in the United Kingdom. *Mycorrhiza* 5: 165-173.

James, E.K. (2000). Nitrogen fixation in endophytic and associative symbiosis. *Field Crops Research* 65: 197-209.

James, E.K., Olivares, F.L., Baldani, J.I., and Döbereiner, J. (1997). *Herbaspirillum,* an endophytic diazotroph colonizing vascular tissue in leaves of *Sorghum bicolor* L. Moench. *Journal of Experimental Botany* 48: 785-797.

Jarstfer, A.G. and Sylvia, D.M. (1995). Aeroponic culture of VAM fungi. In A. Varma and B. Hock (eds.), *Mycorrhiza* (pp. 426-441). Berlin: Springer-Verlag.

Jasper, D.A., Abbott, L.K., and Robson, A.D. (1992). Soil disturbance in native ecosystems—The decline and recovery of infectivity of VA mycorrhizal fungi. In D.J. Read, D.H. Lewis, A.H. Fitter, and I.J. Alexander (eds.), *Mycorrhizas in ecosystems* (pp. 151-155). Wallingford, UK: CAB International.

Johnson, N.C., Graham, J.H., and Smith, F.A. (1997). Functioning of mycorrhizal associations along the mutualism-parasitism continuum. *New Phytologist* 135: 575-585.

Jumpponen, A. (2001). Dark septate endophytes—Are they mycorrhizal? *Mycorrhiza* 11: 207-211.

Jumpponen, A., Mattson, K.G., and Trappe, J.M. (1998). Mycorrhizal functioning of *Phialocephala fortinii:* Interactions with soil nitrogen and organic matter. *Mycorrhiza* 7: 261-265.

Jumpponen, A. and Trappe, J.M. (1998a). Dark-septate root endophytes: A review with special reference to facultative biotrophic symbiosis. *New Phytologist* 140: 295-310.

Jumpponen, A. and Trappe, J.M. (1998b). Performance of *Pinus contorta* inoculated with two strains of root endophytic fungus *Phialocephala fortinii:* Effects of resynthesis system and glucose concentration. *Canadian Journal of Botany* 76: 1205-1213.

Klironomos J.N. and Hart, M.M. (2002). Colonization of roots by arbuscular mycorrhizal fungi using different sources of inoculum. *Mycorrhiza* 12: 181-184.

Koval'skaya, N.Y., Lobakova, E.S., and Umarov, M.M. (2001). The formation of artificial nitrogen-fixing symbioses with rape *(Brassica napus* var. *napus)* plants in nonsterile soil. *Microbiology* 70: 606-612.

Kundu, D.K. and Ladha, J.K. (1995). Efficient management of soil and biologically fixed N_2 in intensively-cultivated rice fields. *Soil Biology and Biochemistry* 27: 431-439.

Lalonde, M. and Calvert, H.E. (1979). Production of *Frankia* hyphae and spores as an infective inoculant for *Alnus* species. In J.C. Gordon, C.T. Wheeler, and D.A. Perry (eds.), *Symbiotic nitrogen fixation in the management of temperate forest* (pp. 95-110). Corvallis: USDA Forest Research Laboratory, Oregon State University.

le Tacon, F., Alvarez, I.F., Bouchard, D., Henrion, B., Jackson, R.M., Luff, S., Parlade, J.I., Pera, J., Stenström, E., Villeneuve, N., and Walker, C. (1992). Variations in field response of forest trees of nursery ectomycorrhizal inoculation in Europe. In D.J. Read, D.H. Lewis, A.H. Fitter, and I.J. Alexander (eds.), *Mycorrhizas in ecosystems* (pp. 119-134). Wallingford, UK: CAB International.

Lupwayi, N.Z., Olsen, P.E, Sande, E.S., Keyser, H.H., Collins, M.M., Singleton, P.W., and Rice, W.A. (2000). Inoculant quality and its evaluation. *Field Crops Research* 65: 259-270.

Malam Issa, O., Stal, L.J., Défarge, C., Couté, A., and Trichet, J. (2001). Nitrogen fixation by microbial cruss from desiccated Sahelian soils (Niger). *Soil Biology and Biochemistry* 33: 1425-1428.

Malik, K.A., Bilal, R., Mehnaz, S., Rasul, G., Mirza, M.S., and Ali, S. (1997). Association of nitrogen-fixing, plant growth-promoting rhizobacteria (PGPR) with kallar grass and rice. *Plant and Soil* 194: 37-44.

Mallesha, B.C., Bagyaraj, D.J., and Pai, G. (1992). Perlite-soilrite mix as a carrier for mycorrhiza and rhizobia to inoculate *Leucaena leucocephala. Leaucaena Research Reports* 13: 32-33.

Mansfeld-Giese, K., Larsen, J., and Bødker, L. (2002). Bacterial populations associated with mycelium of the arbuscular mycorrhizal fungus *Glomus intraradices. FEMS Microbiology Ecology* 41: 133-140.

Marx, D.H., Cordell, C.E., Kenney, D.S., Mexal, J.G., Artman, J.D., Riffle, J.W., and Molina, R.J. (1984). Commercial vegetative inoculum of *Pisolithus tinctorius* and inoculation techniques for development of ectomycorrhizae on bare-root tree seedlings. *Forest Science Monograph* 25: 1-101.

Marx, D.H. and Kenney, D.S. (1982). Production of ectomycorrhizal fungus inoculum. In N.C. Schenck (ed.), *Methods and principles of mycorrhizal research* (pp. 131-146). St. Paul, MN: Americal Phytopathological Society.

Marx, D.H., Ruehle, J.L., and Cordell, C.E. (1991). Methods for studying nursery and field response of trees to specific ectomycorrhiza. *Methods in Microbiology* 23: 383-411.

Menge, J.A. (1983). Utilization of vesicular arbuscular mycorrhizal fungi in agriculture. *New Phytologist* 81: 553-559.

Meyer, J.R. and Linderman, R.G. (1986). Selective influence on populations of rhizosphere or rhizoplane bacteria and actinomycetes by mucorrhizas formed by *Glomus fasciculatum. Soil Biology and Biochemistry* 18: 191-196.

Mikola, P. (1969). Mycorrhizal fungi of exotic forest plantations. *Karstenia* 10: 169-176.

Mikola, P. (1970). Mycorrhizal inoculationin afforestation. *International Review of Forest Research* 3: 123-196.

Miller, R.M. and Jastrow, J.D. (1992). The role of mycorrhizal fungi in soil conservation. In G.J. Bethlenfalvay and R.G. Linderman (eds.), *Proceedings on mycorrhizae in sustainable agriculture* (pp. 29-44). Madison, WI: ASA Special Publication.

Minerdi, D., Bianciotto, V., and Bonfante, P. (2002). Endosymbiotic bacteria in mycorrhizal fungi: From their morphology to genomic sequences. *Plant and Soil* 244: 211-219.

Mogge, B., Loferer, C., Agerer, R., Hutzler, P., and Hartman, A. (2000). Bacterial community structure and colonization petterns of *Fagus sylvatica* ectomycorrhizospheres as determined by in situ hybridization and confocal laser scanning microscopy. *Mycorrhiza* 5: 271-278.

Mugnier, J. and Mosse, B. (1987). Vesicular arbuscular infection in transformed root inducing T-DNA roots grown axenically. *Phytopathology* 77: 1045-1050.

Newsham, K.K., Fitter, A.H., and Watkinson, A.R. (1995). Multi-functionality and biodiversity in arbuscular mycorrhizas. *Trends in Ecology and Evolution* 10: 407-411.

Okon, Y. (1985). *Azospirillum* as a potential inoculant for agriculture. *Trends in Biotechnology* 3: 223-228.

Okon, Y. and Labandera-Gonzalez, C.A. (1994). Agronomic applications of *Azospirillum:* An evaluation of 20 years worldwide field inoculation. *Soil Biology and Biochemistry* 26: 1591-1601.

Olsson, P.A., Chalot, M., Bååth, E., Finlay, R.D., and Söderström, B. (1996). Ectomycorrhiza mycelium reduces bacterial activity in a sandy soil. *FEMS Microbiology Ecology* 21: 81-86.

Perry, D.A., Molina, R., and Amaranthus, M.P. (1987). Mycorrhizae, mycorrhizospheres and reforestation: Current knowledge and research needs. *Canadian Journal of Forest Research* 17: 929-940.

Polyanskaya, L.M., Vedina, O.T., Lysak, L.V., and Zvyagintsev, D.G. (2002). The growth-promoting effects of *Beijerinckia mobilis* and *Clostridium* sp. cultures on some agricultural crops. *Microbiology* 71: 109-115.

Poole, E.J., Bending, G.D., Whipps, J.M., and Read, D. (2001). Bacteria associated with *Pinus sylvestris-Lactarius rufus* ectomycorrhizas and their effects on mycorrhiza formation in vitro. *New Phytologist* 151: 743-751.

Prat, D. (1992). Effect of inoculation with *Frankia* on the growth of *Alnus* in the field. *Acta Oecologia* 13: 463-467.

Ravnskov, S., Larsen, J., Olsson, P.A., and Jakobsen, I. (1999). Effects of various organic compounds on growth and phosphorus uptake of arbuscular mycorrhizal fungus. *New Phytologist* 141: 517-524.

Redecker, D., Thierfelder, H., and Werner, D. (1995). A new cultivation system for arbuscular mycorrhizal fungi on glass beads. *Angewandte Botanik* 69: 189-191.

Richardson, A.E. (2001). Prospects for using soil microorganisms to improve the acquisition of phosphorus by plants. *Australian Journal of Plant Physiology* 28: 897-906.

Rillig, M.C., Wright, S.F., and Eviner, V.T. (2002). The role of arbuscular mycorrhizal fungi and glomalin in soil aggregation: Comparing effects of five plant species. *Plant and Soil* 238: 325-333.

Rojas, A., Holguin, G., Glick, B.R., and Bashan, Y. (2001). Synergism between *Phyllobacterium* sp. (N_2-fixer) and *Bacillus licheniformis* (P-solubilizer), both from a semiarid mangrove rhizosphere. *FEMS Microbiology Ecology* 35: 181-187.

Roper, M.M., Gault, R.R., and Smith, N.A. (1995). Contribution to the N status of soil by free-living N_2-fixing bacteria in a Lucerne stand. *Soil Biology and Biochemistry* 27: 467-471.

Ruiz-Lozano, J.M. and Bonfante, P. (2001). Intracellular *Burkholderia* strain has no negative effect on the symbiotic efficiency of the arbuscular mycorrhizal fungus *Gigaspora margarita. Plant Growth Regulation* 34: 347-352.

Ryan, M.H. and Graham, J.H. (2002). Is there a role for arbuscular mycorrhizal fungi in production agriculture? *Plant and Soil* 244: 263-271.

Schachtman, D.P., Reid, R.J., and Ayling, S.M. (1998). Phosphorus uptake by plants: From soil to cell. *Plant Physiology* 116: 447-453.

Schultz, R.C., Colletti, J.P., and Faltonson, R.R. (1995). Agroforestry opportunities for the United States of America. *Agroforestry Systems* 31: 117-142.

Schwencke, J. and Carù, M. (2001). Advances in actinorhizal symbiosis: Host plant-*Frankia* interactions, biology, and applications in arid land reclamation: A review. *Arid Land Research and Management* 15: 285-327.

Schwintzer, C.R. and Tjepkema, J.D. (1990). *The biology of* Frankia *and actinorhizal plants*. San Diego, CA: Academic Press.

Shivanna, M.B., Meera, M.S., and Hyakumachi, M. (1994). Sterile fungi from zoyziagrass rhizosphere as plant growth promoters in spring wheat. *Canadian Journal of Microbiology* 40: 637-644.

Smith, S.E. and Read, D.J. (1997). *Mycorrhizal symbiosis*. London: Academic Press.

Smolander, A. and Sarsa, M.L. (1990). *Frankia* strains of soil under *Betula pendula:* Behaviour in soil and in pure culture. *Plant and Soil* 122: 129-136.

Socolow, R.H. (1999). Nitrogen management and the future of food: Lessons from the management of energy and carbon. *Proceedings of the National Academy of Sciences USA* 96: 6001-6008.

Sougoufara, B., Diem, H.G., and Dommergues, Y.R. (1989). Response of field-grown *Casuarina equisetifolia* to inoculation with *Frankia* strain ORS 021001 entrapped in alginate beads. *Plant and Soil* 118: 133-138.

Spaink, H.P., Kondorosi, A., and Hooykaas, P.J.J. (eds.) (1998). *The Rhizobiaceae.* Dordrecht, the Netherlands: Kluwer Academic Publishers.

Sprent, J.I. and Parsons, R. (2000). Nitrogen fixation in legume and non-legume trees. *Field Crops Research* 65: 183-196.

Stamford, N.P., Ortega, A.D., Temprano, F., and Santos, D.R. (1997). Effects of phosphorus fertilization and inoculation of *Bradyrhizobium* and mycorrhizal fungi on growth of *Mimosa caesalpiniaefolia* in an acid soil. *Soil Biology and Biochemistry* 29: 959-964.

Stephens, J.H.G. and Rask, H.M. (2000). Inoculant production and formulation. *Field Crops Research* 65: 249-258.

Sundara, B., Natarajan, V., and Hari, K. (2002). Influence of phosphorus solubilizing bacteria on the changes in soil available phosphorus and sugar cane and sugar yields. *Field Crops Research* 77: 43-49.

Sylvia, D.M. (1990). Inoculation of native woody plants with vesicular-arbuscular fungifor phosphate mine land reclamation. *Agriculture, Ecosystems and Environment* 31: 847-897.

Sylvia, D.M. and Jarstfer, A.G. (1992). Sheared-root inocula of vesicular-arbuscular mycorrhizal fungi. *Applied and Environmental Microbiology* 58: 229-232.

Tian, C., He, X., Zhong, Y., and Chen, J. (2002). Effects of VA mycorrhizae and *Frankia* dual inoculation on growth and nitrogen fixation of *Hippophaë tibetana. Forest Ecology Management* 170: 307-312.

Timmusk, S., Nicander, B., Granhall, U., and Tillberg, E. (1999). Cytokinin production by *Paenobacillus polymyza. Soil Biology and Biochemistry* 31: 1847-1852.

Tisdall, J.M. and Adem, H.H. (1990). Mechanised relay cropping in an irrigated red-brown earth in south-eastern Australia. *Soil Use and Management* 6: 21-28.

Toro, M., Azcón, R., and Barea, J.M. (1998). The use of isotopic dilution techniques to evaluate the interactive effects of *Rhizobium* genotype, mycorrhizal fungi, phosphate-solubilizing rhizobacteria and rock phosphate on nitrogen and phosphorus acquisition by *Medicago sativa. New Phytologist* 138: 265-273.

Torrey, J.G. (1978). Nitrogen fixation by actinomycete-nodulated angiosperms. *Bioscience* 28: 586-592.

Triplett, E. (1996). Diazotrophic endophytes: Progress and prospects for nitrogen fixation in monocots. *Plant and Soil* 186: 29-38.

Unkovich, M.J. and Pate, J.S. (2000). An appraisal of recent field measurements of symbiotic N_2 fixation by annual legumes. *Field Crops Research* 65: 211-228.

Unkovich, M.J., Pate, J.S., and Sanford, P. (1997). Nitrogen fixation by annual legumes in Australian Mediterranean agriculture. *Australian Journal of Agricultural Research* 48: 267-293.

van der Heijden, M.G.A., Boller, T., Wiemken, A., and Sanders, I.A. (1998). Different arbuscular mycorrhizal fungal species are potential determinants of plant community structure. *Ecology* 79: 2082-2091.
van der Heijden, M.G.A., Klironomos, J.N., Ursic, M., Moutoglis, P., Streitwolf-Engel, R., Boller, T., Wiemken, A., and Sanders, I.R. (1998). Mycorrhizal fungal diversity determines plant biodiversity, ecosystem variability and productivity. *Nature* 396: 69-72.
Vance, C.P. (1998). Legume symbiotic nitrogen fixation: agronomic aspects, In H.P. Spaink, A. Kondorosi, and P.J.J. Hooykaas (eds.), *The Rhizobiaceae* (pp. 509-530). Dordrecht, the Netherlands: Kluwer Academic Publishers.
Vance, C.P. (2001). Symbiotic nitrogen fixation and phosphorus acquisition. Plant nutrition in a world of declining renewable sources. *Plant Physiology* 127: 390-397.
Vande Broek, A., Dobbelaere, S., Vanderleyden, J., and Vandommelen, A. (2000). *Azospirillum*-plant root interactions: Signaling and metabolic interactions. In E.W. Triplett (ed.), *Prokaryotic nitrogen fixation: A model system for analysis of a biological process* (pp. 761-777). Wymondham, UK: Horizon Scientific Press.
Vandenkoornhuyse, P., Baldauf, S.L., Leyval, C., Straczek, J., and Young, J.P.W. (2002). Extensive fungal diversity in plant roots. *Science* 295: 2051.
Wall, L.G. (2000). The actinorhizal symbiosis. *Journal of Plant Growth Regulation* 19: 167-182.
White, D.P. (1941). Prairie soil as a medium for tree growth. *Ecology* 22: 398-407.
Wilde, H.E. (1944). Mycorrhizae and silviculture. *Journal of Forestry* 42: 290.
Wood, T. and Cummings, B. (1992). Biotechnology and the future of VAM commercialization. In M.F. Allen (ed.), *Mycorrhizal functioning* (pp. 468-487). London: Chapman and Hall.
Zhang, F., Dashti, N., Hynes, R.K., and Smith, D.L. (1996). Plant growth promoting rhizobacteria and soybean (*Glycine max* L. Merr.) nodulation and nitrogen fixation at suboptimal root zone temperatures. *Annals Botany* 77: 453-459.

Chapter 2

Evaluation of the Functional Group of Microorganisms As Bioindicators on the Rhizosphere Microcosm

Ulisses Brigatto Albino
Galdino Andrade

INTRODUCTION

Environmental disturbance can be assessed through the behavior of the microorganism population in the soil. Many methods can be used to assess the environmental impact in the soil. Some groups, such as rhizobia symbiotic microorganisms (Tao et al., 1992; Lakzian et al., 2002) and mycorrhizal fungi (Requena et al., 1997), are widely used as bioindicators. However, the soil is a microcosm where all groups of microorganisms interact directly or indirectly with each other, with the plants and/or the soil. In this sense, the present chapter deals with functional groups of microorganisms as bioindicators to assess soil disturbances or biotic fertility.

The functional groups of soil microorganisms considered in this study are those that participate in a nutrient recycle and organic matter turnover, acting directly in one or more biochemical cycles.

All biogeochemical cycles are important to recycle matter and maintain life on the planet. However, some cycles, such as those of carbon, nitrogen, phosphorus, and sulfur, can be considered more important because they are involved with nutrients that plants require in greater quantities for their nutrition and growth. Disturbances at any stage of these cycles can be detected through their effects on the plant growth. Correlations between plant growth and numbers of colony forming units (CFU), which makes up each functional group, are important tools for the assessment of the soil nutrient cycle dynamics.

These assessments are based on the transformation of compounds due to the metabolic activity of the microorganisms involved. Population assessment by the CFU number is highly questionable, because the number of microorganisms able to grow in culture medium is only a small percentage of the total microbiota present in the soil. However, high qualitative data can be obtained by applying traditional methods when only a sample of the total soil microbiota that makes the compound turnovers important for plant growth is being assessed. According to the nutrient availability, these microorganisms may increase or reduce the number of individuals that make up the communities of cultured microorganisms.

In the past decade many methods of molecular biology were developed, which allowed estimating unculturable microorganisms from environmental samples. Molecular biology methods, such as PCR (polymerase chain reaction), DGGE (denaturing gradient gel electrophoresis), ARDRA (amplified ribosomal DNA restriction analysis), and sequencing of nucleic acid fragments are used to identify unculturable microorganisms in soil and water samples (Stephen et al., 1999; Stach et al., 2001). These techniques permit an estimate of the biodiversity in all environments. However, the extraction of nucleic acids from soil or water samples has not yet been optimized. The variation of the amount of nucleic acid extracted from these samples is very large. The percentage of C + G (cytosine and guanine) opens the possibility for quantification; however, the precision of this method is very low. This method is based on the bisbenzimidazol chemical characteristic of linking with adenosine and thiamine, allowing the separation of both nucleotides by centrifugation with cesium chloride. However, all these techniques need a large amount (around 50 μg) of nucleic acid extracted. It is very difficult to extract this amount from the soil and the results are replicated many times (Tiedje et al., 1999).

Some parameters of variation such as pH, temperature, biocide contamination, and humidity levels can change the microbial activity completely (Griffiths et al., 1997). In this way, the microorganisms whose DNA sequences were obtained from soil should be changed if the conditions are not the same. Nitrogen fixers inhibit nitrogenase activity when high amounts of ammonium or nitrate are present in the soil and start living as saprophytes (Gordon et al., 1981; Martensson and Torstensson, 1996).

The fact that most microorganisms are considered nonculturable is not true; many of these are culturable but live in a small number of cells in the soil. Sometimes, the drawbacks occur in the sample methods used, age of plant, or period that the microorganisms live in soil (Garland et al., 2001; Jansen et al., 2002). The huge variation of nucleic acid sampling in the soil plus the large number of variables that have an influence on molecular technique show that it is only an estimate method, similar to the traditional techniques which have been used for a long time to study the ecology of soil microorganisms.

THE FUNCTIONAL GROUPS

General Communities

The groups of heterotrophic bacteria, filamentous fungi, and actinomycetes are considered to be general soil communities.

Heterotrophic Bacteria

The heterotrophic bacterial community, which is formed by gram-positive and gram-negative bacteria, is important for the assessment of groups participating in other cycles and having heterotrophic activity. They use mainly carbohydrates as a source of carbon and energy, and any alteration in the sugar level of the exudates can influence their CFU number in the community. Microorganisms such as nitrogen-fixing bacteria and mycorrhizal fungi, which can immobilize carbohydrates or alter any other factor that influences plant photosynthetic rate, can directly interact with the community of heterotrophic bacteria. These bacteria have a fairly high growth rate due to high metabolic activity and show quick variations in CFU number over time. Therefore, this group can be regarded as highly sensitive to environmental alterations when compared to other slower-growing groups such as the actinomycetes and fungi.

Actinomycetes

This microorganism community has a slower growth rate than that of the bacteria and is normally involved in the degradation of more complex carbohydrates and proteins such as starch and cellulose. Variations in the concentration of these compounds can alter the number of CFU in the soil community.

Filamentous Fungi

Fungi such as actinomycetes degrade protein and carbohydrate macromolecules such as cellulose and starch. They are also able to degrade other macromolecules and make nutrients available for the several soil microbiota groups. The size of this and of the actinomycetes communities varies very little in the soil, as their growth rates are relatively slow. The changes in their CFU number may be observed over longer time periods, such as weeks or months.

Carbon Cycle Functional Microorganism Groups

Plants demineralize the atmospheric carbon dioxide (CO_2) by using photosynthesis to form carbohydrates that can be polymerized into reserve substances such as starch or other cell components (such as cellulose and lignin). Carbon dioxide can also be used as a carbon source for the biosynthesis of organic molecules, such as amino acids, esters, lipids, fatty acids, and nucleotides. CO_2 is obtained from the atmosphere by autotrophic organisms, and all the other nutrients are absorbed from the soil solution by the roots and transformed into organic molecules, which form plant components through biosynthesis.

During the plant life, a portion of the organic matter produced is constantly released by the roots through the exudates and lysates and is quickly absorbed by the rhizosphere microbiota. Another part comes from the deposit of dead leaves, twigs, fruit, and flowers, which accumulate on the soil surface, and form an organic litter layer. The decomposition of plant materials in the soil occurs in the following sequence:

1. Cell content is released (proteins, nucleic acids, lipids, starch, etc.), quickly hydrolyzed, and absorbed by the soil microbiota.
2. The more complex molecules such as cellulose and hemicellulose are more slowly hydrolyzed into sugar units and used as carbon sources by heterotrophic microorganisms.
3. The remaining difficult-to-degrade molecules, such as lignin and other compounds, accumulate and will start a humus-formation process.

Nitrogen Cycle Functional Microorganism Groups

Nitrogen is fixed by a small group of gram-positive and gram-negative bacteria, which can interact with plant roots in a symbiotic or free-life way. The nitrogen-fixing groups are responsible for the addition of NH_3^+ to the soil, either directly in the root or in the rhizosphere, depending on the association established between the bacteria and the root. The ammonia molecule can enter amino acid biosynthesis or can be transformed into nitrate (NO_3^-), which is another biologically active nitrogen source for amino acid and protein biosynthesis in plant metabolism. NO_3^- can also be used as the final electron acceptor by the fermenting routes to obtain energy for some facultative anaerobic microorganisms.

Regarding the nitrogen demineralization processes, proteins released in the soil are broken down by proteolytic microorganisms into peptides and amino acids, which become available to the soil microbiota. These amino acids can be used by the soil microbiota as a nitrogen, carbon, or energy source. In addition to the bacteria, fungi, and actinomycetes, the free-life protozoa have an important ecological function in the rhizosphere as they take part in the mineralization and nitrogen immobilization in this system. The protozoa, which prey on bacteria, use the bacteria proteins as a nitrogen source. Nitrogen compounds such as urea and nitrates are produced in some metabolic routes and excreted to the soil solution. On the other hand, the cell components immobilize the nitrogen derived from the bacteria, which will be passed to other trophic levels by preying.

The nitrogen cycle is formed by various stages that have not been commented on here, but any step of the nitrogen cycle can be assessed through proper experimental procedures.

Sulfur Cycle Functional Microorganism Groups

The biological processes involved in sulfur transformation occur basically in anaerobic processes, but some aerobic processes are also observed.

The form of biologically active sulfur for plants is SO^{4-}, and many forms of the element (S, H_2S, SO_3^-) are found in the soil. The autotrophic organisms produce sulfured amino acids such as cystine and cysteine from sulfate, which is demineralized in this process. As previously mentioned in this chapter, the plant makes the nutrient turnover with dead tissue deposits and root products released in the soil. The plant material decomposition releases proteins that are broken down by the microbial proteases into amino acids, including the sulfured ones. The sulfured amino acids can be incorporated by the soil microbiota or degraded with release of H_2S. The heterotrophic microorganisms can also oxidize the elemental sulfur and produce sulfite or thiosulfate, which can be transformed into sulfate.

The microorganism community can be assessed at each stage of the sulfur cycle by these transformations. The stages from sulfur demineralization to the H_2S formation are important for this assessment as they involve the proteolytic activity and sulfur mineralization. Another important stage is sulfate formation since sulfate availability can alter plant growth.

Phosphorus Cycle Functional Microorganism Groups

Phosphorus is essential for plant growth and the mycorrhizal fungi are directly involved in the supply of this element to plants. The total phosphorus quantity in the soil is relatively high, but very little is available to the plants.

The inorganic phosphate solubilization carried out by microorganisms is unspecific, and it occurs when there is excretion of products, which act as solubilizing agents. These products are not induced by the presence of the substrate and do not possess specific binding sites for lyses, either. The molecules released by the microorganisms are formed by their cell metabolism, and include carbon dioxide and organic and inorganic acids.

Heterotrophic microorganisms produce carbon dioxide and also organic acids in the catabolism route of their respiration process or in the fermentative route when under anaerobic conditions. Nitric acid is formed in the soil solution in the presence of nitrates, which are derived from ammonia in the nitrification or mineralization of amino acids. Other inorganic acids such as hydrogen sulfide and sulfuric acid can also be formed by sulfur cycle microorganism groups. Several functional microorganism groups that act in other biogeochemical cycles are active in the solubilization of inorganic phosphates. These relationships are very important in the environmental impact analyses, as they affect the soluble phosphate concentration available for the mycorrhizae-plant system.

Mycorrhizal fungi transport soluble phosphates in soil solution to the roots but remove plant carbohydrates. It is well-known that most plants depend on mycorrhizal activity for their nutrition and are dependent on them for growth. Therefore, any factor that influences mycorrhizal activity will have a direct effect on plant growth. On the other hand, mycorrhizal colonization of the roots and plant carbohydrate absorption decrease the quantity of sugars in the exudates, causing a direct impact on the heterotrophic microorganism community of the rhizosphere, affecting the bacteria, fungi, and actinomycete communities.

The interaction among mycorrhizal fungi and phosphate-solubilizing microorganisms is direct, as these fungi need soluble phosphates to absorb and transport to the root. In a soil with low phosphate content, any decrease in the phosphate-solubilizing activity will diminish the phosphorous available for mycorrhizal activity.

The determination of the inorganic-phosphate-solubilizing microorganisms and mycorrhizal fungi communities are important bioindicators of the phosphorus cycle as any disturbance in these two functional groups influences the plant growth directly.

METHODS TO ASSESS THE SOIL FUNCTIONAL MICROORGANISM GROUPS

Collection of Soil Samples

Soil is typically heterogeneous both vertically and horizontally and influences the microbiota both quantitatively and qualitatively. The sampling process, which includes sample collection and size (number), is a very important factor for success in assessment.

The first step is to determine whether the soil under study comes from the field or from pots. The second step is to determine the sample size (number) and quantity of rhizosphere soil in each sample to make the sampling process representative. Collection should be made with adequate material and the samples stored at approximately 7°C. Storage time should not be longer than 12 hours from collection to dilution and plating.

Determining the Number of Colony Forming Units

Collect soil samples, weigh a 10 g subsample and suspend in 90 mL tubes with a ¼ Ringer's solution and keep on chipped ice. For sample dilution (tenfold), prepare a series of eight 9 mL tubes with ¼ Ringer's solution to obtain a final 10^{-8} dilution. Starting with the soil suspension tube, shake in vortex and transfer 1 mL to the next tube successively until the last tube in the series is used, homogenizing the samples in vortex before each transference.

It should be considered that the number of CFU in each soil functional group is different. Normally, bacteria have around 10^8 CFU per gram of soil while fungi and actinomycetes have 10^4 CFU per gram of soil. However, these numbers can vary with soil type or other environmental factors (moisture, pH, organic matter, temperature). After agitating the tube with the required dilution in vortex, inoculate 0.1 mL in two petri dishes. For each plate, sterilize a glass spatula in a flame after dipping it in ethanol. Let the spreader cool briefly and then touch to the agar of an inoculated plate, away from the inoculum, to further cool the spreader. Then, spread the inoculum by moving the spreader in an arc fashion on the surface of the agar while rotating the plate. Continue until the inoculum has been absorbed into the agar. Repeat for each inoculum dilution.

PROTOCOLS

Ringer's Solution

NaCl	$8.50\ g \cdot L^{-1}$
KCl	$0.25\ g \cdot L^{-1}$
$CaCl_2$	$0.30\ g \cdot L^{-1}$
$NaHCO_3$	$0.20\ g \cdot L^{-1}$
Distilled water	1,000 mL

1. Sterilize in autoclave (121°C for 20 minutes).
2. Agitate until the salts are completely dissolved. To obtain the ¼ solution, take 250 mL of the Ringer's solution and complete to one liter.

General Communities

The 1/10 series dilution technique was used as described previously to determine the number of bacteria, actinomycetes, and filamentous fungi CFU. According to the population size, the following dilutions are chosen for inoculation: bacteria (10^{-7} and 10^{-8}), fungi and actinomycetes (10^{-3} and 10^{-4}).

Culture Media—Bacteria

Tryptic Soy Agar (TSA)

Casein pancreatic digest	17.0 $g \cdot L^{-1}$
soybean meal papaic digest	3.0 $g \cdot L^{-1}$
NaCl	5.0 $g \cdot L^{-1}$
K_2HPO_4	2.5 $g \cdot L^{-1}$
Dextrose	2.5 $g \cdot L^{-1}$
Agar	15.0 $g \cdot L^{-1}$
Distilled water	1,000 mL
pH	7.0

1. Sterilize in autoclave (121°C for 20 minutes)
2. Place approximately 20 mL in each petri dish.
3. Inoculate the 10^{-7} and 10^{-8} dilutions and incubate at 28°C for 48 hours.

Culture Media—Fungi

Potato Dextrose Agar

Potato infusion	4.0 $g \cdot L^{-1}$
Dextrose	20.0 $g \cdot L^{-1}$
Bacto agar	15.0 $g \cdot L^{-1}$
Distilled water	1,000 mL
pH	7.0

1. Sterilize in autoclave (115°C for 30 minutes)

2. Place approximately 20 mL in each petri dish.
3. Inoculate the 10^{-3} and 10^{-4} dilutions and incubate at 28°C for 3 to 4 days.

Actinomycetes

Starch Casein Agar (Kuster and Williams, 1964)

Starch	10.0 $g \cdot L^{-1}$
Casein hydrolysate	0.3 $g \cdot L^{-1}$
KNO_3	2.0 $g \cdot L^{-1}$
NaCl	2.0 $g \cdot L^{-1}$
K_2HPO_4	2.0 $g \cdot L^{-1}$
$MgSO_4 \cdot 7H_2O$	0.05 $g \cdot L^{-1}$
$CaCO_3$	0.02 $g \cdot L^{-1}$
$FeSO_4$	0.01 $g \cdot L^{-1}$
Bacto agar	15.0 $g \cdot L^{-1}$
Distilled water	1,000 mL
pH	7.0

1. Sterilize in autoclave (121°C for 20 minutes).
2. Place approximately 20 mL in each petri dish.
3. Inoculate the 10^{-3} and 10^{-4} dilutions and incubate at 28°C for 3 to 4 days.

Carbon Cycle Functional Group Microorganisms

The 1/10 serial dilution technique is used as previously described to determine the CFU number of the carbon cycle functional microorganism groups. The dilutions for inoculation are chosen according to the size of the populations used for inoculation. However, a previous quantification of the microorganisms in the studied soil is suggested.

Starch Hydrolizers (Pontecorvo et al., 1953)

Soluble starch	10.0 $g \cdot L^{-1}$
Levedure extract	0.1 $g \cdot L^{-1}$
$MgSO_4 \cdot 7H_2O$	0.5 $g \cdot L^{-1}$
KH_2PO_4	1.5 $g \cdot L^{-1}$
NH_4NO_3	0.1 $g \cdot L^{-1}$
NaCl (0.85 percent solution)	50 mL

Soil extract*	950 mL
Bacto agar	15.0
pH	7.0

*Soil Extract:

1. Suspend the soil in distilled water (1:1).
2. Agitate well.
3. Decant.
4. Filter through filter paper.
5. Sterilize in autoclave (121°C for 20 minutes).
6. Place approximately 20 mL in each petri dish.
7. Inoculate the 10^{-4} and 10^{-5} dilutions and incubate at 28°C for 4 days.

To observe the starch degradation halo, add an iodine solution to the surface of the medium on top of the formed colonies, remove the excess, and count the halo forming colonies.

Cellulolytic (Wood, 1980)

Carboximethyl cellulose	5.0 $g \cdot L^{-1}$
NH_4NO_3	1.0 $g \cdot L^{-1}$
Saline solution (0.8 percent)	50 mL
Soil extract*	950 mL
Bacto agar	5.0
pH	7.0

1. Sterilize in autoclave (121°C for 20 minutes).
2. *Soil Extract as described on amilolitic media.
3. Place approximately 20 mL in each petri dish.
4. Inoculate the 10^{-3} and 10^{-4} dilutions and incubate at 28°C for 4 days.

Develop

1. Add 1M NaCl solution to the medium surface for 5 minutes.

Eliminate

1. Add a 0.1 percent Congo red solution for 30 minutes.
2. Wash with distilled water until halos around the colonies are observed.
3. Count the halo forming colonies.

Proteolytic (Wood, 1980, modified by Andrade)

Hydrolyzed casein	10.0 $g \cdot L^{-1}$
Levedure extract	0.1 $g \cdot L^{-1}$
K_2HPO_4	1.5 $g \cdot L^{-1}$
$MgSO_4 \cdot 7H_2O$	0.5 $g \cdot L^{-1}$
Saline solution (0.8 percent)	50 mL
Soil extract*	950 mL
Bacto agar	15.0
pH	7.0

1. *Soil Extract as described on amilolitic media.
2. Sterilize in autoclave (121°C for 20 minutes).
3. Place approximately 20 mL in each petri dish.
4. Inoculate the 10^{-3} to 10^{-4} dilutions and incubate at 28°C for 4 days.

Develop

1. Add 0.1N HCl solution.
2. Count the halo forming colonies.

Nitrogen Cycle Functional Group Microorganisms

The 1/10 serial dilution technique was used as previously described to determine the CFU number of the nitrogen cycle functional microorganism groups. Some groups will be determined by the most probable number (MPN) technique, which consists of diluting each sample at least in duplicate and counting the number of positive and negative cases for each dilution. The total number of cells is calculated from the McGrady tables (Postgate, 1969). The dilutions for inoculation depend on the size of the populations. However, a previous quantification of the microorganisms in the studied soil is suggested.

Free-Living Nitrogen-Fixing Bacteria

Nitrogen-Free Malate (Nfb) (Döbereiner and Day, 1976)

KH_2PO_4	0.4 $g \cdot L^{-1}$
K_2HPO_4	0.1 $g \cdot L^{-1}$
$MgSO_4 \cdot 7H_2O$	0.2 $g \cdot L^{-1}$
NaCl	0.1 $g \cdot L^{-1}$

$CaCl_2$	0.02 $g \cdot L^{-1}$
$FeCl_3$	0.01 $g \cdot L^{-1}$
$MoO_4Na \cdot 2H_2O$	0.002 $g \cdot L^{-1}$
Sodium malate	5.0 $g \cdot L^{-1}$
Bromothymol blue (0.5 percent alcohol solution)	5 mL
Agar	15.0 mL
pH	7.0

1. Sterilize in autoclave (121°C for 20 minutes).
2. Place approximately 20 mL in each petri dish.
3. Inoculate the 10^{-4} and 10^{-6} dilutions and incubate at 28°C for 7 days under microaerophyllic conditions.

The colonies that fix nitrogen will alkalinize the culture medium (blue) due to ammonia production. Count the colonies with a blue halo and calculate the CFU number.

Burk's Culture Medium (Wilson and Knight, 1952)

Solution A

K_2HPO_4	6.4 $g \cdot L^{-1}$
KH_2PO_4	1.6 $g \cdot L^{-1}$
Distilled water	1,000 mL

Solution B

NaCl	2.0 $g \cdot L^{-1}$
$MgSO_4 \cdot 7H_2O$	2.0 $g \cdot L^{-1}$
$CaSO_4 \cdot 2H_2O$	0.5 $g \cdot L^{-1}$
Distilled water	1,000 mL

Solution C

$NaMoO_4 \cdot 2H_2O$	0.01
$FeSO_4$	0.03
Distilled water	1,000 mL

Composition

Solution A	100 mL
Solution B	100 mL

Solution C	100 mL
Dextrose	5 mL
Bacto agar	15 mL
Distilled water	700 mL
pH	7.0

1. Sterilize in autoclave (115°C for 30 minutes).
2. Place approximately 20 mL in each petri dish.
3. Inoculate the 10^{-4} to 10^{-6} dilutions and incubate at 28°C for 7 days.
4. Count the number of colonies and calculate the number of CFU.

Ammonifiers (Saratchandra, 1978)

Hydrolyzed casein	10.0 $g \cdot L^{-1}$
Yeast extract	0.1 $g \cdot L^{-1}$
K_2HPO_4	1.0 $g \cdot L^{-1}$
$MgSO_4 \cdot 7H_2O$	0.1 $g \cdot L^{-1}$
$FeSO_4 \cdot 7H_2O$	0.01 $g \cdot L^{-1}$
$MnSO_4 \cdot 4H_2O$	0.01 $g \cdot L^{-1}$
Phenol red	0.02 $g \cdot L^{-1}$
Distilled water	1,000 mL
pH	6.5

1. Prepare at least three replicates of each sample for each dilution (e.g., dilutions from 10^{-1} to 10^{-5}; total, 15 tubes).
2. Place approximately 4 mL of culture medium in each tube.
3. Sterilize in autoclave (121°C for 20 minutes).
4. Inoculate the tubes containing culture medium with 1 mL of the required dilutions and incubate at 28°C for 5 days.
5. Record the tubes with ammonia (orange to pink color) or unaltered as negative.
6. Estimate the ammonifying microorganism population using the most likely number tables (Postgate, 1969).

Protozoans (Darbyshire et al., 1974)

1. Prepare soil extract (see cellulolitic).
2. Take one 24-well with lid tissue culture plate for each soil sample.
3. Fill the wells with 1.8 mL of soil extract and transfer 0.2 mL for dilution to 10^{-6}. Four replications should remain at the 10^{-6} dilution.
4. Incubate at 25°C or at normal soil temperature from the place samples where collected.

After the third day look for the presence or absence of flagellates, ciliates, and amoebas by homogenizing the suspension, collect approximately 0.2 mL, place on a slide, and cover with a slide cover. Observe using 20 to 100 times magnification under a phase-contrast microscope (recommended), or under an interference contrast microscope (highly recommended).

To calculate the number of cells in each group, record the positive and negative results and calculate the MPN according to McGrady's tables (Postgate, 1969).

Sulfur Cycle Functional Group

Anaerobic Sulfur Mineralizing (Pochon and Tardieux, 1962)

K_2HPO_4	0.5 $g{\cdot}L^{-1}$
NH_4Cl	1.0 $g{\cdot}L^{-1}$
$CaCl_2$	0.1 $g{\cdot}L^{-1}$
L-Cystein	1.0 $g{\cdot}L^{-1}$
Fe-NH4 citrate	0.5 $g{\cdot}L^{-1}$
Na lactate at 60 percent	5.0 mL
Soil extract	20.0 mL
Distilled water	100.0 mL
pH	7.0

1. Sterilize in autoclave (121°C for 20 minutes).
2. Place 12 mL of this medium in screw-top test tubes; prepare the dilutions with at least two replicates for each dilution.
3. Inoculate 1 mL of the soil suspension with the same number of replicates and also with the culture medium without the sulfur source.
4. Incubate at 28ºC and examine every 3 days for 29 days.
5. Record the positive and negative results and calculate the MPN according to McGrady's table (Postgate, 1969).

Autotrophic Sulfur Oxidizers

Starkey's Culture Medium (Kuenen and Tuovinen, 1981)

K_2HPO_4	3.5 $g{\cdot}L^{-1}$
$(NH_4)_2SO_4$	0.3 $g{\cdot}L^{-1}$
$MgSO_4{\cdot}7H_2O$	0.5 $g{\cdot}L^{-1}$
$FeSO_4{\cdot}7H_2O$	0.018 $g{\cdot}L^{-1}$
$CaCl_2$	0.25 $g{\cdot}L^{-1}$

Flowers of sulfur	5.0 g·L^{-1}
Distilled water	1,000 mL
pH	7.0

1. Sterilize in autoclave (121°C for 20 minutes).

Elemental sulfur served as the sulfur source in this medium. One mL of rhizosphere-soil-diluted inoculum is added to tubes with 10 mL of culture medium and incubated under anaerobic conditions. The sulfur oxidizer tubes and controls are incubated at 28°C in the dark to exclude phototrophic sulfur bacteria. Initial measurements are taken after four weeks. The presence of H_2SO_4 is checked weekly for six weeks by monitoring the pH changes. Record the positive and negative results and calculate the MPN according to McGrady's table (Postgate, 1969).

Sulfate-Reducing Bacteria: Lapage Culture Medium (Pfennig et al, 1981)

Solution 1

K_2HPO_4	0.5 g·L^{-1}
NH_4Cl	1.0 g·L^{-1}
Na_2SO_4	1.0 g·L^{-1}
$CaCl_2 \cdot 2H_2O$	0.1 g·L^{-1}
$MgSO_4 \cdot 7H_2O$	2.0 g·L^{-1}
Na lactate (70 percent)	3.5 g·L^{-1}
Yeast extract	1.0 g·L^{-1}
Distilled water	980 mL

Solution 2

$FeSO_4 \cdot 7H_2O$	0.5 g·L^{-1}
(Preferably acidic with sulfuric acid)	
Distilled water	10 mL

Solution 3

Ascorbic acid	0.1 g
Na thioglicolate	0.1 g·L^{-1}
Distilled water	10 mL
pH	7.0

1. Sterilize in autoclave (121°C for 20 minutes).
2. Fill screw-top tubes to two-thirds capacity with Lapage culture medium, and 1 mL of rhizosphere-soil-diluted inoculum with at least two replicates for each dilution. The caps must be tightened to exclude the diffusion of atmospheric oxygen during the six-week incubation period.
3. Check every three days for black precipitate at the bottom of the tube, which indicates the presence of FeS.
4. Record the positive and negative results and calculate the MPN according to McGrady's table (Postgate, 1969).

Phosphorus Cycle Functional Group Microorganisms: Inorganic Phosphate Solubilizing (Sylvester-Bradley et al., 1982, Modified)

Solution A

KNO_3	0.1 $g \cdot L^{-1}$
Dextrose	10.0 $g \cdot L^{-1}$
Yeast extract	5.0 $g \cdot L^{-1}$
$MgSO_4 \cdot 7H_2O$	0.2 $g \cdot L^{-1}$
$CaCl_2$	0.02 $g \cdot L^{-1}$
NaCl	0.1 $g \cdot L^{-1}$
Micronutrients*	2.0 mL
FeEDTA**	4.0 mL
CaCl (10 percent, solution B)	100 mL
K_2HPO_4 (10 percent, solution C)	50 mL
Bacto agar	15.0 $g \cdot L^{-1}$
Distilled water	844 mL

Solution B

$CaCl_2$	10.0 $g \cdot L^{-1}$
Distilled water	100.0 mL

Solution C

K_2HPO_4	5.0 $g \cdot L^{-1}$
Distilled water	50.0 mL

**Micronutrients Solution*

$Na_2MoO_4{\cdot}2H_2O$	0.20 $g{\cdot}L^{-1}$
$MnSO_4{\cdot}2H_2O$	0.24 $g{\cdot}L^{-1}$
H_3BO_3	0.28 $g{\cdot}L^{-1}$
$CuSO_4{\cdot}5H_2O$	0.008 $g{\cdot}L^{-1}$
$ZnSO_4{\cdot}7H_2O$	0.024 $g{\cdot}L^{-1}$
Distilled water	200 mL

***FeEDTA Solution*

Add 6.07 g of Na EDTA to 6.17 g of $FeSO_4{\cdot}7H_2O$ in 900 mL distilled water heated to 80°C. Agitate until dissolved and complete volume to 1,000 mL; pH = 7.0

1. Sterilize in autoclave (121°C for 20 minutes).
2. Correct the pH of the A, B, and C solutions and autoclave separately.
3. Check the pH of the B and C solutions before mixing with solution A.
4. Place in Petri dishes.
5. Make the serial dilution, inoculating 0.1 mL of each dilution in duplicate.
6. Incubate at 28°C for 3 weeks and check to see if a solubilization halo has formed around the colonies.
7. Calculate the number of colony forming units.

Arbuscular Mycorrhizal Fungi

Mycorrhizae Infected Root Staining Method (Phillips and Hayman, 1970)

1. Wash the roots in running water to remove soil particles and cut into 1 cm pieces. Submerge the roots in 10 percent KOH for 1 hour at 90°C in a water bath or in autoclave with steam flow. Eliminate the KOH and add 10 percent H_2O_2 alkaline for 1 hour (H_2O_2 is only used when the roots are very dark).
2. Eliminate the H_2O_2 and wash five times in water.
3. Add 0.1N HCl for 3 to 5 minutes at room temperature.
4. Eliminate the 0.1N HCl and wash 2 or 3 times in running water.
5. Add lactic acid trypan blue (0.05 percent) solution for 5 to 10 minutes at 90°C in a water bath or in autoclave with steam flow.
6. Eliminate the excess stain and preserve the roots in lactic acid.

Method to Quantify Root Mycorrhizae Infection (Giovanetti and Mosse, 1980)

The roots stained by the previously described method are quantified under a stereoscopic microscope. The root fragments (1 to 2 cm) are distributed randomly on a reticulate (1.27 cm) 14 × 11 cm plate. Quantification is carried out by counting the number of intercessions which have MA fungi structures in the roots. The results are expressed in percentage of mycorrhizae-infected roots.

FUTURE PERSPECTIVES

Functionality aspects are more important than biodiversity in natural or sustainable agriculture systems. Some questions could be raised concerning biodiversity. The first question that should be asked is this: What is more important to the earth—the number of species that compose the functional group or the transformation power of one group? On the other hand, other questions could be asked: What is the importance of one species in the biological dynamic system? What is the capacity of one species to influence nutrient cycling? What does a species represent within the biological dynamic? What importance can one species have in nutrient cycling? These questions could lead us to conclude that we need to rethink our view of the soil microcosm and extend our understanding of the biological processes and interactions that occur in the soil-plant system, assuming that these processes are a whole and each functional group is only a small fraction of the whole. Only in this way can we improve the determination of the environmental impact of any disturbance effect on the soil microbial community, not only on one specific group of microorganisms. The several functional groups that take part in different stages of the carbon, phosphorus, nitrogen, and sulfur biogeochemical cycles should be assessed, looking for the correlation between them and a response in plant growth.

Microbial diversity is important for other objectives, such as searching for specific microbial phenotypes to use in the food or pharmaceutical industry. Its importance in the environment should also be investigated, as molecular biology does not permit assessment of microbial interaction mechanisms in the soil microcosm.

CONCLUSIONS

The studies of environmental impact can be carried out through the evaluation of the correlations of the CFU number of the microorganism populations that live in the environment. In the soil, the interactions between microorganisms and plants possess high metabolic activity and result in the cycling of nutrients. In this way, it is possible to divide the microorganism community into functional groups, so that one species can participate, in different situations, in more than one functional group of a certain biogeochemical cycle. In studies of environmental impact assessment, the functional group can be an important tool with which to assess disturbance in the natural environment.

It is necessary to continue to pursue a more holistic view of the complexity of interactions existing in the environment. It is worthless to study deeply only one group of microorganisms. A widening of our understanding of microbial interactions is possible not only through new methods but also through the human mind. By doing so we will be able to understand the miracle of life maintenance on our planet.

REFERENCES

Darbyshire J.F., Wheatley R.E., Greaves M.P., and Inkson R.H.E. (1974). A rapid micromethod for estimating bacterial and protozoan populations in soil. *Rev. Ecol. Biol. Soil. 11: 465-475.*

Döbereiner J. and Day J.M. (1976). Associative symbiosis in tropical grasses: Characterization of microorganisms and dinitrogen fixing sites. In *Proceedings of the 1st International Symposium on Nitrogen Fixation II* (pp. 518-538). Pullman: Washington State University Press.

Garland J.L., Cook K.L., Adams J.L., and Kerkhof, L. (2001). Culturability as an indicator of succession in microbial communities. *Microbial Ecology* 42: 150-158.

Giovanetti M. and Mosse B. (1980). An evaluation of techniques for measuring vesicular-arbuscular mycorrhizal infections in roots. *New Phytol.* 84: 489-500.

Gordon J.K., Shah V.K., and Brill W.J. (1981). Feedback inhibition of nitrogenase. *J. Bacteriol.* 148: 884-888.

Griffiths B.S., Ritz K., and Wheatley R.E. (1997). Relationship between functional diversity and genetic diversity in complex microbial communities. In H. Insam and A. Rangger (eds.), *Microbial communities, functional versus structural approaches* (pp. 1-9). Berlin: Springer.

Jansen P.H., Yates P.S., Grinton B.E., Taylor P.M., and Sait M. (2002). Improved culturability of soil bacteria and isolation in pure culture of novel members of the divisions Acidobacteria, Actinobacteria, Proteobacteria and Verrucomicrobia. *Appl. Environ. Microbiol.* 68: 2391-2396.

Kuenen J.G. and Tuovinen O.H. (1981). The genera of *Thiobacillus* and *Thiomicrospira*. In M.P. Starr and H. Stolp (eds.), *The prokaryotes: A handbook on habitats, isolation and identification of bacteria* (pp. 1023-1036). New York: Springer Verlag.

Kuster E. and Williams S.T. (1964). Selection of media for isolation of streptomycetes. *Nature* 202: 928-929.

Lakzian A., Murphy P., Turner A., Beynon J.L., and Giller K.E. (2002). *Rhizobium leguminosarum* bv. *viciae* populations in soils with increasing heavy metal contamination: Abundance, plasmid profiles, diversity and metal tolerance. *Soil Biol. Biochem.* 34: 519-529.

Martensson A.M. and Torstensson L. (1996). Monitoring sewage sludge using heterotrophic nitrogen fixing microorganisms. *Soil Biol. Biochem.* 28: 1621-1630.

Pfennig N., Widdel F., and Trüper H.G. (1981). The dissimilatory of sulfate-reducing bacteria. In M.P. Starr and H. Stolp (eds.), *The prokaryotes: A handbook on habitats, isolation and identification of bacteria* (pp. 1005-1022). New York: Springer Verlag.

Phillips D.A. and Hayman D.S. (1970). Improved procedures for clearing roots and staining parasitic and vesicular-arbuscular mycorrhizal fungi for rapid assessment of infection. *Trans. Br. Mycol. Soc.* 55: 158-161.

Pochon J. and Tardieux Y. (1962). *Techniques d'analyse en microbiologie du sol.* Paris: De la Tourelle.

Pontecorvo G., Roper J.A., Hemons L.M., MacDonalds K.D., and Buffon A.W.J. (1953). The genetics of *Aspergillus nidulans. Adv. Gen.* 5: 141-238.

Postgate J.R. (1969). Viable counts and viability. In J.R. Norris and D.W. Ribbins (eds.), *Methods in microbiology,* Vol. 1 (pp. 611-618). New York: Academic Press.

Requena N., Jimenez I., Toro M., and Barea J.M. (1997). Interactions between plant-growth-promoting rhizobacteria (PGPR), arbuscular mycorrhizal fungi and *Rhizobium* spp. in the rhizosphere of *Anthyllis cytisoides,* a model legume for revegetation in mediterranean semi-arid ecosystems. *New Phytol.* 136: 667-677.

Saratchandra S.V. (1978). Nitrification activities and the changes in the population of nitrifying bacteria in soil prefused with two different H-ion concentrations. *Plant Soil* 50: 99-111.

Stach J.E.M., Batle S., Clapp J.P.,and Bums R.G. (2001). PCR-SSCP comparison of 16S rDNA sequence diversity in soil DNA obtained using different isolation and purification methods. *FEMS Microbiol. Ecol.* 36: 139-151.

Stephen J.R., Chang Y.J., Gan Y.D., Peacock A., Pfiffner S.M., Barcelona M.J., White D.C., and Macnaughton S.J. (1999). Microbial characterization of a JP-4 fuel-contaminated site using a combined lipid biomarker/polymesase chain reaction-denaturing gradient gel electrophoresis (PCR-DGGE)-based approach. *Environm. Microbiol.* 1: 231-241.

Sylvester-Bradley R., Akasawa N., La Torranca S., Magalhães F.M.M., Oliveira L.A., and Pereira R.M. (1982). Levantamento quantitativo de microrganismos solubilizadores de fosfatos na rizosfera de gramíneas e leguminosas forrageiras na Amazonia. *Acta Amazonica* 12: 15-22.

Tao H., Brewin N.J., and Noel K.D. (1992). *Rhizobium leguminosarum* CFN42 lipopolysacharide antigenic changes induced by environmental conditions. *J. Bact.* 174: 2222-2229.

Tiedjc J.M., Asuming-Brempong S., Nusslein K., Marsh T.L., and Flynn S.J. (1999). Opening the black box of soil microbial diversity. *Appl. Soil Ecol.* 13: 109-122.
Wilson P.W. and Knight S.C. (1952). *Experiments in bacterial physiology*. Minneapolis, MN: Burgess.
Wood P.J. (1980). Specify in the interactions of direct dyes with polysaccharides. *Carbohydrate Res.* 85: 271-287.

Chapter 3

Tripartite Relationship of *Rhizobium*, AMF, and Host in Growth Promotion

Nanjian Raman
Thangaswamy Selvaraj

INTRODUCTION

The increase in fertilizer use, especially nitrogen, has played a major role in recent productivity gains. However, increasing the amount of nitrogen applied decreases its use efficiency, and a large amount of it is left behind, causing environmental problems. Nitrogen, even though it is abundant in the atmosphere, is the most limiting nutrient for crop yields in many soils, while phosphorus is the most limiting nutrient under natural conditions in the tropics and the second most limiting plant nutrient for farmers. Biological nitrogen fixation by free-living microorganisms in the soil or those associated with plants has a great potential to contribute to food production. Biological nitrogen fixation uses energy derived from photosynthesis and does not accumulate excess nitrogen to cause pollution. *Rhizobium,* a genus of gram-negative soil bacteria, is capable of establishing nitrogen-fixing symbiosis within roots of plants belonging to the Leguminosae family. Legume nodules export symbiotically fixed nitrogen as amino acids, amides, or ureides apoplastically via xylem into the shoot of the host plant. Assimilates from photosynthesis are imported into nodules via phloem in the form of sucrose. Export of nitrogen compounds from the infected cells and import of energy-rich assimilates are essential requirements for efficient biological N_2 fixation. The legume host supplies carbon and energy to the nitrogen-fixing bacteria, which in turn reduce nitrogen into ammonia and export it to the plants (Bergerson and Turner, 1967; Miller et al., 1988). The mutual symbiotic association between a plant and a fungus, mycorrhizae, is also playing major role in plant growth. A diverse group of soil fungi spanning the Zygomycetes, Ascomycetes, and Basidiomycetes (Harley and Smith, 1983) form mycorrhizae. Arbuscular mycorrhizae are formed between members of the order Glomales of Zygomycetes and the species of angiosperms which also includes the most important agricultural crops. The arbuscular mycorrhizal fungi (AMF) are ubiquitous and are found in diverse habitats ranging from the arctic to the tropics, arid to aquatic environments, and stable plant communities to highly disturbed ecosystems. AMF are vital for uptake and accumulation of ions from soil and translocation to hosts because of their high metabolic rate and strategically diffuse distribution in upper soil layers. In fact, the fungus serves as a highly efficient extension of the host root system. The fungi derive most of their needed organic matter from their symbiotic niche in the roots and, in turn, help their host plants achieve better growth through enhanced phytochrome levels (Raman and Mahadevan, 1996) in absorption of phosphorous and other mobile elements from soil, impart tolerance to heavy metals, and afford protection against disease, salinity, drought, and temperature (Raman and Sambandan, 1998).

The term *plant growth promotion* is used to describe the increased plant growth and crop yield that occurs upon inoculating roots with certain specific root colonizing organisms (Kloepper et al., 1980; Gupta et al., 1998). Mycorrhizal fungi are of great value in promoting water transport, phosphorus uptake, plant growth, and high yield. *Rhizobium* species are able to fix the atmospheric nitrogen in the nodules of legumes plants, by which they help in the N_2 uptake by plants. Microbial interactions involving AMF are particularly interesting in the case of *Rhizobiun* species since both microorganisms colonize the root system of a common host plant, a legume (Barea and Azcon-Aguilar, 1983; Hayman, 1986). The interactions between both endophytes show nutritional as well as nonnutritional host-mediated effects (Harris et al., 1985; Brown and Bethlenfalvay 1987; Brown et al., 1988). However, information is scarce concerning combinations of *Rhizobium* with different AM fungal species. Such information would be important from the basic point of view and also for practical purposes, since synergism between AM colonization and nodulation can result in decisive nutrient uptake and nitrogen fixation by the common host legume.

Considering the importance of the tripartite relationship of *Rhizobium,* AMF, and host in growth promotion, this chapter deals with the *Rhizobium* and AMF interactions.

RHIZOBIUM *AND AMF INTERACTIONS*

The agriculture productivity in tropical and subtropical countries is often minimal because of low soil nitrogen and phosphate availability (Freire, 1984). Exploitation of legume-*Rhizobium* symbiosis can provide an economic source of available nitrogen (Ahmad et al., 1981). The AM fungi can improve phosphorous uptake and ultimately plant growth and yield (Mosse, 1973). Phosphorous (P) deficiency is an important limiting factor in nitrogen (N) fixation and legume production (Freire, 1984; Jakobsen, 1985).

Inoculation of AM fungi and rhizobia with legumes gives benefit from P uptake and generally has greater N_2 fixation, nodule mass, N, P content, and yield than legumes inoculated with only rhizobia (Barea and Azcon-Aguilar, 1983). However, proper selection of AM fungi and rhizobial strains is an important step because in a tripartite association (AM fungi, rhizobia, and the host) there should be compatibility to form an effective symbiosis (Ames et al., 1991).

Various researchers have shown that the inoculation of legumes with AM fungi improves their growth and yield (Islam and Ayanaba, 1981; Ganry et al., 1982). Badr El-Din and Moawad (1988) indicated that dual inoculation with AM fungus and rhizobia enhanced the growth of legumes. Bagyaraj et al. (1979) and De la Cruz et al. (1988) have shown that there is no significant difference in yield of legumes when inoculated with *Rhizobium* alone. However, Pacovsky, Bethlenfalvay, and Paul (1986) and Pacovsky, Paul, and Bethlenfalvay (1986) have claimed that by changing the *Bradyrhizobium* strain, mycorrhizal plants have shown improvement in the yield of soybean. None of the researchers coselected the compatible pairing of AM fungus-rhizobia prior to their use as dual inocula. Dual inoculations increased pod dry weight by 25 percent and more than 16 percent in single inoculations of AM fungi and rhizobia, respectively, in black and sandy clay loam soil types. Thiagarajan et al. (1992) have demonstrated the field testing of coselected strains of rhizobia and AM fungi on cowpea and shown that the growth in a given soil type can be maximized by inoculation with compatible *Rhizobium*–AM fungus pairing.

Positive effects of dual colonization of soybean roots by AM and diazotrophic rhizobia have been studied by many investigators (Barea and Azcon-Aquilar, 1983; Pacovsky, Paul, and Bethlenfalvay, 1986; Gianinazzi-Pearson et al., 1981). Both rhizobia and AM fungi are active in root cortical cells and, hence, it can be assumed that the presence of one microsymbiont will affect the activity of the other (Azcon et al., 1991). Most studies have concentrated on the indirect relationships between AM fungi and rhizobia in which successful symbiosis has been measured as an increased uptake of nutrient elements, P in particular, by the plant (Ames and Bethlenfalvay, 1987) and an increased N_2 fixation and nodule mass (Fredeen and Terry, 1988).

AM infection stimulated Zn uptake in plants. In presence of both AM and rhizobial strains, Zn concentrations in leaves of mycorrhizal fungi were greater at all harvests. Concentrations of Cu were significantly higher in

AM plants than in non-AM plants at the initial stage of growth, but not at flowering, and they increased again later (Vejsadova et al., 1992). The mutual relationship among the AM fungus *Bradyrhizobium* and soybean were strongly dependent on the growth stages of the plants (Bethlenfalvay et al., 1984).

Skrdleta et al. (1984) indicated that in spite of the rhizobial activity, the presence of AM fungus was significant in soybean. Manjunath et al. (1984) showed the improved nodulation in *Leucaena leucocephala* when dual inoculations with *Glomus fasciculatum* and *Rhizobium* were made. AM fungi increase phosphorus uptake of plants growing in phosphorus-deficient soils (Manjunath and Bagyaraj, 1984) mainly through increased surface area of absorption. Therefore there has been increasing interest regarding the possible use of AMF and *Rhizobium* together to improve growth and nodulations in legumes. These results bring out a synergistic or additive interaction between *G. fasciculatum* and *Rhizobium* with consequential effects on nodulation, nitrogen fixation, and phosphorus uptake in *Leucaena*. Legumes have repeatedly been shown to require high levels of phosphorus for effective growth, nodulation, and nitrogen fixation (Gibson, 1967).

It was believed that *Phaseolus* was either less effective at photosynthesis or G transport than other legumes and so the decreased supply of carbohydrates to the root limited N_2 fixation (Graham, 1981) or AM colonization (Daft and El-Giahmi, 1976). Host-endophyte interactions (Hohenberg et al., 1982), excessive H_2 production by nodules (Pacovsky et al., 1984) and in commercially important cultivars, and lower nitrogenase activity (Graham, 1981) or decreased N transport have also been reported for the reduced levels of N_2 fixation in beans, mycorrhizae (Gianinazzi-Pearson and Gianinazzi, 1978). Lack of host response to colonization by diazotrophic bacteria or endomycorrhial fungi may be due to factors that derive from the host (Hungria and Neves, 1986), the symbiosis (Iruthayathas and Vlassak, 1982), or the environment (Graham and Temple, 1984). Even under different environmental conditions, the genetic potential of various *Phaseolus* cultivars, *Rhizobium* strains (Israel, 1981), and the interaction between host and endophyte (Minchin et al., 1978) are the predominant influences governing the expression of nitrogen fixation.

The role of isozymes in the assimilation, transfer, and transport of P is still uncertain (Gianinazzi-Pearson and Gianinazzi, 1978), but it is clear that the presence of an AM fungus does alter the P physiology and metabolism of a host (Capaccio and Callow, 1982). The increase in polyphosphate hydrolase in AM roots is interesting because it raises the possibility that the fungal endophyte may enhance phosphate availability in the rhizosphere. Typically, plants do not contain their own polyphosphate hydrolase (Bieleski, 1973); rather, they rely on the activity of soil microorganisms to release free phosphate from mineral or organic P reserves. These results, then, must be approached with caution, since other phosphatases can also apparently release free phosphate from polyphosphoric acid in a nonspecific fashion. Earlier reports of Gianinazzi-Pearson and Gianinazzi (1976, 1978) confirmed that the acid, not alkaline phosphatases of the mycorrhizae, is both specifically induced in the presence of *Glomus* and is sensitive to the level of phosphate in the environment. These results indicate that there are alterations in extracellular acid phosphates (Goldstein et al., 1988) and in the cell wall and membrane-bound acid isozymes (Lefebvre et al., 1990) in response to P starvation. Acid phosphatases may play an even more fundamental role in the establishment of mycorrhizal symbioses. Under P starvation, plant cells will produce a major amount of acid phosphatase with a pronounced preference for phosphoenolpyruvate (Duff et al., 1989). This inducible enzyme can apparently "bypass" the inorganic phosphate and adenylate-requiring steps of glycolysis, allowing G metabolism to proceed in P-deprived cells.

In Woody Legumes

Desertification is a complex and dynamic process (Allen, 1988, 1989; Smith and Gianinazzi-Pearson, 1988; Lopez-Bermudez and Albaladejo, 1990) that obviously has a negative environmental impact. The process claims several million hectares annually (Francis and Thornes, 1990; Morgan et al., 1990). Physicochemical and biological properties, such as soil aggregation and microbial functioning, are also degraded. Microbial number and/or activities are reduced by the desertification process, and this reduction becomes critical for the reestablishment of vegetation cover (Sylvia, 1990). In particular, AM fungi and *Rhizobium* have been reported to be key elements for plant establishment under xeric and nutrient-unbalanced conditions (Herrera et al., 1993).

The scarcity of available P and the imbalance of trace elements in desertified ecosystems actually limit legume establishment and N_2 fixation (Barea et al., 1990). Moreover, woody legumes usually exhibit a considerable degree of dependence on mycorrhizae to thrive in stressed situations (Roskoski et al., 1986; Carpenter and Allen, 1988; Habte et al., 1988; Jasper et al., 1989; Sylvia, 1990). Considering that restoration of vegetation cover in degraded sites is the first step in slowing down desertification processes in semiarid ecosystems, it follows that a revegetation strategy based on the management of arbuscular mycorrhizae and *Rhizobium* appears worthwhile.

After biotrophic mutualistic colonization of plant roots, the microsymbiont develops an extraradical mycelium that links the roots and the soil environment and helps the plant use soil nutrients more efficiently, particularly those that diffuse slowly toward the root surface, such as phosphate and trace elements (Smith and Gianinazzi-Pearson, 1988; Barea et al., 1992). In addition, the symbiosis enhances the ability of the plant to become established and cope with stress situations (nutrient deficiency, drought, trace element imbalance, soil disturbance) which are typical in desertification situations (Stahl et al., 1988; Barea et al., 1990).

In the case of rhizobial bacteria, it is also well-known that the relationships between woody legumes and rhizobia are rather promiscuous and often nonspecific (Roskoski et al., 1986; Olivares et al., 1988). Moreover, rhizobia isolated from legume trees can nodulate herbaceous legumes and vice versa (Olivares et al., 1988). There is, in fact, a great deal of genetic diversity among the rhizobia that are able to nodulate a given tree legume (Batzli et al., 1992), just as there is a great deal of genetic diversity among mycorrhizal fungi (Smith and Gianinazzi-Pearson, 1988), a situation which is expressed in the wide spectrum of symbiotic host responses.

Mycorrhizal plants are more efficient in phosphate uptake than nonmycorrhizal plants are in low-phosphate soils (Mosse, 1973). In such soils, a satisfactory nodulation and nitrogen fixation depend on P-solubilizing bacteria and mycorrhizal fungi for more P. These phosphobacteria were able to produce plant-growth-regulating substances, including auxins, gibberellins, and cytokinins (Barea et al., 1990); such substances may be involved in the previously mentioned interactions. Ferrera-Cerrato and Villerias (1985) have reported the possible results in the growth and nodulation of leguminous plants inoculated with these microorganisms. It has been found that the effects of endomycorrhiza can be improved with the addition of low levels of phosphate fertilizers (Gardezi et al., 1988). These circumstances suggest the need to study the possibilities of inoculation with both microorganisms in soils deficient in nitrogen and with problems of phosphorus assimilation. Nitrogen content increased more in plants inoculated with *Glomus* sp. than in those inoculated only with *Rhizobium;* the largest nitrogen content corresponded to the treatment with both *Glomus* and *Rhizobium* (Ferrera-Cerrato and Villerias, 1985).

In Crop Legumes

It has often been reported that mycorrhizal infection also increases the concentration of nutrients other than P in plant tissues, but it is not clear whether this enhancement of nutrient uptake is merely a consequence of improved P supply. Murphy et al. (1981) concluded that a high P concentration in plant tissues is sometimes associated with reduced concentrations of other nutrients, particularly micronutrients. Janos (1987) and Stribley (1987) observed a direct positive effect of AM on Zn and Cu nutrition. Pacovsky, Bethlenfalvay, and Paul (1986) and Pacovsky, Paul, and Bethlenfalvay (1986) also documented a reduction in Mn and Fe uptake following AM formation, whereas the P uptake was more.

Ahiabor and Hirata (1994) showed the importance of mycorrhizal infection for the growth of legume plants confirming other reports of favourable legume response to AM fungi inoculation, whether in sterilized or unsterilized soils (Daft and El-Giahmi, 1976; Manjunath et al., 1984). Unlike cowpea and groundnut, colonization of pigeon pea roots was relatively low, which differs from the results of Manjunath and Bagyaraj (1984). AM formation induced a higher concentration of P in the shoot dry matter of all three legumes than the 0.1 percent P observed in the nonmycorrhizal plants (Mengel and Kirkby, 1987). The P concentration in groundnut was more than double this value. This enhanced P nutrition is most likely the cause of both the improved shoot growth and yield in these crops. Their findings agree with the proposal for cowpea (Manjunath and Bagyaraj, 1984) and other legumes (Carling et al., 1978)—that the effects of AM increasing plant growth and nodulation are explained by enhanced P nutrition. In cowpea and groundnut, these increases could also be attributed to the high concentration of Ca in the shoots and with pigeon pea K could also be implicated. The level of Ca observed in such plants falls within the upper limits of the concentration range (0.5 to 3 percent) generally found in the dry matter of higher plants (Mengel and Kirkby, 1987).

According to Jarell and Beverly (1981) and Cooper (1984), an increase in both shoot weight and shoot nutrient concentration is evidence of a treatment effect. Increases in P concentration following AM inoculation were also reported by other authors (Neilsen and Jensen 1983; Raju et al., 1990). However, Neilsen and Jensen (1983) observed a decrease in Ca concentration in the dry matter of *Medicago sativa* when the plants were inoculated with *Glomus caledonium*. Comparing mycorrhizal and nonmycorrhizal plants, the results for Ca have been inconsistent (Mosse, 1957; Gerdemann, 1964); the concentrations of Ca in AM plants were higher, lower, or not significantly different from those in corresponding plant parts of nonmycorrhizal controls. The high concentrations of K and Ca might be a consequence of mycorrhizal improvement in P supply. AM inoculation either had no effect on or decreased Mg and N concentrations in shoots of all three legumes in both soils.

Mycorrhizal enhancement of shoot micronutrient concentrations was rare in all plants. Negative responses (signifying a dilution effect) were observed in certain cases, especially with Zn in subsoil of cowpea and pigeon pea and also with Mn in the former. High concentrations of P in plant shoots (as commonly observed) are some times associated with reduced concentrations of micronutrients (Murphy et al., 1981). In agreement with the findings of Raju et al. (1990), Cu concentrations did not change with mycorrhizal treatment in any of the plants.

Mycorrhizal plants are known to grow more actively as a result and have a higher ash content than those not infected. This is because in AM plants the fungus stimulates respiration and additional release of H^+ ions from the host root for use in exchange reactions with the soil, leading to increased absorption of salts (Routein and Dawson, 1943).

In either soil, the legumes responded to *Glomus* sp. better than *Gigaspora margarita*. Apart from the possibility that the two fungi have inherently different characteristics, one reason for their contrasting performances may be the different preculture host plants used (Simpson and Daft, 1990). The synergistic interaction between AMF *(Glomus clarum* or *G. mosseae)* and *Rhizobium leguminosarum* enhanced the productivity in *Lens culinaris*.

RHIZOBIUM, *AMF, AND BIOCHEMICAL ACTIVITIES*

The interface between the arbuscular membrane and the plant cell has very high enzymatic activity (Jeanmaire et al., 1988). Most extra activities (synthesis, metabolic processes) occur at the interface between the fungus and the plant cell (Gianinazzi, 1991) and specific proteins, endomycorrhizins, are produced during cellular interactions in AM (Wyss et al., 1990). It is generally accepted that any interaction among cells involves the exchange of chemical signals (Halverson and Stacey, 1986). Some elicitors have been identified in the *Rhizobium*-legume symbiosis (Spaink et al., 1991). Since the different signals (chemicals) involved have to be synthesized, the relevant metabolic processes should be stimulated after establishing contact between host and fungus.

Thomas (1991) observed pronounced changes in proline level in *Lolium perenne* subjected to drought. In leaf bases, proline accumulated gradually during drought to reach about three times the content in control plants. In laminae, accumulation of proline only after severe drought coincided with dehydration, scorching cessation of leaf extension, and very low water potential (below –2M Pa). It therefore seems likely that proline accumulation in laminae is either an injury response or occurs just before injury. This is supported by Kemble and MacPherson (1954) who found greater accumulation of proline in leaf blades, which was probably a consequence of the artificial stress technique. Proline may have accumulated because compartmentation between production (in the cytoplasm or chloroplasts) and oxidation (in the mitochondria) was disrupted, as proposed by Stewart and Bogges (1978). Riazi et al. (1985), working with barley, also found more proline in bases than in laminae, and that proline accumulates in bases under moderate stress and was not a consequence of cell death. Giovennetti et al. (1991) found that tobacco plants inoculated with *Glomus monosporum* showed better tolerance to root rot pathogen *Thielaviopsis basicola* than did nonmycorrhizal seedlings. There were no appreciable differences in free amino acid composition among the different treatments, except for proline and arginine, which were higher in the mycorrhizal plants than in other treatments. In *T. basicola*–infected plants, the free amino acids level increased in the initial stages and decreased in later stages compared with the control plants.

Protein content was much higher in mycorrhizal than in nonmycorrhizal root extracts, in agreement with Dumas et al. (1994) and Gianinazzi-Pearson and Gianinazzi (1978) for tobacco and onion. A 2.6-fold higher protein content was found in mycorrhizal than in nonmycorrhizal red clover roots. Other reports have not shown such a large difference in both types of roots as have been described so far, and further deep studies of this aspect are needed. Perhaps this difference is a consequence of factors such as higher metabolic activity in VA (vesicular-arbuscular)-colonized root cells and the presence of internal and external fungal mycelium. It is difficult to speculate because our actual knowledge of VA fungal proteins is very limited, and we do not know if the new proteins are of fungal or plant origin. Five polypeptides with 16, 17, 18, 22, and 30 kDA were found only in AM roots and were considered to be VA mycorrhizins (Pacovsky, 1989). Wyss et al. (1990) have also found some polypeptides in both the low and high molecular weight ranges that were immunoprecipitated with an antiserum against soluble nodulins or membrane-bound nodulins. Some polypeptides, at both low and high molecular weight, were more abundant in P2, P6, and P8 or were exclusive to P4, P5, and P7 mycorrhizal roots; others were less abundant (P1 and P3) in mycorrhizal than in nonmycorrhizal roots. The similarity in the degree of nodulation in both mycorrhizal and nonmycorrhizal roots allows us to discard the possibility of a bacterial origin for these polypeptides. After mycorrhizal infection, changes in polypeptide accumulation fell into three categories: decreasing, increasing, or synthesis of new proteins (Martin and Hilbert, 1991). We do not know the role of the polypeptides in the symbiosis. But after isoelectric focusing (IEF) analysis and staining for activity of superoxide dismutase (SOD), an SOD isozyme appeared only in mycorrhizal root extracts. Legocki and Verma (1979) identified nodulin-35, a protein identified from soybean root nodules, having 35,000 molecular weight, developed by different strains of *Rhizobium japonicum,* irrespective of their effectiveness in fixing atmospheric nitrogen. This protein is not detected in uninfected plants and bacteroids or in free-living *Rhizobium* and appears to be synthesized by the plant during the formation of nodules. The *nodABC* genes, present in all rhizobial species, are the only *nod* genes strictly required for the synthesis of lipochitin oligosaccharide (Spaink et al., 1991). Samara et al. (1997) have demonstrated the symbiosis-related polypeptides (14,18, 21, 28, 35 KDa) during the establishment of *Glomus mosseae* in *Pisum sativum* roots.

The interaction between AM fungus and plant cell takes place at both extracellular and intracellular levels. Jeanmaire et al. (1988) showed high enzymatic activity in the interface between the cells of root and endomycorrhizal hyphae. According to Gianinazzi (1991) most metabolic processes occur at the interface between the fungus and plant cell. The mycorrhizal roots possess higher pectin esterase and endopolymethyl galacturonase activities than the nonmycorrhizal roots (Garcia-Romera et al., 1991).

Nemec and Meredith (1981) found that *Glomus etunicatum*–inoculated *Citrus limon* leaves had higher total amino acids than control. Krishna and Bagyaraj (1983) found proteins and amino acids increased in *G. fasciculatum*–inoculated *Arachis hypogaea* roots. Dumas et al. (1994) observed higher protein content in AMF–inoculated *Nicotiana tabacum* and onion *(Allium cepa)* roots than in control. There were no appreciable differences in free amino acid composition among the different treatments, except proline and arginine, which were higher in the mycorrhizal plants than in all other treatments. In *T. basicola*–plants, the free amino acids level increased in the initial stages and decreased in later stages compared with the control plants. A high amount of protein was shown in *G. mosseae*–infected *Trifolium pratense* roots (Arines et al., 1993).

Studies on the lipid metabolism of *Rhizobium* and AM fungi are very limited. Cooper and Losel (1978) showed that roots infected by *G. mosseae* contained significantly more lipid than did uninfected roots. The mycorrhizal roots of *Trifolium repens* and *Lolium perenne* contained larger amounts of total lipids than did nonmycorrhizal roots.

ENZYME ACTIVITIES

Higher peroxidase activity has been demonstrated in mycorrhizal roots than in nonmycorrhizal roots (Spanu and Bonfante-Fasolo, 1988). This agrees with the results on SOD activity, in that the activities of both enzymes are associated with the defense of plants against oxidative stress produced by generation of active oxygen species (Rabinowitch and Fridovich,

1983; Fridovich, 1986). More interesting is the presence of a new SOD isozyme associated with the symbiosis (Palma et al., 1992). This may operate the interface. Both nodulation (Becana et al., 1989) and environmental stress (Tsang et al., 1991) are associated with increased SOD activity. Increased peroxidase is one of the most widespread biochemical activities in diseased and injured plant tissues. Peroxidase has an important function in secondary cell wall biosynthesis by polymerizing hydroxy and methoxy cinnamic alcohols into lignin (Grisebach, 1981) and also is associated with deposition of suberin in plant cell walls around the lesions (Wu, 1973). Spanu and Bonfante-Fasolo (1988) measured the cell wall bound peroxidase in *Allium porrum* during root growth and development of *G. versiforme*. At initial stages of fungal infection, the enzyme activity was maximum and decreased in well-established and highly colonized roots. Pacovsky et al. (1990) studied peroxidase activity in *Phaseolus vulgaris* infected by *G. etunicatum,* and found that peroxidase activity increased in the mycorrhizal plants.

The capacity of AM plants to utilize complex inorganic phosphate was first demonstrated by Daft and Nicolson (1964). The activity of some mycorrhizal fungi in culture to utilize some phytates (Theodorou, 1968, 1971), and the presence of an active p-nitrophenyl phosphatase in *Fagus sylvatica* mycorrhizas (Woolhouse, 1969) has already been demonstrated. MacDonald and Lewis (1978) cytochemically demonstrated the presence of acid phosphatase in *G. mosseae*. Acid phosphatase was found in lysing and growing AM fungal structures. The presence of AM-specific alkaline phosphatase activity in *Allium cepa* and *Platanus occidentalis* plants inoculated with *G. mosseae* has been reported by Gianinazzi-Pearson and Gianinazzi (1976). Qualitative changes of root-soluble phosphatase activity were observed during AM infection.

Gianinazzi-Pearson and Gianinazzi (1978) demonstrated the presence of soluble alkaline phosphatase specific to AM fungal infection in enzyme extracts from onion roots inoculated with *G. mosseae*. Maximum activity occurred when the infection was still in its early stage (100 percent arbuscular). Gianinazzi-Pearson and Gianinazzi (1983) found the increased alkaline phosphatase activity in *G. mosseae* and *Glomus* spp. inoculated soybean *(Glycine max)* roots of nonphosphate-fertilized plants of 50 days old more than it did in phosphate-fertilized nonmycorrhizal plants. Smith and Gianninazzi-Pearson (1987) found higher alkaline phosphatase activity of *G. mosseae* than in the infected roots of *Allium cepa*. Dodd et al. (1987) compared the effect of individual inoculation of *G. geosporum, G. monosporum,* and *G. mosseae* on acid phosphatase activity in roots of *Brassica rapa, Triticum aestivum,* and *Allium cepa*. They found higher acid phosphatase activity in plants infected with *G. geosporum* and *G. mosseae* than in *G. monosporum*–inoculated plants. Anita et al. (1988) found high acid phosphatase activity in *G. fasciculatum*–inoculated roots of *Trigonella* sp. Polyphosphatase activity increased in the mycorrhizal plants.

The greater part of the nitrogen incorporated into plants is taken up as nitrate from the soil. It has not been demonstrated yet whether the mycorrhizal fungi themselves reduce nitrate to nitrite and ammonia, and synthesize an amino acid which is then transferred to the host by active transport (Allen, 1982; Ames et al., 1983). Proteins and/or amino acids could be made available for the plant with each degradation of an arbuscule within the root. An alternative is that nitrate could be directly transferred to the plants, which then reduces it and supplies the symbiotic fungus. The key enzyme for nitrate assimilation is nitrate reductase. Sequence comparison and/or DNA-DNA hybridizations indicated that the PCR amplificates did indeed come from the organisms from which DNA had been isolated. A DNA hybridization experiment with the digoxigenin-labeled PCR segment and DNA isolated from about 0 to 5 million *Glomus* spores confirmed that this mycorrhizal fungus possesses a nitrate reductase gene (Kaldorf et al., 1994).

GROWTH HORMONES

Roots colonized by AM fungi are often thicker and carry fewer root hairs. Such changes in morphology are expected to be under phytohormonal control. Abscisic acid (ABA) was found to be considerably enhanced in both roots and shoots of AM plants as compared to nonmycorrhizal controls. In contrast, the levels of auxins and cytokinins *(zeatin riboside)* were essentially the same in colonized and noncolonized plants (Danneberg et al., 1992). It is unlikely that ABA is originally synthesized by the maize cells, since the fungi would then have to enrich this phytohormone from the cytoplasm by active transport. It is more probable that the fungus exerts control over the morphology of the roots and ABA plays an important role in this. ABA is involved in the regulation of solute fluxes within plants, which could also happen in AM symbiosis. Cytokinins have been described for the flux *(Linum usitatissimum)* symbiosis (Druge and Schonbeck, 1992). Raman et al. (1994) found an enhanced indole acetic acid (IAA) level in *G. mosseae* and *Rhizobium*–inoculated *Prosopis juliflora. Glomus fasciculatum* and *Rhizobium* inoculation in tannery-effluent-treated *P. juliflora* resulted in increased NPK (nitrogen, phosphorus, and potassium) and useful macro- and microelements, increased levels of hormones, enzymes, amino acids, protein, lipid, and phenol (Raman and Selvaraj, 1998). It was suggested that rhizobial infection might be analogous to the cytokinin-auxin stimulation, causing endomitotic cortical cells already present in the root to undergo mitosis upon the initiation of nodules (Torrey, 1961). In addition to these cytological observations, it was reported that cytokinins were released into the medium by cultures of both *Rhizobium japonicum* and *R. leguminosarum* (Giannatiasis and Coppula, 1969; Phillips and Torrey, 1970, 1972).

It is generally believed that both auxin and cytokinin are required for cell division in plants (Das et al., 1976). It is not unreasonable to suggest that both cytokinin and auxin have an important role in root nodule development. High levels of auxin in root nodules have already been demonstrated (Thimann, 1936; Pate, 1958). The presence of high cytokinin activity in pea root nodules was clearly demonstrated. The presence of cytokinins in the root nodules of *Phaseolus vulgaris* infected by *Rhizobium phaseoli* was also reported (Puppo et al., 1974). Cytokinins play a critical role in nodule development, particularly in relation to the meristematic nature of the nodule, in other words, its cell division activity and also to the distinctive polyploid state of the dividing nuclear population (Torrey, 1961). The cytokinins produced by *R. leguminosarum* were different (Syono and Torrey, 1976). According to Phillips and Torrey (1972), *R. leguminosarum* produces a different cytokinin than that produced by *R. japonicum;* from chromatographic mobilities one would assign the activity to isopentenyladenine (2-iP) or isopentenyladenosine (IPA). Giannatasis and Coppula (1969) have reported that *R. leguminosarum* release IPA and two unknown cytokinins into the medium.

The host tissue, especially the meristematic region of the nodule, is the source of zeatin and its derivatives, while the invading bacteria and bacteroids produce 2-iP and its derivatives (zeatin and its ribosides and ribotide were the major free cytokinins occurring in the uninfected root apex (Syono and Torrey, 1976).

CALCIUM AND PHOSPHORUS REQUIREMENTS

Ca requirements for unstressed cells of *Rhizobium* and *Bradyrhizobium* are quite low. Vincent and Humphrey (1968) and Munns (1970) noted a calcium requirement of only 25 to 50 μM. Calcium appears significantly more important in cells exposed to low pH. Beck and Munns (1984, 1985) noted that P-limited cells or cells grown at low pH needed Ca for P mobilization in the cell. Calcium also plays a role in cell wall integrity. Ca-depleted cells of *R. leguminosarum* are swollen, lack rigidity, and express an additional somatic antigen normally blocked by side chains of lipopolysaccharide O antigen (Vincent and Humphrey, 1968; De Maagd et al., 1989). Calcium requirements for nodulation are increased by low pH. With alfalfa plants grown at pH 6.0, Munns (1970) achieved abundant nodulation with 0.1 to 0.2 mM Ca, whereas when the pH was dropped to pH 5.2, 6.4 mM Ca was needed to achieve comparable results. Calcium-dependent cell surface components affect attachment of *Rhizobium* to root hair cells (Smit et al., 1987, 1989; Caetano-Anolles et al., 1989). Addition of 10 mM calcium increased *nod* gene induction activity by five- to tenfold at pH 4.5 in white and subterranean clover (Richardson et al., 1988a). Calcium also enhanced *nod* gene expression at pH 4.8 to 5.2 (Richardson et al., 1988b).

P deficiency for rhizobia in soil and rhizosphere is a real possibility, especially under acid conditions where solution P may be precipitated in the presence of aluminum. Slow-growing strains appear more low-P tolerant than fast growers, with *R. meliloti,* in particular, P requiring (Beck and Munns, 1984, 1985). Rhizobia grown at luxury levels of P can accumulate this substance to levels as high as 2.4 percent (Cassman, Munns, and Beck, 1981; Cassman, Whitney, and Fox, 1981; Beck and Munns, 1984), using this reserves on transfer to P-limited media. *Rhizobium meliloti* strains may be an exception to this, requiring high calcium levels for P mobilizaton (Beck and Munns, 1985).

Phosphorus appears essential for both nodulation and nitrogen fixation. Significant delays have been reported in the infection of primary soybean roots by P-limited cells of *Bradyrhizobium japonicum* (Mullen et al., 1988), although in some host-strain combinations P stress subsequently resulted in enhanced nodule mass and shoot N. Nodules are strong sinks for phosphorus and range in P content from 0.72 to 1.12 percent (Hart, 1989). This is a higher level than found in roots, shoots, or even young mature leaves at all but luxury levels of P. As a consequence, fixation-dependent plants will require more of this element than those supplied combined nitrogen (Cassman, Munns, and Beck, 1981; Cassman, Whitney, and Fox, 1981; Israel, 1987). Nodulation, nitrogen fixation, and specific nodule activity can all be directly related to P supply (Graham and Rosas, 1979; Jakobsen, 1985).

HEAVY METALS

Among soil microorganisms, mycorrhizal fungi are the only ones providing a direct link between soil and roots, and can therefore be of great importance in heavy metal availability and toxicity to plants (Leyval et al., 1997). The toxicity of metals in soil depends upon their bioavailability, defined as their ability to be transferred from a soil compartment to a living organism (Juste, 1988). According to Berthelin et al. (1995), metal bioavailability is a function not only of their total concentration but also of physiochemical (e.g., pH, Eh, organic matter, clay content) and biological (e.g., biosorption, bioaccumulation, and solubilization) factors.

High metal concentrations in soil are toxic to bacteria and fungi. Metal tolerance in soil microorganisms has been studied in the context of removing metals from polluted soils and also to provide a biological understanding of the adaptation of living organisms to extreme environments (Gadd, 1990a,b). In mine spoils heavily polluted with metals (up to 8.3 percent Zn and 863 $\mu g \cdot g^{-1}$ Cd). Gildon and Tinker (1981) found 35 percent colonized roots of clover (growing naturally on these soils). Diaz and Honrubia (1993) used *Medicago sativa* to show that mine spoils and waste sediments containing Zn and Pb (up to 236 and 456 $mg \cdot kg^{-1}$ DTPA-extractable Zn and Pb, respectively) had a mycorrhizal infection potential, although the number of AM fungal spores was lower than in an adjacent soil not altered by mining activity.

All organisms, including microorganisms, can achieve resistance to heavy metals by "avoidance" when the organism is able to restrict metal uptake or by "tolerance" when the organism survives in the presence of high internal metal concentrations (Joho et al., 1985; Baker, 1987; Turnau et al., 1996). The first mechanism involves, for example, reduced uptake or increased efflux, formation of complexes outside cells, and organic acid release. In the second situation, metals are chelated intracellularly through the synthesis of ligands such as metallothioneins, polyphosphates, and/or compartmentation within vacuoles. Heavy-metal-tolerant and -sensitive individuals can be distinguished by their growth performance on metal-contaminated substrates.

Metal tolerance has been tested regarding spore germination ability in sand amended with Cd solution and in heavy-metal-contaminated soils (Weissenhorn et al., 1993, 1994). AM fungi cannot be cultivated without a host plant. So metal tolerance cannot be really be evaluated through fungal growth on axenic media. Using a compartmented system separating roots and extraradical hyphae, Joner and Leyval (1997) showed that hyphal length was unaffected by up to as much as 100 $mg \cdot kg^{-1}$ Cd added to the soil of the hyphal compartment and that juvenile spores were observed. Sporulation and spore germination of nonmycorrhizal fungi are also more affected than hyphal development by Cd (Babich and Stotzky, 1977; Ross, 1982), suggesting that distinctions should be made between concentrations that inhibit spore germination and those that inhibit fungal growth. Increased Cd tolerance of a mixed fungal culture was observed in a soil amended with Cd salts, but only one year after amendment of Cd (Weissenhorn et al., 1994). Increased metal tolerance has also been observed for AM fungi isolated from a soil naturally high in metals (Leyval et al., 1995).

the resulting toxic symptoms (Bradley et al., 1982). A similar explanation has been proposed for ectomycorrhizal fungi (Denny and Wilkins, 1987). The importance of vesicular-arbuscular mycorrhizae varies between different hosts and fungal species. In some cases, protection from toxicity is apparent (Gildon and Tinker, 1983) whereas in some other cases, increased metal uptake may reduce growth (Killham and Firestone, 1983). Two species of ectomycorrhizal fungi, *Amanita muscaria* and *Paxillus involutus,* increased zinc tolerance of *Betula* sp. (Brown and Wilkins, 1985b), the latter fungus also ameliorated aluminium toxicity toward Norway spruce when nonmycorrhizal and located in the rhizosphere (Wilkins and Hodson, 1989).

PLANT UPTAKE OF METALS

The effect of AM fungi on plant uptake of metals is not always clear. At high metal concentrations, some reports show higher uptake of metals by mycorrhizal plants (Gildon and Tinker 1983; Killham and Firestone 1983; Weissenhorn and Leyval, 1995). Others found reduced concentrations in plants or in shoots due to mycorrhizal colonization (Weissenhorn et al., 1995). In two pot experiments with maize in metal-contaminated soil, mycorrhizal colonization either increased plant biomass and decreased Cd, Cu, Zn, and Mn concentrations in shoots and roots or had no effect on growth and metal uptake, depending on root density, plant growth conditions, and mycorrhizal inoculum (Weissenhorn et al., 1995). Loth and Hofner (1995) observed that mycorrhizal colonization led to a higher uptake of Cu, Zn, and Cd in oat roots from a highly contaminated soil but a reduced translocation to the aerial part. Inoculation of *Festuca rubra* and *Calamagrostis epigejos,* grown in pots filled with sand amended with Zn sulphate at high concentration with *G. fasciculatum,* alleviated the negative effect of Zn on plant growth but had no effect on the Zn concentration in shoots and roots (Dueck et al., 1986).

Mycorrhizal colonization reduced the shoot concentration of Cd and Zn in field-grown maize and lettuce when the soil had high concentrations of available metals (Schuepp et al., 1987). In a soil with low metal concentrations, AM fungi decreased Cd uptake but increased Zn uptake in some plants. These results suggest that the effect of AM fungi on plant heavy metal uptake is metal specific and depends on the concentration. El-Kherbawy et al. (1989) showed that the AM effect on plant metal uptake also depends on soil pH. With increasing soil pH, the DTPA-extractable metals decreased, but at the same time AM fungi increased Cd, Zn, and Mn uptake. In both cases, mycorrhizal colonization enhanced alfalfa growth (Biro et al., 2000; Tsimilli-Michael et al., 2000).

To differentiate fungus and host plant effects, plant containers with different compartments separating roots and extraradical hyphae have proven very useful (Burkert and Robson, 1994; Guo et al., 1996; Joner and Leyval, 1997). Transport of Zn in extraradical hyphae from a sandy soil to clover was shown by Burkert and Robson (1994). In the experiments of Guo et al. (1996) up to 37, 33, and 44 percent of the total Cd, Cu, and Zn uptake, respectively, by bean plant was attributed to mycorrhizal transfer via hyphae from a root-free compartment. Joner and Leyval (1997) studied the transfer of Cd by extraradical hyphae of *G. mosseae* from a sandy loam to clover using ^{100}Cd.

PLANT-GROWTH-PROMOTING RHIZOBACTERIA

The role of other microorganisms such as plant-growth-promoting rhizobacteria (PGPR) as modifiers of soil fertility and facilitators of plant establishment has been considered by Linderman (1992), McInroy and Kloepper (1994), and Glick (1995). Several PGPR have been described by different authors to promote emergence of host plants (de Freitas and Germida, 1990; Turner and Backman, 1991) and have been named emergence-promoting rhizobacteria (EPR). Some authors have reported that PGPR can influence plant development not only directly (hormone production, P solubilization, or asymbiotic nitrogen fixation) but also indirectly through modifications to the activity of other plant-microbe interactions, such as the mycorrhizal or the *Rhizobium* symbioses (Meyer and Linderman, 1986; Pololenko et al., 1987; Garbaye, 1994; Linderman,1994) or by inducing changes in the microbial population balance, for example, exerting biological control against plant pathogens (Weller and Thomashow, 1994). Thus, since the final effect of soil microorganisms is the result of interactions among the different soil microbial components involved, the study of the antagonistic or synergic effects of PGPR, AMF, or *Rhizobium* when coinoculated (Meyer and Linderman, 1986; Azcon-Aguilar and Barea, 1992; Fitter and Garbaye, 1994; Puppi et al., 1994; Barea et al., 1996), is important for improvement of plant performance.

COEXISTENCE OF RHIZOBIUM *AND AM SYMBIOSES*

There are number of similarities between *Rhizobium* and AM symbioses. The physical similarities include the development of specialized membranes and interface compartments around the microbial/fungal symbionts. There are also similarities in the molecular components of these membranes and interfacial matrixes (Harrison, 1997). It has been shown that the monoclonal antibodies which recognize the glycoproteins of the peribacteroid membrane and the peribacteroid space of pea nodules also cross-react with the components of the periarbuscular membrane and interface compartments in mycorrhizal pea roots (Perotto et al., 1994). Similarly, in soybean, antibodies prepared to soybean bacteroid membranes recognize in vitro translated proteins from mycorrhizal root RNA (Wyss et al., 1990). The most striking similarity which has been demonstrated was that a proportion of pea nodulation mutants are also mycorrhizal mutants and are unable to complete mycorrhizal associations (Duc et al., 1989). Following this report, mutants unable to form nodulation and mycorrhizal symbioses have been identified in alfalfa (Bradbury et al., 1991), bean (Shirtliffe and Vessey, 1996), faba bean (Duc et al., 1989), and *M. truncatula* (Sagan et al., 1995). The existence of *myc/Nod* mutants defines genes that are essential for both symbioses, and it has been speculated that rhizobia exploited signals and pathways from the mycorrhizal symbiosis to develop the symbiosis with legumes (Gianizazzi-Pearson, 1996). The nature of mutated *myc/Nod* loci are unknown, but the identification mutants in the model legume *M. truncatula* offers a possibility to clone these genes (Harrison, 1997). Lipochitooligosaccharides act as signals in plant microbe interactions but may also act as general plant growth regulators (De Jong et al., 1992; Rohrig et al., 1995). Lipochitooligosaccharides are signals in the *Rhizobium*-legume symbiosis (Denarie et al., 1996) and have also been demonstrated to promote cell division of protoplasts in the absence of auxin and cytokinin and can stimulate auxin-inducible gene expression (Rohrig et al., 1996). Because the AM fungi have cell walls from which chitooligosaccharides could be readily derived, they might also be considered as potential signaling molecules (Harrison, 1997).

The first recognizable reaction of the plant to *Rhizobium* infection is the curling of the root hair. A cell wall is then broken down and the bacterium is able to enter. An infection thread develops which penetrates the inner cortex of the root. The infected cortex cells increase in size and divide to form a sphere surrounded by uninfected cells and outer fibrous layers. Within the infected cells, the rhizobia divide and enlarge to form bacteroids which are separated in the plant cytoplasm by the peribacteroid membrane (Meller and Werner, 1990). Rhizobia are able to elicit the formation of a new organ, the nodule, on the roots of leguminous plants. Legumes secrete flavonoids and isoflavonoids, which are able to activate the rhizobial regulatory nodulation protein NodD in a host-specific way (Schlaman et al., 1992). The activated nod protein is able to induce the expression of the other bacterial nodulation genes. It has been shown that many nodulation genes involved in the biosynthesis are secretion of Nod metabolites, which function as signals for the host plant. The basic structure of the signal molecules is a β-1,4 linked *N*-acetylglucosamine oligosaccharide of which the nonreducing terminal sugar residue is *N*-acylated, so that they are described as lipo-chitin oligosaccharides (Bloemberg et al., 1997).

The initial events of AM colonization include hyphal growth onto the surface of the root followed by appressorium formation. Appressorium formation has been observed only on plant roots, suggesting that a specific plant signal is necessary for this differentiation event. The signal could potentially be chemical in nature. Flavonoid/isoflavonoid compounds have been demonstrated to act as signals in other plant-microbe interactions, and since they are components of root exudates they have been examined extensively for possible signal function in the mycorrhizal symbiosis. A wide range of these compounds display both stimulatory and inhibitory effects on spore germination and hyphal growth (Chabot et al., 1992). In addition to the signals for the initial stages of the association it is expected that there will also be signals for later stages of the interactions. The nature of the signals is also currently unknown. Oligisaccharins act as signal molecules in both plant-pathogen and plant-symbiont interactions, as well as endogenous signals for plant development (Fry et al., 1993). Considering their widespread occurrence in the plant kingdom, they could be signals in mycorrhizal association. Mycorrhizal fungi possess both cellulose and endopolygalacturonase activity that would be capable of releasing oligosaccharides or oligosaccharins from the plant cell walls (Bonfante and Perotto, 1995). Lipochitooligosaccharides are signals in the *Rhizobium* symbiosis (Denarie et al., 1996). Because the AM fungi have cell walls from which chitooligosaccharides could be readily derived, they might also be considered as potential signal molecules. Recent investigations indicate that a nodulin-26-like gene may be activated also during mycorrhiza interactions, even in nonlegumes (Roussel, 1995; Gianinazzi-Pearson, 1997), and that an unusual leghemoglobin gene is activated both in nodules and in mycorrhiza of faba bean (Fruhling et al., 1997).

CONCLUSIONS AND FUTURE PERSPECTIVES

It is evident that more information about the molecular and genetic bases of mycorrhizal and rhizobial nodule symbioses will improve our understanding of root behavior toward microbes and provide insight into the evolutionary relationships between plant processes involved in the two types of interactions. Research on plant genes and molecules involved in AM interactions is still in its infancy. The interactions between host plants cells and microsymbionts that lead to the expression of the various nodulin genes are totally unknown. To induce most of the nodulins essential for functional symbiosis, an additional series of signal exchanges between plant cells and differentiating bacteriods must be required. Activation of plant defense responses does not appear to be a key event in colonization processes in symbiotic fungal development. However, comparisons with nodulation symbiosis, the use of plant mutants as experimental systems, and constantly evolving molecular biology techniques are now opening the way to identifying plant molecules that play an essential role in mechanisms involved in symbiotic interactions during root colonization by AM fungi. Multidisciplinary approaches are necessary to study the mechanisms and various processes of the coexistence of the AM and *Rhizobium* in legume. Resources, such as mycorrhizal mutants, are particularly valuable for many areas of study, including physiological investigations, biochemical studies in search of signals, and molecular and genetic approaches aimed at cloning the genes essential for the symbioses. Molecular characterization of the symbiosome membrane, the gas diffusion barrier, and the function of cell organelles such as mitochontria and plastids are all important research topics that need further investigations. Clearly an intensive area of research for the coming decade is to elucidate the various signals and signal transduction pathways operating during nodule organogenesis and interaction in the root system with AMF. Increased efforts to connect the biochemical and genetic analysis are especially needed. Over the past ten years, many aspects of the molecular signaling between rhizobia and leguminous plants have been elucidated. A wide variety of rhizobial lipochitin oligosaccharides (LCO) that have been shown to be major determinants of the host specificity of nodulation have been identified. Almost nothing is known about the mechanism of perception of LCO by plants. Microballistic targeting of signal molecules is likely to be of great importance in the study of LCO perception and the mechanism of root nodule formation and interaction with AMF. Our knowledge of the biosynthesis of rhizobial LCO will be a powerful tool in the search for novel chitinlike signal molecules in plants and animals. The morphological and metabolic changes in roots resulting from the establishment of AMF symbiosis are associated with modifications in gene expression in both symbionts. Endomycorrhizins which have been designated as symbiosis-related (SR) polypeptides have been observed in AM symbiosis. Targeted approach on the SR polypeptides will improve our knowledge of molecular mechanisms governing AM symbiosis.

REFERENCES

Adjimani J.P. and Emery T. (1987). Characteristic responses of three tropical legumes to the inoculation of two species of VAM fungi in Andosol soils with different fertilities. *Mycorrhiza* 5: 63-70.

Ahiabor B.D. and Hirata H. (1994). Characteristic responses of three tropical legumes to the inoculation of two species of VAM fungi in Andosol soils with different fertilities. *Mycorrhiza* 5: 63-70.

Ahmad M.H., Eaglesham A.R.J., and Hassouna S. (1981). Examining the potential for inoculant raise with cowpeas in West African soils. *Trop. Agric* (Trinidad) 58: 325-335.

Allen E.B. (1989). The restoration of disturbed arid landscapes with special reference to mycorrhizal fungi. *Journal of Arid Environment* 17: 279-286.

Allen M.F. (1982). Influence of vesicular arbuscular mycorrhizae on water movement through *Bouteloua gracilis* (J.BK.) Lag.ex Steud. *New Phytologist* 91:191-196.

Allen M.F. (1988). Below ground structure: A key to reconstructing a productive arid ecosystem. In E.B. Allen (ed.), *The reconstruction of disturbed arid lands* (pp. 113-115). Boulder, CO: Westview Press,

Ames R.N. and Bethlenfalvay G.J. (1987). Localized increase in nodule activity but no competitive interaction of cowpea rhizobia due to pre-establilshment of vesicular-arbuscular mycorrhiza. *New Phytologist* 106: 207-215.

Ames R.N., Reid C.P.P., Porter L.K., and Cambardella C. (1983). Hyphal uptake and transport of nitrogen from two ^{15}N labeled sources by *Glomus mosseae,* a vasicular arbuscular mycorrhizal fungus. *New Phytologist* 95: 381-396.

Anita K., Singh V.P., and Mukerji K.G. (1988). Studies on the phosphotase on mycorrhizal and non-mycorrhizal *Trigonella* roots. In A. Mahadevan, N. Raman, and K. Natarajan (eds.), *Mycorrhiza for green Asia* (pp. 125-127). Madras: University of Madras.

Arines J., Palma J.M., and Vilarino A. (1993). Comparison of protein patterns in non-mycorrhizal and vesicular-arbuscular mycorrhizal roots of red clover. *New Phytologist* 123: 763-768.

Azcon R., Rubio R., and Barea J.M. (1991). Selective interactions between different species of mycorrhizal fungi and *Rhizobium meliloti* strains and their effects on growth, N_2 fixation (^{15}N) and nutrition of *Medicago sativa* L. *New Phytologist* 117: 399-404.

Azcon-Aguilar C. and Barea J.M. (1992). Interactions between mycorrhizal fungi and other rhizosphere microorganisms. In M.J. Allen (ed.), *Mycorrhizal functioning: An integrative plant fungal process* (pp. 163-198). New York: Chapman Hall.

Babich H. and Stotzky G. (1977). Sensitivity of various bacteria, including actinomycetes, and fungi to cadmium and the influence of pH on sensitivity. *Applied Environmental Microbiology* 33: 681-695.

Badr El-Din S.M.S. and Moawad H. (1988). Enhancement of nitrogen fixation in lentil, faba bean and soyabean by dual inoculation with rhizobia and mycorrhiza. *Plant and Soil* 108: 375-381.

Bagyaraj D.J., Manjunath A., and Patil R.B. (1979). Interaction between a vesicular arbuscular mycorrhiza and *Rhizobium* and their effects on soyabean in the field. *New Phytologist* 82: 141-145.

Baker A.J.M. (1987). Metal tolerance. *New Phytologist* 106: 93-111.

Barea J.M., Azcon R., and Azcon-Aguilar C. (1992). Vesicular-arbuscular mycorrhizal fungi in nitrogen-fixing systems. *Methods in Microbiology* 24: 391-416.

Barea J.M. and Azcon-Aguilar C. (1983). Mycorrhiza and their significance in nodulating nitrogen-fixing plants. In N.C Brady (ed.), *Advances in agronomy* (pp. 1-54). New York: Academic Press.

Barea J.M., Azcon-Aguilar C., and Azcon R. (1996). Interactions between mycorrhizal fungi and rhizosphere micro-organisms within the context of sustainable soil-plant systems. In A.C. Gange and V.K. Brown (eds.), *Multitrophic interactions in terrestrial systems* (pp. 65-77). Cambridge, UK: Blackwell Science. Symposium of the British Ecological Society.

Barea J.M., Salamanca C.P., and Herrera M.A. (1990). The role of VA mycorrhiza at improving nitrogen fixation by woody legumes in arid zones. In D. Werner and P. Muller (eds.), *Fast growing trees and nitrogen fixing trees* (pp. 303-311). Stuttgart, Germany: Gustav Fisherverlag.

Bartnicki-Garcia S. (1973). Fundamental aspects of fungal morphogenesis. In J.O. Shworth and J.E. Smith (eds.), *Microbial differention* (pp. 245-267). Cambridge: Cambridge University Press.

Batzli J.M., Graves W.R., and Van Berkum P. (1992). Diversity among rhizobia effective with *Robinia pseudoacacia* (L.). *Applied Environmental Microbiology* 58: 2137-2143.

Becana M., Paris F.J., Sandalio L.M., and Del Rio L.A. (1989). Enzymes of superoxide dismutase in nodules of *Phaseolus vulgaris* L., *Pisum sativum* L. and *Vigna unguiculata* (L.) Walp. *Plant Physiology* 90: 1286-1292.

Beck D.P. and Munns D.N. (1984). Phosphate nutrition of *Rhizobium* sp. *Applied Environmental Microbiology* 47: 278-282.

Beck D.P. and Munns D.N. (1985). Effects of calcium to the phosphorus nutrition of *Rhizobium meliloti*. *Soil Science Society of America J* 49: 334-337.

Bergerson F.J. and Turner G.L. (1967). Nitrogen fixation by bacteroid fraction of breis of soyabean nodudes. *Biochemica Biophyca Acta* 141: 507-515.

Berthelin J., Munier-Lamy C., and Leyval C. (1995). Effect of microorganisms on mobility of heavy metals in soils. In P.M. Huang, J. Berthelin, J.M. Bollag, W.B. McGill, and A.L. Page (eds.), *Metals, other organics and microbial activities: Environmental impacts of soil component interactions,* Volume 2 (pp. 3-17). Boca Raton, FL: Lewis.

Bethlenfalvay G.J., Dakessian S., and Pacovsky R.S. (1984). Mycorrhizae in a southern California desert: Ecological implications. *Canadian Journal of Botany* 62: 519-524.

Bieleski R.L. (1973). Phosphate pools, phosphate transport and phosphate ability. *Annual Review of Plant Physiology* 24: 225-252.

Biro B., Koves-Pechy K., Voros I., Takacs T., Eggenberger P., and Strasser R.J. (2000). Interactions between *Azospirillum* and *Rhizobium* nitrogen fixer and arbuscular mycorrhizal fungi in the rhizosphere of alfalfa in steril, AMF- free of normal soil conditions. *Applied Soil Ecology* 15: 159-168.

Bloemberg G.V., Thomas-Oates J.E., Lugtenberg J.J., and Spaink H.P. (1997). Biosynthesis and host specificity of rhizobial lipo-chitin oligosaccharide signal molecules. In G. Stacey and N.T. Keen (eds.), *Plant-microbe interactions,* Volume 2 (pp. 1-26). London: Chapman and Hall.

Bonfante P. and Perotto S. (1995). Strategies of arbuscular mycorrhizal fungi when infecting host plants. *New Phytologist* 130: 3-21.

Bradbury S.M., Petrson R., and Bowley S. (1991). Interaction between three alfalfa nodulation genotypes and two *Glomus* species. *New Phytologist* 119: 115-120.

Bradley R., Burt A.J., and Read D.J. (1981). Mycorrhizal infection on resistance to heavy metals toxicity in *Calluna vulgaris. Nature* 292: 335-337.

Bradley R., Burt A.J., and Read D.J. (1982). The biology of mycorrhiza in the Ericaceae. VII. The role of mycorrhizal infection in heavy metal resistance. *New Phytologist* 91: 197-209.

Brown M.S. and Bethlenfalvay G.J. (1987). *Glycine-Glomus-Rhizobium* symbiosis. VI. Photosynthesis in nodulated, mycorrhizal, or N and P-fertilized soyabean plants. *Plant Physiology* 85: 120-123.

Brown M.S., Thamasurakul S., and Bethlenfalvay G.J. (1988). The *Glycine-Glomus-Bradyrhizobium* symbiosis. VIII. Phosphorus-use efficiency of CO_2 and N_2 fixation in mycorrhizal soyabean. *Physiologia Plantarum* 74: 159-163.

Brown M.T. and Wilkins D.A. (1985a). Zinc tolerance of *Amanita* in *Paxillus. Transactions of British Mycological Society* 367-369.

Brown M.T. and Wilkins D.A. (1985b). Zinc tolerance of mycorrhizae *Beutula. New Phytolo*gist 99: 101-106.

Burkert B. and Robson A. (1994). ^{65}Zn uptake in subterranean clover (*Trifolium subterraneum* L.) by three vesicular-arbuscular mycorrhizal fungi in a root-free sandy soil. *Soil Biology and Biochemistry* 26: 117-1124.

Caetano-Anolles G., Lagares A., and Favelukes G. (1989). Adsorption of *Rhizobium meliloti* to alfalfa roots: Dependence on divalent cations and pH. *Plant and Soil* 117: 67-74.

Capaccio L.C.M. and Callow J.A. (1982). The enzymes of polyphosphate metabolism in vesicular-arbuscular mycorrhizas. *New Phytologist* 91: 81-91.

Carling D.E., Richle W.G., Brown M.F., and Johnson D.R. (1978). Effects of a vesicular-arbuscular mycorrhizal fungus on nitrate reductase and nitrogenase activities in nodulating and non-nodulating soyabeans. *Phytopathology* 68: 1590-1596.

Carpenter A.T. and Allen M.F. (1988). Responses of *Hedysarum borelae* Nutt. to mycorrhizas and *Rhizobium:* Plant and soil nutrient changes in a disturbed shrub-steppe. *New Phytologist* 109: 125-132.

Cassman K.G., Munns D.N., and Beck D.P. (1981). Phosphorus nutrition of *Rhizobium japonicum:* Strain differences in phosphate storage and utilization. *Soil Science Society of America, J.* 45: 517-520.

Cassman K.G., Whitney A.S., and Fox R.L. (1981). Phosphorus requirements of soyabean and cowpea as affected by mode of nutrition. *Agraon. J.* 73: 17-22.

Chabot S., Bel-Rhlid R., Chenebert R., and Piche Y. (1992). Hyphal growth promotion in vitro of the VA mycorrhizal fungus, *Gigaspora margarita* Becker & Hall, by the activity of structurally specific flavonoid compounds under CO_2-enriched conditions. *New Phytologist* 122 : 461-467.

Cooper K.M. (1984). Physiology of VA mycorrhizal association. In C.L. Powell and D.J. Bagyaraj (eds.), *VA mycorrhizae* (pp. 155-186). Boca Raton, FL: CRC Press.

Cooper K.M. and Losel D. (1978). Lipid physiology of vesicular-arbuscular mycorrhiza. Composition of lipids in roots of onion, clover and ryegrass infected with *Glomus mosseae*. *New Phytologist* 80: 143-151.

Daft M.J. and El-Giahmi A.A. (1976). Studies on nodulation and mycorrhizal peanuts. *Annual Applied Biology* 83: 273-276.

Daft M.J. and Nicolson T.H. (1964). Effect of *Endogone* mycorrhiza on plant growth. II. Influence of inoculum concentration on growth and infection in tomato. *New Phytologist* 68: 953-963.

Danneberg G., Latus C., Zimmer W., Hundeshagen B., Schneider-Poetsch H., and Bothe H. (1992). Influence of vesicular-arbuscular mycorrhiza on phytohormone balances in maize (*Zea mays* L.). *Journal of Plant Physiology* 141: 33-39.

Das N.K., Patel K., and Skoog F. (1976). Extraction of kinetin and Indole acetic acid excised from tobacco pith tissue. *Physiology of Plant* 9: 640-648.

de Freitas J.R. and Germida J.J. (1990). Plant growth promoting rhizobacteria for winter wheat. *Canadian Journal of Microbiology* 36: 265-272.

De Jong A.J., Cordewener J., LoSchiavo F., Terzi M., Vendekerchhove J., van Kammen A., and de Vries S.C. (1992). A carrot somatic embryo mutant is rescued by chitinase. *Plant Cell* 4: 42-433.

De la Cruz R.E., Manalo M.Q., Aggangan N.S., and Tombalo D.J. (1988). Growth of three legume trees inoculated with VA mycorrhizal fungi and *Rhizobium*. *Plant and Soil* 108: 111-115.

De Maagd R.A., de Rijk R., Mulders I.H.M., and Lugtenberg B.J.J. (1989). Immunological characterization of *Rhizobium leguminosarum* outer membrane antigens by use of polyclonal and monoclonal antibodies. *Journal of Bacteriology* 171: 1136-1142.

Denarie J., Debelle F., and Prome J.C. (1996). *Rhizobium* lipo-chitooligosaccharide nodulation factors: Signaling molecules mediating recognition and morphogenesis. *Annual Review of Biochemistry* 65: 503-535.

Denny H.J. and Wilkins D.A. (1987). Zinc tolerance in *Betula* spp. III. Variation in response to zinc among ectomycorrhizal amelioration of zinc toxicity. *New Phytologist* 106:533-544.

Diaz G. and Honrubia M. (1993). Infectivity of mine soils from South-East Spain. 2. Mycorrhizal population in levels in spoilt sites. *Mycorrhiza* 4: 85-88.

Dixon R.K. (1988). Response of ectomycorrhizal *Quercus rubra* to soil cadmium, nickel and lead. *Soil Biology and Biochemistry* 20: 555-559.

Dixon R.K. and Buschena C.A. (1988). Response to ectomycorrhizal *Pinus banksiana* and *Picea glauca* to heavy metals in soil. *Plant* and *Soil* 105: 265-271.

Dodd J.C., Burton C.C., Burnus R.G., and Jeffries P. (1987). Phosphatase activity associated with rhizosphere of plants infected with vesicular-arbuscular mycorrhizal fungi. *New Phytologist* 7: 163-172.

Druge U. and Schonbeck F. (1992). Effect of vesicular arbuscular mycorrhizal infection on transpiration, photosynthesis and growth of flax (*Linum usitatissimum* L.) in relation to cytokinin levels. *Journal of Plant Physiology* 141: 40-48.

Duc G., Trouvelot A., Goaninazzi-Pearson V., and Gianinazzi S. (1989). First report of non-mycorrhizal mutants (Myc$^-$) obtained in pea (*Pisum sativum.* L.) and Fababean (*Vicia faba* L.). *Plant Science* 60: 215-222.

Dueck T.A., Visser P., Ernst W.H.O., and Schat H. (1986). Vesicular arbuscular mycorrhizae decrease zinc toxicity to grasses growing in zinc polluted soil. *Soil Biology and Biochemistry* 18: 331-333.

Duff S.M.G., Lefebvre D.D., and Plaxton W.C. (1989). Phosphate starvation inducible "by passes" of adenylate and phosphate dependent glycolytic enzymes in *Brassica nigra* suspension cells. *Plant Physiology* 90: 1275-1278.

Dumas G.E., Guillaume P., Tahiri A.A., Gianinazzi-Pearson V., and Gianinazzi S. (1994). Changes in polypeptide patterns in tobacco roots by *Glomus* spp. *Mycorrhiza* 4: 215-221.

El-Kherbawy M., Angle J.S., Heggo A., and Chaney R.L. (1989) Soil pH, rhizobia, and heavy metal uptake of alfalfa (*Medicago sativa* L.). *Biology and Fertility of Soil* 8: 61-65.

Ferrera-Cerrato R. and Villerias S.J. (1985). The VA endomycorrhiza and its effect of the development of three arboreous legumes. *Proceedings of the Sixth North American Conference on Mycorrhizae,* Bend, Oregon State University, Corvallis, USA. pp 328.

Fitter A.H. and Garbaye J. (1994). Interactions between mycorrhizal fungi and other soil organisms. *Plant Soil* 159: 123-132.

Francis C.F. and Thornes J.B. (1990). Matoral: Erosion and reclamation. In J. Albaladejo, M.A. Stocking, and E. Diaz (eds.), *Soil degradation and rehabilitation in Mediterranean environmental conditions* (pp. 87-115). Murcia, Spain: Consejo Superior de Investigaciones Cientificas.

Fredeen A.L. and Terry N. (1988). Influence of vesicular-arbuscular mycorrhizal infection and soil phosphorus level on growth and carbon metabolism of soya bean. *Canadian Journal of Botany* 66: 2311-2316.

Freire J.J.R. (1984). Important limiting factors in soil for the *Rhizobium*-legume symbiosis. In M. Alexander (ed.), *Biological nitrogen fixation and its ecological basis* (pp. 75-98). New York: Plenum Press.

Fridovich I. (1986). Superoxide dismutases. In A. Meister (ed.), *Advances in enzymology related areas of molecular biology* (pp. 61-97). New York: John Wiley and Sons.

Fruhling M., Roussel H., Gianinazzi-Pearson V., Puhler A., and Perlick A.M. (1997). The *Vicia faba* legheomoglobin gene *VfLb*29 is induced in root nodules and in roots colonized by the arbuscular mycorrhizal fungus *Glomus fasciculatum. Mol. Plant-Microbe. Int.* 10: 124-131.

Fry S.C., Aldington S., Hetherington P.R., and Aitken J. (1993). Oligosaccharides as signals and substrates in the plant cell wall. *Plant Physiology* 103: 1-5.

Gadd G.M. (1986a). The response of fungi towards heavy metals. In R.A. Herbert and G.A. Codd (eds.), *Microbes in extreme environments,* Volume 83 (p. 110). London: Academic Press.

Gadd G.M. (1986b). The uptake of heavy metal by fungi and yeasts: The chemistry and physiology of the process and applications for biotechnology. In H. Eccles and S. Hunt (eds.), *Immobilisation of ions by biosorption* (pp. 135-147). Chichester: Ellis Horwood Ltd.

Gadd G.M. (1990a). Fungi and yeasts for metal binding. In M. Ehrlich and C.L. Brierley (eds.), *Microbial mineral recovery* (pp. 249-279). New York: McGraw-Hill.

Gadd G.M. (1990b). Metal tolerance In C. Edwards (ed.), *Microbiology of extreme environments* (pp. 178-210). Philadelphia: Milton Keynes Open University Press.

Gadd G.M. (1992a). Microbial control of heavy metal pollution. In J.C. Fry, G.M. Gadd, R.A. Herbert, C.W. Jones, and J. Watson-Craik (eds.), *Microbial control of environmental pollution* (pp. 59-88). Cambridge: Cambridge University Press.

Gadd G.M. (1992b). Molecular biology and biotechnology of microbial interctions with organic and inorganic heavy metal compounds. In R. A. Herbert, R.J. Sharp (eds.), *Molecular biology and biotechnology of extremophiles* (pp. 225-257). Glassgow: Blackie and Sons.

Gadd G.M. and Griffiths A.J. (1978). Microorganisms and heavy metal toxicity. *Microbial Ecology* 4: 303-317.

Gadd G.M. and Griffiths A.J. (1980). Effect of copper on morphology of *Aureobasidium pullulans. Transactions of British Mycological Society* 74: 387-392.

Galiazzo F., Ciriolo M.R., Carri M.T., Civitareale L., Marmocchi F., and Marmocchi G. (1991). Activiation and induction by copper of Cu/Zn superoxide dismutasein *Saccharomyces cereviseae. European Journal of Biochemistry* 19: 545-549.

Ganry F., Diem H.G., and Dommergues Y.R. (1982). Effect of inoculation with *Glomus mosseae* on nitrogen fixation by field growth soya beans. *Plant and Soil* 68: 321-329.

Garbaye J. (1994). Helper bacteria: A new dimension to the mycorrhizal symbiosis. *New Phytologist* 128: 197-210.

Garcia-Romera I., Garcia-Garrido J.M., Martinez-Molina E., and Ocampo J.A. (1991). Production of pectolytic enzymes in lettuce root colonized by *Glomus mosseae. Soil Biology and Biochemistry* 23: 597-601.

Gardezi A.K., Ferrera-Cerrato R., and Lara-Fernandez V. (1988). Effect of double inoculation of *Rhizobium* sp. and VA-endomycorrhizae on *Acacia cyanophylla* in an andisol in Mexico. *NFTRR* 6: 562-575.

Gerdemann J.W. (1964). The effects of mycorrhiza on the growth of maize. *Mycologia* 54: 342-349.

Gianinazzi S. (1991). Vesicular-arbuscular (endo)-mycorrhizas: cellular, biochemical and genetic aspects. *Agriculture Ecosystems and Environment* 35: 105-109.

Gianinazzi-Pearson V. (1996). Plant cell responses to arbuscular mycorrhiza fungi: Getting to the roots of the symbiosis. *Plant Cell* 8: 1871-1883.

Gianinazzi-Pearson V. (1997). Have common plant systems co-evolved in fungal and bacterial root symbioses? In A. Legocki, H. Bothe, and A. Puhler (eds.), *Biological fixation of nitrogen for ecology and sustainable agriculture* (pp. 321-324). NATO ASI Series Vol. G39. Berlin: Springer-Verlag.
Gianinazzi-Pearson V., Fordeau J.C., Asimi S., and Gianinazzi S. (1981). Source of additional phosphorus absorbed by soil by vesicular-arbuscular mycorrhizal soybeans. *Physiol. Veg.* 19: 33-44.
Gianinazzi-Pearson V. and Gianinazzi S. (1976). Enzymatic studies on the metabolism of vesicular-arbuscular mycorrhiza. I. Effect of mycorrhiza formation and Phosphorus nutrition on soluble phosphatase activities in onion roots. *Physiol. Veg.* 14: 833-841.
Gianinazzi-Pearson V. and Gianinazzi S. (1978). Enzymatic studies on metabolism of vesicular arbuscular mycorrhiza II. Soluble alkaline phosphatase specific to mycorrhizal infection in onion roots. *Physiological Plant Pathology* 12: 45-53.
Gianinazzi-Pearson V. and Gianinazzi S. (1983). The physiology of vsicular arbuscular mycorrhizal roots. *Plant and Soil* 71: 197-209.
Giannatiasis M. and Coppula S. (1969). Isolomento de citochinine et *Rhizobium leguminosarum* Frank. *Giorn. Bot. Ital.* 103: 11-17.
Gibson A.H. (1967). Carbon dioxide limitations of plant growth in tube culture, with special reference to legume-nodulation studies. *Austr. Journal of Biological Science* 20: 837-842.
Gildon A. and Tinker P.B. (1981). A heavy metal tolerant strain of a mycorrhizal fungus. *Transactions of British Mycological Society* 77: 648-649.
Gildon A. and Tinker P.B. (1983). Interactions of vesicular-arbuscular mycorrhizal infection and heavy metals in plants. The effects of heavy metals on the development of vesicular-arbuscular mycorrhizas. *New Phytologist* 95: 247-261.
Giovennetti M., Tiso. G., Torry. D., and Zazzerini A. (1991). Histological and biochemical interactions between vesicular-arbuscular mycorrhizae and *Thielaviopsis basicola* in tobbaco plants. *Journal of Phytopathology* 131: 265-274.
Glick B.R. (1995). The enhancement of plant growth by free- living bacteria. *Canadian Journal of Microbiology* 41: 109-117.
Goldstein A.H., Baertlein D.A., and McDaniel R.G. (1988). Phosphate starvation inducible metabolism in *Lycopersicon esculentum.* I. Excretion of acid phosphatage by tomato plants and suspension cultured cells. *Plant Physiology* 87: 711-715.
Graham P.H. (1981). Some problems of nodulation and symbiotic nitrogen fixation in *Phaseolus vulgaris.* A review. *Field Crops Research* 4: 93-112.
Graham P.H. and Rosas J.C. (1979). Phosphorus fertilization and symbiotic nitrogen fixation in common bean. *Agron. J.* 71: 925 - 926.
Graham P.H. and Temple S.R. (1984). Selection for improved nitrogen fixation in *Glycine max* (L.) Merr. and *Phaseolus vulgaris* L. *Plant and Soil* 82: 315- 327.
Greco M.A., Hrab D.I., Magner W., and Kosman D.J. (1990). Cu, Zn superoxide dismutase and copper deprivation and toxicity in *Saccharomyces cereviseae. Journal of Bacteriology* 172: 317-323.
Grisebach H. (1981). Lignin. In E.E. Conn (ed.), *The biochemistry of plants. Secondary plant products* (pp. 457-478). New York: Academic Press.

Guo Y., George E., and Marschner H. (1996). Contribution of an arbuscular mycorrhizal fungus to the uptake of cadmium and nickel in bean and maize plants. *Plant and Soil* 184: 195-205.

Gupta A., Saxena A.K., Gopal M., and Tilak K.V.B.R. (1998). Effect of plant growth promoting rhizobacteria on competitive ability of introduced *Bradyrhizobium* sp. *(Vigna)* for nodulation. *Microbiol. Res.* 153 : 113.

Habte M., Fox R.L., Aziz T., and El-Swaify V. (1988). Interaction of vesicular-arbuscular mycorrhizal fungi with erosion in an oxisol. *Applied Environmental Microbiology* 54: 945-950.

Halverson L.J. and Stacey G. (1986). Signal exchange in plant-microbe interactions. *Microbiological Reviews* 50: 193-225.

Harley J.L. and Smith S.E. (1983). *Mycorrhizal symbiosis.* London and New York: Academic Press.

Harris D., Pakovsky R.S., and Paul E.A. (1985). Carbon economy of soybean-*Rhizobium-Glomus* associations. *New Phytologist* 101: 427-440.

Harrison M.J. (1997). The arbuscular mycorrhizal symbiosis. *Plant-Microbe Interactions* 3: 1-34.

Hart A.L. (1989). Nodule phosphorus and nodule activity in white clover. *New Zealand Journal of Agricultural Research* 32: 145-149.

Hayman D.S. (1986). Mycorrhizae of nitrogen-fixing legumes. *MIRCEN Journal* 2: 121-145.

Herrera M.A., Salamanca C.P., and Barea J.M. (1993). Introduction of woody legumes with selected arbuscular mycorrhizal fungi and rhizobia to recover desertified mediterranean ecosystems. *Applied Environmental Microbiology* 59: 129-133.

Hohenberg J.S., Munns D.N., and Tucker C.L. (1982). *Rhizobium* host specificities in *Phaseolus coccineus* L. and *Phaseolus vulgaris* L. *Crop Science* 22: 455-459.

Hungria M. and Neves M.C.P. (1986). Ontogenia da fixacao biologica do nitrogenio em *Phaseolus vulgaris. Pseq. Agropec. Bras.* 21: 715-730.

Iruthayathas E.E. and Vlassak K. (1982). Symbiotic specificity in nodulation and nitrogen fixation between winged bean and *Rhizobium. Scientia Horticulture* 16: 313-322.

Islam R. and Ayanaba A. (1981). Effect of seed inoculation and pre-infecting cowpea *(Vigna unguiculata)* with *Glomus mosseae* on growth and seed yield of the plants under field conditions. *Plant and Soil* 61: 341-350.

Israel D.W. (1981). Cultivar and *Rhizobium* strain effects on nitrogen fixation and remobilization by soyabean. *Agron. J.* 73: 509-516.

Israel D.W. (1987). Investigation of the role of phosphorus in symbiotic nitrogen fixation. *Plant Physiology* 84: 835-840.

Jakobsen I. (1985). The role of phosphorus in nitrogen fixation by young pea plants. *Physiological Plantarum* 64: 190-196.

Janos D.P. (1987). VA mycorrhizas in humid tropical ecosystems. In G.R. Sufir (ed.), *Ecophysiology of VA mycorrhizal plants* (pp. 107-134). Boca Raton, FL: CRC Press.

Jarell W.M. and Beverly R.B. (1981). The dilution effect in plant nutrition studies. *Adv. Agron.* 34: 197-224.

Jasper D.A., Abbott L.K., and Robson A.D. (1989). Acacias respond to additions of phosphorus and to inoculation with VA mycorrhizal fungi in soils stockpiled during mineral sand mining. *Plant and Soil* 115: 99-108.

Jeanmaire C., Dexheimer J., and Gerard F. (1988). Approche, au niveau cellulaire, du fonctionnement des endomycorhizes a vesicules et arbuscules. *Cytologia* 53: 19-32.

Joho M., Imai M., and Murayama T. (1985). Different distribution of Cd^{2+} between Cd-sensitive and Cd-resistant strains of *Saccharomyces cerevisiae. Journal of General Microbiology* 131: 53-56.

Joner E.J. and Leyval C. (1997). Uptake of ^{100}Cd by roots and hyphae of a *Glomus mosseae/Trifolium subterraneum* mycorrhiza form soil amended with high and low concentrations of cadmium. *New Phytologist* 135: 353-360.

Juste C. (1988). Appreciation de la mobilite e de la biodispoinibilite des elements en traces du sol. *Sci. Sol.* 26: 103-112.

Kaldorf M., Zimmer W., and Bothe H. (1994). Genetic evidence for the occurrence of assimilatory nitrate reductase in arbuscular mycorrhizal and other fungi. *Mycorrhiza* 4: 37-42.

Kemble A.R. and MacPherson H.T. (1954). Liberation of amino-acids in perennial ryegrass during wilting. *Biochemistry Journal* 58: 46-49.

Killham K. and Firestone M.K. (1983). Vesicular-arbuscular mycorrhizal mediation of grass response to acidic and heavy metal depositions. *Plant and Soil* 72: 39-48.

Kloepper J.W., Leong J., Teintz M., and Schroth, M.N. (1980). Enhanced plant growth by siderophores produced by plant growth promoting rhizobacteria. *Nature* 286: 885.

Krishna K.R. and Bagyaraj D.J. (1983). Changes in free amino nitrogen and protein fractions of groundnut caused by inoculation with VA-mycorrhiza. *Annals of Botany* 51: 399-401.

Lefebvre D.D., Duff S.M.G., Fife C.A., Julien-Inalsingh C., and Plaxton W.C. (1990). Response to phosphorus deprivation in *Brassica nigra* suspension cells. *Plant Physiology* 93: 504-511.

Legocki R.P. and Verma D.P.S. (1979). A nodule specific plant protein (Nodulin-35) from soyabean. *Science* 205: 190-193.

Leyval C., Singh B.R., and Joner E.J. (1995). Occurrence and infectivity of arbuscular mycorrhizal fungi in some Norwegian soils influenced by heavy metals and soil properties. *Water Air Soil Pollution* 84: 203-216.

Leyval C., Turnau K., and Haselwandter K. (1997). Effect of heavy metal pollution on mycorrhizal colonizaiton and function: Physiological, ecological and applied aspects. *Mycorrhiza* 7: 139-153.

Linderman R.G. (1992). Vesicular-arbuscular mycorrhizae and soil microbial interactions. In G.J. Bethlenfalvay and R.G. Linderman (eds.), *Mycorrhizae in sustainable agriculture* (pp. 65-77). Madison: ASA Special Publication.

Linderman R.G. (1994). Role of VAM fungi in biocontrol. In F.L. Pfleger and R.G. Linderman (eds.), *Mycorrhizae and plant health* (pp. 1-26). St. Paul, MN: APS Press.

Lopez-Bermudez F. and Albaladejo J. (1990). Factors ambientales de la degradacion del suelo enel area mediterranea. In J. Albaladejo, M.A. Stocking, and E. Diaz (eds.), *Soil degradation and rhabilitation in Mediterranean environmental conditions* (pp. 15-45). Murcia, Spain: Consejo Superior de Investigaciones Cientificas.

Loth F.G. and Hofner W. (1995). Einfluss der VA-Mykorrhiza auf die Schwermetallufnahme von Hafer (*Avena sativa* L.) in Abhangigheir vom Kontaminationsgrad der Boden. *Z. Pflanzenernaehr Bodenk* 158: 339-345.

MacDonald R.M. and Lewis M. (1978). The occurrence of some acid phosphatases and dehydrogenases in the vesicular-arbuscular mycorrhizal fungus *Glomus mosseae*. *New Phytologist* 80: 135-141.

Manjunath A. and Bagyaraj D.J. (1984). Response of pigeonpea and cowpea to phosphate and dual inoculation with vesicular-arbuscular mycorrhiza and *Rhizobium. Tropical Agricultural Trinidad* 61: 48-52.

Manjunath A., Bagyaraj D.J., and Gopala Gowda H.S. (1984). Dual inoculation with VA mycorrhiza and *Rhizobium* is beneficial to *Leucaena. Plant and Soil* 78: 445-448.

Marschner H. (1995). *Mineral nutrition of higher plants*. London: Academic Press.

Martin E. and Hilbert J.L. (1991). Morphological, biochemical and molecular changes during ectomycorrhizal development. *Experientia* 47: 321-331.

McInroy J.A. and Kloepper J.W. (1994). Studies on indigenous endophytic bacteria of sweet corn and cotton. In F. O'Gara, D.N. Dowling, and B. Boesten (eds.), *Molecular ecology of rhizosphere microorganisms: Biotechnology and the release of GMOs* (pp. 19-28). Winheim: VCH Verlagsgesellschaft.

Mehra R.K. and Winge D.R. (1991). Metal ion resistance in fungal molecular mechanism and their related expression. *Journal of Cellular Biochemistry* 45: 30-40.

Meller R.B. and Werner D. (1990). Legume nodule biochemistry and function. In P.M. Gresshoff (ed.), *Molecular biology of symbiotic nitrogen fixation* (pp. 111-129). Boca Raton, FL: CRC Press,.

Mengel K. and Kirky E.A. (1987). Phosphorus in physiology. In K. Mengel and E.A. Kirky (eds.), *Principles of plant nutrition,* Fourth Edition (pp. 414-420). Bern, Switzerland: International Potash Institute.

Meyer J.R. and Linderman R.G. (1986). Response of subterranean clover to dual inoculation with vesicular-arbuscular mycorrhizal fungi and a plant growth promoting bacterium. *Pseudomonas putida*. *Soil Biology and Biochemistry* 18: 185-190.

Miller R.W., McRac D.G., Al-Jobore A., and Berndt W.B. (1988). Respiration supported nitrogenase activity of isoloted *Rhizobium meliloti* bacteroids. *Journal of Cell Biochemistry* 38: 35-49.

Minchin F.R., Summerfield R.J., and Eaglesham A.R.J. (1978). Plant genotype *Rhizobium* strain interactions in cowpea (*Vigna unguiculata* L. Walp). *Tropical Agriculture (Trinidad)* 55: 107-115.

Minney S.F. and Quirk A.V. (1985). Growth and adaptation of *Saccharomyces cerevisiae* at different cadmium concentrations. *Microbios* 42: 37-44.

Morgan R.P.C., Rickson R.T., and Wright W. (1990). Regeneration of degraded soils. In J. Albaladejo, J. Stocking, and E. Diaz (eds.), *Soil degradation and re-*

habilitation in Mediterranean environmental conditions (pp. 69-85). Murcia, Spain: Consejo Superior de Investigaciones Cientificas.

Mosse B. (1957). Growth and chemical composition of mycorrhizal and non-mycorrhizal apples. *Nature* (London) 179: 922-924.

Mosse B. (1973). Advances in the study of vesicular-arbuscular mycorrhiza. *Annual Review of Phytopathology* 11: 171-196.

Mullen M.D., Israel D.W., and Wollum A.G. (1988). Effects of *Bradyrhizobium japonicum* and soyabean (*Glycine max* (L.) Merr.) phosphorus nutrition on nodulation and dinitrogen fixation. *Applied Environmental Microbiology* 54: 2387-2392.

Mullen M.D., Wolf D.C., Beveridge T.J., and Bailey G.W. (1992). Absorption of heavy metals by the soil fungi *Aspergillus niger* and *Mucor rouxii*. *Soil Biology and Biochemistry* 24: 129-135.

Munns D.N. (1970). Nodulation of *Medicago sativa* in solution culture. I. Acid sensitive steps. *Plant and Soil* 28: 129-146.

Murphy L.S., Ellis R. Jr., and Adriano D.C. (1981). Phosphorus micronutrient interaction effects on crop production. *Journal of Plant Nutrition* 3: 593-613.

Murphy R.J. and Levy J.F. (1983). Production of copper oxalate in some copper tolerant fungi. *Transactions of the British Mycological Society* 81: 165-168.

Neilsen J.D. and Jensen A. (1983). Influence of vesicular-arbuscular mycorrhizal fungi on growth and uptake of various nutrients as well as uptake ratio of fertilizer P for lucerne *(Medicago sativa)*. *Plant and Soil* 70: 165-172.

Nemec S. and Meredith F.I. (1981). Amino acid content of leaves on mycorrhizal and non-mycorrhizal citrus rootstocks. *Annals of Botany* 47: 351-358.

Olivares J., Herrera M.A., and Bedmar E.J. (1988). Woody legumes in arid and semi-arid zones: The *Rhizobium-Prosopis chilensis* symbiosis. In D.P. Beck and L.A. Materon (eds.), *Nitrogen fixation legumes in Mediterranean agriculture* (pp. 65-72). Dordrecht, the Netherlands: ICARDA and Martinus Nijhoff.

Ono B.I., Ohue H., and Ishihara F. (1988). Role of cell wall in *Saccharomyces cerevisiae* mutants resistant to Hg^{2+}. *Journal of Bacteriology* 170: 5877-5882.

Pacovsky R.S. (1989). Carbohydrate, protein and amino acid status of *Glycine-Glomus-Bradyrhizobium* symbioses. *Physiologia Plantarum* 72: 733-746.

Pacovsky R.S., Bayne H.G., and Bethlenfalvay G.J. (1984). Symbiotic interactions between strains of *Rhizobium phaseoli* and cultivars of *Phaseolus*. *Plant and Soil* 128: 101-105.

Pacovsky R.S., Bethlenfalvay G.J., and Paul E.A. (1986). Comparison between phosphorus fertilized and mycorrhizal plants. *Crops Science* 26: 151-156.

Pacovsky R.S., De Silva P., Carvalho M.T.V., and Tsai S.M. (1990). Increased nutrient assimilation and enzyme activities in field beans inoculated with *Glomus etunicatum*. *Proc, 8th North American Conference on Mycorrhizae,* Jackson, Wyoming, p. 230.

Pacovsky R.S., Paul E.A., and Bethlenfalvay G.J. (1986). Response of mycorrhizal and P fertilized soyabeans to nodulation by *Bradyrhizobium* on ammonium nitrate. *Crops Science* 26: 151-156.

Palma J.M., Longa M.A., Del Rio L.A., and Arines J. (1992). Superoxide dismutase in VA-mycorrhizal red clover plants. *Physiologia Plantarum* 86: 147-152.

Pate J.S. (1958). Studies of growth substances of legume nodules using paper chromatography. *Australian Journal of Biological Science* 11: 516-526.

Peberdy J.F (1990). Fungal cell walls—A review. In P.J. Kuhn, A.P.J. Trinci, M.J. Yung, M.W. Cosey, and L.E. Copping (eds.), *Biochemistry of cell walls and membranes in fungi* (pp. 5-30). Berlin: Springer-Verlag.

Perotto S., Brewin N., and Bonfante P. (1994). Colonization of pea roots by the mycorrhizal fungus *Glomus versiforme* and by *Rhizobium* bacteria: Immunological comparison using monoclonal antibodies as probes for plant cell surface components. *Molecular Plant Microbial Interaction* 7: 91-98.

Phillips D.A. and Torrey J.G. (1970). Cytokinin production by *Rhizobium japonicum. Physiology of Plant* 23: 1057-1063.

Phillips D.A. and Torrey J.G. (1972). Studies on cytokinin production by *Rhizobium. Plant Physiology* 49: 11-14.

Pololenko D.R., Scher F.M., Kloepper J.W., Singleton C.A., Lalibert E.M., and Zaleska I. (1987). Effects of root colonizing bacteria on nodulation of soybean roots by *Bradyrhizobium japonicum. Candian Journal of Microbiology* 33: 498-503.

Puppi G., Azcon R., and Hoflich G. (1994). Management of positive interactions of arbusular mycorrhizal fungi with essential groups of soil microorganisms. In S. Gianinazzi and H. Shuepp (eds.), *Impact of arbuscular mycorrhizas on sustainable agriculture and natural ecosystems* (pp. 201-215). Basel: Birkhauser-Verlag.

Puppo A., Riguad J., and Bafthe J. (1974). Sur la presence de cytokinines dans legume nodules de *Phaseolus vulgaris* L. *C. R. Acad. Sci. Soc. D.* 279: 2029-2032.

Rabinowitch H.D. and Fridovich I. (1983). Superoxide radicals, superoxide dismutase and oxygen toxicity in plants. *Phytochemical Photobiology* 37: 843-859.

Raju P.S., Clark R.B., Ellis J.R., and Maranville J.W. (1990). Mineral uptake and growth of *Sorghum* colonized with VA mycorrhiza at varied soil phosphorus levels. *Journal of. Plant and Nutrition* 13: 843-859.

Raman N. and Mahadevan, A. (1996). Mycorrhizal research—A priority in agriculture. In K.G. Mukerji (ed.), *Concepts in Mycorrhizal research* (pp. 41-75). Dordrecht, the Netherlands: Kluwer Acadamic Publishers.

Raman N. and Sambandan K. (1998). Distribution of VAM fungi in tannery effluent polluted soils of Tamil Nadu, India. *Bull. Environ. Contamin. Toxicol.* 60:142-150.

Raman N. and Selvaraj T. (1998). Remediation of tannery effluent polluted soils using VAM fungus and *Rhizobium* inoculated *Prosopis juliflora. International Conference on Environmental Contamination, Toxicology and Health.* Hong Kong: Hong Kong Baptist University, September 23-25, 1998, pp. 33-35.

Raman N., Ravi I., and Gnanaguru M. (1994). Enhancement of Indole-3-acetic acid in nodules of *Prosopis juliflora* inoculated with *Gl. mosseae* and *Rhizobium. Indian J Microbiol.* 34: 33-45.

Raymond K.N., Muller G., and Matzanke B.F. (1984). Complexation of iron by siderophores. A review of their solution and structural chemistry and biological function. *Topics in Current Chemistry* 123: 49-102.

Riazi A., Matsuda K., and Arslan A. (1985). Water stress induced changes in concentrations of proline and other solutes in growing regions of young barley leaves. *Journal of Experimental Botany* 36: 1716-1725.

Richardson A.E., Simpson R.J., Djordjevic M.A., and Rolfe B.J. (1988a). Effects of pH Ca and Al on the exudation from clover seedlings of compounds that induce the expression of nodulation genes in *Rhizobium trifolii*. *Plant and Soil* 109: 37-47.

Richardson A.E., Simpson R.J., Djordjevic M.A., and Rolfe B.J. (1988b). Expression of nodulation genes in *Rhizobium leguminosarum* var *trifolii* is affected by low pH and by Ca and Al ions. *Applied Environmental Microbiology* 54: 2541-2548.

Rohrig H., Schmidt J., Walden R., Czaja I., Lubenov H., Wieneke U., Schell J., and John M. (1996). Convergent pathways for lipochitooligosaccharide and auxin signaling in tobacco cells. *Proceedings of National Academy of Science, USA* 93: 13389-13392.

Rohrig H., Schmidt J., Walden R., Czaja I., Miklaswvics E., Wieneke U., Schell J., and John M. (1995). Growth of tobacco protoplasts stimulated by lipo-chitooligosaccharides. *Science* 269: 841-843.

Roskoski J.P., Pepper. I., and Pardo E. (1986) Inoculation of leguminous trees with rhizobia and VA mycorrhizal fungi. *Forest Ecology and Management* 16: 57-68.

Ross I.S. (1982). Effect of copper, cadmium and Zinc on germination and mycelial growth in *Candida albicans*. *Transactions of the British Mycological Society* 78: 543-545.

Roussel H. (1995). Etude sur les raciness de plantes endomycorhizees du gene codant la noduline 26. DEA des Sciences de la Vie et de la Sante, Biochimie, Biologie Cellulaire et Moleculaire. Dijon, France: Universite de Franche-Comte, Besancon et de Bourgogne.

Routein J.B. and Dawson R.F. (1943). Some interrelationships of growth, salt absorption, respiration and mycorrhizal development in *Pinus echinata* (Mill). *American Journal of Botany* 30: 440-451.

Sagan M., Morandi D., Tarenghi E., and Duc G. (1995). Selection of nodulation and mycorrhizzal mutants in the model plant *Medicago truncula*ta (Gaertn.) after g-ray mutagenasis. *Plant Science* 11: 63-71.

Salt D.E., Blaylock M., Kumar N.P.B.A., Dushenkov V., Ensley B.D., Chet I., and Raskin I. (1995). Phytoremediation: A novel strategy for the removal of toxic metals from the environment using plants. *Biotechnology* 13: 468-474.

Samara A., Dumas-Gaudot E., and Gianinazzi S. (1997). Detection of symbiosis—Related polypeptides during the early stages of the establishment of arbuscular mycorrhizae between *Glomus mosseae* and *Pisum sativum* roots. *New Phytologist* 135: 711-722.

Schlaman H.R.M., Okker R.J.H., and Lugtenberg B.J.J. (1992). Regulation of nodulation gene expression by NodD in *Rhizobia*. *Journal of Bacteriology* 174: 5177-5182.

Schuepp H., Dehn B., and Sticher H. (1987). Interaktionen zwischen VA-Mykorrhizen und Schwermetallbelastungen. *Angew Botany* 61: 85-95.

Shirtliffe S.J. and Vessey J.K. (1996). A nodulation (Nod+/Fix−) mutant of *Phaseolus velgaris* L. has nodule-like structures lacking peripheral vascular bundles (Pvb−) and is resistant to mycorrhizal infection (Myc−). *Plant Science* 118: 209-220.

Shumate S.E. and Strandberg G.W. (1985). Accumulation of metals by microbial cells. In M. Moo-Young, C.N. Robinson, and J.A. Howell (eds.), *Comprehensive biotechnology,* Volume 4 (pp. 235-247). New York: Pergamon Press.

Simpson. D. and Daft M.J. (1990). Spore production and mycorrhizal development in various tropical crop hosts infected with *Glomus clarum. Plant and Soil* 12: 171-178.

Skrdleta V., Nemcova M., Lisa L., and Gaudinova A. (1984). Root respiration and potential N_2 fixation in Pea in relation to the source of nitrogen nutrition. *Rostl. Uyr.* 30: 1171-1178.

Smit G., Kijne J.W., and Lugtenberg B.J.J. (1987). Involvement of both cellulose fibrils and Ca^{2+} dependent adhesin in the attachment of *Rhizobium leguminosarum* to pea root tips. *Journal of Bacteriology* 169: 4294-4301.

Smit G., Kijne J.W., and Lugtenburg B.J.J. (1989). Roles of flagella lipopolysaccharide, and a Ca^{2+} dependent cell surface protein in attachment of *Rhizobium leguminosarium* to pea har tips. *Journal of Bacteriology* 171: 569-572.

Smith S.E. and Gianinazzi-Pearson V. (1987). Enzymatic separation of VA mycorrhizal fungus from roots: Fungal characteristics. *Proceedings of the 7th North American Conference of Mycorrhizae* (p. 221). Gainesville, Florida.

Smith S.E. and Gianinazzi-Pearson V. (1988). Physiological interactions between symbionts in vesicular-arbuscular mycorrhizal plants. *Annual Review of Plant Physiology and Plant Molecular Biology* 39: 221-244.

Spaink H.P., Sheeley D.M., Van Brussel A.A.N, Glushka J., York W.S., Tak T., Geiger O., Kennedy E.P., Reonhold V.N., and Lugtenberg B.J.J. (1991). A novel highly unsaturated fatty acid moiety of lipo-oligosaccharide signals determines host specificity of *Rhizobium. Nature* 354: 125-130.

Spanu P. and Bonfante-Fasolo P. (1988). Cell wall bound peroxidase activity in roots of mycorrhizal *Allium porrum. New Phytologist* 109: 119-124.

Stahl P.D., Williams S.E., and Christensen V. (1988). Efficacy of native vesicular-arbuscular mycorrhizal fungi after severe soil disturbance. *New Phytologist* 110: 347-354.

Stewart C.R. and Bogges S.F. (1978). Metabolism of 5-H proline by barley leaves and its use in measuring the effects of water stress and proline oxidation. *Plant Physiology* 61: 654-657.

Stribley D.P. (1987). Mineral nutrition. In G. Safir (eds.) *Ecophysiology of VA mycorrhizal plants* (pp. 59-70). Boca Raton, FL: CRC Press.

Sutter H.P., Jones E.B.J., and Walchli O. (1983). The mechanism of copper tolerance in *Poria placenta* (Fr.) Cke and *Poria callion* (Pers.) Fr. *Material und Organismen* 18: 243-263.

Sylvia D.M. (1990). Inoculation of native woody plants with vesicular-arbuscular mycorrhizal fungi for phosphate mine land reclamation. *Agricultural Ecosystem and Environment* 31: 253-261.

Syono K. and Torrey J.G. (1976). Identification of cytokinins of root nodules of the garden pea, *Pisum sativum* (L.). *Plant Physiology* 57: 602-606.

Theodorou C. (1968). Inositol phosphates in needles of *Pinus radiata* (D.Don.) and the phytase activity of mycorrhizal fungi. *Proc. Ninth International Congress on Soil Science, Adelaide* 3: 483-493.

Theodorou C. (1971). The phytase activity of the mycorrhizal fungus *Rhizopogon roseolus. Soil Biology and Biochemistry* 3: 89-90.

Thiagarajan T.R., Ames R.N., and Ahmed M.H. (1992). Response of cowpea *(Vigna unguiculata)* to inoculation with co-selected vesicular-arbuscular mycorrhizal fungi and *Rhizobium* strains in field trails. *Canadian Journal of Microbiology* 38: 573-579.

Thimann K.V. (1936). On the physiology of the formation of nodules on legume roots. *Proceedings of Natural Academy of Science U.S.A.* 22: 511-514.

Thomas H. (1991). Accumulation and consumption of solutes in swards of *Lolium perenne* during drought and after rewatering. *New Phytologist* 118: 35-48.

Torrey J.G. (1961). Kinetin as trigger for mitosis in mature endomitotic plant cells. *Experimental Cell Research* 23: 281-299.

Trevors J.T., Stratton G.W., and Gadd G.M. (1986). Cadmium transport resistance and toxicity in algae, bacteria and fungi. *Canadian Journal of Microbiology* 32: 447-464.

Tsang W.T., Bowler C., Herouart D., Van Camp W., Villarroel R., Genetello C., Van Montagu M., and Inze D. (1991). Differential regulation of superoxide dismutases in plants exposed to environmental stress. *The Plant Cell* 3: 783-792.

Tsimilli-Michael M., Eggenberger P., Biro B., Koves-Pechy K., Voros I., and Strasser R.J. (2000). Synergitic and antagonistic effects of arbuscular mycorrhizal fungi and *Azospirillum* and *Rhizobium* nitrogen fixers on the phosynthetic activity of alfalfa, probed by the polyphasic chlorophyll a florescence transient O-J-I-P. *Applied Soil Ecology* 15: 169-182.

Turnau K., Kottke I., and Dexheimer (1996). Toxic element filtering in *Rhizopogon roseolus/Pinus sylvestris* mycorrhizas collected from calamine dumps. *Mycological Research* 100: 16-22.

Turner J.T. and Backmam P.A. (1991). Factors relating to peanut yield increases following *Bacillus subtilis* seed treatment. *Plant Disease* 75:347-353.

Vargas A.A.T. and Graham P.H. (1988). *Phaseolus vulgaris* cultivar and *Rhizobium* strain variation in acid-pH tolerance and nodulation under acid conditions. *Field Crops Research* 19: 91-101.

Vejsadova H., Siblikova D., Hrselova H., and Vancura V. (1992). Effects of the VAM fungus *Glomus* sp. on the growth and yield of soybean inoculated with *Bradyrhizobium japonicum. Plant and Soil* 140: 121-125.

Vincent J.M. and Humphrey B.A. (1968). Modification of the antigenic surface of *Rhizobium trifolii* by a deficiency of calcium. *Journal of General Microbiology* 54: 397-405.

Weissenhorn I., Glashoff A., Leyval C., and Berthelin J. (1994). Differential tolerance to Cd and Zn of arbuscular mycorrhizal (AM) fungal spores isolated from heavy metal-polluted and unpolluted soils. *Plant and Soil* 167: 189-196.

Weissenhorn I. and Leyval C. (1995). Root colonization of maize by a Cd-sensitive and a Cd-tolerant *Glomus mosseae* and cadmium uptake in sand culture. *Plant and Soil* 175 : 233-238.

Weissenhorn I., Leyval C., and Belgy G. (1995). Arbuscular mycorrhizal (AM) fungi from heavy-metal uptake by maize (*Zea mays* L.) in pot culture with contaminated soils. *Mycorrhiza* 5: 245 - 251.

Weissenhorn I., Leyval C., and Berthelin J. (1993). Cd-tolerant arbuscular mycorrhizal (AM) fungi from heavy-metal polluted soils. *Plant and Soil* 157: 247-256.

Weller D.M. and Thomashow L.S. (1994). Current challenges in introducing beneficial microorganisms into the rhizosphere. In F. O'Gara, D.N. Dowling, and B. Bousten (eds.), *Molecular ecology of rhizosphere microorganisms: Biotechnology and the release of GMOs* (pp. 1-13). Winheim: VCH Verlagsgesellschaft.

Wilkins D.A. and Hodson M.J. (1989). The effects of aluminium and *Paxillus involutus* Fr. on the growth of Norway spruce *Picea abcea* (L) Karst. *New Phytologist* 113: 225-232.

Winklemann G. (1992). Structures and functions of fungal siderophores containing hydroxamate and compexane types iron binding ligands. *Mycological Research* 96: 529-534.

Woolhouse H.W. (1969). Difference in the properties of the acid phosphatases of plant roots and their significance in the evolution of edaphic ecotypes. In I.H. Rorison (ed.), *Ecological aspects of the mineral nutrition of plants* (pp. 357-380). Oxford: Blackwell.

Wu L.C. (1973). Changes in enzymes of mung bean seeds germinated on mycelial macerates of *Rhizoctonia solani*. *Physiological Plant Pathology* 3: 16-27.

Wyss P., Mellor R.B., and Wiemken A. (1990). Vesicular-arbuscular mycorrhizas of wild type soyabean and non-nodulating mutants with *Glomus mosseae* contain symbiosis-specific polypeptides (mycorrhizins), immunologically cross-reactive with nodulins. *Planta* 182: 22-26.

Chapter 4

Biological Fertilizers for Sustainable Rice Production

S. Kannaiyan
K. Kumar

INTRODUCTION

Nitrogen is an important nutrient that is essential for crop growth. The atmosphere contains about 80 percent nitrogen by volume in a free state. Most of the elemental nitrogen that finds its way into the soil is due to its fixation by a certain specialized group of microorganisms. The natural supply of nitrogen comes from irrigation water, rainfall, and atmosphere, which is utilized by soil microflora and recycled to crops during decomposition. Biological nitrogen fixation (BNF) is considered to be an important process that determines nitrogen balance in soil ecosystem.

Nitrogen inputs through BNF support sustainable and environmentally sound agricultural production. BNF conserves soil nitrogen resources. Soil fertility and soil tilth could be improved by practices such as green manuring with nitrogen-fixing legumes, and companion or subsequent non-legume crops can also benefit from BNF. The selection of crop sequences and the integration of farm inputs for maximal returns are all components of effective soil management.

The search for alternatives or supplements to fossil-fuel-based inorganic nitrogen fertilizers has been brought about not only by the likelihood of future price increases for chemical fertilizers but also by the need to maintain

---

The author is thankful to the Indian Council of Agricultural Research, New Delhi, Department of Biotechnology, New Delhi, Tamil Nadu Council for Science and Technology, Tamil Nadu, BARC, Bombay, IRRI, Philippines, EEC, Belgium, British Council, Madras and London, FAO, Rome, Italy and USDA, USA for the financial support for BNF research. The financial assistance extended by ICAR, New Delhi for continuation of research on biological nitrogen fixation and biofertilizers technology under National Agricultural Technology Project-Team of Excellence on Biofertilizers for rice-based cropping system is gratefully acknowledged.

long-term soil productivity and ecological sustainability. The value of nitrogen-fixing legumes in improving and sustaining soil fertility is very important, and higher yields of legumes can be achieved by the application of biofertilizers. The process of BNF and its potential for increasing protein production by legume inoculation is one of the means of providing improved nutrition in developing countries.

Biofertilizers are a cost-effective, eco-friendly, and renewable source of plant nutrients to supplement chemical fertilizers in sustainable agricultural systems in India. Chemical fertilizers are the major supplier of nutrients besides organic and green manures. Despite sizable increases in the use of chemical fertilizers over the years, the gap between nutrient removal and replenishment is significant in India. The role of biofertilizers in sustainable agriculture assumes special significance, particularly in the present context of the very high cost of chemical fertilizers. The production and application of biofertilizers for leguminous crops, oilseeds, rice, millets, and other important crops, along with forest nursery plants, are very common in India.

AZOLLA *BIOFERTILIZER FOR RICE CROP*

Azolla is a free-floating water fern which, in symbiotic association with the cyanobacterium *Anabaena azollae,* contributes substantial amounts of biologically fixed nitrogen to the rice crop (see Photos 4.1 and 4.2). The ni-

PHOTO 4.1. Temperature-tolerant strain of *Azolla microphylla,* SK-TNAU (S. Kannaiyan–Tamil Nadu Agrculture University).

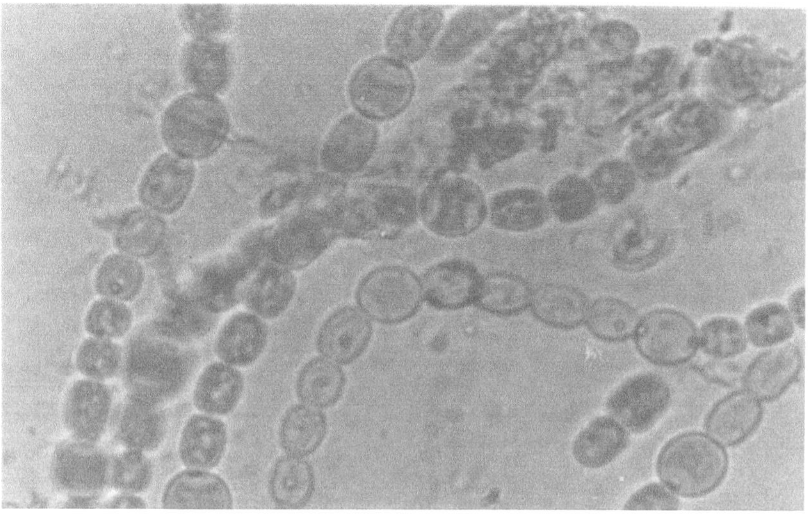

PHOTO 4.2. Algal symbiont of *Azolla-Anabaena azollae.*

trogen-fixing water fern *Azolla* has been used as a biofertilizer for rice crops in India, China, Vietnam, Thailand, the Philippines, Korea, Sri Lanka, Bangladesh, Pakistan, Nepal, Burma, Indonesia, Brazil, and West Africa (Kannaiyan, 1987). The technology of the nitrogen-fixing water fern *Azolla* has been attempted by these countries, but the full potential of this biological system has not been exploited. The *Azolla-Anabaena* association is a "living floating nitrogen manufacturing factory" which utilizes the energy from photosynthesis to fix atmospheric nitrogen amounting to 150 to 200 $kg \cdot ha^{-1}$ per year. *Azolla* is used as biofetilizer for rice crops, and it could contribute 40 to 60 kg $N \cdot ha^{-1}$ per crop (Kannaiyan, 1992). The important factor in using *Azolla* as a biofertilizer for rice crops is its quick decomposition in soil and efficient availability of its nitrogen to rice plants (Kannaiyan, 1990). In tropical rice soils the applied *Azolla* mineralizes rapidly and its nitrogen is made available to the rice crop in a very short period.

An increase in the grain yield of rice by the application of *Azolla* as green manure, dual crop, or both with rice was equivalent to 30 to 40 kg $N \cdot ha^{-1}$, and this has been well documented by several investigators in various countries, such as China (Liu Chung Chu, 1987), the Philippines (Mabbayad, 1987; Watanabe, 1987), India (Singh, 1977; Kannaiyan, 1987), Thailand (Loudhapasitiporn and Kanareugsa, 1987), Sri Lanka (Kulasooriya et al., 1987), the United States (Talley et al., 1977), and West Africa (Diara et al., 1987). The use of *Azolla* in rice fields has considerably increased the soil microbial as well as enzyme activity (Kumar et al., 1995), which in turn maintains soil fertility. It is an important low-cost input that plays a vital role in improving soil quality management in sustainable rice farming (Kannaiyan, 1992, 1993b, 1995). Gopalaswamy and Kannaiyan (1998) have investigated the potential of *Azolla* hybrids and reported that inoculation of *Azolla* hybrids significantly improved the soil organic carbon as well as increased the rice yield. Dual cropping of *Azolla* is a most effective and ecofriendly practice that plays a major role in sustainable rice farming (Kannaiyan, 1992, 1993c, 1995).

NITROGEN-FIXING BLUE GREEN ALGAE

The species of blue-green algae (BGA) that are known to fix atmospheric nitrogen are classified into three groups:

1. Heterocystous—aerobic forms,
2. Aerobic unicellular forms, and
3. Nonheterocystous, filamentous, microaerophilic forms.

The blue-green algal culture's composite inoculum consisting of *Nostoc, Anabaena, Calothrix, Tolypothrix, Plectonema, Aphanothece, Gloeocapsa, Oscillatoria, Cylindrospermum, Aulosira,* and *Scytonema* have been used for inoculation in rice (Kannaiyan, 1985). These N_2-fixing cultures have better adaptability in wetland rice.

Algal Inoculation to Rice Crops

Inoculation of BGA is highly desirable during the dry season. Application of dried soil-based blue-green algal flakes at the rate of 10 $kg \cdot ha^{-1}$ is recommended for rice crops. At the time of inoculation of BGA, 500 g of single superphosphate may be mixed with 10 kg of soil-based BGA, which encourages better growth and establishment in rice fields. A composite starter culture consisting of more than one species or genera is preferable. Fast-growing species that can tolerate more extremities are preferable to highly susceptible ones. The algal species should have a fair amount of tolerance to such agricultural chemicals as pesticides and weedicides. The algal inoculant is to be applied ten days after transplantation of the rice crop. Continuous application of blue-green algae to rice fields for at least three or four consecutive seasons sustains the high crop yield even with reduced levels of commercial nitrogen fertilizer input in ensuing cropping seasons (Kannaiyan, 1993c).

The beneficial effect of algal inoculation on the grain yield of rice has been well documented at a number of locations (Venkataraman, 1972; Kannaiyan, 1985). Algal inoculation to rice crop was effective in different agroclimatic conditions, and soil types (Goyal and Venkataraman, 1971). Blue-green algal inoculation with composite cultures was found to be more effective than single-culture inoculation. A technology for mass-scale production of composite culture of blue-green algae in rice field conditions was developed by Kannaiyan (1985), and the soil-based BGA inoculum could survive for more than two years (Kannaiyan, 1990). At many sites where algal inoculation was used for three to four consecutive cropping seasons, the inoculated algae established well and the effect persisted for subsequent rice crops (Kannaiyan, 1990).

Saline-Tolerant Cyanobacteria

Cyanobacteria react to salt stress either through osmotic adjustment or by secretion of excess polysaccharide (Saxena and Kaushik, 1991). Salt tolerance in cyanobacteria is mediated by an osmoregulatory mechanism that involves the accumulation of carbohydrates, amino acids, and quaternary ammonium compounds (Reed and Stewart, 1988). The salt adaptation strategy of cyanobacteria includes two main processes, namely (1) maintenance of low internal contents of inorganic ions by active export mechanisms, and (2) synthesis and accumulation of osmoprotective compounds corresponding to the osmotic potential of the surrounding medium. The first process is achieved by Na^+/H^+ antiporters that are coupled to cytochrome oxidase. Several compounds such as sucrose, trehalose, glucosyl glycerol, and glycine acting in the second process were identified in salt-loaded cyanobacteria (Reed and Stewart, 1988). The physiological basis of salt tolerance may be due to the extrusion of Na^+ and maintenance of low intracellular Na^+, accumulation of K^+, or ionic balance between K^+ and Na^+ (Apte and Thomas, 1986).

Increasing evidence suggests that respiration plays a primary role in the adaptation of cyanobacteria to higher concentrations of salt (Wastyn et al., 1987). In *Anacystis nidulans,* increasing salt concentration led to decreased photosynthesis and respiration (Vonshak et al., 1988). When cyanobacterial cells were exposed to salt stress, a short-term increase in the cellular sodium concentration was noticed, which was due to a temporary increase in the permeability of the plasma membrane during the first few seconds of exposure to high salt and led to the immediate inhibition of the photosynthetic and respiratory system.

With increased sodium chloride concentration, the filaments of *Anabaena doliolum,* a salt-sensitive freshwater cyanobacterium, was shorter, with fewer heterocysts, which was indirect evidence of reduced nitrogen fixation, as well as a decrease in relative protein content and chlorophyll a content (Rai and Abraham, 1993). *Synechocystis* sp. PCC 6803, when grown in different salt concentrations, exhibited optimal growth with maximal photosynthesis per cell volume, chlorophyll, and phycocyanin at 34.2 mM NaCl (Schubert et al., 1993). Anand et al. (1994) have reported that while *Chroococcus minor* and *Oscillatoria salina* grew up to 3 percent salinity whiter, *Nostoc piscinale, Tolypothrix tenuis,* and *Gloeocapsa polydermatica* failed to grow beyond 1.5 percent salinity. The cells of *Anabaena* strain L-31, when exposed to 350 mM sucrose, 10 mM KNO_3, osmotic stress induced proteins (OSPS) were stimulated which resulted in significant osmoprotection (Iyer et al., 1994).

ACID-TOLERANT CYANOBACTERIA

In India, out of a total of 157 million ha of cultivable area, 49 million ha of land are acidic in nature. Moore (1963) reported the occurrence of many Cyanophyceae in acid soils. Many cyanobacterial cultures were isolated from acid soils of Kerala having a pH of 3.8 (Aiyer, 1965; Madhusoodanan and Dominic, 1999). The survival of cyanobacteria under acid soil rice field condition was studied by Jurgensen and Davey (1968). Later Sardeshpandey and Goyal (1981) observed that the native isolates survived better under acid soils than the introduced cultures did. Singh et al. (1997) isolated many acid-tolerant strains of cyanobacteria from the acid soils of Nagaland.

Recently, several efficient cyanobacterial cultures were isolated from the acid soils of Tamil Nadu. The predominant genera are *Anabaena* (25 percent), *Westiellopsis* (17.5 percent), *Nostoc* (15.0 percent), *Oscillatoria,* and *Phormidium* (25.0 percent), and the rest were unicellular algae *Microcystis* (5 percent). The acid-tolerant cyanobacterial culture *Westiellopsis* AT-TGK-4A7 performed better both under normal and acid pH levels, such as 4, 5, and 6. The effect of inoculation of acid-tolerant cyanobacterial cultures on rice variety ASD 20 was studied. The study revealed that the inoculation of composite cultures of the acid-tolerant cyanobacteria constituting *Anabaena* AT-TGK-5C10, *Nostoc* AT-TGK-5A4, *Oscillatoria* AT-TGK-5C9, *Westiellopsis* AT-TGK-4AT, and *Westiellopsis* AT-TGK-5A9 was found to be better in increasing the growth and yield of ASD 20, along with excreting maximum amounts of ammonia into the floodwater.

Immobilization of Cyanobacteria

A carrier-based immobilized cyanobacterial inoculant was developed to maintain better quality. The polyurethane (PU)-foam-, sugar-cane-waste, and paper-waste-immobilized cyanobacteria were the components of the inoculant (see Photos 4.3 and 4.4). Rice husk and soil mixture act as a carrier for these immobilized cyanobacteria (see Figure 4.1). PU-foam-immobilized cyanobacterial cultures recorded higher growth, nitrogenase activity, ammonia excretion, and heterocyst frequency than free-living cultures. Sugarcane wastes (SCW) and paper waste (PW) solid matrixes were used for immobilization of cyanobacteria in combination with rice husk and rice soil for improving the quality of cyanobacterial biofertilizers. Kannaiyan et al. (1997) showed that the immobilized cyanobacteria treated with the systemic fungicide Bavistin stimulated nitrogenase activity and inhibited the GS activity. The cyanobacterial cultures immobilized in PU foam increased the quality and viability of cyanobacteria and could survive for a period of two to three years. The cyanobacterial culture is well protected inside the pores of the PU foam and also increases the nitrogen use efficiency (NUE) of rice, owing to its continuous photoproduction of ammonia in the rice field. The PU foam immobilized with mixed culture recorded higher ammonia excretion, higher growth performance, and biochemical constituents of rice seedlings (Anitha and Kannaiyan, 1999).

PHOTO 4.3. Mass production of PU-foam-immobilized cyanobacterial inoculum in cement tanks under open-air conditions.

PHOTO 4.4. Close view of PU-foam-immobilized *Anabaena azollae.*

N_2-FIXING STEM-NODULATING SESBANIA *BIOFERTILIZER FOR RICE*

Biological nitrogen fixation by leguminous green manure crops in symbiotic association with *Rhizobium* is a low-cost input for rice crop. The cultivation of nitrogen-fixing legume green manure crop would be an efficient way of improving soil fertility status and sustainability (Kannaiyan, 1993b). Many *Sesbania* species are widely used in Asia, of which *S. aculeata* and *S. speciosa* are considered to be the most predominant species grown in India (Kannaiyan, 1989). These two are more commonly grown as N_2-fixing green manures because of their vigorous growth habits, higher rates of N_2 fixation, and general ability to withstand a wide range of soil conditions, namely waterlogging, salinity, and alkalinity (Kannaiyan and Kalidurai, 1995). The role of N_2-fixing green manures and other associated systems that contribute nitrogen in rice field ecosystems and their impact on rice crops is given in Figure 4.2. The stem-nodulating tropical legume *S. rostrata* (see Photo 4.5) has shown better performance as a potential N_2-fixing green manure crop in rice farming systems due to its fast growth rate, N_2-fixing efficiency, and tolerance for flooding (Kalidurai and Kannaiyan,

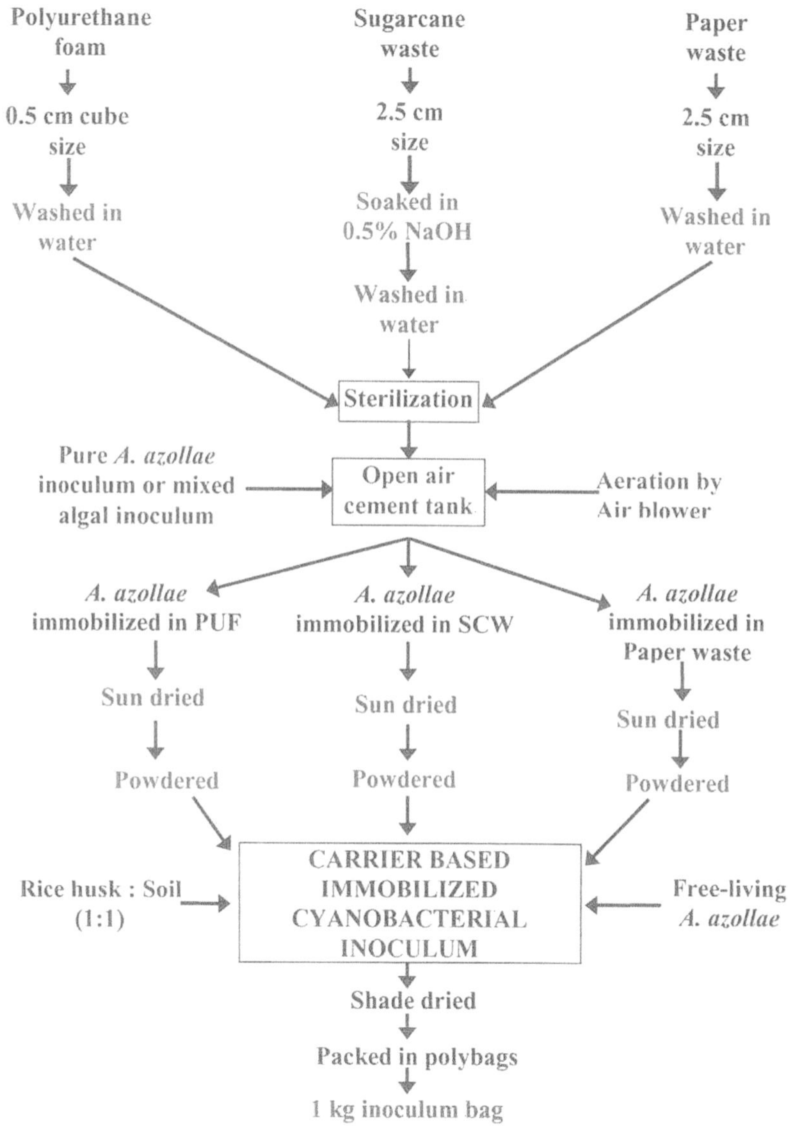

FIGURE 4.1. Mass production of carrier-based immobilized cyanobacterial inoculants. (*Source:* Kannaiyan, 1996.)

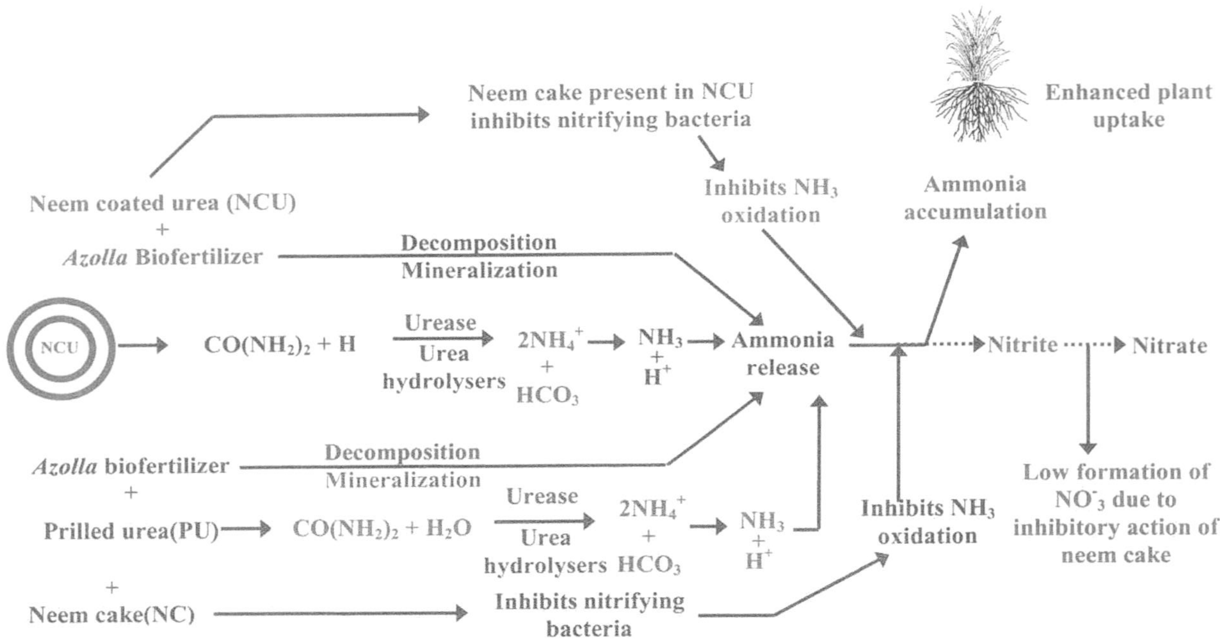

FIGURE 4.2. Role of neem cake in the efficient utilization of nitrogen by rice crop. (*Source:* Kannaiyan, 1999.)

PHOTO 4.5. Stem nodules in *Sesbania rostrata.*

1989, 1991). The use of stem-nodulating *S. rostrata* and root-nodulating *S. aculeata* as green manure has not only increased the yield of rice but also improved the soil microbial population and soil enzymes (Kannaiyan and Kalidurai, 1989; Kalidurai and Kannaiyan, 1991; Kannaiyan, 1993b).

Sesbania rostrata-Azorhizobium caulinodans symbiosis is highly species specific. *Azorhizobium caulinodans* neither forms effective root nodules in other *Sesbania* species nor does it produce stem and root nodules in any of the stem-nodulating *Aeschynomene* species (Ladha et al., 1990). The occurrence of azorhizobia in seeds of *S. rostrata* has been reported by Ladha et al. (1988) and Balasubramani et al. (1992). The transmission of *A. caulinodans* in stem-nodulating *S. rostrata* by a beetle, *Mylabris pustulata,* was reported by Balasubramani and Kannaiyan (1993).

SESBANIA *AS BIOFERTILIZER FOR RICE*

The utilization of nitrogen-fixing leguminous green manure crops for rice production would be a viable component of integrated nutrient management to enrich soil organic matter (Jiao, 1983) and to attain higher grain yield (Ladha et al., 1988; Kalidurai and Kannaiyan, 1991). Blending urea with oil cakes, particularly neem cake or neem seed extract, was shown to be a promising practice for nitrogen use efficiency. Neem cakes that contain nimbin, nimbosterin, nimbidin, and azadiractin have been reported to inhibit the nitrification process by inhibiting the nitrifying bacteria, *Nitrosomonas* and *Nitrobacter* (Khandelwal et al., 1977; Kumar et al., 1995). Field experiments conducted with fertilizer nitrogen coupled with *S. rostrata* and *S. aculeata* have recorded higher grain yield of rice. Fertilizer nitrogen coated with neem cake along with incorporation of *S. rostrata,* or with *S. aculeata,* have further increased the grain yield of rice (Kannaiyan, 1993b).

Incorporation of nitrogen-fixing green manures encourages the soil microbial population substantially. The application of N_2-fixing green manures not only induces the total soil microbial population but also preferentially stimulates the N_2-fixing bacteria. Application of N_2-fixing green manures on soil enzymes was investigated, and the induction of enzymes in rice soil was found. Soil enzymes and soil microbial populations are important components in rice soil for maintaining soil fertility and sustainability (Kannaiyan, 1993b; Sundaravarathan and Kannaiyan, 1998).

NEEM CAKE AND N USE EFFICIENCY

Neem cake is widely used as an organic fertilizer for several crops. But it seems to have good nitrification inhibition capacity, due to the presence of the alkaloids solanin, meliantrool, azadiractin, nimbin, and nimbidin, which possess a furan ring system responsible for inhibition (Jotwani and Srivastava, 1981). When neem cake is mixed with urea, the rate of inhibition increases with increasing proportion of neem cake. The nitrification inhibition period of neem cake is around 10 to 25 days. It retains ammonium N for longer periods, and higher NH_4^+ –N contents are noticed in the soil with neem-cake-coated urea application (Subbiah and Kothandaraman, 1980). The application of nitrification inhibitors along with urea in lowland rice could inhibit nitrification and suppress the transformation of ammoniacal N, thereby accumulating more ammoniacal N for effective utilization by rice crops (Manickam et al., 1976; De Datta, 1986). The increase in rice yields due to the application of dicyandiamide (DCD)-mixed urea and neem-cake-mixed urea was reported (Sharma and Sinha, 1977). In the experiments conducted at Tamil Nadu Agricultural University, Coimbatore, the grain and straw yields of rice were enhanced significantly due to the application of neem cake and DCD. Of the two nitrification inhibitors, neem cake recorded maximum total N content in plant, grain, and straw yield when added during the incorporation of N_2-fixing green manure plants, namely *Sesbania rostrata* and *Azolla microphylla* compared to DCD. Among different treatments with different combinations of green manures, fertilizer N, and nitrification inhibitors, *S. rostrata* incorporation (20 $t \cdot ha^{-1}$) with fertilizer N (75 $kg \cdot ha^{-1}$) and neem cake (250 $kg \cdot ha^{-1}$) registered the highest nitrogen accumulation, grain, and straw yield in IR 64 rice. It has been apparently shown that both nitrification inhibitors neem cake and DCD have influenced the grain yield of rice considerably. The beneficial effect on IR 64 rice, as a result of the addition of nitrification inhibitors during the incorporation of *A. microphylla* and *S. rostrata,* was attributed to the significant biological and chemical changes brought by the inhibitors in the rice soil, which eventually influenced the availability of nitrogen to the rice crop. Similar effects of nitrification inhibitors in increasing the yield of lowland rice and reducing N loss have been well documented (Manickam et al., 1976; Kamire and Sonar, 1979; Kumar et al., 1995). Recently, Kannaiyan (1996) more clearly elucidated the inhibitory action of neem cake on the N loss mechanism in rice soil, eventually increasing the N use efficiency of *Azolla* biofertilizer and inorganic fertilizer N by rice crops (see Figure 4.3).

Application of neem cake along with *S. rostrata* incorporation resulted in a significant reduction in the accumulation of nitrate N in flooded water and the population of nitrifying bacteria in rice soil (Sundaravarathan and Kannaiyan, 1998). Influence of *S. rostrata* resulted in a greater increase in evolution of CO_2 and CH_4 than application of fertilizer N alone (Sundaravarathan and Kannaiyan, 1998). An attempt was made to develop *S. rostrata* mutants using gamma irradiation and screening for increased growth, biomass production, nodulation, and nitrogen fixation. Gamma irradiation of *S. rostrata* seeds at 50, 55, 60, 65, and 70 kR reduced the plant height compared to the wild parent. Gamma irradiation significantly increased the number of compound leaves per plant, branches per plant, total biomass, stem nodulation, and stem nodule nitrogenase activity compared to the wild parent. However, gamma irradiation had no significant influence on root nodulation and root nodule nitrogenase activity of *S. rostrata* (Sundaravarathan and Kannaiyan, 1998).

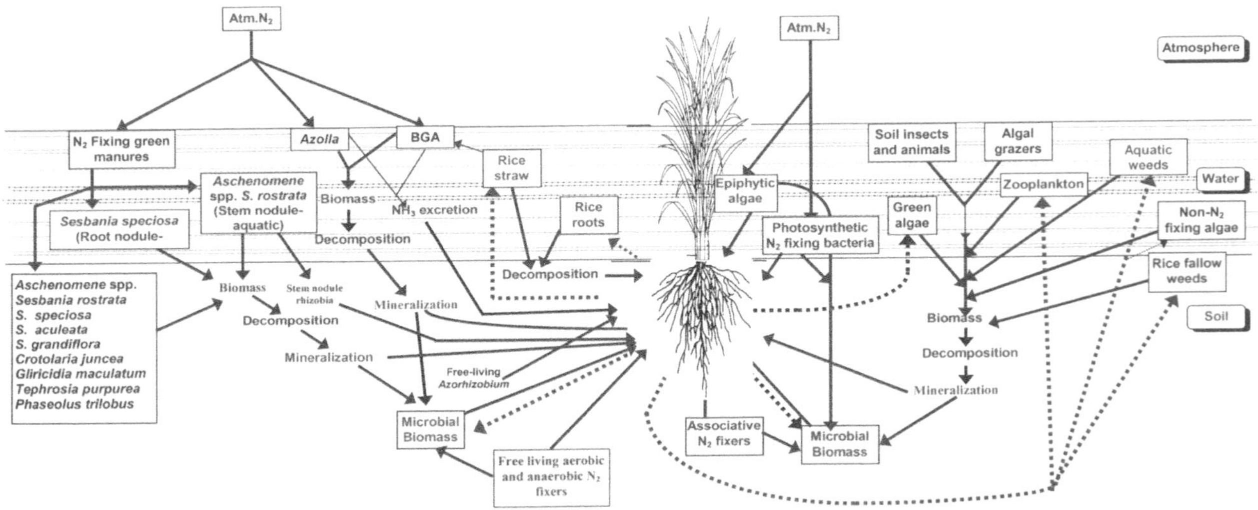

FIGURE 4.3. Natural bioresources input for rice crop nutrition. (*Source:* Kannaiyan, 1999.)

UREA SUPER GRANULES AND N USE EFFICIENCY

Deep placement of fertilizer in reduced soil has been considered the most efficient method to increase fertilizer N efficiency in lowland rice and is becoming increasingly relevant since the introduction of such modified urea materials as urea super granules (USG). The large size urea granule commonly referred to as the urea super granule can be placed in the soil by hand or with mechanical operator. The advantage of deep placement of urea super granules is that various losses such as ammonia, volatilization, denitrification, runoff, and leaching can be minimized (Mikkelsen et al., 1978; Savant et al., 1983; Kannaiyan, 1995). An additional advantage is that this mode of fertilization does not seem to interfere with dinitrogen fixation in floodwater by *Azolla* and blue-green algae (Roger et al., 1980).

Several years of agronomic data from the International Network on Soil Fertility and Fertilizer Evaluation for Rice (INSFFER) have shown that deep placement is generally more efficient than the traditional split application of urea (De Datta, 1986). Sharma (1995) has documented the positive influence of USG on rice soil based on the field experiments conducted at different locations.

When nitrogen-fixing green manures such as *Azolla* and *Sesbania rostrata* and such biofertilizers as blue-green algae are used, the application of neem-cake-coated urea and deep placement of USG proved better (Kannaiyan, 1993b). The application of different forms of urea along with *Azolla* inoculation has also been evaluated in the field experiments conducted during the early and late wet seasons of 1984-1985 at Tamil Nadu Agricultural University (Latha et al., 1988). Significant increase in grain yield was obtained over 75 kg $N \cdot ha^{-1}$ as prilled urea with *Azolla* as a dual crop, due to the application of 75 kg $N \cdot ha^{-1}$ as urea super granule or neem-coated urea along with *Azolla* inoculation as dual crop in IR 50 rice during early wet season.

The increased yield obtained due to the application of urea super granule and neem-coated urea with *Azolla* as a dual crop may be attributed to the continuous availability of nitrogen for rice crop. From our results, it is evident that the nitrogen-fixing green manure plants, namely *Azolla* and *Sesbania rostrata,* and biofertilizers such as blue-green algae could be effectively utilized by incorporation with different forms of urea, namely neem-coated urea and urea super granule, to increase the rice production under lowland systems (Kumar et al., 1995).

SOIL ENZYME ACTIVITY

Soil enzymes or soil biocatalysts not only play a major role in the soil chemical and biochemical environment but also affect the rate at which nutrients become available to crop plants as well as to other forms of life in soil. Soil enzyme activity is considered an index of microbial activity as well as soil fertility (Burns, 1982). Activity of such soil enzymes as urease and dehydrogenase were positively correlated with increasing organic carbon, total nitrogen, and total phosphorus content (Barush and Mishra, 1984). An increased activity of certain soil enzymes due to the application of green manures has been reported. Frankenberger and Tabatabai (1991) observed significant correlation of L-asparaginase activity with organic carbon and total nitrogen.

The incorporation of *Azolla* has resulted in increased soil urease activity (Subramani and Kannaiyan, 1989; Kannaiyan and Kalidurai, 1995). The addition of neem cake, carbofuran, rice straw, and *Azolla* increased the phosphatase activity in flooded soils. Significant increase in the activity of soil enzymes, namely dehydrogenase, phosphatase, cellulase, and amylase, due to the incorporation of N_2-fixing green manures such as *S. rostrata, S. speciosa,* and *A. microphylla,* has been noticed (Kannaiyan and Kalidurai, 1995).

INDUCTION OF NODULE-LIKE STRUCTURES IN RICE ROOTS FOR N_2 FIXATION

The role of azorhizobia in associative N_2 fixation in rice and higher levels of *A. caulinodans* populations in the rhizosphere of rice has been reported by Ladha et al. (1988). In rice, maize, and *Sorghum,* seedlings grown in IRRI(–) NO_3 medium were inoculated with 48 h old *A. caulinodans* ZB-SK-5 in combination with the treatment of 2,4-D at 0.5, 1.5, and 3.0 ppm, and 5.0 ppm of benzylamino purine (BAP) and kinetin, and an increase in shoot growth and biomass production was found (Zahara Begum, 1996). The formation of nodule-like structures was noticed with the inoculation of *A. caulinodans* and 2,4-D treatment at different levels at 0.5, 1.5, and 3.0 ppm and also with BAP and kinetin at 5.0 ppm level (see Photo 4.6). Re-isolation of *A. caulinodans* from the crushed nodule-like structure was achieved, and inoculation of the same in young *S. rostrata* plants produced nodules on the root and stem. The influence of different inoculum levels on the population dynamics of the rhizoplane showed that the population of *A. caulinodans* increased in the rhizoplane region of the rice roots (var. ADT-36) at an inoculum level of 10^9 cells per mL. The population dynamics were

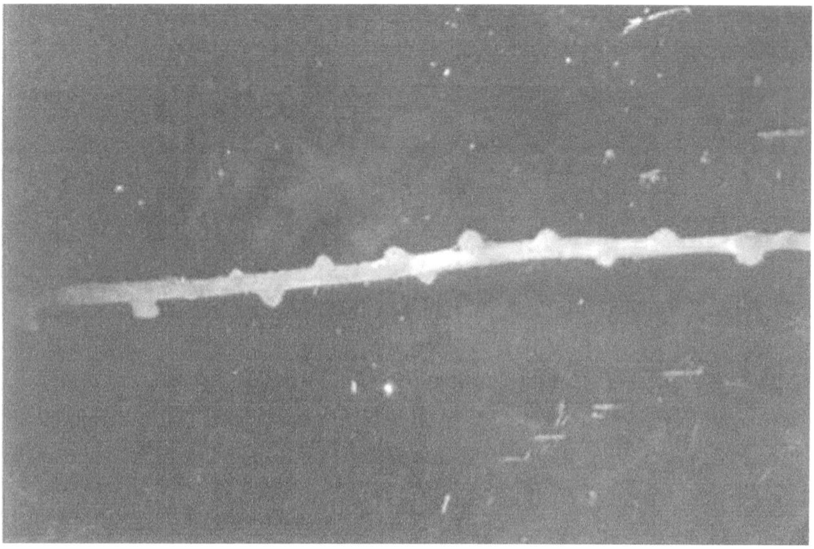

PHOTO 4.6. Nodule-like structures formed in rice roots due to inoculation of *A. caulinodans* with 2.4-D.

high in the rice rhizoplane and rhizosphere region when inoculated into soil with *A. caulinodans* in rice, maize, and sweet potato. Chandrasekar and Kannaiyan (1995) reported the formation of small nodule-like structures in the roots of rice, maize, and *Sorghum* when treated with 2,4-D/NAA at 3.0 and 7.0 ppm and inoculation with *A. caulinodans*. The bacterium *A. caulinodans* is a unique one that fixes nitrogen under free-living condition and is also tolerant to combined nitrogen at low concentrations. The bacterium could also tolerate flooded conditions, and *A. caulinodans* also produces the pectic and cellulolytic enzymes with good penetration capacity, and in general it fixes nitrogen in rice root systems at microaerophilic conditions (2 to 3 percent O_2 level). The nitrogenase enzyme that is responsible for nitrogen fixation was detected from the root nodule structures of rice, maize, sorghum, cumbu, and ragi (Amutha and Kannaiyan, 1999).

The nitrogen-fixing rate in the nodular structures of cereal crops is low, but the colonization of the bacterium could be induced by biotechnological manipulation of the rice plant. Flavonoids are a class of phenolics secreted by plant roots as part of the regulator system for colonization of *A. caulinodans*. Four families of flavonoids, the flavones, flavonols, flavonones, and isoflavonones, have been shown to have *nod*-gene-inducing activity. The naringenin-a-specific flavonoid is known to stimulate the colonization of *A. caulinodans* in rice roots (Kumar et al., 1996). It is a signal molecule for attracting the population of *A. caulinodans*. Recently it has been established that the flavonoid naringenin could induce *A. caulinodans* to colonize the xylem of rice as shown in Photo 4.7 (Gopalaswamy et al., 2000).

The induction of nodulation in cereal crops has excellent scope in terms of nitrogen transfer to rice crops for higher productivity, and this finding has a potential practical significance in cereal crop nitrogen fixation. It is a strong sign to intensify research in this direction of endophytic bacterial nitrogen fixation inside the root of the rice plant, which would reduce fertilizer consumption in wetland rice in sustainable rice farming. Rice-root-associated nitrogen fixation through nodular structures in rice is like a mini fertilizer factory in the rice root itself. The new approach on nodule induction in rice and other cereal crops would be a breakthrough for intensive colonization and N_2 fixation by diazotrophic bacteria in rice root. To move further in this direction, it is possible to introduce new genes into rice through transformation by plant genetic manipulation, which would create an excellent opportunity for incorporating N_2-fixation capability to form nodules in rice (Kannaiyan, 1998).

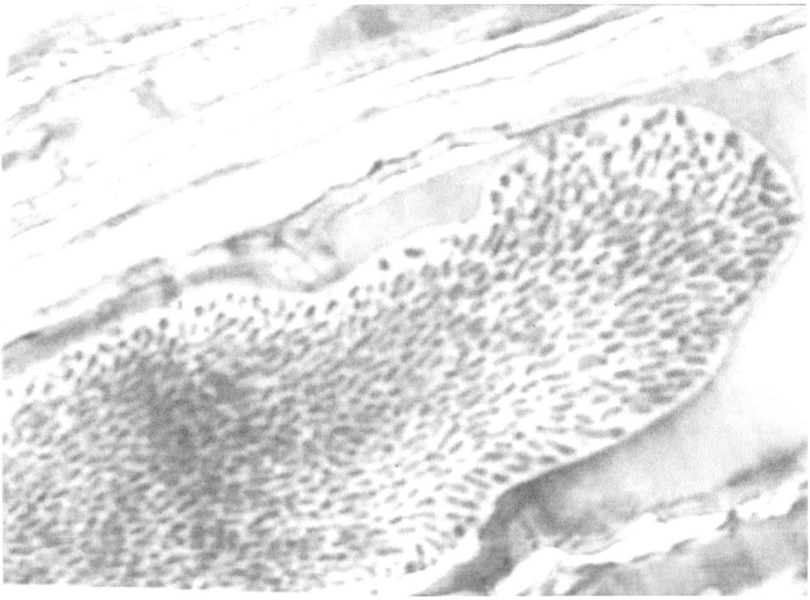

PHOTO 4.7. Colonization of *A. caulinodans* in xylem vessels of rice.

AZOSPIRILLUM *BIOFERTILIZER FOR RICE*

Field experiments conducted in different places in India have indicated the positive response of rice cultivars to *Azospirillum* inoculation (Subba Rao, 1979; Rao et al., 1983). Experiments conducted at IRRI, the Philippines, showed that inoculation with *Azospirillum lipoferum* promoted early tillering and also the growth of rice, and significantly increased the filling rate of grain and the grain weight per plant at harvest (Watanabe and Lin, 1984; Nayak et al., 1986). Response of rice to inoculation of *Azospirillum lipoferum* has been reported by several researchers (Rao et al., 1983; Kumar and Balasubramanian, 1986; Gopalaswamy and Vidhyasekaran, 1989; Jeyaraman and Purushothaman, 1988). However, the response of rice to *Azospirillum* inoculation under lowland rice field conditions was variable, depending on the initial nitrogen status of the soil and the varieties of the rice. Gopalaswamy and Vidhyasekaran (1989) have reported variation due to *Azospirillum* inoculation in different rice varieties.

Beneficial effects of *Azospirillum* inoculation have been reported on maize, sorghum, sunflower, bhendi, cotton, and pearl millet (Dhanabal et al.,1978); sorghum (Subba Rao et al., 1980, 1986; Okon and Labandera-Gonzalez, 1994); pearl millet (Subba Rao et al., 1986; Tilak and Subba Rao, 1987); maize (Tilak et al., 1982; Fulchieri and Frioni, 1994); and cotton (Purushothaman and Gunasekaran, 1981).

PHOSPHOBACTERIAL BIOFERTILIZER

Many fungi *(Aspergillus, Penicillium, Mucor, Fusarium, Candida),* bacteria *(Bacillus, Pseudomonas, Micrococcus, Flavobacterium),* and actinomycetes *(Streptomyces)* are involved in phosphorus solubilization. The population of phosphate-dissolving microorganisms (PDM) is more likely in the rhizosphere than in the nonrhizosphere, and the higher population of PDM in the rhizosphere is of great relevance to plants, especially in P- deficient soil as it helps in the solubilization of phosphorus.

In India, by developing improved techniques for isolation of rock-phosphate-solubilizing microorganism, new efficient bacteria and fungi such as *Pseudomonas striata, Bacillus polymyxa, Aspergillus awamori,* and *Penicillium digitatum* were selected for preparation of a carrier-based inoculant.

In the former USSR, the use of phosphobacteria has been found to increase yield of plants by 5 to 10 percent. The increase in crop yield has been attributed to

1. increased soluble phosphate nutrition in plants,
2. synthesis of growth-promoting substances, and
3. production of antibiotic-like compounds.

Sharma and Singh (1971) observed that phosphobacteria along with bonemeal application increased the grain yield of rice and also the nitrogen and phosphorus contents of plants. Significant yield increase by application of rock phosphate and many phosphate-solubilizing microorganisms have been reported in wheat, paddy, and soybean (Gaur and Ostwal, 1972). Inoculation of rice with *Pseudomonas striata* significantly increased the yield when soil was amended with rock phosphate and super phosphate in a 1:1 ratio (Gaur, 1990). Anthoni Raj et al. (1994) reported a positive response to phosphobacteria inoculation in lowland rice, even though the extent of benefit varied with different P levels tested. Thamizhvendhan and Subramanian (2000) observed that 50 percent of super phosphate with phosphobacteria registered higher microbial populations, available phosphorus content, and yield of rice, and indicated that it was comparable with the application of 50 percent rock phosphate with phosphobacteria.

FUTURE PERSPECTIVES

A quantitative understanding of the ecological factors that control the fate and performance of BNF systems in crop fields is essential for promotion and successful adoption of these technologies. We can no longer afford the time and expense of trial-and-error experimentation, which has been the hallmark of experimental agricultural sciences in the past. Future research will have to be directed toward approaches that generate information which is transferable from one site to another. Scientists from different biophysical and socioeconomical fields should work together to identify and eliminate farmers' limitations in using BNF technology. Because of technological and socioeconomical limiting factors, the agronomic potential of BNF is still largely underutilized in rice cultivation. Methodological progress and comprehensive evaluation of BNF in rice fields are still needed to develop and test agricultural practices that take advantage of BNF in the rice ecosystems.

Research is needed to understand the mechanisms of N gains and losses and to identify and refine appropriate soil and crop management practices to maximize soil N and legume BNF use in lowland rice-based cropping systems.

The economics of labor cost, agricultural inputs, and gross returns of the farmer under different legume/cereal combinations and varying socioeconomic conditions must be addressed to identify a profitable combination for specified areas.

Genetic enhancement of BNF agents used in rice fields is far less developed than are those agents of grain legumes. The improvement of BNF systems by traditional breeding and modern biotechnology is absolutely necessary. The various attempts to achieve effective nodulation of nonleguminous field crops using *Azorhizobium* may also yet to prove beneficial under field conditions. This will require suitable measures to be taken to allow expression of nitrogenase activity. Similar problems for persistence of nodules and continued activity still remain. Despite these severe challenges, we are convinced that the eventual application of this developing technology in one form or other is now more likely than not. By her brilliant examples represented in the legume-*Rhizobium* and the plant-*Frankia* symbioses, nature has provided the models that must be matched, at least in overall function. However, these new technologies with very different species of plants may not necessarily mimic these naturally evolved symbioses. The next ten years will tell whether this confidence is justified. It is also conceivable that the technique of paranodulation with desirable bacteria may find new applications outside the N_2-fixing area.

CONCLUSIONS

Biological nitrogen fixation is an important process in rice farming systems because it is an inexpensive source of nitrogen for increasing the productivity of crops. The nitrogen-fixing biological systems such as *Azolla,* cyanobacteria, and legume green manures were able to adapt well under wetland rice field ecological conditions and also to fix considerable amounts of nitrogen. The biomass of these biological systems decomposes rapidly in rice soil and supplies the nitrogen to the rice crop. Besides nitrogen addition, these biofertilizers or biomanures also contribute significant amounts of phosphorus, potassium, sulfur, zinc, iron, molybdenum, and other micronutrients. The organic acids released during the mineralization process of the biomass of biomanures would accelerate the phosphorus availability in rice soil. These biological fertilizers or biomanures are known to influence the total soil microbial population including bacteria, fungi, and actinomycetes, and also substantially activate the population dynamics of such nitrogen-fixing free-living bacteria as *Azotobacter* and *Azospirillum,* and also the free-living symbiotic bacteria such as *Rhizobium*. It is evidently clear now that the application of biological fertilizers was greatly involved in the accumulation of soil enzymes, which directly reflects on the soil fertility index. The use of biological fertilizers would reduce the cost of chemical fertilizers involved in crop production. The effective utilization of biological fertilizers for crops will not only provide economic benefits to the farmers but also improve and maintain the soil fertility and sustainability in the natural soil ecosystem.

REFERENCES

Aiyer R.S. (1965). Comparative algological studies in rice fields of Kerela state. *Agric. Res. J. Kerela.* 3(1) : 100-104.

Amutha K. and Kannaiyan S. (1999). Effect of 2-4D, NAA and inoculation with *Azorhizobium caulinodans* on induction of lateral roots and para nodules in rice. *Indian J. Microbial.* 39: 125-128.

Anand N., Hopper R.S.S., and Jagatheswari G. (1994). Response of certain blue-green algae (cyanobacteria) in salinity. In A.K. Kashyap and H.D. Kumar (eds.), *Recent Advances in Phycology* (pp. 125-129). Varanasi, India: Rastogi Publications.

Anitha K.G. and Kannaiyan S. (1999). Immobilized and carrier mixed cyanobacterial cultures as biofertilizer for rice *J. Microb. World* 1(2): 33-37.

Anthoni Raj S., Gopalaswamy G., and Abdul Kareem A. (1994). Effect of graded levels of phosphorus and phosphobacteria inoculation on rice yields. *Madras Agric. J.,* 81: 457-458.

Apte S.K. and Thomas J. (1986). Membrane electrogenesis and sodium transport in filamentous nitrogen fixing cyanobacteria. *Eur. J. Biochem.* 154: 395-401.

Balasubramani G. and Kannaiyan S. (1993). Transmission of *Azorhizobium caulinodans* in stem nodulating *Sestania rostrata* by *Mylabris pustulata. J. Soil Biol. Ecol.* 13: 75-77.

Balasubramani G., Krishnaveni S., and Kannaiyan S. (1992). Inhibition of *Azorhizobium* and *Rhizobium* by seed leachates. *World J. Microbiol. Biotechnol.* 8: 542-543.

Barush M. and Mishra R.R. (1984). Dehydrogenase and urease activities in rice field soils. *Soil Biol. Biochem.* 16: 423-424.

Burns R.C. (1982). Enzyme activity in soil. Location and possible role in microbial ecology. *Soil Biol. Biochem.* 14: 107-108.

Chandrasekar M.V. and Kannaiyan S. (1995). Induction of nodular structures by *Azorhizobium caulinodans. Tamil Nadu Agric. Univ. Newslett.* 24(12): 3.

De Datta S.K. (1986). Improving nitrogen fertilizer efficiency in lowland rice in tropical Asia. *Fert. Res.* 9: 171-186.

Dhanabal N., Purushothaman D., and Nadanam M. (1978). Effect of seed inoculation with *Spirillum lipoferum* on pearl millet and *Sorghum. Food Fong. Agrl.* 10: 85-88.

Diara H.F., Van Brandt H., Diop A.M., and Van Hove C. (1987). *Azolla* and its use in rice culture in West Africa. In *Azolla utilization: Proc. workshop on Azolla use* (pp. 147-152). Fuzhou, Fujian, China. Los Banos, Philippines: International Rice Research Institute.

Frankenberger, W.T. Jr. and Tabatabai M. A. (1991). L-asparaginase activity in soils. *Biol. Fertil. Soils* 11: 6-12.

Fulchieri M. and Frioni L. (1994). *Azospirillum* inoculation on maize *(Zea mays):* Effect of yield in a field experiment in Central Argentina. *Soil. Biol. Biochem.* 26: 921-924.

Gaur A.C. (1990). *Phosphate solubilising microorganisms as biofertilizer*. New Delhi, India: Orega Scientific Publishers.

Gaur A.C. and Ostwal K.P. (1972). Influence of phosphate dissolving bacilli on yield and phosphate uptake of wheat crop. *Indian J. Exp. Biol.* 10: 393-394.

Gopalaswamy G. and Kannaiyan S. (1998). Effect of inoculation of *Azolla* hybrids on soil organic matter and rice yield. In S. Kannaiyan (ed.), *Cyanobacterial biofertilizer for rice crop* (pp. 214-217). Coimbature, Tamil Nadu, India: Tamil Nadu Agricultural University.

Gopalaswamy G., Kannaiyan S., O'Callaghan J., Davey M.R., and Cocking E.C. (2000). The xylem of rice *(Oryza sativa)* is colonized by *Azorhizobium caulinodans*. *Proc. R. Soc. Lond.* B 267: 103-107.

Gopalaswamy G. and Vidhyasekaran P. (1989). Response of different rice cultivars to *Azospirillum inoculation. Cur. Sci.* 58(13): 752-754.

Goyal S.K. and Venkataraman G.S. (1971). Response of high yielding rice varities to algalization. II. Interaction of soil types with algal inoculation. *Phykos* 10 (1-2): 32-33.

Iyer V., Fernandes T., and Apte S.K. (1994). A role for osmotic stress induced proteins in the osmotolerance of nitrogen fixing cyanobacterium, *Anabaena* sp. strain L-31. *J. Bacteriol.* 176 (18) :5868-5870.

Jeyaraman S. and Purushothaman S. (1988). Biofertilizers efficiency in low land rice. *Intl. Rice Res. Newsletter* 13: 24-25.

Jiao B. (1983). Utilizing green manure for raising soil fertility in China. *Soil Science.* 135: 65-69.

Jotwani M.G. and Srivastava K.P. (1981). Neem insecticide of the future: Chemistry, toxicology and future strategy. *Pesticide* 15(2) : 12-14.

Jurgensen M.F. and Davey C.B. (1968). Nitrogen fixing blue green algae in acid forest and nursery soils. *Can. J. Microbiol.* 14: 1179-1183.

Kalidurai M. and Kannaiyan S. (1989). Effect of *Sesbania rostrata* on nitrogen uptake and yield of low land rice. *J. Agron. Crop Sci.* 163: 284-289.

Kalidurai M. and Kannaiyan S. (1991). *Sesbania* as a biofertilizer for rice. *Bioresource Technol.* 36: 141-145.

Kamire K.K. and Sonar K.R. (1979) Mineralization of urea-N with 'Neem' *(Azadirachta indica)* and 'Karanj' *(Pongamia glabrarent)* cakes. *J. Maharashtra Agric. Univ.* 4: 10-13.

Kannaiyan S. (1985). *Studies on the algal application for low land rice crop*. Coimbature, Tamil Nadu, India: Tamil Nadu Agricultural University.

Kannaiyan S. (1987). *Azolla biotechnology*. Coimbature, Tamil Nadu, India: Tamil Nadu Agricultural University.

Kannaiyan S. (1989). Nitrogen fixing potential of stem nodulating *Sesbania rostrata*. In *Organic recycling in Asia and Pacific*. Coimbature, Tamil Nadu, India: Tamil Nadu Agricultural University.

Kannaiyan S. (1990). *Biotechnology of biofertilizer for rice crop*. Coimbature, Tamil Nadu, India: Tamil Nadu Agricultural University.

Kannaiyan S. (1992). *Azolla biofertilizer technology for rice*. Coimbature, Tamil Nadu, India: Tamil Nadu Agricultural University.

Kannaiyan S. (1993a). Algal biofertilizer technology for rice. *Ind. Phyco. Rev.* 2: 73-82.

Kannaiyan S. (1993b). *Biofertilizers for rice.* Coimbature, Tamil Nadu, India: Tamil Nadu Agricultural University.

Kannaiyan S. (1993c). Nitrogen contribution by *Azolla* to rice crop. *Proc. Indian Natl. Science. Acad.* 59(5&6): 21-31.

Kannaiyan S. (1995). Biofertilizers for rice status and scope. In S. Kannaiyan (ed.), *Rice management biotechnology* (pp. 237-265). New Delhi, India: Assoc. Pub. Co.

Kannaiyan S. (1996). EEC project report on the use of immobilized cyanobacteria for continuous ammonia fertilization in rice production. Presented in the First Review Committee meeting of EEC Project at the Univ. of Seville, Spain, p. 1-29.

Kannaiyan S. (1998). Induction of root nodule structures in rice for nitrogen fixation. *CADRI Newsletter* 11(11): 1-3.

Kannaiyan S. (1999). Ammonia Excretion by the polyurethane foam immobilized cyanobacteria for rice. In T. Fatma (ed.), *Cyanobacterial and algal metabolism and environmental biotechnology* (pp. 19-29). New Delhi, India: Narosa Publishing House.

Kannaiyan S., Aruna S.J., and Merina Prem Kumari S. (1997). Immobilized cyanobacteria as a biofertilizer for rice crop. *J. Appl. Phycol* 7: 1-8.

Kannaiyan S. and Kalidurai M. (1989). Studies on the nitrogen fixing potential of stem nodulating *Sesbania rostrata.* In J.P. Varghese (ed.), *In vitro approaches for plant breeding* (pp. 91-98). Kerala, India: Ashram Press.

Kannaiyan S. and Kalidurai M. (1995). Nitrogen fixing potential of stem nodulating *Sesbania rostrata* and its effect on rice yield. In S. Kannaiyan (ed.), *Rice management biotechnology* (pp. 173-195). New Delhi, India: Assoc. Pub. Co.

Khandalwal K.L., Singh D.P., and Kapoor K.K. (1977). Mineralization of urea coated with neam extract and response of wheat. *Indian J. Agric. Sci.* 47: 267-270.

Kulasooriya S.A., Hirmburegama W.K., and Abeysekara S.W. (1987). Use of *Azolla* in Sri Lanka. In *Azolla Utilization. Proc. Workshop on Azolla use* (pp. 131-140). Fuzhou, Fujian, China. Los Banos, Phillipines: International Rice Research Institute.

Kumar K. and Balasubramanian (1986). Field response of rice to *Azospirillum* biofertilizer. *Curr. Res.* 15: 74-76.

Kumar K., Kannaiyan S. and Cocking E.C. (1996). Studies on the colonization of *Azorhizobium caulinodons* in rice roots and their role in symbiotic nitrogen fixation. Paper presented at 37th Ann. Conference Association of Microbiologists of India, held at IIT, Chennai Tamil Nadu, India.

Kumar K., Thangaraju M., and Kannaiyan S. (1995). Use of nitrification inhibitors and nitrogen use efficiency in lowland rice soil. In S. Kannaiyan (ed.), *Rice management biotechnology* (Chapter 8). New Delhi, India: Assoc. Publ. Co.

Ladha J.K., Pareek R.P., and Becker M. (1990). Stem nodule symbiosis and its unusual properties. In P.M. Gresshoff, L.E. Roth, G. Stacey, and W.L. Newton (eds.), *N_2 fixation: Achievements and objectives* (pp. 633-640). New York: Chapman and Hall.

Ladha J.K., Watanabe I., and Saono S. (1988). Nitrogen fixation by leguminous green manure and practices for its enhancement in tropical low land rice. In *Green manure in rice farming* (pp. 165-183). Los Banos, Philippines: International Rice Research Institute.

Latha K.R., Kannaiyan S., and Subramanian S. (1988). Uptake of nitrogen in different growth phases of rice through *Azolla* biofertilizer. *Acta Agron. Hungarica* 37 (3-4) : 239-243.

Liu Chung Chu (1987). Re-evaluation of *Azolla* utilization in agricultural production. In *Azolla Utilization. Proc. on the Workshop on Azolla use* (pp. 67-76). Fuzhou, Fujian, China. Los Banos, Phillipines: International Rice Research Institute.

Loudhapasitiporn L. and Kanareugsa C. (1987). *Azolla* use in Thailand. In *Azolla Utilization Proc. on the Workshop on Azolla use* (pp. 119-122). Fuzhou, Fujian, China. Los Banos, Phillipines: International Rice Research Institute.

Mabbayad B.B. (1987). The *Azolla* program of the Philippines. In *Azolla Utilization. Proc. on the Workshop on Azolla use* (pp. 101-108). Fuzhou, Fujian, China. Los Banos, Phillipines: International Rice Research Institute.

Madhusoodanan P.V. and Dominic T.K. (1999). Cyanobacteria from extreme acidic environment *Curr. Sci.,* 77(8): 1021-1023.

Manickam T.S., Selvaseelan A., and Krishnamurthy K.K. (1976). Nitrification of urea treated with nitrification inhibitors, coated fertilizers and slow acting fertilizers. *Fert. News,* 21 (9):64-65.

Mikkelsen D.S., De Dutta S.K., and Obcemae W.N. (1978). Ammonia volatilization losses from flooded rice soils. *Soil Sci. Soc. Am. J.* 442: 725-730.

Moore A.W. (1963). Occurrence of non-symbiotic nitrogen fixing microorganisms in nitrogen soils. *Plant Soil* 19: 385-395.

Nayak D.N., Ladha J.K., and Watanabe I. (1986). The fate of marker *Azospirillum lipoferum* inoculated into rice and its effect on growth, yield and N_2 feeding and ^{15}N dilution techniques. *Biol. Fertil. Soils* 2: 7-14.

Okon Y. and Labandera-Gonzalez C.A. (1994). Agronomic applications of *Azospirillum:* An evaluation of 20 years worldwide field inoculation. *Soil Biol. Biochem.* 12: 1591-1601.

Purushothaman D. and Gunasekaran S. (1981). *Azospirillum,* a biofertilizer for cotton. *TNAU Newsletter* 11:4.

Rai A.K. and Abraham G. (1993). Salinity tolerance and growth analysis of the cyanobacterium, *Anabaena doliolum. Bull. Environ. Cont. and Toxicol.* 51(5): 724-731.

Rao V. R., Nayak D.N., Charyulu P.B.D., and Adhya T.K. (1983). Yield response of rice to root inoculation with *Azospirillum. J. Agric. Sci.* 100: 689-691.

Reed R.H. and Stewart W.D.P. (1988). The responses of cyanobacteria to salt stress. In L.J. Rogers and J.R. Gallon (eds.), *Biochemistry of the algae and cyanobacteria* (pp. 217-237). Oxford, UK: Clarendon Press.

Roger P.A., Kulasooriya S.A., Third A.C., and Craswell E.T. (1980). Deep placement of nitrogen fertilizer application, compatible with algal nitrogen fixation in wetland rice soils. *Plant Soil* 57: 137-142.

Sardeshpandey J.S. and Goyal S.K. (1981). Distributional pattern of blue green algae in rice field soils of Konkan region of Maharastra State. *Phykos* 20 (1&2): 102-106.

Savant N.K., Craswell E.T., and Kiamond R.E. (1983). Use of urea supergranules for wetland rice: A review. *Fert. News* 28(8): 27-35.

Saxena S. and Kaushik B.D. (1991). Polysaccharides (biopolymers) from halotolerant cyanobacteria. *Indian J. Exp. Biol.* 30: 433-434.

Schubert H., Fulda S., and Hagemann M. (1993). Effects of adaptation of different salt concentration on photosynthesis and pigmentation of the cyanobacterium, *Synechocystis* sp. PCC 6803. *J. Plant Physiol.* 142: 291-295.

Sharma J.P. and Singh M. (1971). Response of rice to phosphatic and nitrogenous fertilizers with and without phosphobacterin culture. *Indian J. Agron.* 16: 15-18.

Sharma N.N. and Sinha U.P. (1977). Neem cake blending of urea for raising efficiency of applied nitrogen to wheat. *Indian. J. Agron.* 22: 106-107.

Sharma S.K. (1995). Urea super granules—Its performance and prospects for rice production. In S. Kannaiyan (ed.), *Rice management biotechnology* (pp. 117-134). New Delhi, India: Assoc. Publ. Co.

Singh N.I., Dorycanta H., Devi G.A., Singh N.S., and Singh M.S. (1997). Blue-green algae from rice field soils of Nagaland. *Phykos* 36(182): 115-120.

Singh P.K. (1977). The use of *Azolla pinnata* as green manure for rice. *Inte. Rice Res. Newsletter* 2(7): 7.

Subba Rao N.S. (1979). Biofertilizers in Indian agriculture problems and prospects. *Fert. News* 24: 84-90.

Subba Rao N.S., Tilak K.V.B.R., Singh C.S., and Nair P.V. (1986). Yield and nitrogen gains of *Sorghum* as influenced by *Azospirillum brasilense*. In S.P. Wani (ed.), *Cereal nitrogen fixation* (pp. 69-74). Patancheru, India: Inte. Crop Res. Inst. for Semi Arid Tropics.

Subba Rao N.S., Tilak K.V.B.R., and Singh S. (1980). Root nodulation studies on *Aeschynomene aspera. Plant Soil* 56: 491-494.

Subbiah S. and Kothandaraman G.V. (1980). Effect of neem cake blending of urea on ammonium and nitrate N content in soil under field conditions. *J. Indian. Soc. Soil. Sci.* 28: 136-137.

Subramani S. and Kannaiyan S. (1989). Effect of the water fern *Azolla* on enzyme activity in soil under rice cultivation. *Microbiology* 58(1): 118-121.

Sundaravarathan S. and Kannaiyan S. (1998). Decomposition of *Azolla microphylla* and *Sesbania rostrata* with fertilizer N and neem cake under submerzed soil conditions. *Indian J. Microbiol.* 38: 89-93.

Talley S.N., Talley B.J., and Rains W.D. (1977). Nitrogen fixation by *Azolla* in rice fields. In H. Alexander (ed.), *Genetic engineering for nitrogen fixation* (pp. 250-282). New York: Plenum Publishing Company.

Thamizhvendhan R. and Subramanian M. (2000). Effect of phosphobacteria with graded levels of phosphate fertilizers on lowland rice. *Crop Res.* 19: 194-197.

Tilak K.V.B.R., Singh C.S., and Subba Rao N.S. (1982). *Azospirillum* and *Azotobacter* inoculation. Effect on yield of maize and *Sorghum. Soil Biol. Biochem.* 14: 417-418.

Tilak K.V.B.R. and Subba Rao N.S. (1987). Association of *Azospirillum brasilense* with pearl millet *(Pennisetum americanum). Biol. Ferti. Soils* 4: 97-102.
Venkataraman G.S. (1972). *Algal biofertilizers and rice cultivation.* New Delhi, India: Today and Tomorrow's Printers.
Vonshak A., Guy R., and Guy M. (1988). The response of the filamentous cyanobacterium, *Spirulina platensis* to salt stress. *Arch. Microbiol.* 150: 417-420.
Wastyn M., Achatz A., Trnka M., and Peschek G.A. (1987). Immunological and special characterisation of partly purified cytochrome oxidase from the cyanobacterium, *Synechocystis* 6714. *Biochem, Biophys. Res. Commun.* 149: 102-111.
Watanabe I. (1987). Fertility and fertilizer evaluation for rice. In *Azolla Utilization. Proc. on the Workshop on Azolla use* (pp. 197-208). Fuzhou, Fujian, China. Los Banos, Philippines: International Rice Research Institute.
Watanabe I. and Lin C. (1984). Response of wetland rice to inoculation with *Azospirillum lipoferum* and *Pseudomonas* sp. *Soil Sci. Nutr.* 30: 117-124.
Zahara B. (1996). Studies on the free-living association of nitrogen fixing bacterium *Azorhizobium caulinodans* in the root systems of rice, maize and sweet potato and its effect on growth: N_2 fixing activity and biomass production of plants. Master's thesis, Tamil Nadu Agricultural University, Coimbatore, Tamil Nadu, India.

Chapter 5

Mycorrhiza Helper Bacteria: Their Ecological Impact in Mycorrhizal Symbiosis

Robin Duponnois

INTRODUCTION

Degradation of nature and agroecosystems, resulting from anthropogenic impacts, is associated with disturbances in plant communities but also with a loss of physicochemical and biological soil properties such as plant nutrient availability, organic matter, and microbial activities (Barea and Jeffries, 1995). In recent years, increasing evidence has suggested that the microbial communities of soil and plants play a key ecological role in the development of sustainable agriculture. Among the microorganisms living in the rhizosphere of plants, mycorrhizal fungi are usually considered essential components of sustainable soil-plant systems (Bethlenfalvay and Linderman, 1992; Hooker and Black, 1995; Van der Hejden et al., 1998). This mycorrhizal symbiosis is being actively studied for its beneficial effect on plant growth and its potential use in agriculture and forestry. Mycorrhizal fungi improve uptake of plant minerals, such as phosphorus (Sanders and Tinker, 1971; Bolan, 1991; Plenchette and Fardeau, 1998), micronutrients (Cooper, 1984; Kucey and Janzen, 1987; Bürkert and Robson, 1994), and nitrogen (Barea et al., 1991). Mycorrhizal symbiosis can also enhance water absorption (Sieverding, 1991; George et al., 1992) and improve plant health with antagonism against some pathogens (Dehne, 1982; Duponnois and Cadet, 1994). In addition, mycorrhizal fungi play a significant role on soil structure in the formation and stability of soil aggregates (Hamel et al., 1997; Wright and Upadhyaya, 1998).

The establishment of these symbiotic relationships generally occurs in the presence of myriad microorganisms, and abundant literature demonstrates that mycorrhizal relationships are highly influenced by some of these microorganisms (Rambelli, 1973; Bowen and Theodorou, 1979;

Bowen, 1980; De Oliveira, 1988; De Oliveira and Garbaye, 1989; Garbaye, 1991). Some of these interactions are competitive with fungal symbionts and inhibit the mycorrhizal development, but in other cases some can be beneficial to the mycorrhizal infection process.

The purpose of this chapter is to describe mycorrhizal interactions with rhizosphere microflora (mycorrhizosphere effect) and to analyze some results in which positive interactions occurred between mycorrhizal fungi and rhizobacteria (mycorrhiza helper bacteria). Moreover, discussions are largely limited to interactions with bacteria and ectomycorrhizal fungi where literature is more abundant.

THE MYCORRHIZOSPHERE EFFECT

Roots provide water and mineral nutrients to the plant and release nutrients for a wide range of microorganisms (Curl and Truelove, 1986). These nutrients, such as exudates (nonmetabolic leakage), secretions (metabolic release), plant mucilages, and lysates sloughed from roots are used by many microorganisms and, consequently, this exudation of carbohydrates by the roots significantly increases microbial activity. This portion of the soil influenced by the root activity has been recognized as the rhizosphere by Hiltner (1904) (Curl and Truelove, 1986) (see Photo 5.1). Mycorrhizae usually modify plant physiological processes by altering nutrient balance of the plant tissues, partitioning carbon in mycorrhizal plants, or changing phytohormone production (Allen et al., 1980; Graham et al., 1981; Barea and Azcon-Aguilar, 1982). For example, mycorrhizal plants usually transfer more assimilates to the roots than do nonmycorrhizal plants (Eissenstat et al., 1993). This is the result of the carbon needs of the mycorrhizal fungus and of the higher respiration rate of the mycorrhizal roots than of the nonmycorrhizal roots (Kucey and Paul, 1982; Fitter, 1991). Just as root exudates provide most of the low molecular weight carbon compounds, which are easily used by microorganisms in the soil, mycorrhizal symbiosis has a significant effect on the bacterial community composition in the rhizosphere. This fungal effect is commonly called the "mycorrhizosphere effect" (Linderman, 1988). The microbial communities of the mycorrhized roots differ largely from those of the uninfected roots and of the surrounding soil (Katznelson et al., 1962; Garbaye and Bowen, 1989; Garbaye, 1991) and specific relationships occur between mycorrhizal fungi and mycorrhizosphere microflora. First of all, this mycorrhizosphere effect has been demonstrated with culturable bacteria such as rhizobia or fluorescent pseudomonads at a quantitative level.

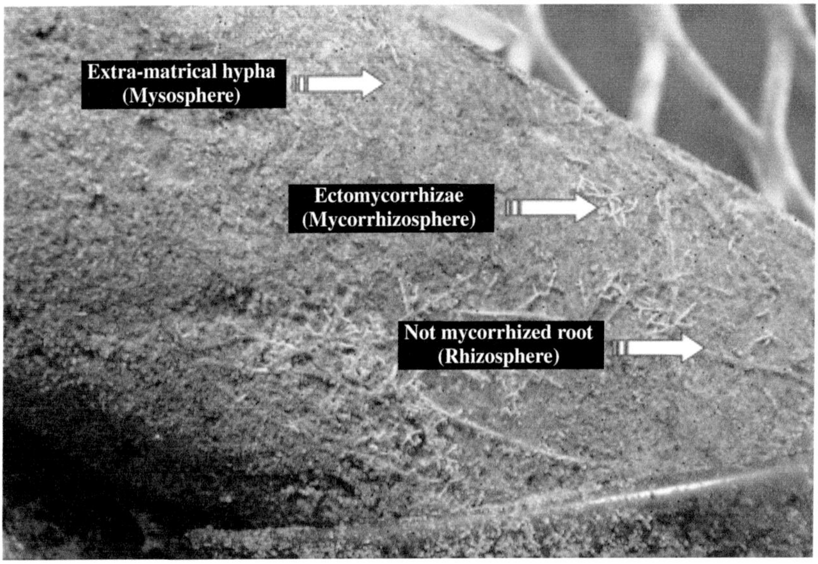

PHOTO 5.1. Biological compartments resulting in the development of ectomycorrhizas (fungal symbiont: *Pisolithus alba;* host plant: *A. holosericea*).

Table 5.1 shows some examples of positive interactions between rhizobia and mycorrhizal fungi with tropical leguminous plants (*Acacia* species). It is well established that many N_2-fixing trees and shrubs are dependent on mycorrhizae to absorb mineral nutrients required for plant growth and efficient N_2 fixation (Cornet and Diem, 1982; Cornet et al., 1982 ; Bâ et al., 1996; Duponnois et al., 2001). It has also been demonstrated that mycorrhizal symbiosis could influence the distribution of rhizobia along the root systems. André et al. (2003) have shown that besides a stimulatory effect of the mycorrhizosphere of *Acacia raddiana/Glomus intraradices* on the development of two rhizobial isolates (*Sinorhizobium teranga* ORS 1009 and *Mesorhizobium plurifarium* ORS 1096), *G. intraradices* colonization induces two different dynamics of each bacterial strain in sand-grown plants. ORS 1096 seems to be more competitive than ORS 1009 in sand, but mycorrhizal infection increases the competitiveness of ORS 1009. Specific relationships could occur during the development of the tripartite symbiosis at the physiological and molecular levels (Van Rhijn et al., 1997; Blilou et al., 1999). This mycorrhizal effect has been enlarged to rhizobial populations. According to Fulele-Laurent et al. (2002), ectomycorrhizal fungus (*Pisolithus alba* strain IR100) inoculation induced different dynamics of rhizobia populations and different polymerase chain reactions/restriction

TABLE 5.1. Influence of mycorrhizal symbiosis on the rhizobial infection process.

		No. of nodules per plant		Total dry wt of nodules per plant (mg)		
Plant species	Fungal symbiont	Control[a]	Inoc.[b]	Control[a]	Inoc.[b]	References
Acacia seyal	*Glomus* sp.	17.8 a[c]	16.1 a	37.2 a	113.6 b	Duponnois and Cadet, 1994
Acacia raddiana	*G. intraradices*	3.0 a	11.0 b	—	—	André et al., 2003
A. holosericea	*G. intraradices*	6.0 a	15.4 b	2.1 a	13.3 b	Duponnois and Plenchette, 2003
A. auriculiformis	*Pisolithus alba*	8.3 a	17.7 b	2.9 a	55.2 b	Duponnois and Plenchette, 2003
A. eriopoda	*P. alba*	1.8 a	9.8 b	1.6 a	10.1 b	Duponnois and Plenchette, 2003
A. mangium	*Pisolithus* sp.	1.8 a	17.5 b	—	—	Duponnois et al., 2002
A. holosericea	*P. alba*	8.1 a	24.6 b	11.2 a	58.1 b	Duponnois and Plenchette, 2003
A. holosericea	*Pisolithus* sp.	8.1 a	12.7 b	11.2 a	26.1 b	Duponnois and Plenchette, 2003
A. holosericea	*Scleroderma dictyosporum*	8.1 a	25.6 b	11.2 a	52.2 b	Duponnois and Plenchette, 2003
A. holosericea	*S. verrucosum*	8.1 a	17.6 b	11.2 a	33.1 b	Duponnois and Plenchette, 2003

[a]Control: not inoculated treatments; [b]Inoc.: inoculated treatments; [c]For each parameter (no. of nodules and total weight of nodules per plant), data in the same row followed by the same letter are not significantly different according to the Student "t" test ($p < 0.05$).

fragment length polymorphism (PCR-RFLP) profiles were recorded (see Table 5.2).

In general, mycorrhizosphere activity exerts a significant stimulating effect on the populations of fluorescent pseudomonads. According to Frey et al. (1997), the fluorescent pseudomonad population was significantly higher in the mycorrhizosphere of Douglas fir seedlings ectomycorrhized with *Laccaria bicolor* (8.91 10^4 CFU·g^{-1} dry weight of soil) than in the bulk soil (2.40 10^4 CFU·g^{-1} dry weight of soil). This stimulating effect of the mycorrhizosphere has been recovered by Founoune et al. (2002) where the fluorescent pseudomonad populations were about four times greater in the mycorrhizosphere of *A. holosericea* seedlings mycorrhized with *Pisolithus alba* than in the roots. All these results are in accordance with those obtained with several mycorrhizal plant species such as hybrid larch, Sitka spruce (Grayston et al., 1994), and hazel trees mycorrhized with truffles (Mamoun and Olivier, 1989). Moreover, the mycorrhizal infection modifies the structure of fluorescent pseudomonad populations at metabolic and genotypic levels. With the biolog systems, Frey et al. (1997) have shown that most isolates of *P. fluorescens* (62.8 percent) from the soil belonged to one biovar (Biovar V), whereas most of the isolates (50.4 percent) from the

TABLE 5.2. Number and percentages of rhizobia isolated from *A. holosericea* root systems per PCR-RFLP groups after eight months' culture in an undisinfected soil.

	Total number and percentages of isolates collected from	
PCR-RFLP groups	**Control (20 isolates)[a]**	***P. alba* IR100 (14 isolates)**
I	2 (10)[b]	4 (29)
II	5 (25)	2 (14)
III	1 (5)	0 (0)
IV	5 (25)	0 (0)
V	3 (15)	5 (36)
VI	1 (5)	0 (0)
VII	1 (5)	0 (0)
VIII	0 (0)	1 (7)
IX	0 (0)	1 (7)
X	2 (10)	1 (7)

[a]Total number of rhizobia analyzed by PCR-RFLP from each treatment.
[b]Total number and percentage of isolates of each PCR-RFLP group from each treatment and compartment.

mycorrhizosphere belonged to another biovar (Biovar I), and 90.6 percent of the isolates in the mycorrhiza were assigned to Biovar I. The same kinds of modifications have been recorded with another method (siderotyping) (Founoune et al., 2002). All the strains isolated from the mycorrhizosphere belonged to siderovar 1, whereas in the mycorrhiza compartment 75 percent were assigned to siderovar 1 and 25 percent to siderovar 2. The diversity among bacterial strains was also affected by mycorrhizal symbiosis. Fulele-Laurent et al. (2002) showed that the ectomycorrhizal fungus *Pisolithus alba* strongly modified the distribution of fluorescent pseudomonad genotypes along the root systems and in the cultural soil of *Acacia holosericea* seedlings (Fulele-Laurent et al., 2002) (see Figure 5.1).

Culture-independent methods, such as fatty acid extraction (Cavigelli et al., 1995; Ibekwe and Kennedy, 1998) and polymerase chain reaction-denaturing gradient gel electrophoresis (PCR-DGGE) (Muyzer et al., 1993; Muyzer and Smalla, 1998), are increasingly used for microbial community analysis. With such methods, Marshner et al. (2001) have demonstrated that

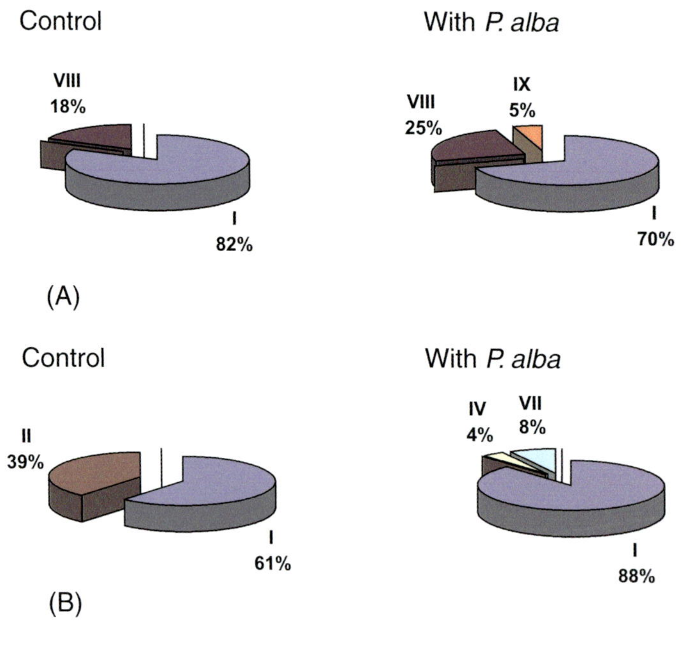

FIGURE 5.1. Percentages of fluorescent pseudomonads isolated from the soil adherent to the root systems (A) and from the soil (B) per PCR/RFLP groups of *Acacia holosericea* seedlings inoculated or not with *Pisolithus*. (*Source:* Adapted from Fulele-Larent et al., 2002.)

eubacterial community compositions examined by the PCR-DGGE of 16 S rDNA changed with time and were root zone specific of maize inoculated or not with *Glomus mosseae* or *G. intraradices*. In Senegal, Duponnois (data not published) has found that mycorrhizosphere bacterial community of *A. holosericea/P. alba* was rather different from that of rhizosphere using RISA (rRNA intergenic spacer analysis).

EVIDENCE

Beside the significant effect of mycorrhizae on bacterial communities (see Table 5.3), some soil bacteria can promote or inhibit the establishment of mycorrhizal symbiosis (Bowen and Theodorou, 1979; Garbaye and Bowen, 1989). De Oliveira (1988) isolated some bacterial strains that enhanced mycorrhizal infection of beech seedlings by the ectomycorrhizal fungus *Hebeloma crustuliniforme* under controlled conditions. Garbaye and Bowen (1989) postulated, for the first time, that some bacteria could be

TABLE 5.3. Mycorrhiza helper bacteria in other mycorrhizal fungus/host plant associations.

	Host plant	Fungal symbiont	References
Ectomycor			
	Fagus sylvatica	*Hebeloma crustuliniforme*	De Oliveira, 1988
	Pinus radiata	*Rhizopogon luteolus*	Garbaye and Bowen, 1989
	P. menziesii	*L. laccata* S238N	Duponnois and Garbaye, 1991
	Pinus sylvestris	*L. laccata* S238	Rozycki et al., 1994
	Eucalyptus diversicolor	*L. fraterna*	Dunstan et al., 1998
	Pinus sylvestris	*Lactarius rufus*	Poole et al., 2001
	Picea abies	*Amphinema byssoides*	Geri et al., 2000
Endomycor			
	Trifolium parviflorum	*Glomus* sp.	Mosse, 1962
	Lycopersicon esculentum	*G. fasciculatum*	Bagyaraj and Menge, 1978
	Trifolium subterraneum	Not determined	Meyer and Linderman, 1986
	Ipomoea batata	*Glomus clarum*	Paula et al., 1992
	Herbaceous species	Not determined	Von Alten et al., 1993

closely associated with mycorrhizae and could interact with the "host" fungus (i.e., bacteria in contact with fungal mycelium [Photo 5.2]) or embedded in the outer mantle of the ectomycorrhiza). Bacteria were isolated from surface-sterilized ectomycorrhizae of *Rhizopogon luteolus* with *Pinus radiata*. Researchers found that most of them belonged to the fluorescent pseudomonad group and that 80 percent of the bacteria stimulated mycorrhiza formation, whereas 20 percent had a neutral or inhibitory effect. The same approach has been followed by Garbaye et al. (1990), Duponnois and Garbaye (1991), and Duponnois (1992) with another biological model: Douglas fir (*Pseudotsuga menziesii* Mirb. Franco) and *Laccaria laccata* Scop. Ex Fr.. In a greenhouse experiment, among 45 bacterial isolates tested, 14 bacteria stimulated the ectomycorrhizal establishment of *L. laccata* with Douglas fir. The best stimulating effect was recorded with a strain of *Bacillus subtilis,* which increased the ectomycorrhizal rate (percent of short roots mycorrhized with *L. laccata*) from 67.3 to 97.3 percent. This MHB effect has been observed in different environmental conditions (in vitro, greenhouse, and nursery conditions) (Duponnois, 1992) and indicates the relative competitiveness of these introduced bacteria against the indige-

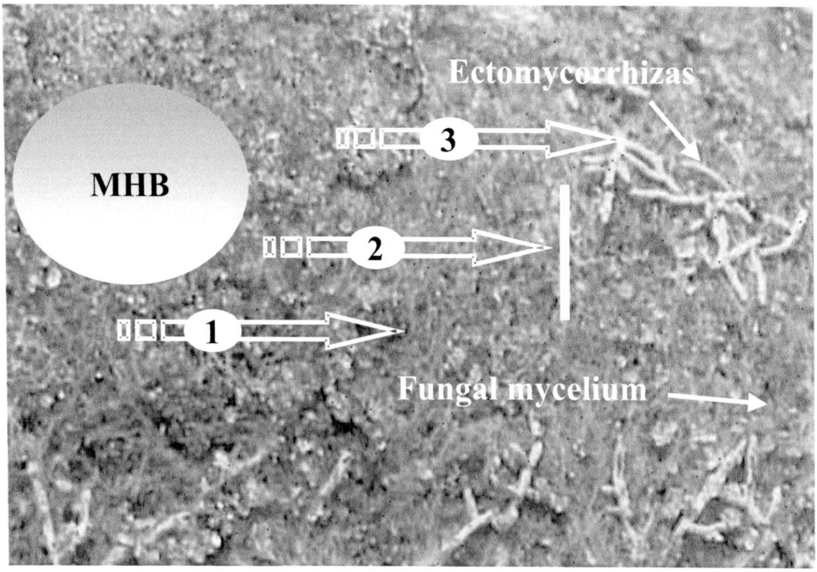

PHOTO 5.2. Representation of the possible ways of action of MHBs on the mycorrhizal establishment. 1: Effect of MHBs on fungal growth; 2: effect of MHBs on root-fungus recognition; 3: effect of MHBs on the receptivity of the root.

nous microflora. This same kind of experiment has been realized with a tropical tree species *(Acacia holosericea)* associated with *Pisolithus alba* (strain IR100) (Founoune et al., 2002). Twenty-four strains of fluorescent pseudomonads isolated from the mycorrhizosphere and the mycosphere were tested for their effect on mycorrhizae formation in greenhouse conditions. The stimulation of ectomycorrhizal establishment has been recorded with 17 isolates. The higher stimulation effect was observed with HR1 (mycorrhizal establishment: 32 percent) compared to the control without bacteria (14.2 percent).

BIOLOGICAL PROPERTIES

Beside the promoting effect of MHBs on ectomycorrhizal establishment, another property of these bacteria has been demonstrated through several experiments: the fungal specificity of MHBs. One bacterial isolate (*Pseudomonas fluorescens* BBc6), isolated from sporocarps of *Laccaria laccata,* could stimulate mycorrhizae formation by *Laccaria* species but inhibit symbiosis establishment by other fungi (Garbaye and Duponnois, 1992). This result has been observed in a wide range of experimental conditions (see Table 5.4). In contrast, there was a lack of bacterial specificity for host plants. This bacterial isolate BBc6 promoted mycorrhizae formation with *L. laccata* S238 and host plants such as *Picea abies, Pinus nigra, Pinus sylvestris, Pseudotsuga menziesii,* and an angiosperm *Quercus robur* (Garbaye et al., 1992). Dunstan et al. (1998) showed that mycorrhizae formation by *Laccaria fraterna* with *Eucalyptus diversicolor* increased significantly, by up to 296 percent, in treatments coinoculated with *P. fluorescens* BBc6. However, in treatments coinoculated with *L. laccata* (strain E766) and BBc6, mycorrhizal development was significantly inhibited. These conclusions on the fungus specificity of MHBs are not supported by the results of Duponnois and Plenchette (2003). This study focussed on the effects of an MHB, *Pseudomonas monteilii* strain HR 13, on the mycorrhization of (1) an Australian *Acacia, A. holosericea,* by several ectomycorrhizal fungi or one endomycorrhizal fungus *Glomus intraradices,* and of (2) several Australian *Acacia* species by *Pisolithus alba* strain IR 100 under greenhouse conditions. A stimulating effect of HR13 on the ectomycorrhizal establishment has been recorded with all the fungal isolates (strains of *Pisolithus* and of *Scleroderma*). The same effect of bacteria was recorded on frequency of endomycorrhizal colonization of *A. holosericea* seedlings by *G. intraradices* with vesicles and hyphae frequencies. In conclusion, although *P. monteilii* HR13 was isolated from *P. alba,* it stimulated mycorrhizae development of *A. holosericea* with two species of *Scleroderma* and, more

TABLE 5.4. Effect of a MHB (*P. fluorescens* BBc6) on mycorrhizae formation (percent of mycorrhized short roots) of Douglas fir seedlings inoculated with ectomycorrhizal fungi in different experimental conditions.

		Mycorrhiza formation (%)	
Experimental condition	**Fungal isolate**	**Control (without bacteria)**	**+ BBc6**
In vitro	*Laccaria laccata* S238	12.0 a[a]	30.5 b
	Hebeloma cylindrosporum		
	Paxillus involutus		
	Coenococcum geophilum		
Greenhouse	*L. laccata* S238	32.1 a	89.6 b
	L. bicolor 993	78.6 a	97.9 b
	L. bicolor S3	5.2 a	19.6 a
	L. bicolor A4B3xA1B2	21.7 a	45.7 b
	Telephora terrestris	18.0 a	4.2 b
Nursery	*L. laccata* S238	75.2 a	93.6 b
	L. bicolor D101	55.9 a	89.8 b
	L. proxima	38.7 a	43.4 a
	H. cylindrosporum	89.6 a	57.8 b
	P. involutus	35.1 a	14.9 a

[a]Data in the same row followed by the same letter are not significantly different according to the Student "t" test ($p < 0.05$).

surprisingly, with the arbuscular mycorrhizal fungus *G. intraradices*. However, bacterial inoculant HR13 has significantly promoted the ectomycorrhizal colonization for all Australian *Acacia* species tested in this experiment *(A. auriculiformis, A. eriopoda, A. holosericea, A. mangium,* and *A. platycarpa)*. The not-plant-specific effect of MHBs is demonstrated but the fungus specificity of MHBs seems to be restricted to the system with *L. laccata* S238 and its helper bacteria.

MECHANISMS INVOLVED IN THE MHB EFFECT

The main hypothesis that concern the mechanisms underlying the MHB effect have been reviewed by Garbaye (1994) and are presented in Photo 5.2. MHBs could act on mycorrhizal formation at different steps:

1. On the saprophytic fungal growth in the soil
2. On the mechanisms of recognition between the fungal symbiont and the host plant roots
3. On the host plant receptivity to the fungal infection

Actually, few works have been done to elucidate the mechanisms involved in the MHB effect on mycorrhizal development. To date, most of the data concern the interactions between MHBs and fungal symbiont growth. The bacterium could help the growth of the fungus in its saprophytic stage in the rhizosphere soil and on the root surface. Duponnois (1992) showed that bacteria adhered to the hyphae, which suggested close relationships between both microorganisms (see Photo 5.3). Using a test with a poor medium to simulate through an in vitro system the environmental conditions which are encountered by the fungus in the soil, Duponnois (1992) found a highly significant correlation between the ability of the bacteria to promote or inhibit fungal growth and their influence on mycorrhiza formation. More recently, Founoune et al. (2002) observed that fluorescent pseudomonads isolated from the mycorrhizosphere and mycosphere had a positive influence on mycorrhiza formation between *P. alba* and *A. holosericea*. This result suggests that the MHB are not strictly rhizobacteria. This hypothesis is in accordance with that of Frey-Klett et al. (1997) who found that the helper bacteria (BBc6) of the mycorrhizal symbiosis between *Laccaria bicolor* and Douglas fir is associated with the mycelium of the fungus in the soil. Moreover, Sen et al. (1994) have demonstrated that BBc6 was attached to the hyphal wall of *L. bicolor* in axenic experiments. Duponnois and Garbaye (1990) found that in nutrient-limiting conditions, helper bacteria can act through at least two mechanisms: direct trophic effect and detoxication of the medium enriched in fungal metabolites. Bacteria can excrete some organic acids (i.e., citric and malic acids) that were metabolized by fungal isolates, but they can also break down some toxic molecules (i.e., polyphenols compounds which are toxic for mycorrhizal fungi as *Paxillus involutus*). Moreover, MHB can produce some volatile compounds that inhibit or stimulate fungal growth (Garbaye and Duponnois, 1992). Carbon dioxide could have a significant effect in these interactions. It has been previously shown that depending on CO_2 concentration can inhibit or promote the growth of different fungi (Imolehin and Grogan, 1980; Le Tacon et al., 1983). Other compounds could be involved such as ethylene, ammonia, amines, alcohols, sulfur compounds, or low-molecular-weight fatty acids (Duponnois, 1992).

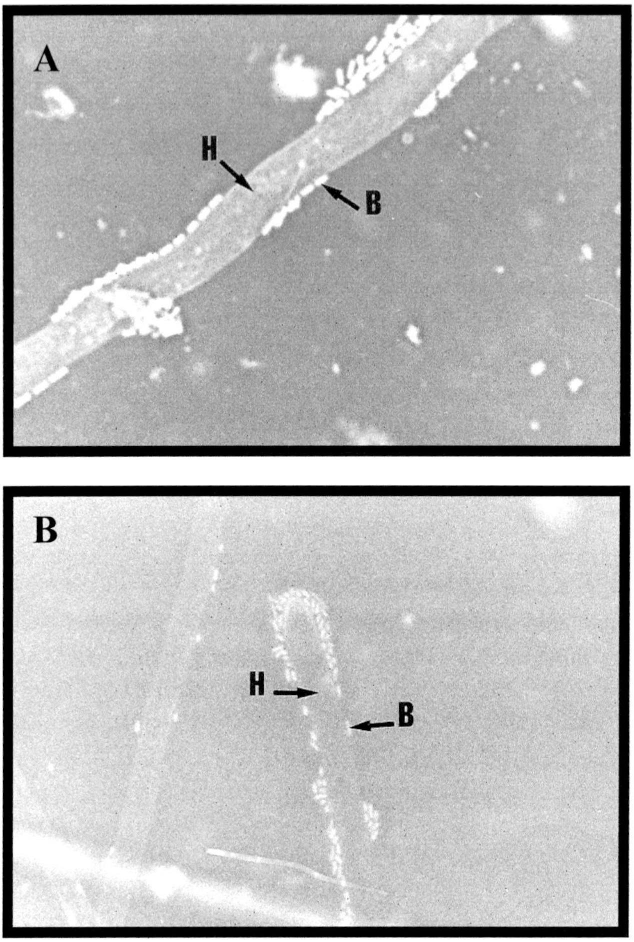

PHOTO 5.3. Hyphal colonization by mycorrhiza helper bacteria on *Paxillus involutus* (A) and *Hebeloma crustuliniforme* (B). (H: Hyphae; B: Bacteria.)

APPLICATION

The fast and important mycorrhizal establishment that determines the success of fungal inoculation depends on abiotic factors (pH, temperature, fertility, moisture) (Slankis, 1974) but also on such biotic factors as soil microbial communities. According to the MHB properties that have been previously described, three main benefits can be expected by using MHBs in controlled mycorrhization in forest nurseries:

1. They reduce the need for soil fumigation with the fungus specificity of MHBs.
2. They optimize the mycorrhizal establishment with the inoculated fungal isolate.
3. They reduce the quantity of fungal inoculum mixed to the soil.

To date, the main results have been obtained with the Douglas fir—*Laccaria bicolor* S238N symbiotic system using the MHB *Pseudomonas fluorescens* BBc6. In a fumigated soil with methyl bromide, BBc6 stimulated mycorrhiza formation from around 60 percent (percent of short roots mycorrhized with *L. bicolor* S238N) to 87 percent. This experiment was carried out with a peat-vermiculite fungal inoculum. The same positive effect of BBc6 was recorded when this bacterial isolate was entrapped together with the fungus in alginate beads (Duponnois, 1992). After four months culture, BBc6 increased ectomycorrhizal infection from 42 percent to 75 percent. The MHB effect was also observed with a range of bacterial inoculation doses from 10^5 to 10^9 CFU per m^2. This result contrasts markedly with most of the data with PGPRs (plant-growth-promoting-rhizobacteria) used as rhizosphere biocontrol agents where plant protection against pathogens is generally linked with the inoculation dose.

Moreover, with these two kinds of fungal inoculum (peat-vermiculite or alginate beads), Duponnois (1992) found that BBc6 inoculation permits to reduce the dose of fungal inoculum ($1\ L{\cdot}m^{-2}$ to $0.5\ L{\cdot}m^{-2}$) while ensuring the same mycorrhizal index. Finally, in a nondisinfected soil, BBc6 inhibits the development of indigenous ectomycorrhizal fungi, which could decrease mycorrhiza formation with introduced fungal isolate. In conclusion, MHB could be used as an alternative to soil fumigation. All these results are summarized in Table 5.5.

CONCLUSIONS AND PERSPECTIVES

All the results described in this section show that MHBs are major components of mycorrhizal symbiosis and must be included in all ecological studies on mycorrhizal fungi. They interact with most steps of mycorrhizal development. As controlled mycorrhization of tree species is a tool to rehabilitate degraded soils, their effect has to be studied as well as the mycorrhizal symbiosis sensu stricto. However, our knowledge is still fragmentary to describe this bacterial effect. For instance, all these studies show that

TABLE 5.5. Effect of an MHB (*Pseudomonas fluorescens* BBc6) on the ectomycorrhization of Douglas fir seedlings with *Laccaria bicolor* S238N in bare root forest nurseries.

	L. bicolor **mycorrhizal index (%)**	
Experiments	**Control**	**+BBc6**
Mycelium dose		
0.5 $L \cdot m^{-2}$	45 a[a]	89 b
1 $L \cdot m^{-2}$	66 a	91 b
Bacterial dose		
10^6 $CFU \cdot m^{-2}$	60 a	76 b
10^8 $CFU \cdot m^{-2}$	60 a	86 b
10^{10} $CFU \cdot m^{-2}$	60 a	88 b
Specificity		
Unidentified ectomycorrhizal fungus	62 a	28 b

[a]Data in the same row followed by the same letter are not significantly different according to the one-way analysis of variance ($p < 0.05$).

MHBs could have a great ecological importance in degraded areas through reaforestation programs, especially in tropical areas. Further reasearches must be undertaken to

1. elucidate the mechanisms involved in these interactions,
2. determine the influence of MHB on the dynamic of mycorrhizal symbiosis associated with a host plant, and
3. describe the impact of MHB on the functioning of mycorrhizal symbiosis with different fungal symbionts, or host plants in different environmental conditions.

All of these ecological aspects must be reached with a close collaboration between scientific disciplines from pedology to plant physiology to bacteriology. In addition, these bacteria could be a great biological indicator to describe mycorrhizal communities (diversity and functionality).

REFERENCES

Allen M.F., Moore T.S. Jr., and Christensen M. (1980). Phytohormone changes in *Bouteloua gracilis* infected by vesicular-arbuscular mycorrhizae: I. Cytokinin increases in the host plant. *Canadian Journal of Botany,* 58: 371-374.

André S., Duponnois R., and Neyra M. (2003). Arbuscular mycorrhizal symbiosis changes the colonization pattern of *Acacia tortilis* spp. *raddiana* rhizosphere by two strains of rhizobia. *Microbial Ecology,* 45: 137-144.

Bâ A.M., Dalpé Y., and Guissou T. (1996). Les glomales d'*Acacia holosericea* et d'*Acacia mangium. Bois et Forêts des Tropiques,* 250 : 5-18.

Bagyaraj J.M. and Menge J.A. (1978). Interactions between a VA mycorrhiza and *Azotobacter* and their effects on rhizosphere microflora and plant growth. *New Phytologist,* 80: 567-573.

Barea J.M. and Azcon-Aguilar C. (1982). Production of plant growth regulating substances by the vesicular-arbuscular mycorrhizal fungus *Glomus mosseae. Applied Environmental and Microbiology,* 43: 810-813.

Barea J.M., El-Atrach F., and Azcon R. (1991). The role of VA mycorrhizas in improving plant N acquisition from soil as assessed with ^{15}N. In C. Flitton (ed.), *The use of stable isotopes in plant nutrition* (pp. 677-808). Vienna: Soil Fertility and Environmental Studies, Joint IAEA, FAO Division.

Barea J.M. and Jeffries P. (1995). Arbuscular mycorrhizas in sustainable soil plant systems. In B. Hock and A. Varma (eds.), *Mycorrhiza: Structure, function, molecular biology and biotechnology* (pp. 521-559). Heidelberg: Springer-Verlag.

Bethlenfalvay G.J. and Linderman R.G. (1992). Mycorrhizae in sustainable agriculture. Special publication 54. Madison, WI: ASA.

Blilou I., Ocampo J.A., and Gracia-Garrido J.M. (1999). Resistance of pea roots to endomycorrhizal fungus or *Rhizobium* correlates with enhanced levels of endogenous salicylic acid. *Journal of Experimental Botany,* 50: 1663-1668.

Bolan N.S. (1991). A critical review on the role of mycorrhizal fungi in the uptake of phosphorus by plant. *Plant and Soil,* 134: 189-207.

Bowen G.D. (1980). Misconceptions, concepts and approaches in rhizosphere biology. In D.C. Ellwood, J.N. Hedger, M.J. Latham, J.M. Lynch, and J.M. Slater (eds.), *Contempary microbial ecology* (pp. 283-304). London: Academic Press.

Bowen G.D. and Theodorou C. (1979). Interactions between bacteria and ectomycorrhizal fungi. *Soil Biology and Biochemistry,* 11: 119-126.

Bürkert B. and Robson A. (1994). ^{65}Zn uptake in subterranean clover (*Trifolium subterraneum* L.) by three vesicular-arbuscular mycorrhizal fungi in a root free sandy soil. *Soil Biology and Biochemistry,* 26: 1117-1124.

Cavigelli M.A., Robertson G.P., and Klug M.J. (1995). Fatty acid methyl ester (FAME) profiles as measures of soil microbial community structure. *Plant and Soil,* 170: 99-113.

Cooper K.M. (1984). Physiology of VA mycorrhizal associations. In C.L. Powell and D.G. Bagyaraj (eds.), *VA mycorrhiza* (pp. 155-158). Boca Raton, FL: CRC Press.

Cornet F. and Diem H.G. (1982). Etude comparative de l'efficacité des souches de *Rhizobium* d'*Acacia* isolées de sols du Sénégal et effet de la double symbiose *Rhizobium-Glomus mosseae* sur la croissance de *Acacia holosericea* et *A. raddiana. Bois et Forêts des Tropiques,* 198: 3-15.

Cornet F., Diem H.G., and Dommergues Y.R. (1982). Effet de l'inoculation avec *Glomus mosseae* sur la croissance d'*Acacia holosericea* en pépinière et après transplantation sur le terrain. In *INRA Les Mycorhizes: Biologie et utilisation* (pp. 287-293). Dijon: INRA.

Curl E.A. and Truelove B. (1986). *The rhizosphere.* New York: Springer-Verlag.

De Oliveira V.L. (1988). Interactions entre les microorganismes du sol et l'établissement de la symbiose ectomycorhizienne chez le hêtre (*Fagus silvatica* L.) avec *Hebeloma crustuliniforme* et *Paxillus involutus*. Doctoral thesis, University of Nancy, France.

De Oliveira V.L. and Garbaye J. (1989). Les microorganismes auxiliaires de l'établissement des symbioses ectomycorhiziennes. *European Journal of Forest Pathology,* 19: 54-64.

Dehne H.W. (1982). Interaction between vesicular-arbuscular mycorrhizal fungi and plant pathogens. *Phytopathology,* 72: 1115-1119.

Dunstan W.A., Malajczuk N., and Dell B. (1998). Effects of bacteria on mycorrhizal development and growth of container grown *Eucalyptus diversicolor* F. Muell. seedlings. *Plant and Soil,* 201: 241-249.

Duponnois R. (1992). Les bactéries auxiliaires de la mycorhization du Douglas (*Pseudotsuga menziesii* (Mirb.) Franco) par *Laccaria laccata* souche S238. Doctoral thesis, University of Nancy, France.

Duponnois R. and Cadet P. (1994). Interactions of *Meloidogyne javanica* and *Glomus* sp. on growth and N_2 fixation of *Acacia seyal. Afro-Asian Journal of Nematology,* 4: 228-233.

Duponnois R., Founoune H., and Lesueur D. (2002). Influence of the controlled dual ectomycorrhizal and rhizobial symbiosis on the growth of *Acacia mangium* provenances, the indigenous symbiotic microflora and the structure of plant parasitic nematode communities. *Geoderma,* 109: 85-102.

Duponnois R. and Garbaye J. (1990). Some mechanisms involved in growth stimulation of ectomycorrhizal fungi by bacteria. *Canadian Journal of Botany,* 68: 2148-2152.

Duponnois R. and Garbaye J. (1991). Mycorrhization helper bacteria associated with the Douglas fir—*Laccaria laccata* symbiosis: Effects in aseptic and in glasshouse conditions. *Annales des Sciences Forestières,* 48: 239-251.

Duponnois R. and Plenchette C. (2003). A mycorrhiza helper bacterium (MHB) enhances ectomycorrhizal and endomycorrhizal symbiosis of Australian *Acacia* species. *Mycorrhiza,* 13: 85-91.

Duponnois R., Plenchette C., and Bâ A.M. (2001). Growth stimulation of seventeen fallow leguminous plants inoculated with *Glomus aggregatum* in Senegal. *European Journal of Soil Biology,* 37: 181-186.

Eissenstat D.M., Graham J.H., Syvertsen J.P., and Drouillard D.L. (1993). Carbon economy of sour orange in relation to mycorrhizal colonization and phosphorus status. *Ann Bot* 71: 1-10.

Fitter A.H. (1991). Costs and benefits of mycorrhizas: Implications for functioning under natural conditions. *Experimentia,* 47: 350-355.

Founoune H., Duponnois R., Meyer J.M., Thioulouse J., Masse D., Chotte J.L., and Neyra M. (2002). Interactions between ectomycorrhizal symbiosis and fluorescent pseudomonads on *Acacia holosericea:* Isolation of mycorrhiza helper bacteria (MHB) from a soudano-Sahelian soil. *FEMS Microbiology Ecology,* 41: 37-46.

Frey P., Frey-Klett P., Garbaye J., Berge O., and Heulin T. (1997). Metabolic and genotypic fingerprinting of fluorescent pseudomonads associated with the Douglas fir *Laccaria bicolor* and Douglas fir. *Applied and Environmental Microbiology,* 63: 1852-1860.

Frey-Klett P., Pierrat J.C., and Garbaye J. (1997). Location and survival of mycorrhiza helper *Pseudomonas fluorescens* during establishment of ectomycorrhizal symbiosis between *Laccaria bicolor* and Douglas fir. *Applied and Environmental Microbiology,* 63: 139-144.

Fulele-Laurent E., Duponnois R., and Chotte J.L. (2002). Ectomycorrhization of *Acacia holosericea* by *Pisolithus:* Effect on nodulation and on the rhizospheric bacteria. 17th WCSS, August 14-21, 2002, Bangkok, Thailand.

Garbaye J. (1991). Biological interactions in the rhizosphere. *Experientia,* 47: 370-375.

Garbaye J. (1994). Helper bacteria: A new dimension to the mycorrhizal symbiosis. *New Phytologist,* 128: 197-210.

Garbaye J. and Bowen G.D. (1989). Stimulation of ectomycorrhizal infection of *Pinus radiata* by some microorganisms associated with the mantle of ectomycorrhizas. *New Phytologist,* 112: 383-388.

Garbaye J., Churin J.L., and Duponnois R. (1992). Effects of substrate disinfection, fungicide treatments and mycorrhization helper bacteria (MHB) on ectomycorrhizal formation of pedunculate oak inoculated with *Laccaria laccata* in two bare-root nurseries. *Biology and Fertility of Soils,* 13: 55-57.

Garbaye J. and Duponnois R. (1992). Specificity and function of mycorrhization helper bacteria (MHB) associated with the *Pseudotsuga menziesii-Laccaria laccata* symbiosis. *Symbiosis,* 14: 335-344.

Garbaye J., Duponnois R., and Wahl J.L. (1990). The bacteria associated with *Laccaria laccata* ectomycorrhizas or sporocarps: Effect on symbiosis establishment on Douglas-fir. *Symbiosis,* 9: 267-273.

George E., Haussler K., Vetterlein G., Gorgus E., and Marshner H. (1992). Water and nutrient translocation by hyphae of *Glomus mosseae. Canadian Journal of Botany,* 70: 2130-2137.

Geri B., Rupnik M., and Kraigher H. (2000). Isolation and identification of mycorrhization helper bacteria in Norway spruce, *Picea abies* (L.) Karst. *Annales Rei Botanicae,* 40: 65-70.

Graham J.H., Leonard R.T., and Menge J.A. (1981). Membrane mediated decrease in root exsudation responsible for phosphorus inhibition of vesicular-arbuscular mycorrhiza formation. *Plant Physiol,* 68: 548-552.

Grayston S.J., Campell C.D., and Vaughan D. (1994). Microbial diversity in the rhizospheres of different tree species. In C.E. Pankhurst (ed.), *Soil biota: Management in sustainable farming systems* (pp. 155-157). Adelaide: CSIRO.

Hamel C., Dalpé Y., Furlan V., and Parent S. (1997). Indigenous populations of arbuscular mycorrhizal fungi and soil aggregate stability are major determinants of leek (*Allium porrum* L.) response to inoculation with *Glomus intraradices*

Schank and Smith or *Glomus versiforme* (Karsten) Berch. *Mycorrhiza,* 7: 187-196.

Hiltner R. (1904). Uber neure Erfahrugen und Probleme auf dem Gebiet der Bodenbakteriologie und unter besonderer Berücksichtingung der Gründüngung und Brache. *Arb Deutsche Landw Ges,* 98: 59-78.

Hooker J.E. and Black K.E. (1995). Arbuscular mycorrhizal fungi as components of sustainable soil-plant systems. *Crit. Rev. Biotechnol.,* 15: 201-212.

Ibekwe, A.M. and Kennedy, A.C. (1998). Fatty acid methyl ester (FAME) profiles as tool to investigate community structure of two agricultural soils. *Plant and Soil,* 206: 151-161.

Imolehin E.D. and Grogan R.G. (1980). Effect of oxygen, carbon dioxide and ethylene on growth, sclerotial production, germination and infection by *Sclerotinia minor. Phytopathology,* 70: 1158-1161.

Katznelson H., Rouatt J.W., and Peterson E.A. (1962). The rhizosphere effect of mycorrhizal and non mycorrhizal roots of yellow birch seedlings. *Canadian Journal of Botany,* 40: 377-382.

Kucey R.M.N. and Janzen H.H. (1987). Effects of VAM and reduced nutrient availability on growth and phosphorus and micronutrient uptake of wheat and field beans under greenhouse conditions. *Plant and Soil,* 104: 71-78.

Kucey R.M.N. and Paul E.A. (1982). Carbon flow, photosynthesis, and N_2 fixation in mycorrhizal and nodulated faba beans. *Soil Biology and Biochemistry,* 14: 407-412.

Le Tacon F., Skinner F.A., and Mosse B. (1983). Spore germination and hyphal growth of a vesicular-arbuscular mycorrhizal fungus, *Glomus mosseae* (Gerdeman and Trappe), under decreased oxygen and increased carbon dioxyde concentrations. *Canadian Journal of Microbiology,* 29: 1280-1285.

Linderman R.G. (1988). Mycorrhizal interactions with the rhizosphere microflora: The mycorrhizosphere effect. *Phytopathology,* 78: 366-371.

Mamoun M. and Olivier J.M. (1989). Dynamique des populations fongiques et bactériennes de la rhizosphère des noisetiers truffiers. II. Chélateur du fer et répartition taxonomique chez les *Pseudomonas* fluorescents. *Agronomie,* 9: 345-351.

Marshner P., Crowley D.E., and Lieberei R. (2001). Arbuscular mycorrhizal infection changes the bacterial 16 S rDNA community composition in the rhizosphere of maize. *Mycorrhiza,* 11: 297-302.

Meyer J.R. and Linderman R.G. (1986). Response to subterranean clover to dual inoculation with vesicular-arbuscular mycorrhizal fungi and a plant growth promoting bacterium, *Pseudomonas putida. Soil Biology and Biochemistry,* 18: 185-190.

Mosse B. (1962). The establishment of vesicular-arbuscular mycorrhiza under aseptic conditions. *Journal of General Microbiology,* 27: 509-520.

Muyzer G., De Waal E.C., and Uitterlinden A.G. (1993). Profiling of complex microbial populations by denaturing gradient gel electrophoresis analysis of polymerase chain reaction-amplified genes coding for 16 S rRNA. *Applied Environmental Microbiology,* 59: 695-700.

Muyzer G. and Smalla K. (1998). Application of denaturing gradient gel electrophoresis (DGGE) and temperature gradient gel electrophoresis (TGGE) in microbial ecology. *Antonie van Leeuwenhoek,* 73: 127-141.

Paula M.A., Urquiaga S., Siqueira J.O., and Dobereiner J. (1992). Synergistic effects of vesicular-arbuscular mycorrhizal fungi and diazotrophic bacteria on nutrition and growth of sweet potato *(Ipomoea batatas). Biology and Fertility of Soils,* 14: 61-68.

Plenchette C. and Fardeau J.C. (1998). Prélèvement du phosphore par les racines et les mycorhizes. *C. R. Académie des Sciences,* 4: 117-123.

Poole E.J., Bending G.D., Whipps J.M., and Read D.J. (2001). Bacteria associated with *Pinus sylvestris-Lactarius rufus* ectomycorrhizas and their effects on mycorrhiza formation in vitro. *New Phytologist,* 151: 743-751.

Rambelli A. (1973). The rhizosphere of mycorrhizae. In G.C. Marks and T.T. Kozlowski (eds.), *Ectomycorrhizae, their ecology and physiology* (pp. 299-349). New York: Academic Press.

Rozycki H., Kampert M., Strzelczyk E., Ly C.Y., and Perry D.A. (1994). Effect of different soil bacteria on mycorrhizae formation in Scots pine (*Pinus sylvestris* L.) in vitro studies. *Folia Forest. Polon,* 36: 92-102.

Sanders F.E. and Tinker P.B. (1971). Mechanism of absorption of phosphate from soil by endogone mycorrhizas. *Nature,* 233: 278-279.

Sen R., Nurmiaho-Lassila E.L., Haahtela K., and Korhonen T. (1994). Attachment of *Pseudomonas fluorescens* strains to the cell walls of ectomycorrhizal fungi. In C. Azcón-Aguilar and J. M. Barea (eds.), *Mycorrhizas in integrated systems from genes to plant development* (p. 15). Luxembourg: EUR 16728 EN. European Commission.

Sieverding E. (1991). Vesicular-arbuscular mycorrhiza management in tropical agrosystems. Eschborn, Germany: Deutsche Gesellschaft für Technische Zusammenarbeit (GTZ).

Slankis V. (1974). Soil factors influencing formation of mycorrhizae. *Annual Review of Phytopathology,* 12: 437-457.

Van der Hejden M.G.A., Klironomos J.N., Ursic M., Moutoglis P., Streitwolf-Engel R., Boller T., Wiemken A., and Sanders I.R. (1998). Mycorrhizal fungal diversity determines plant biodiversity ecosystem variability and productivity. *Nature,* 396: 69-72.

Van Rhijn P., Fang Y., Galili S., Shaul O., Atzon N., Wininger S., Eshed Y., Lum M., Li Y., To V., et al. (1997) Expression of early nodulin genes in alfalfa mycorrhizae indictes that signal transduction pathways used in forming arbuscular mycorrhizae and *Rhizobium*-induced nodules may be conserved. *Plant Biology,* 94: 5467-5472.

Von Alten H., Linderman A., and Schönbeck F. (1993). Stimulation of vesicular-arbuscular mycorrhiza by fungicides or rhizosphere bacteria. *Mycorrhiza,* 2: 167-173.

Wright S.F. and Upadhyaya A. (1998). A survey of soils for aggregate stability and glomalin produced by hyphae of arbuscular mycorrhizal fungi. *Plant and Soil,* 198: 97-107.

Chapter 6

Plant-Growth-Promoting Rhizobacteria As Biofertilizers and Biopesticides

Manas Ranjan Banerjee
Laila Yesmin
Joseph Kevin Vessey

INTRODUCTION

Plants absorb nutrients and moisture from soil through their root systems. As soil consists of minerals, organic matter, air, and water, it acts not only as a medium for plant growth but also as a home for numerous microbes. When plants grow in soil they also release a wide range of organic compounds. These substances and sloughed-off root tissues attract millions of microbes around the root zone of plants and create a region of vigorous microbial activity. There are millions of tiny bacteria, actinomycetes, yeast, protozoa, and fungi in each gram of soil. Among the microorganisms that reside in soil, bacteria are found to be the most abundant groups both in number and variety. Over 1 billion bacteria may be contained in 1 gram of fertile topsoil. They live on the roots of plants or in the volume of soil that is adjacent to and influenced by the plant roots (rhizosphere). These microbes interact with plants' roots and the nutrients in the soil. Some of them have no effect on plants, but others have either beneficial or detrimental effects. Utilization of these interactions by soil microbiologists and biotechnologists has resulted in renewed interest in the use of microbial inoculants in agriculture, forestry, and environmental management. Use of microbial inoculants (biofertilizers, phytostimulators, and biopesticides) as a strategy to ensure competitive yields and protection of crops in agriculture represents an attractive and environmentally friendly alternative to further applications of synthetic fertilizers and chemical pesticides. This method of farming, often referred to as sustainable agriculture, seeks agricultural practices that are proenvironment for maintaining the long-lasting ecological equilibrium of the soil system and ultimately beneficial to human health (see Figure 6.1).

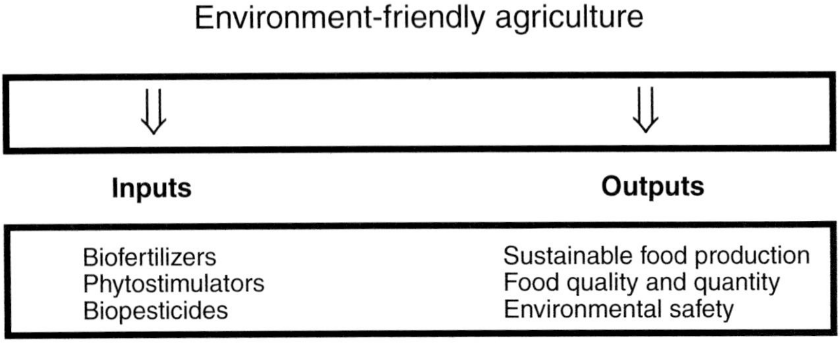

FIGURE 6.1. Role of biofertilizers and biopesticides in sustainable agriculture.

Among the microbial inoculants, bacteria from the plant's rhizosphere (rhizobacteria) have received considerable attention over the past two decades with respect to plant growth promotion. Rhizobacteria that interact with plant roots and soil and, by doing so, provide a beneficial effect on plant growth are generally called plant-growth-promoting rhizobacteria (PGPR). The acronym *PGPR* was first introduced by Kloepper and Schroth (1978), and they defined the term as the rhizobacteria that colonize plant roots and stimulate plant growth. It was used to describe a subset of rhizobacteria that induce increased plant growth after seed inoculation. However, the term has broad meaning in the sense that the presence of PGPR might not result in signs of visible plant growth but contribute to plant yield by reducing or suppressing disease incidents. It may be more practical to say that PGPR help plant performance either directly via visible growth promotion or indirectly via disease control. Due to these important characteristics of PGPR, in recent years rhizosphere microbial ecology and microbial inoculation became a subject of enormous interest to many soil microbiologists and plant scientists. Table 6.1 provides the list of bacteria currently commercially used as microbial inoculants for improving plant performance.

This chapter focuses on the alternatives for using PGPR to produce good harvest yield. Growth promotion as used here implies both visible plant growth as well as crop protection to produce regular yield results. Because of the broad range of topics covered, and although there is evidence of PGPR having biofertilizing or seedling disease control effects on forest tree species (e.g., Elo et al., 2000; Shishido et al., 1999), the focus in this chapter will be on crop plants.

TABLE 6.1. Currently used bacteria as bioinoculants for improved crop performance.

Inoculant type	Crop inoculated	Modified traits	Bacterial identity	Use
Naturally occurring	Rice	None	*Anabaena-Azollae*	Biofertilizer
Naturally occurring	Cereals	None	*Azospirillum brasilense*	Biofertilizer
Naturally occurring	Cereals	None	*Azospirillum lipoferum*	Biofertilizer
Naturally occurring	Forage and grain legumes	None	*Bradyrhizobium japonicum*	Biofertilizer
Naturally occurring	Forage and grain legumes	None	*Rhizobium* spp.	Biofertilizer
Naturally occurring	Nonlegume trees (e.g., *Alnus*)	None	*Frankia* spp.	Biofertilizer
Naturally occurring	Various crops	None	*Bacillus subtilis*	Biopesticide
Naturally occurring	Various crops	None	*Bacillus thuringiensis*	Biopesticide
Naturally occurring	Various crops	None	*Pseudomonas fluorescens*	Biopesticide
Naturally occurring	Various crops	None	*Pseudomonas* spp.	Biopesticide
Genetically modified	Various crops	Deletion of transfer genes	*Agrobacterium radiobacter*	Biocontrol of crown gall
Genetically modified	Forage and grain legumes	Introduction of *nif* gene	*Bradyrhizobium japonicum*	Increased N_2 fixation
Genetically modified	Legumes	Introduction of *nif* and/or *dct* genes	*Rhizobium meliloti*	Increased N_2 fixation
Genetically modified	Various crops	Deletion of ice nucleation gene	*Pseudomonas syringae*	Biocontrol of frost damage
Genetically modified	Various crops	Addition of endotoxin from *Bacillus thuringiensis*	*Clavibacter* spp.	Biocontrol of insect damage

BIOFERTILIZERS

One of the important factors that controls plant growth and development is the adequate supply of necessary plant nutrients or food from soil. Nutrition must be in a form that plants can absorb immediately as they require. At present, addition of nutrients to soil in the form of chemical fertilizer dominates the means of increasing food supply in modern agriculture. However, instead of chemical means, use of natural sources is not new to farmers. For thousands of years people have been using lime, marl, ashes, bones, manure, mud, and legumes to enhance soil nutrients. Utilization of microorganisms in the field to increase soil fertility is comparatively new in that respect. Nevertheless, the agronomic importance of plant-microbial interactions has been widely known for many years, especially the beneficial effects of *Rhizobium* species on legume crops. In recent years a tremendous increase in interest has been noticed among the researchers on this topic both in basic and applied aspects. It is known that soil microorganisms play a very active role in nearly all the biochemical transformations occurring in the soil environment. As an example, microbial decomposition liberates organically bound nutrients into soil in a form that plants can absorb. In particular, they take an active role in the maintenance of soil fertility as a result of their involvement in nutrient cycling (e.g., carbon, nitrogen), which are essential for plant growth. Soil microorganisms can increase the amount of nutrients present in the soil. Thus, these microorganisms, which improve the fertility status of the soil and contribute to plant growth, have been termed *biofertilizers* and are receiving worldwide attention for use as microbial inoculants in agriculture.

What is a biofertilizer? In the scientific literature and on the World Wide Web, the term has a great number of meanings, for everything from plant extracts to chemical fertilizers with organic components (e.g., vitamins) to unspecified mixes of microbial organisms. Vessey (2003) defined *biofertilizer* as

> a substance which contains living microorganisms which, when applied to seed, plant surfaces, or soil, colonizes the rhizosphere or the interior of the plant and promotes growth by increasing the supply or availability of primary nutrients to the host plant.

Likewise, Subba Rao (1981) also defined the term *biofertilizers* or more appropriately *microbial inoculants* as preparations containing live or latent cells of efficient strains of nitrogen-fixing, phosphate-solubilizing, or cellulolytic microorganisms used for application to seed, soil, or composting areas with the objective of increasing the numbers of such microorganisms and accelerating certain microbial processes to augment the extent of the availability of nutrients in a form which can easily be taken up by plants. In an applied sense, the microbial inoculants are carrier-based preparations containing beneficial microorganisms in a viable state intended for seed or soil application and designed to improve soil fertility and help plant growth by increasing the number and biological activity of desired microorganisms in the root environment. These definitions focus on living microorganisms and thereby separate biofertilizers from organic fertilizers and animal or green manures. Some rhizospheric fungi (e.g., arbuscular mycorrhizae; Bethlenfalvay, 1993) can be biofertilizers, but the focus of this chapter is on PGPR, so we will not be addressing fungi as biofertilizers.

ENDOPHYTIC AND RHIZOSPHERIC RELATIONSHIPS BETWEEN PGPR AND THEIR HOSTS

Two types of relationships can exist between PGPR and their hosts: endophytic and rhizospheric. In most cases, the former living arrangement is considered more complex than the latter. In endophytic relationships, the bacteria infect and colonize the plant's apoplast (i.e., nonliving spaces such as intercellular regions). In rhizospheric relationships, the bacteria colonize the zone of plant influence adjacent to the root (i.e., the rhizosphere) or actually attach to the root surface (i.e., the rhizoplane).

Endophytic relationships are considered more sophisticated or complex than rhizospheric relationships because to colonize the interior of a plant the bacteria must either evade or suppress the host plant's defense systems. Since the host is availing itself of a bacterial infection, the infection and colonization processes must be quite highly regulated to ensure that the bacterium getting access to the plant is beneficial and not pathogenic. The regulation of infection and development of the symbiosis is quite well understood for rhizobia-legume symbioses (Gualtieri and Bisseling, 2000).

Endophytic PGPR have been found in all parts of host plants (seeds, stems, fruit, leaves, and roots). They may be found in normal, nonspecialized apoplastic space, such as intercellular spaces in parenchyma tissue (Dong et al., 1997) or xylem vessels (James et al., 2001). However, a debate is under way on the suitability of xylem apoplast as a region for colonization by endophytes with its proponents (James and Olivares, 1998) and its opponents (McCully, 2001). In contrast to simply colonizing normal apoplastic regions, the beneficial bacteria may colonize anatomical features specialized to host the microsymbiont, such as the nodules of legume-rhizobia, *Parasponia*-rhizobia, and *Alnus-Frankia* symbioses, or the gland tissue of *Gunnera-Nostoc* symbioses. For these more specialized symbioses, the means of infection of the PGPR are relatively well understood. However, for the endophytic relationships where no specialized structures are formed, the means of infection and the bacterial habitat within the plant host is not as well understood. Some endophytes have cell-wall-degrading enzymes (e.g., cellulases and pectinases; Verma et al., 2001) that aid in the penetration into the interior of the plant. Other endophytes may use third-party vectors. For example, the sugarcane mealybug and mycorrhizal fungus are both possible vectors in the infection of sugarcane by *Gluconacetobacter diazotrophicus* (Franke et al., 2000; Isopi et al., 1995). There is some controversy that the much referred to "crack entry" by endophytes at the junction of lateral roots with primary roots may be artifactual (McCully, 2001) and not nearly as common in nature as reported in the scientific literature.

Given that rhizospheric PGPR do not have to infect the host plant, their relationship with the host is often considered less specific than those of endophytes. Nonetheless, the rhizospheric milieu is very different from that of the bulk soil in terms of pH, water potential, O_2 levels, nutritional components, and many other physical and chemical characteristics associated with the influence of the plant roots on this soil zone. As a result of these differences, a certain degree of specificity for bacteria present in the rhizosphere exists. An even greater degree of specificity exists for bacteria that attach to the rhizoplane of the host plant (see Photo 6.1). Although detail on this sort of specificity is generally lacking, the means of attachment of *Azospirillum*

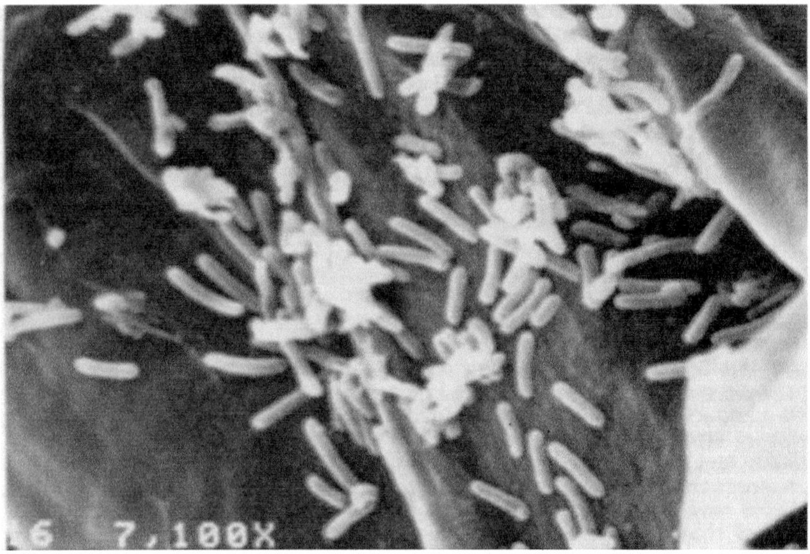

PHOTO 6.1. Rod-shaped bacterial cells exhibiting aggregates on root rhizoplane. (*Source:* L. F. Elliot, C. M. Gilmour, J. M. Lynch, and D. Tittemore, Bacterial colonization of plant roots, in R. L. Todd and J. E. Giddens, eds., *Microbial-plant interactions.* Copyright 1984, American Society of Agronomy. Used by permission.)

sp. have been quite well characterized (Steenhoudt and Vanderleyden, 2000). It must also be recognized that the rhizoplane itself is not uniform and colonization of bacteria on the root surfaces is not consistent across the entire root (e.g., Ma et al., 2001).

MODE OF ACTION OF PGPR AS BIOFERTILIZERS

A great number of papers published in the literature indicate a stimulation of plant growth by a beneficial microorganism but do not address the mode of action used. Such papers are of limited usefulness in increasing our understanding of biofertilizing PGPR. However, many studies do investigate mode of action, and these mechanisms can be divided into four broad areas:

1. Enhanced N supply to the host by N_2 fixation
2. Enhanced supply of some other plant nutrient (e.g., P, S, Fe)
3. Phytohormone production leading to increases in root surface area
4. Enhancement of other beneficial bacterial or fungal symbioses with the host

In many cases, the beneficial effect of the PGPR is mediated by a combination of two or more of the listed modes of action.

Increasing N Availability Through N_2 Fixation

Although rhizobia that infect and fix N_2 in legumes are the most well studied and longest commercially exploited of the PGPR (Fred et al., 1932), given the many excellent and comprehensive reviews that exist on the topic (e.g., Schultze and Kondorosi, 1998), the focus of this section will be on nonsymbiotic N_2 fixers (associative N_2 fixers) (see Table 6.2).

For any PGPR to contribute significant amounts of fixed N_2 in, on, or near a plant, certain criteria must be met. The bacterium must provide an adequate flux of O_2 for respiration, but not an excessive flux of O_2 that would inhibit nitrogenase; it must have a sufficient supply of carbon substrate to meet the high respiratory demand to fuel nitrogenase activity; and it must have a means of supplying the fixed N_2 to the plant. Of these three criteria, the most is known about the first criterion (Marchal and Vanderleyden, 2000), and much less about the latter two. Plant-growth-promoting bacteria fix N_2 in microaerobic and aerobic environments (see Table 6.2) and avoid excessive concentrations of O_2 through high respiration consumption of O_2

TABLE 6.2. Bacteria whose ability of fix N_2 is believed to be partly responsible for their promotion of plant growth.

PGPR	Relationship to host	O_2 condition for N_2 fixation	Host crops
Azospirillum sp.	Rhizospheric and endophytic	Microaerobic	Maize Rice Wheat
Azoarcus sp.	Endophytic	Microaerobic	Kallar grass Sorghum Rice
Azotobacter sp.	Rhizospheric	Aerobic	Maize Wheat
Paenibacillus polymyxa	Rhizospheric	Microaerobic	Wheat
Burkholderia sp.	Endophytic	Microaerobic	Rice Sugarcane
Cyanobacteria[a]	Rhizospheric	Aerobic	Rice Wheat
Gluconacetobacter diazotrophicus	Endophytic	Microaerobic	Sorghum Sugarcane
Herbaspirillum sp.	Endophytic	Microaerobic	Rice Sorghum Sugarcane

Source: Adapted and updated from Vessey, 2003.
[a]Predominantly of the genera *Anabaena* and *Nostoc.*

(e.g., *Azotobacter*) by colonizing sites with low partial pressures of O_2 (e.g., submerged rice roots; rhizospheric mucilage), or by forming structures to decrease the flux rate of O_2 (e.g., heterocysts in cyanobacteria; mucilage production in *G. diazotrophicus*).

Despite the evidence for associative N_2 fixation being involved in the promotion of plant growth in some associations (Omar et al., 1996; Malik et al., 1997; Baldani et al., 2001; Mrkovacki and Milic, 2001; Hashem, 2001; Hurek et al., 2002) and much excitement during the 1970s and 1980s regarding the potential of diazotrophs as PGPR, actual performance in the field has been disappointing. One of the most promising candidates in this area was *Azospirillum brasilense*. However, it is now recognized that the bacterium's growth-promoting effects originate from mechanisms other than N_2 fixation (Dobbelaere et al., 2001). Still today, many bacteria being investigated as PGPR are diazotrophic, yet their stimulation of plant growth is not related to their ability to fix N_2 (e.g., *Azoarcus* sp. [Hurek et al., 1994], *Beijerinckia* sp. [Baldani et al., 1997], *Klebsiella pneumoniae* [Riggs et al., 2001], *Pantoea agglomerans* [Riggs et al., 2001]). Vessey (2003) speculated that either there is something about the ability to fix N_2 that makes the organisms well adapted to living in the rhizosphere (e.g., low mineral N levels) or, when attempting to isolate PGPR, researchers commonly use culture media well suited to diazotrophs (e.g., minimal N media), which may account for the preponderance of diazotrophic bacteria being identified as putative PGPR.

Increasing the Availability of P, S, and Fe to the Host Plant

Some PGPR stimulate the growth of their host plants by increasing the availability of certain inorganic nutrients in the rhizosphere. These bacteria are involved in the solubilization of phosphate, the oxidation of sulfur, and the chelation of iron.

Solubilization of Phosphates

Phosphorus (P) is often limiting to crop production in tropical soils. However, even in temperate and higher organic matter soils, P can often be limiting to plant growth because it is predominantly in insoluble forms, and plants can only absorb P in its soluble ionic forms, $H_2PO_4^-$ and HPO_4^{2-}. Many bacteria isolated from rhizospheres have the ability to solubilize unavailable forms of P in culture by excretion of organic acids and phosphatases (Kim et al., 1998). There is great potential to utilize such bacteria as PGPR (Richardson, 2001). Evidence for P solubilization as a mode of action of PGPR is indicated for *Rhizobium* sp. and *Bradyrhizobium japonicum* and radish (Antoun et al., 1998); *R. leguminosarum* bv. *phaseoli* and maize (Chabot et al., 1998); *Enterobacter agglomerans* and tomato (Kim et al., 1998); *Bacillus* sp. and various crop species (Pal, 1998); *Azotobacter chroococcum* and wheat (Kumar and Narula, 1999); *Bacillus circulans* and *Cladosporium herbarum* and wheat (Singh and Kapoor, 1999); and *Pseudomonas chlororaphis* and *P. putida* and soybean (Cattelan et al., 1999). However, simply the ability to solubilize P in culture does not mean that the bacterium will act as a PGPR. Only two of five P-solubilizing isolates from the rhizosphere of soybean promoted growth of soybean seedlings (Cattelan et al., 1999). Likewise, *Bacillus* and *Xanthomonas* isolates from rapeseed rhizosphere could stimulate the growth of the plants but did not increase the P content of the host plants (de Freitas et al., 1997).

Oxidation of Sulfur

Sulfur (S) is the fourth major plant nutrient after N, P, and K, and is one of the 16 most essential elements required by all life-forms. Sulfur is a vital component of proteins and enzyme systems in the plant body; however, plants can only take up S as the sulfate ion (SO_4^{2-}). If agricultural lands are S deficient for growing crops, farmers use sulfate fertilizers (e.g., ammonium sulfate, potassium sulfate). These sulfate fertilizers are produced in chemical factories and, as sulfate is soluble in water, these fertilizers are liable to cause leaching loss and thus may cause potential groundwater contamination. As our environmental awareness is increasing, more and more farmers are using elemental S fertilizer in their fields instead. Elemental S is a less expensive, high analysis fertilizer, minimizing handling cost, and is not directly soluble in water thus reducing leaching loss and avoiding potential groundwater contamination. Moreover, elemental S is available as a by-product of the oil and gas industry. The only concern of using elemental S is that it is not directly available to plants and must be oxidized to sulfate form for plant uptake. This S oxidation as follows is largely mediated by soil microorganisms and in natural conditions it may take 18 to 24 months, depending on the number of S oxidizers and conducive environmental conditions.

$$\text{Elemental S } (S^0) \rightarrow \text{thiosulfate } (S_2O_3^{2-}) \rightarrow \text{tetrathionate } (S_4O_6^{2-}) \rightarrow \text{trithionate } (S_3O_6^{2-}) \rightarrow \text{sulfite } (SO_3^{2-}) \rightarrow \text{sulfate } (SO_4^{2-})$$

Many types of soil microorganisms are capable of oxidizing S, but these microorganisms represent less than 1 percent of the total soil microbial population. Among those are bacteria (e.g., *Thiobacillus* sp.), fungi (*Fusarium* sp.), and actinomycetes (*Streptomyces* sp.), but bacteria are believed to be the most active S oxidizers. The nature and activity of S-oxidizing microorganisms in soils has been a controversial and promising topic. Although the fundamental concept of the enhanced elemental S oxidation by the appropriate soil bacterial isolates was proven (Grayston and Germida, 1991; Banerjee, 1995) the success of utilizing the S-oxidizing plant-growth-promoting rhizobacteria in different agroclimatic condition is yet to be determined. Nevertheless, Banerjee and Yesmin (2000) found that biological seed treatment with naturally occurring S-oxidizing PGPR of high S-demanding crops (e.g., canola) has tremendous potential to enhance crop production with lower input cost in the Canadian prairies. For example, *Thiobacilli* (sulfur-oxidizing bacteria) were used in the preparation of "biosuper" in Australia by mixing rock phosphate and sulfur. The sulfuric acid produced in the mixture dissolves the rock phosphate and thereby enhances phosphorous nutrition of plants. Whether this kind of biofertilizer is worthwhile and economical depends on the availability of sulfur and rock phosphate for this purpose and the utility of the product on the farm for increasing yields of crop plants.

Chelation of Ferric Iron

Ferrous iron (Fe^{2+}) is more soluble but less common compared to the ferric (Fe^{3+}) ion in soil solution at neutral pH (Taiz and Zeiger, 1998). In the ferric form, iron easily precipitates into insoluble iron-oxide forms. Aside from acidifying the rhizosphere to make iron more available, some plants also excrete iron chelators (phytosiderophores) that bind Fe^{3+}, stopping it from being oxidized into insoluble forms. These phytosiderophores-Fe^{3+} complexes can then be absorbed into root cells (von Wiren et al., 2000).

Some rhizospheric bacteria also produce siderophores that chelate Fe^{3+}, but this activity is more often associated with biocontrolling activities (e.g., Manthey et al., 1994; also see the section called Siderophore Production in the upcoming Biopesticides section) rather than biofertilizing activity. Nonetheless, there is evidence that these microbial siderophore-Fe^{3+} complexes can be absorbed by some plant species. Some researchers suggest that this type of absorption is very important in the iron nutrition of plants, especially in calcareous soils (Masalha et al., 2000); others feel that it plays only a marginal role (Bar-Ness et al., 1991; Glick, 1995). The importance of siderophore production by rhizospheric bacteria and its potential to promote plant growth has yet to be definitively determined.

Phytohormone Production

Undoubtedly the most common mode of action of PGPR is the alteration of phytohormone levels, which in turn changes root growth and morphology such that the host plant has greater root surface area. Many studies and models of nutrient uptake in plants indicate the extremely important role of root surface area as a factor influencing the capacity of plants to absorb inorganic nutrients (Barber and Cushman, 1981). Therefore, the most common method of PGPR stimulation of plant growth is actually an indirect one: production of phytohormones which increases root surface area, which in turn enables the plant to absorb more nutrients to enhance growth of the entire plant.

The effect of phytohormone production by PGPR on host plants is evident in a number of root parameters, namely greater root mass, more root branching, thinner roots, and more root hairs (see Photo 6.2). PGPR-medi-

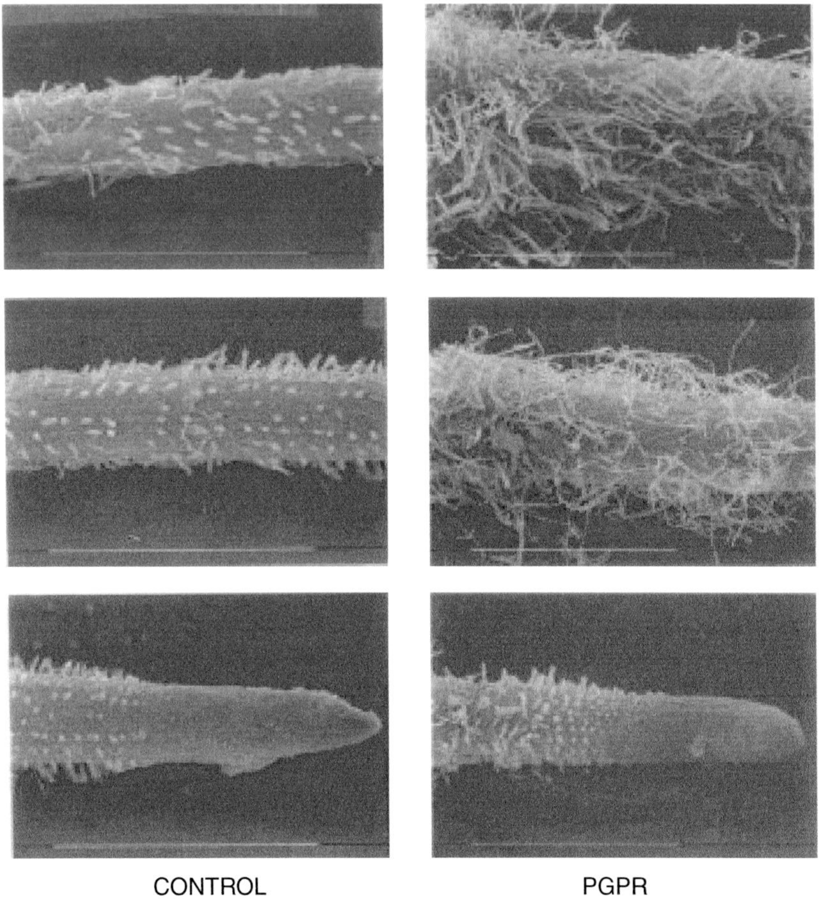

PHOTO 6.2. Scanning electron microscopic (SEM) view of canola root after 96 hours of seed inoculation. PGPR-inoculated roots showed higher densities of larger root hairs. PGPR enhanced the root hair formation and root length. Root hairs are longer and thicker than those of controls. (*Source:* Banerjee, unpublished.)

ated increases in root mass are common (e.g., Vessey and Buss, 2002); however, even if root mass does not increase or it increases only marginally, the increase in specific root length (i.e., root length per unit root mass) observed in many studies means greater root surface area and a greater volume of soil volume explored for nutrients. For example, Molla et al. (2001) showed that *Azospirillum brasilense* Sp7 caused a 63 percent increase in soybean root dry weight but more than a sixfold increase in specific root length, and more than a tenfold increase in total root length.

Table 6.3 lists hormones (IAA, cytokinin, and gibberellic acid) and a hormone-influencing substance (ACC deaminase) produced by certain PGPR and responsible for growth promotion in host plants. The most common of these substances is IAA (indole-3-acetic acid), a phytohormone involved in root initiation, cell division, and cell enlargement (Salisbury, 1994). Effects of IAA-producing PGPR are most commonly identified as increases to root length (e.g., Antoun et al., 1998; Barazani and Friedmen, 1999; Mehnaz et al., 2001; Patten and Glick, 2002).

TABLE 6.3. Hormones or hormone-altering substances produced by PGPR shown to stimulate the growth of their host plants.

Hormonal influence	**PGPR**	**Host**
IAA production	*Aeromonas veronii*	Rice
	Agrobacterium sp.	Lettuce
	Alcaligenes piechaudii	Lettuce
	Azospirillum brasilense	Wheat
	Bradyrhizobium sp.	Radish
	Comamonas acidovorans	Lettuce
	Enterobacter cloacae	Rice
	Enterobacter sp.	Sugarcane
	Pseudomonas putida	Rape
	Rhizobium leguminosarum	Radish
ACC deaminase production	*Alcaligenes* sp.	Rape
	Bacillus pumilus	Rape
	Enterobacter cloacae	Rape
	Pseudomonas cepacia	Soybean
	Pseudomonas putida	Mung bean
	Pseudomonas sp.	Rape
	Variovorax paradoxus	Rape
Cytokinin production	*Paenibacillus polymyxa*	Wheat
		Pine
	Rhizobium leguminosarum	Rape and lettuce
Gibberellin production	*Bacillus* sp.	Alder
		Pine

Source: Adapted and updated from Vessey, 2003.

ACC (1-aminocyclopropane-1-carboxylate) is the precursor of ethylene in its biosynthetic pathway. ACC deaminase is an enzyme produced by some PGPR (see Table 6.3) that decreases the production of ethylene. Ethylene is unique among the phytohormones as a gas and an inhibitor of root growth in many species (Salisbury, 1994). Glick and colleagues (Burd et al., 1998; Glick et al., 1998; Grichko and Glick, 2001) have provided quite convincing evidence that PGPR which secrete ACC deaminase stimulate plant growth by inhibiting ethylene production in the host plant, which leads to an increase in root length.

Two other plant hormones have been implicated in the means by which PGPR stimulate plant growth. Cytokinins promote cell divisions, cell enlargement, and tissue expansion in certain plant parts, and gibberellic acid (GA) is a class of phytohormone commonly shown to cause extension of plant tissues, particularly stem tissue (Salisbury, 1994). Cytokinin production by PGPR has recently been considered responsible for the stimulation of plant growth in a number of species (Table 6.3; Timmusk et al., 1999; Bent et al., 2001; de Salamone et al., 2001). Although only reported for woody species, GA has been recently implicated as responsible for PGPR stimulation of growth in alder (Gutierrez-Manero et al., 2001) and pine (Probanza et al., 2002). Given these recent findings with cytokinins and GA, and given the occurrence of many other plant growth regulating substance (e.g., brassins; Salisbury, 1994), it is highly likely that the production of plant-growth-regulating substance will be identified as the mode of action of other PGPR in the future.

Stimulating Other Plant-Microbe Symbiosis: Helper Bacteria

Biofertilizing PGPR sometimes manifest their beneficial effects on host plants as third parties, aiding or stimulating the beneficial effect of other plant-bacterial or plant-fungal symbioses. These PGPR are sometimes referred to as "helper" bacteria. Although the vast majority of studies investigating PGPR as aids of other host-symbiont relationships involve either legume-rhizobia symbioses or plant-fungi symbioses, coinoculation of PGPR or mixed-culture consortia may be another approach worth pursuing. Chiarini et al. (1998) reported that while testing *Burkholderia cepacia* for its ability to colonize roots and promote the growth of sorghum alone or in combination with *Enterobacter* sp. or *Pseudomonas fluorescens,* all three strains were able to colonize the root system of sorghum but only *B. cepacia* and *P. fluorescens* promoted plant growth in single-strain inoculation. Dual-strain inocula were no more effective than single ones, and in one case the dual-strain inoculum (*B. cepacia* and *P. fluorescens*) did not have any significant effect on plant growth, in contrast to the separate inoculation of both strains. Although the aforementioned study did not find success with combined inoculation, from the competition and survivability of introduced strain's standpoint the authors believe that this area of research might have a tremendous potential of commercialization of PGPR strains. Likewise, Schroth and Becker (1990) suggested that mixtures of microorganisms might be more efficacious than single strains in promoting plant growth. Most tests with PGPR have been done with single strains; however, there is some evidence that combinations (bacterium-bacterium or bacterium-fungus) have proven beneficial in plant growth promotion than single strains alone (Cook et al., 1988; Park et al., 1988). Obviously, in-depth investigations are needed to compare pure culture single strains versus complementary mixed strains of microorganisms to form synergistic consortia that might have greater potential to give a consistent performance with better competitive ability under different environmental and growth conditions (Yesmin and Banerjee, 2001).

Stimulation of Legume-Rhizobia Symbioses

Since the late 1970s, when Singh and Subba Rao (1979) reported positive effects of *A. brasilense* inoculation on nodulation and growth of soybeans, many more studies have identified the synergistic effects of helper bacteria on legume rhizobia symbioses (see Table 6.4).

The most commonly implicated mode of action of helper bacteria is phytohormone production (usually IAA) (e.g., Molla et al., 2001; Vessey and Buss, 2002). The phytohormones increase root length in a similar manner to that discussed in the previous section. The increase in root length results in more sites for infection and nodule development. However, simply the production of a phytohormone by a PGPR does not mean that it will stimulate the legume-rhizobia symbiosis. For example, Cattelan et al. (1999) isolated from the rhizosphere of soybean a number of strains which produced IAA and ACC deaminase, but these strains had no positive effects on nodulation.

Another means by which helper bacteria appear to stimulate legume-rhizobia symbioses involves the production of signaling molecules involved in the development of the plant-microbe symbioses. Flavonoids are chemical signals secreted by legumes that act as chemoattractants to the rhizobia and also turn on bacterial genes associated with the infection and nodulation process (i.e., the *nod* genes). The helper bacteria *Azospirillum brasilense* and *Pseudomonas fluorescens* have been implicated in the stim-

TABLE 6.4. Plant-growth-promoting rhizobacteria that stimulate legume-rhizobia symbioses and the maximum percentage change on aspects of the symbioses due to inoculation.

Crop	PGPR	Nodulation increase (%)	N Benefit[a] (%)	Reference
Soybean	*Bacillus* sp. *Pseudomonas* sp.	57	NM[b]	Li and Alexander, 1988
Alfalfa	*Pseudomonas syringae*	141	194	Knight and Langston-Unkefer, 1988,
Peanut	*Azospirillum lipoferum*	47	NM	Raverkar and Konde, 1988
Lentil	*P. putida*	46	228	Chanway et al., 1989
Common bean	*A. brasilense*	17	NM	Burdman et al., 1996
Common bean	*Bacillus* sp.	87	115	Srinivasan et al., 1996
Soybean	*Aeromonas hydrophila*	73	140	Zhang et al., 1996, 1997
Pea	*P. fluorescens*	298	NM	Andrade et al., 1998
Soybean	*Serratia proteamaculans*	191	92	Dashti et al., 1997, 1998
Soybean	*S. liquefaciens*	231	111	Dashti et al., 1997, 1998
Soybean	Isolates GW3205 & LN3212	189	NM	Cattelan et al., 1999
Red clover	*Pseudomonas* sp.	210	124	Marek-Kozaczuk and Skorupska, 2001
Soybean	*A. brasilense*	730	NM	Molla et al., 2001
Chickpea	*Pseudomonas* sp.	15	41	Goel et al., 2002
Soybean	*Serratia sp.*	44	6	Pan et al., 2002
Soybean	*B. cereus*	16	12	Vessey and Buss, 2002

Source: Adapted and updated from Vessey, 2003.
[a]Measured as nitrogenase activity or N accumulation.
[b]NM = Not measured.

ulation of flavonoid production in the common bean and pea symbioses, respectively (Burdman et al., 1996; Andrade et al., 1998). Pan et al. (2002) tested the effects of the helper bacteria of the *Serratia* genus in combination with genistein, a flavonoid signal in the soybean-*Bradyrhizobium* symbiosis. They found that there were no synergistic or additive effects of the helper bacteria in combination with genistein on nodulation in soybean.

The final mode of action implicated in helper bacteria of legume-rhizobia symbioses involves strain competition in the rhizosphere. Ghosh et al. (2001) have evidence that a toxin released by *Pseudomonas syringae* may be responsible for the stimulation of the alfalfa-*Sinorhizobium* symbiosis. Likewise, B vitamins secreted by *Pseudomonas* sp. may be responsible for the stimulation of the red clover-*Rhizobium* symbiosis (Marek-Kozaczuk and Skorupska, 2001).

Stimulation of Plant-Fungal Symbioses

Recently PGPR have been shown to sometimes stimulate beneficial plant-fungal symbioses involving both arbuscular mycorrhizae (AM) and ectomycorrhizae. These fungi are well-known to benefit growth by making more nutrients (especially P) available to the host plants.

Ectomycorrhizae are more commonly associated with woody plants than with herbaceous species. Numerous bacterial genera have been isolated from the plant-ectomycorrhizal rhizosphere (e.g., Bending et al., 2002), and some isolates increase ectomycorrhizal abundance and plant growth under gnotobiotic conditions. The fluorescent pseudomonades appear to be a group of bacteria often identified as helper bacteria (e.g., Founoune et al., 2002). This area of research is still quite new and the mechanisms underlying the stimulation of host-ectomycorrhizal associations by helper bacteria are not yet understood.

Numerous bacterial genera including *Bacillus*, *Azospirillum*, and *Enterobacter* have been found to stimulate the level of infection of AM of the *Glomus* genus and the growth and P content of their hosts (Kim et al., 1998; Ratti et al., 2001; Srinath et al., 2003). Interestingly, the bacteria in these studies are also P-solubilizing bacteria, and this feature may be related to their mode of action as helper bacteria. However, phytohormone production by the PGPR may also aid the plant-fungal relationship. Shabayev et al. (1996) found that nitrogen fixation, photosynthesis, and yield increases in soybeans inoculated with mixed cultures of *Bradyrhizobium japonicum* and *Pseudomonas fluorescens* as well as *Glomus mosseae* were due to growth-promoting substances.

The relationships between rhizospheric bacteria and plant-mycorrhizae associations are not always positive. Coinoculation of *Pseudomonas* sp. with both plant-AM (Germida and Walley, 1996) and plant-ectomycorrhizae (Bending et al., 2002) associations have been found to have negative effects on the symbioses. In an interaction study with *Pseudomonas* sp. and the AM fungi *Glomus clarum* NT4, Walley and Germida (1997) found typically that dual inoculation influenced the magnitude of response associated with any organism applied alone. However, they also suggested that a nonvolatile, diffusible substance might have been produced by their bacterial strains which inhibit the spore germination of the AMF.

PGPR with Multiple Modes of Action

In many cases a PGPR will display more than one of the modes of actions described previously. As mentioned, many PGPR that have the capacity to fix N_2 are also P solubilizing and/or phytohormone producers (e.g., *Azospirillum*). Several studies have specifically addressed the multiple modes of action of PGPR populations associated with a plant. Antoun et al. (1998) surveyed 266 strains of rhizobia and found 83 percent produced siderophores, 58 percent produced IAA, and 54 percent could solubilize phosphorus. Twenty-five percent of all the isolates had positive effects on the growth of radish. Cattelan et al. (1999) screen isolates from soybean rhizosphere for 9 different PGPR modes of action (both biofertilizing and biocontroling). Of 22 isolates that promoted some aspect of soybean growth, 6 were ACC deaminase producers, 4 produced siderophores, 4 produced β-1,3-gluconase (a biopesticide), and 2 were P solubilizers. Likewise, Belimov et al. (2001) found that among five *Pseudomonas* sp. isolates which stimulated the growth of rape, three produced ACC deaminase and were P solubilizers, and the other two had these features plus produced IAA. These very important studies indicate that not only does an individual plant-growth-promoting rhizobacterium often have more than one mode of action but that within the population of PGPR in a plant's rhizosphere, multiple modes of action are working interactively to stimulate plant growth.

BIOPESTICIDES

Chemical pesticides are generally used to control pests and diseases, but the use of these pesticides to control some pests and diseases (e.g., root diseases) can be expensive. They may also play a dominant role in environmental pollution, such as soil and groundwater contamination. The other way of combating these diseases is to develop new resistant cultivars. New cultivar development can be costly and time-consuming. Thus, using biological control agents, especially biopesticides, to fight such diseases and pests is gaining increasing momentum in the agricultural biotechnology research world. Biopesticides or biological pesticides are defined as pesticides resulting from such natural materials as animals, plants, bacteria, and certain minerals. For instance, garlic, mint, and baking soda each have pesticidal applications and are therefore considered biopesticides. The U.S.

Enivronmental Protection Agency (EPA) categorized biopesticides into three main divisions (www.epa.gov/pesticides/biopesticides):

1. *Microbial pesticides*—These contain a microorganism (bacterium, fungus, alga, virus, or protozoan) as the active ingredient. The most widely known microbial pesticides are derived from the bacteria *Bacillus thuringiensis,* or Bt, and can control specific types of insects in potatoes, cabbages, or other crops.
2. *Plant pesticides*—These are pesticidal materials which plants produce from genetic material that has been added to the plant. For instance, researchers may take the gene for the Bt pesticidal protein and introduce the gene into the plant's own genetic material. Consequently, the plant itself generates the substances that control the pest.
3. *Biochemical pesticides*—These are naturally occurring materials which destroy pests by nontoxic mechanisms. Biochemical pesticides may include substances that interfere with growth or mating, such as plant growth regulators, or substances that repel or attract pests, for example, pheromones. Therefore, microbial pesticides are microorganisms, including bacteria, that help to control insects, weeds, fungi, and bacteria that cause plant diseases.

Curl and Truelove (1986) divided biological control into direct and indirect methods. The direct methods include inoculation of parasitized plant parts or soil with antagonistic microorganisms through seed inoculation with bacterial suspensions; treatment of vegetative structures (roots and cuttings or transplants) by dipping in suspensions of biocontrol agents; and direct application of bacterial suspensions (biocontrol agent) to soils. The indirect methods include agricultural practices using organic amendments, fertilization, crop rotation, and pesticides to enhance biological control; and plant breeding programs developed to produce resistant crop varieties. In the scope of this chapter we mainly discuss the PGPRs that may directly be used as potential microbial pesticides—in particular, the bacterial biopesticides. As mentioned earlier, being antagonistic to many pests, different strains of *Bacillus thuringiensis* have been used extensively as biopesticides. At present many other bacteria have also been used as active ingredients in different bioproducts and are gaining significance in the market, especially as biofungicides. Table 6.5 shows examples of some bacteria other than Bt used as biopesticides. A survey with a wide range of different materials and organisms that can be classified as biopesticides illustrated that although in the immediate future biopesticides may continue to be limited mainly to niche and specialty markets, there is great potential for long-

TABLE 6.5. Examples of some bacterial bioproducts being used in different parts of the world.

Bacteria	Target organism	Trade name	Manufacturer	Category
Pseudomonas fluorescens	Fire blight *(Erwinia amylovora)*	BlightBanA506	Plant Health Technologies, USA	Bactericide
Pseudomonas cepacia	Damping off diseases	Blue Circle	Stine Microbial Products, USA	Fungicide
Pseudomonas syringae	Blight Leaf spot	BioSave	Ecoscience Corp., USA	Fungicide
Pseudomonas cepacia	Root lesion nematode, spiral nematode	DENY	CCT Corp., USA	Nematocide
Serratia entomophila	New Zealand grass grab	Invade	AgResearch, New Zealand	Insecticide
Bacillus papillae	Japanese beetle larvae	Milky Spore-Bacilluspopillae	Arizona Biological Control Inc., USA	Insecticide
Agrobacterium radiobacter	Crown gall diseases	NOGALL	Bio-Care Technology Pty Ltd., Australia	Bactericide
Pseudomonas solanacearum	Bacterial wilt	PSSOL	Natural Plant Protection (NPP), France	Bactericide
Pseudomonas fluorescens	Blotching disease of mushrooms *(Pseudomonas tolassii)*	Victus	Sylvan Spawn Laboratory, USA	Bactericide
Bacillus sphaericus	Culex mosquito larvae	VectoLexCG	Fomerly Abbott Laboratories, USA	Insecticide
Bacillus subtilis	Bacteria, fungus, and viral diseases	System 3	Helena Chemical Company, USA	Bactericide, fungicide, viricide
Streptomyces griseoviridis K61	Seed rot, root and stem rot, wilt, and damping off	Mycostop	Kemira agro Oy, Finland	Fungicide
Bacillus subtilis	*Alternaria* sp. *Aspergillus* sp. *Fusarium* sp. *Rhizoctonia* sp.	Epic	Gustafson, Inc., USA	Fungicide
Bacillus subtilis	Damping off diseases	Stimulex	Levington Horticulture Ltd. UK	Fungicide
Bacillus subtilis	Fire blight Grey mould Late blight Potato blight Powdery mildew	Serenade QST713	AgraQuest Inc., USA	Fungicide
Bacillus subtilis	*Alternaria* spp. *Phytophthora* spp. *Colletotrichum* spp.	Rotor	Appliedchem (Thailand) Co. Ltd., Thailand	Fungicide
Pseudomonas aureofaciens Tx-1	Dollar spot Anthracnose Phythium Pink snow mold	Spot less	Eco Soil Systems, USA	Fungicide
Pseudomonas chlororaphis 63-28	*Pythium* spp. *Rhizoctonia solani* *Fusarium oxysporum*	AtEze	Eco Soil Systems, USA	Fungicide
Bacillus subtillis	*Aspergillus* *Fusarium* *Rhizoctonia*	Kodiak	Gustafson, Inc., USA	Fungicide

TABLE 6.5. *(continued)*

Bacteria	Target organism	Trade name	Manufacturer	Category
Pseudomonas fluorescens	Bloching diseases of mashroom *(Pseudomonas tolassii)*	Conquer	Mauri Laboratories, Australia	Bactericide
Bacillus subtilis	Fungal and mold diseases	MBI600	MicrobioGroup Ltd., UK	Fungicide
Agrobacterium radiobacter	Crown gall diseases	Galltrol-A	AgBioChem Inc., USA	Bactericide
Bacillus subtilis	*Alternaria* spp. *Fusarium* spp. *Rhizoctonia spp.* *Sclerotinia* spp. *Streptomyces scabies* *Verticillium* spp.	Rhizo Plus/ Rhizo-plus Konz	KFZB Biotechnik GmbH, Germany	Fungicide and actinomyceticide

term development and growth, both in their own merit and in providing leads in other areas of pest management (Copping and Menn, 2000). According to the EPA, there were approximately 175 registered biopesticide active ingredients and 700 products at the end of year 1998.

MODE OF ACTION OF PGPR AS BIOPESTICIDES

Not all PGPR can be considered biofertilizers. Rhizobacteria that promote plant growth by control of deleterious organisms are biopesticides, but not biofertilizers. It is fascinating to note that some PGPR appear to promote growth by acting as both biofertilizer and biopesticide. For example, strains of *Burkholderia cepacia* have been shown to have biocontrol characteristics in *Fusarium* spp. but also can stimulate growth of maize under iron-poor conditions via siderophore production (Bevivino et al., 1998). However, microbial pesticides in general can be subdivided into three main categories:

1. Biofungicides that biocontrol fungal diseases
2. Bioinsecticides that biocontrol insect pests
3. Bioherbicides that biocontrol weed plants

Biopesticides, be it biofungicides or bioinsecticides or bioherbicides, are achieving some recognition as alternatives to conventional pesticides. Although their full potential has not yet been reached, their role in replacing or complementing the reduction of application of conventional pesticides in the environment will be significant.

The microbial pesticides, or biopesticidal PGPR, use several different processes to control the targeted plant pathogens. Kloepper (1997) described some of these mechanisms by which rhiozobacteria exhibit their biocontrolling power against pathogens. Among these, competition, antibiosis, induced systemic resistance, siderophore production, cyanide production, and parasitism warrant further discussion. However, more than one mechanism may also operate at the same time to mediate the biocontrol effect. Nevertheless, understanding the mode of action of these biopesticidal agents and how they interact with the different plant pathogens will give direction toward developing strategies to control and manage pests and diseases.

Competition

Some biocontrol agents act through competition for limited resources, and via this competition the growth of the pathogenic population can be prohibited, suppressing the incidence or severity of disease (Van Driesche and Bellows, 1996). Some PGPR are capable of utilizing nutrients and energy materials either from host plant exudates, from habitat, or from both that are needed for growth of the pathogenic microbes. In many cases, competition for carbon is considered a well-known phenomenon. For example, ethanol user *Pseudomonas putida* uses soybean seed exudates (Paulitz, 1991), and *Enterobacter cloacae* uses cottonseed exudates (Nelson and Maloney, 1992), but both of them reduce the growth of sporangia of *Pythium ultimum*. This kind of competition is also believed to occur for other macro- and micronutrients. In iron-deficient soil, fluorescent *Pseudomonas* spp. produce siderophores or pyoverdine siderophores that make Fe less abundant for other microbes and thereby suppress disease. Competition for nutrients between harmless and pathogenic microbes could also occur as microbial associations and different modes of action interact. Harmless *Fusarium oxysporum* strains and the PGPR strain of fluorescent *Pseudomonas* spp. together were documented to provide biological control of fusarium wilts (Alabouvette and Lemanceau, 1999). In this case, they might compete for two different nutrients, carbon and iron, and thereby improve biocontrol efficiency. Many naturally occurring competitive microbes were isolated from the rhizosphere and are currently used as biopesticides. *Pseudomonas aeruginosa* is used against *Pythium splendans* which causes damping off of tomato (Buysens et al., 1994), *P. putida* is used against *Fusarium oxysporum* (fusarium wilt) for carnation (Lemanceau et al., 1992) and *Pythium ultimum* (damping off) for pea and soybean (Paulitz, 1991). However, success of introduced bacteria as a biopesticide depends on proper and rapid colonization to adequate population level to be competitive against the plant pathogens.

Antibiosis

Many rhizosphere microorganisms in soil are found to produce different metabolites that are toxic to target pathogens but not the host plants. These microbes compete via the production of antimicrobial substances that prevent pathogen growth directly, rather than through preventive utilization of limiting resources. Over the past four decades *Bacillus thuringiensis* has been widely used as a safe biopesticide for controlling agricultural pests and diseases because it produces pesticidal crystal proteins. Disease suppression of host plant by antibiosis is an antagonistic mechanism, similar to competition for nutrients or siderophore production. Foldes et al. (2000) isolated *Bacillus subtilis* IFS-01 strain, which produces natural biocontrol agents (BCAs) that can be used as biopesticides. This *Bacillus* strain produces either a broad-spectrum antimicrobial compound or several compounds with different activities. Fungi and gram-positive bacteria were more susceptible to their antagonistic effects than were gram-negative bacteria. *Burkholderia cepacia* strains isolated from the corn rhizosphere have shown antifungal activity against a number of fungal pathogens (Hebber et al., 1992), and pyrrolnitrin antibiotic has been reported as the main metabolite responsible for antifungal activity (Homma et al., 1989). Bacteria isolated from barley root, *Pseudomonas* Hv37a, produce oomycin A, which control *Pythium ultimum* (Howie and Suslow, 1991), and *Pseudomonas fluorescens,* CHA0 isolated from Swiss soil produce monoacetyl- and 2, 4-diacetyl-phloroglucinol (Voisard et al., 1994), control black root of tobacco (Keel et al., 1990). A single strain of biocontrol agent could produce several toxic metabolites. Disease caused by *Gaeumannomyces graminis* var. *tritici* in wheat and barley was observed to be controlled by *P. fluorescens* 2-79 and *P. aureofaciens* 30-84 via antibiosis-phenazine production (Pierson and Thomashow, 1993). However with the production of phenazine-1-carboxylic acid (PCA), the strain *P. aureofaciens* 30-84 also produces other metabolite-like PCA derivatives and hydrogen cyanide (Pierson and Thomashow, 1993). Production of multiple metabolites is an important phenomenon for biocontrol compared to chemical pesticides, as it makes it difficult for pathogens to develop resistance toward different metabolites. Metabolite production and its activity in many biocontrol agents were found to be influenced by soil biotic and abiotic conditions, including nutrient status (Ownley et al., 1991; Slininger and Jackson, 1992). Inoculation at the time of planting with *Pseudomonas chlororaphis* strain 2E3 inhibited growth of wheat pathogen *Fusarium culmorum* and may also suppress the winter wheat pathogen *Rhizoctonia solani*. Higher emergence of wheat in soil for this PGPR may sometimes be attributed to the suppression of pathogenic microorganisms in soil (Kropp et al., 1996). *Bacillus circulans* no. 4.1 produced high levels of a chitinase enzyme when grown in triptic soy broth (TSB) amended with 0.3 percent colloidal chitin. Wiwat et al. (1999) purified and characterized this enzyme that hydrolyzed colloidal chitin, purified chitin, glycol chitin, carboxymethyl-chitin (CM-chitin), and 4-methylumbelliferyl-beta-D-N,N'-diacetylchitobioside. This information might give insight into the development of genetically modified *Bacillus* spp. as biopesticides. Seligy and Rancourt (1999) used bioreduction and DNA-based assays for antibiotic MIC (minimum inhibitory concentration) and MBC (minimum bactericidal concentration) analysis of *Bacillus*-based commercial insecticides. Resistance to beta-lactams was attributed to beta-lactamase activity in both Bt products and cultures derived from them. The MIC and MBC break points in spore-based assays were also approximated by changes in genome copy, using delta-endotoxin and beta-lactamase genes as probes. These bioassays can be valuable for characterizing and biomonitoring microorganisms in biopesticidal products. Because of its relative ease and inexpensiveness, antibiosis may be one of the most popular modes of action used in the screening of microbial biocontrol agents, but the inherent disadvantage of antibiosis is that antibiosis in vitro is not always equivalent to antibiosis in vivo. The reason may be as stated by Baker and Dickman (1993), that adequate amounts of antibiotics needed to be produced in order to have sufficient biocontrol activity, and they ought to be produced at the optimal time when the plant pathogen is susceptible to the antibiotic-producing agent. We believe that in vitro and in vivo reaction to antibiosis, sufficiency of antibiotic concentrations, and how to control the timing of antibiotic production warrant further investigation to develop the biopesticidal research to its full potential.

Induced Systemic Resistance

Many plants react to pathogenic attack by enabling their natural defense mechanisms. But some microorganisms are thought to act as inducer to host plants' defense mechanisms in the presence of pathogenic microbes in the ecological niche of the host plants, even before the virulence of pathogenic activity occurs and thereby suppresses disease incidence. Induced systemic resistance is thus an indirect biocontrol mechanism that inhibits disease infection through the induction of plant defense against pathogens by nonpathogenic microorganisms. Induced plant defenses may consist of lignification of cell walls by chemical cross-linkages, an addition in cell wall peptides that makes it harder for infection establishment through lysis. Other processes include suberification of tissues (in which plant cell walls are penetrated with the fatty substance suberin, making them more cork-like), and other general defense mechanisms, including production of chitinases and β1,3-glucanases (Van Driesche and Bellows, 1996). *Pseudomonas* sp. via ISR was found to control such diseases as fusarium wilt in carnation (Van Peer et al., 1991), halo blight of bean (Alstrom, 1991), and tobacco necrosis in tobacco (Maurhofer et al., 1994). Moreover, in commercial greenhouse trials, *Pseudomonas fluorescens* strain WCS374 not only suppressed fusarium wilt in radish but also increased radish yield via ISR (Leeman et al., 1995). Induced systemic resistance in cucumber and tomato against cucumber mosaic cucumovirus (CMV) using PGPR strain of *Pseudomonas fluorescens* and *Serratia marcescens* was observed by Raupach et al. (1996). Seed treatment with both PGPR strains significantly and consistently reduced mean numbers of symptomatic plants when CMV was inoculated onto cotyledons. Plants treated with PGPR did not develop initial symptoms 14 days after CMV inoculation and remained symptomless throughout the experimental period. Wei et al. (1996) reported ISR by rhizobacteria for cucumber disease under field situations and found that PGPR-mediated ISR was consistently effective against angular leaf spot and naturally occurring anthracnose *(Colletotrichum obiculare),* along with significant effects on early-season plant growth promotion and yield increase. Zehnder et al. (1997) reported that cucumber plant growth and yields were significantly higher and numbers of cucumber beetles were significantly lower on PGPR-treated cucumber compared to nontreated cucumber on dates when peak beetle populations were present. The PGPR treatment resulted in significantly greater cucumber beetle control than did weekly applications of the insecticide esfenvalerate. They observed that the incidence of cucurbit wilt disease was also significantly lower in PGPR-treated plants. These suggested that PGPR-induced resistance might be more effective than insecticides for control of cucumber beetles and cucurbit wilt disease on cucumber.

Siderophore Production

Different strains of bacteria inhabiting the rhizosphere are capable of suppressing soilborne diseases via secondary metabolite production. Some of these metabolites are directly lethal to plant pathogens (e.g., antibiotics or hydrogen cyanide), but some metabolites help bacteria with acquisition of such nutrients as iron. In fact, in iron-deficient soils certain microorganisms produce metabolites known as siderophores which serve as iron carriers for microbial growth (e.g., pyoveridine, pyochelin). Some rhizobacteria excrete low-molecular-weight, high-affinity ferric-chelating siderophores or microbial iron cofactors that particularly increase their acquisition of iron by binding to membrane-bound siderophore receptors. As siderophores have a high affinity for iron, siderophore producers easily outcompete other microbes in their vicinity by quickly scavenging for iron. The mechanism of siderophore production may also be intimately correlated with the competition for nutrients. Numerous studies have supported biocontrol mechanisms by *Pseudomonas* spp. via siderophore production, for example, suppression of fusarium wilt (Lemanceau and Alabouvette, 1993) and pythium damping off (Loper, 1988). According to Kloepper et al. (1988), the mode of siderophore production is based on the following suppositions:

1. Siderophores produced by certain rhizobacterial strains prevent plant pathogens in vitro and promote plant growth.
2. The addition of dissolved Fe^{3+} to growth media prevents the inhibition of in vitro antagonism and also nullifies bacteria-induced growth promotion.
3. The pseudomonad PGPR (WCS358) that produces siderophores consistently increases plant growth in potato crop, whereas the siderophore-negetative mutant does not show that effect—although the colonization levels are the same in both cases.

The *Pseudomonas putida* WCS358 strain is reported to be extraordinarily capable of absorbing and synthesizing iron to combine for a variety of siderophores (Bitter et al., 1991). Some strains of siderophore producers (e.g., *Pseudomonas fluorescens* WCS374) are also reported to trigger induced resistance in addition of siderophore production (Schippers et al., 1995). In another study, de Brito-Alvarez et al. (1995) demonstrated that the addition of compost to soil increased the incidence in the tomato rhizosphere of bacteria exhibiting antagonism toward *Fusarium oxysporum, Pyrenochaeta lycopersici, Pythium ultimum,* and *Rhizoctonia solani* due to the marked increases in siderophore producers.

For developing effective biological control agents as bioherbicides, Kremer et al. (1990) suggested that PGPR must be screened directly on host seedlings, and must possess various properties including high colonizing ability, specific phytotoxin production, and tolerance to antibiotics produced by other rhizosphere microbes, and they must either synthesize or utilize other bacterial siderophores. Microbial biocontrol agents might utilize one or several mechanisms to reduce pathogenic progression. Whatever the mechanisms used by the introduced species to control disease incidence, the speed of establishment and efficiency of introduced species to attain a desired population size is the key to be effective against target pathogens. Proper colonization by rhiobacteria is often referred as the limiting step in disease control process (Weller, 1988).

Cyanide Production

Different strains of biocontrol agents might respond differently to specific environments for specific metabolite release. Hydrogen cyanide produced by *Pseudomonas fluorescens* CHAO controls black root rot of tobacco caused *Thielaviopsis basicola,* and adequate iron is necessary for the production of hydrogen cyanide (Keel et al., 1989). As mentioned earlier, metabolites produced by biocontrol agents are target specific, but in most cases they are not toxic to the host plant. For example, hydrogen cyanide (HCN) was found to be effective for black root rot of tobacco but showed no influence on take-all of wheat (Voisard et al., 1989; Keel et al., 1992). However, it also depends on the amount and kind of metabolite produced for the specific target, the target's sensitivity to the metabolite, and the host plant tolerance level. Wei et al. (1991) tested a model system of cucumber and anthracnose, caused by *Colletotrichum orbiculare* for ISR, and found that four of the six strains that induced resistance in cucumber produced HCN in vitro. Nevertheless, excessive production of HCN may be lethal to the host plant and could cause phytotoxicity. Overproduction of cyanide at phytotoxic levels by pseudomonads have been shown to reduce growth and yield of potato (Schippers et al., 1987). Reduction of plant growth by different strains of *Pseudomonas* were also shown by Alstrom and Burns (1989) and it was postulated that some siderophore-producing plants increase yield by reducing cyanide production through reduction of iron availability needed for cyanide production (Kloepper, 1997). However, as discussed earlier in several investigations, cyanide production provided evidence of disease suppression and is considered an important mechanism of biological control (Defago et al., 1990; Voisard et al., 1989).

Parasitism

Parasitism is another mode of action by which beneficial microbes suppress plant pathogens. Some beneficial microbes act as parasites to pathogenic microbes, thereby reducing disease incidence in host plants. Negative interactions between beneficial and pathogenic organisms are the basis of parasitism. Different lytic enzymes including chitinase, β-glucanase, cellulase, and protease produced by biocontrol microbes are considered to be liable for feeble pathogenic progression or degradation of the cell wall (Weller and Thomashow, 1993). Support for this hypothesis as a mechanism for biocontrol—especially for bacterial biocontrol—is not clearly established. However, many biocontrol fungi produce different enzymes that are responsible for growth reduction of pathogens and are considered an important mode for biocontrol fungi. Through genetic modifications some bacteria showed improved biocontrol activity, such as recombinant strains of *Escherichia coli* or *Pseudomonas pudita* encoded with a chitinase gene cloned from *Serratia marcescens,* and have been shown to control bean seedling diseases caused by *Rhizoctonia solani* and *Sclerotium rolfsii* (Shapira et al., 1989; Chet et al., 1993). In one example, growth inhibition of *P. ultimatum* by *Enterobacter cloacae* in vitro was related to attachment of *E. cloacae* to fungal hyphae. However, the addition of sugar was found to reduce the bacterial capability to suppress fungal growth (Nelson et al., 1986). Nonetheless, bacterial contribution of disease suppression by parasitism still remains to be established.

VIABILITY AND SHELF LIFE OF PGPR

To reestablish the rhizosphere's ecological balance in a microbial inoculation, the ability of an introduced strain to outperform the native strains in soils may be a primary concern in choosing a suitable strain and is a criterion that can be achieved by thorough laboratory work, field testing, and serological studies. Other considerations may be tolerability of different root temperatures, greater survival in carriers and on seed, and tolerance to pesticides, fertilizers, and other agrochemicals. Thus, for a microbial inoculant to be successful in the real-world environment includes considerations at all stages of inoculum use, from strain selection to the ultimate application in the field.

The success of microbial inoculation for enhanced crop production is greatly influenced by the number of viable cells introduced into the soil (Duquenne et al., 1999). Therefore, it is crucial to determine the duration of bacterial survivability after the bacterial treatment in the final product and to obtain the desired level of bacterial population for the inoculant to be effective. For example, coating bacteria-treated seeds seals the bacteria onto the seeds, prevents cells from drying out, and keeps bacteria alive much longer than bare seeds. There are now several other methods of delivery that are both practical and ecologically sound. But little progress has been made with alternative carriers that might enhance the numerical quality of microbial inoculants (Brockwell and Bottomley, 1995). When broth is directly used for inoculation, large quantities of fresh broth are necessary to obtain good colonization comparable to that obtained by peat-based inoculants. Moreover, various scientists have been of the opinion that both agar and broth cultures are inferior to peat-based inoculants because peat provides natural protection to the rhizobacteria and thus enhances its survival on seeds. Daza et al. (2000) evaluated peat- and perlite-based inoculants and suggested the existence of interactions between carriers and adhesives, and showed that the combination of a sucrose adhesive with the perlite carrier gave better survival of bacteria on seeds. The final product of PGPR can be liquid, powder, solid, suspension, or granular, but the formulation of the final product should maintain viability of the bacteria from the production facility to the end users via dealers and retailers, and through long-term storage (over three months) (Paau, 1988). As formulation of PGPR biofertilizers or biopesticides plays a vital part in the performance consistency by improving or maintaining bacterial viability following application, the congenial formulation should provide a protective environment for the introduced strain, thereby enhancing the potential for survival and successful colonization (Boyetchko et al., 1999). Microbial inoculants, in particular bacterial inoculants, tend to have a very low shelf life or storage life as biological activity in the PGPR may decline rapidly with handling and storage procedures, if not done properly. This is a huge roadblock for commercializing more PGPR. The formulation may be chosen to capture the dormant or metabolically active states that have a longer shelf life without compromising the beneficial effects of biofertilizing or biocontrol. However, storage should be in a cool place (<18°C) without sunlight exposure, as the most critical factors affecting PGPR survival is storage temperature and moisture loss during storage. Therefore, developing new formulations and/or carrier materials and/or testing of compatibility of different existing commercial inoculant carrier materials for use in biological seed treatment or inoculant product development is urgently needed to maintain bacterial viability and extend the shelf life of the inoculant product to the adequate level.

BENEFITS OF PGPR

One of the major worries in today's world is pollution and contamination of soil. The use of chemical fertilizers and pesticides has caused incredible harm to the environment. An answer to this is the biofertilizing or biopesticidal PGPR, an environmentally friendly approach now being used in most countries. In many instances, the yield advantages due to these biofertilizing PGPR may be marginal, but resource-poor countries can hardly afford to overlook any low-cost input that can bring corresponding dividends to the farmer where labor is not as expensive as it is in developed world. Moreover, bacterial biopesticides are cheaper to develop and safer than agrochemicals. Although potential for advantages and benefits of using PGPR in agriculture is vast, as described throughout this chapter, the following beneficial aspects of PGPR is worth mentioning again. The PGPR are natural in origin; avoid potential groundwater contamination from chemical fertilizers and pesticides; may reduce pesticide use by tolerance to stresses like diseases and pests; and enhance crop performance by improving growth and development (seedling emergence, vigor, leaf size, root growth, and development), nutrient uptake, and yield. Undoubtedly, the most successful and effective use of PGPR as microbial inoculant for agricultural crops is based on the symbiotic nitrogen fixation of *Rhizobium* spp. There are a number of commercial inoculant products available in the market for using different strains of *Rhizobium* spp. with such various leguminous crops as pea, soybean, alfalfa, clover, and bean. Many other free-living bacteria such as *Azotobacter, Azospirillum, Bacillus, Clostridium, Pseudomonas* are also capable of biological nitrogen fixation. The PGPR (e.g., *Azotobacter*) can affect the nodulation process in legume plants through the production of enzyme pectinase, which directly affects nodule formation. Many rhizobacteria produce a variety of phytohormones (e.g., auxin, gibberlins, cytokinins) and phytohormone-like materials in the rhizosphere that are indispensable for plant growth and development. For example, pseudomonads are suggested to produce indole-3-acetic acid (IAA) and increase the amount of IAA in plants that affect the root morphology. The PGPR are not only responsible for nitrogen fixation; they also help in other nutrient acqusitions in the plant. They may help in phosphorous solubilization in the rhizosphere and thus improve phosphorous uptake in plants, or they may help in sulfur oxidation, making more sulfate available in soil and thus helping with sulfur uptake in plants. The PGPR like *Pseudomonas* is reported to produce a siderophore called pseudobactin which helps potato plants to inhibit the growth of the pathogen of soft-rot of potato. The addition of pseudobactin to the growth medium not only prevents infection of *Erwinia carotovora* but also substantially depresses the pathogenic fungal population in the potato root region. The biocontrol effect of PGPR on various diseases and pests are not only environment friendly and remarkable but also offers huge commercial potential for using these as biopesticides. For example, commercial products (e.g., Quantum 4000 and Kodiak produced by Gustafson, Inc.) based on *Bacillus subtilis* are used in peanut, cotton, and broad bean, and more than 4 million of hectares of land are treated with these products in the United States alone.

LIMITATIONS OF PGPR

Although PGPR may reveal huge potential for crop production, for a microbial inoculant to be commercially successful it must be economically mass produced and then formulated into a product that is cost-effective, uniform, and readily applicable by the end user (Walter and Paau, 1997). Although much research has gone into identifying and characterizing the potential microbial agent, little has been done on these aspects. According to Glass (1997), several obstacles must be overcome to achieve the successful commercialization of these new generation products. First, microbial products are composed of living organisms; therefore, they must be produced, formulated, and sold in ways such that their viability and biological activity are maintained. Second, microbial products must compete in the marketplace with a huge number of synthetic chemicals, which are better known to the end users (e.g., farmers). Finally, microbial products suffer a bad reputation based on perceived deficiencies with some earlier biological products. Like chemical fertilizer, PGPR should not be expected to make up for every shortcoming of the soils and crops. Several conditions can limit the effectiveness of PGPR. Weather constraint is a major problem for PGPR to work, especially on dry marginal soils. People have little control over air, light, temperature, and moisture. Presently a number of inoculants products are available in the marketplace. Manufacturing and application methods are different for every product. Therefore the effectiveness of these products is difficult to evaluate on a standard scale. These are often associated with overambitious claims made in advertisements and promotions. Also there is no standard efficacy method to evaluate such microbial products. There needs to be concerted efforts and collaborations between government and university researchers, and industry, to enhance the process of discovery and ultimately result in product development to explore the enormous opportunity PGPR has created.

Another limitation is the regulatory aspect of these PGPR either as biofertilizers or biopesticides to be commercialized. Regulatory bodies are well conversant about registering a conventional fertilizers or chemical pesticides, but when they are dealing with microbial products, either biofertilizers or biopesticides, the safety assessments and efficacy requirements may call for totally new approaches. Genetic modifications present advantages related to efficacy, flexibility, and safety, but public perception concerning the risks must be tackled. Practitioners must be ready to promote biofertilizing and biocontrol to farming communities and also deal with their concerns to ensure public support and suitable regulatory process.

FUTURE PERSPECTIVES

This chapter has shown that there is huge potential for the use of PGPR as biofertilizing agents for a wide variety of crop plants in a wide range of climatic and edaphic conditions. It is interesting that so much research in this area has focused on associative diazotrophs when, aside from the *Gluconacetobacter diazotrophicus*-sugarcane symbiosis, there is little evidence for consistent, significant levels of fixed-N supply by rhizospheric or endophytic bacteria to their host plants. As stated earlier, the focus on diazotrophs may be reflective of the selection criteria that researchers use in looking for a biofertilizing PGPR, or indicative that there is something about the nature of diazotrophic bacteria which make them well suited for life in the rhizosphere (e.g., lower levels of mineral N). It is quite likely that the use of inoculants will become routine technology in the future to increase crop production, cure problems with nutrient uptake, and control plant pathogens. But much work is needed urgently to demonstrate that the mass production of inoculants, other than rhizobia, is technologically and economically viable.

The vast majority of studies that investigate the mode of action of biofertilizing PGPR indicate bacterial influences on the plant-growth-regulating substance (e.g., phytohormone production, production of enzymes which decrease phytohormone production by the host, or induction of the host to produce signally substances to other symbionts [i.e., flavonoids]). It is interesting that so many of the beneficial effects of biofertilizing PGPR come from influences on plant growth regulation. Essentially, what is happening in these cases is simply a modification of plant processes, in other words, changes in root architecture and assimilate partitioning patterns. If bacterial actions can result in improved root morphology for nutrient uptake, this begs the question: Why do plants have less than ideal root morphology for nutrient uptake in the first place? No doubt what we are observing is a modification of plant processes for better plant performance as we as humans see fit (i.e., increased growth and yield), but not which the plant might associate with increased fitness. If this is the case, then we must be aware that modification of plant processes to improve plant growth may increase the risk of adverse effects for the plant (e.g., do longer, more slender roots make the plant more susceptible to the infection by certain pathogens?).

Apart from PGPR as biofertilizers, PGPR have a tremendous potential to be successful as biopesticides in the marketplace through more basic and applied research. In the year 1990, from the world agrochemicals estimated sales of $24 billion, world biopesticide sales were estimated to be $120 million, which was less than 0.5 percent of the world market for agrochemicals.

Moreover, most of the biopesticides sales were associated with a single bacterial species of *Bacillus thuringiensis* (Bt). As researchers are showing that this scenario is certain to be changed as we start to exploit more and diversified microbial products as inoculants or active ingredients. Due to rising environmental awareness, the world agrochemical market is static or even shrinking, and the biopesticide sales are estimated to be increasing at 10 to 25 percent per annum (Rodgers, 1993). This boosting energy can be seen in a number of biotechnology companies as well as major agrochemical companies investing in their R&D resources for the commercialization and development of biopesticides.

In many places in the world, fungicide-treated seeds are commonly used as an important element to control plant diseases. These fungicides/insecticides (e.g., Vitavax RS Flowable, Helix, Gaucho, etc.) are formulated as a suspension and used as seed treatment to control seed decay, preemergence damping off, soilborne blackleg, and insect-flea beetles. It is expected that the microbial cultures (e.g., bacteria) might not survive with these pesticides at the recommended doses due to their high toxicity toward living organisms (Yesmin and Banerjee, 2001). Nevertheless, chance of survivability of these microbial agents may be enhanced if applied with reduced rates, although the reduced rate has to be adequate enough to serve its original purpose for what is intended. As environmental concerns about groundwater quality and pesticide exposure in foods grow, biological alternatives are becoming necessary (Walter and Paau, 1997). Thus, developing biological treatments compatible to pesticides or even reducing the amount of these toxic chemicals could be a real boon to the agricultural industry.

In the African continent, there are very few places where fertilizer factories exist, and the average farmer scarcely uses fertilizer nitrogen for crop production. Therefore, agronomists all over the world should be encouraged to develop in the area of organic fertilization of soil, especially biofertilizers. Using crop protection chemicals is also cost prohibitive for the marginal farmers in the continent. The PGPR or PGPR products may be well suited to this formidable task. They offer advantages in terms of low cost of production as well as environmental friendliness. The authors anticipate that PGPR or PGPR-based products may increasingly fill the gaps in the umbrella of synthetic fertilizers or crop protection chemicals in many parts of the world including the developing continents. However, before we see this advance in the PGPR technology to its full extent, our research efforts need to address the following challenges, as described in an online article by Parr et al. (1994):

1. Ensuring a high level of consistency in performance and benefit of PGPR
2. Evaluating the effects of pure culture versus mixed culture inocula
3. Determining the modes of action of inocula on the indigenous soil microflora and on plant growth, yield, quality, and protection
4. Ensuring quality control for producing inocula with respect to inoculum density and activity
5. Evaluating the practical feasibility of using microbial inoculants to enhance and improve the species and genetic diversity of marginal soils
6. Detecting and monitoring the survival and dispersal of PGPR inocula in treated soils

CONCLUSIONS

Aside from the century-long use of rhizobial inoculants on legumes, there has been little extensive commercialization of biofertilizing PGPR. An exception to this is the *Azospirillum* inoculant that is available for a variety of crops in Europe and Africa. No doubt the lack of consistent responses in different host crops and different field sites are reasons limiting the more widespread commercialization of biofertilizing PGPR. Likewise, we will continue to face this limitation until our understanding of rhizosphere ecology and the interactions between the rhizospheric organisms improve. In fact, this knowledge will have to increase before we can be assured that biofertilizing PGPR in inoculants will successfully colonize host rhizospheres and consistently promote the growth of host plants. On the other hand, PGPR as biopesticides have huge potential as environmentally friendly alternatives to chemical pesticides, as in many instances various pathogens and weed populations are developing resistance to conventional pesticides. The PGPR products through consortia or association of different microbes could be an effective and more approachable alternative to chemical pesticides, as several mechanisms could interact to affect a broad spectrum of pathogens and crops. Although these beneficial microorganisms live naturally on the roots and in soils, they must not only survive but also outcompete the negative and neutral microorganisms to be successful as inoculants. Biotechnology research using the tools of PGPR technology can help them do their jobs better. Thus, the potential for PGPR technology either as biofertilizers or as biopesticides, or as both, is huge. Nevertheless, the obstacles to conquer for safe and successful utilization of this technology include widening our understanding of how introduced microorganisms, naturally occurring or genetically modified, interrelate and contend with indigenous plant-soil habitats and how this relationship varies across different soil types as well as different agro-climatic conditions.

REFERENCES

Alabouvette C. and Lemanceau P. (1999). Joint action of microbial for disease control. In F.R. Hall and J.J. Menn (eds.), *Biopesticides: Use and delivery* (pp. 117-135). Totowa, NJ: Humana Press.

Alstrom S. (1991). Induction of disease resistance in common bean susceptible to halo blight bacterial pathogen after seed bacterization with rhizosphere pseudomonads. *Journal of General and Applied Microbiology* 37: 495-501.

Alstrom S. and Burns R.G. (1989). Cyanide production by rhizobacteria as a possible mechanism of plant growth inhibition. *Biology and Fertility of Soils* 7: 232-238.

Andrade G., DeLeij F.A., and Lynch J.M. (1998). Plant mediated interactions between *Pseudomonas fluorescens, Rhizobium leguminosarum* and arbuscular mycorrhizae on pea. *Letters in Applied Microbiology* 26: 311-316.

Antoun H., Beauchamp C.J., Goussard N., Chabot R., and Lalande R. (1998). Potential of *Rhizobium* and *Bradyrhizobium* species as plant growth promoting rhizobacteria on non-legumes: Effect on radishes (*Raphanus sativus* L.). *Plant and Soil* 204: 57-67.

Baker R. and Dickman M.B. (1993). Biological control with fungi. In F. Blaine Metting Jr. (ed.), *Soil microbial ecology* (pp. 275-305). New York: Marcel Dekker, Inc.

Baldani J.I., Caruso L., Baldani V.L.D., Goi S.R., and Döbereiner J. (1997). Recent advances in BNF with non-legume plants. *Soil Biology and Biochemistry* 29: 911-922.

Baldani V.L.D., Baldani J.I., and Döbereiner J. (2001). Inoculation of rice plants with the endophytic diazotrophs *Herbaspirillum seropedicae* and *Burkholderia* spp. *Biology and Fertility of Soils* 30: 485-491.

Banerjee M.R. (1995). Sulfur-oxidizing bacteria as a potential canola plant growth promoting rhizobacteria. In D.L. Gustine and H.E. Flores (eds.), *Phytochemicals and health* (pp. 330-332). Rockville, MD: American Society of Plant Physiologists.

Banerjee M.R. and Yesmin L. (2000). Biological seed treatment by sulfur-oxidizing rhizobacteria for potential canola growth promotion. In *Agronomy abstracts, annual meeting, Soil Science Society of America,* November 5-9, Minneapolis, Minnesota, pp. 257.

Barazani O. and Friedman J. (1999). Is IAA the major root growth factor secreted from plant-growth-mediating bacteria? *Journal of Chemical Ecology* 25: 2397-2406.

Barber S.A. and Cushman J.H. (1981). Nitrogen uptake model for agronomic crops. In I.K. Iskander (ed.), *Modeling wastewater renovation-land treatment* (pp. 382-409). New York, Wiley-Interscience.

Bar-Ness E., Chen Y., Hadar H., Marschner H., and Romheld V. (1991). Siderophores of *Pseudomonas putida* as an iron source for dicot and monocot plants. *Plant and Soil* 130: 231-241.

Belimov A.A., Safronova V.I., Sergeyeva T.A., Egorova T.N., Matveyeva V.A., Tsyganov V.E., Borisov A.Y., Tikhonovich I.A., Kluge C., Preisfeld A., et al. (2001). Characterization of plant growth promoting rhizobacteria isolated from polluted soils and containing 1-aminocyclopropane-1-carboxylate deaminase. *Canadian Journal of Microbiology* 47: 642-652.

Bending G.D., Poole E.J., Whipps J.M., and Read D.J. (2002). Characterisation of bacteria from *Pinus sylvestris-Suillus luteus* mycorrhizas and their effects on root-fungus interactions and plant growth. *FEMS Microbiology Ecology* 39: 219-227.

Bent E., Tuzun S., Chanway C.P., and Enebak S. (2001). Alterations in plant growth and in root hormone levels of lodgepole pines inoculated with Rhizobacteria. *Canadian Journal of Microbiology* 47: 793-800.

Bethlenfalvay G.J. (1993). Mycorrhizae in the agricultural plant-soil system. *Symbiosis* 14: 413-425.

Bevivino A., Sarrocco S., Dalmastri, C., Tabacchioni, S., Cantale, C. and Chiarini L. (1998). Characterization of a free-living maize-rhizosphere population of *Burkholderia cepacia:* Effect of seed treatment on disease suppression and growth promotion of maize. *FEMS Microbiolial Ecology* 27: 225-237.

Bitter W., Marugg J.D., de Weger L.A., Thommassen J., and Weisbeek P.J. (1991). The ferric-pseudobactin receptor PupA of *Pseudomonas putida* WCS358: Homology to TonB dependent *Escherichia coli* receptors and specificity of the protein. *Molecular Microbiology* 5: 647-655.

Boyetchko S., Pedersen E., Punja Z., and Reddy M. (1999). Formulations of biopesticides. In F.R. Hall and J.J. Menn (eds.), *Biopesticides: Use and delivery* (pp. 487-508). Totowa, NJ: Humana Press.

Brockwell J. and Bottomley P. (1995). Recent advances in inoculant technology and prospects for the future. *Soil Biology and Biochemistry* 27: 683-697.

Burd G.I., Dixon D.G., and Glick B.R.(1998). A plant growth-promoting bacterium that decreases nickel toxicity in seedlings. *Applied and Environmental Microbiology* 64: 3663-3668.

Burdman S., Volpin H., Kapulnik Y., and Okon Y. (1996). Promotion of nod gene inducers and nodulation in common bean *(Phaseolus vulgaris)* root inoculated with *Azospirillum brasilense* Cd. *Applied and Environmental Microbiology* 62: 3030-3033.

Buysens S., Poppe J., and Hofte M. (1994). Role of siderophores in plant growth stimulation and antagonism by *Pseudomonas aeruginosa* 7NSK2. In M.H. Ryder, P.M. Stevens, and G.D. Bowen (eds.), *Improving plant productivity with rhizobacteria* (pp. 139-141). Adelaide, Australia: CSIRO Division of Soils.

Cattelan A.J., Hartel P.G., and Fuhrmann J.J. (1999). Screening for plant growth-promoting rhizobacteria to promote early soybean growth. *Soil Science Society of America Journal* 63: 1670-1680.

Chabot R., Beauchamp C.J., Kloepper J.W., and Antoun H. (1998). Effect of phosphorus on root colonization and growth promotion of maize by bioluminescent mutants of phosphate-solubilizing *Rhizobium leguminosarum* biovar *phaseoli*. *Soil Biology and Biochemistry* 30: 1615-1618.

Chanway C.P., Hynes R.K., and Nelson L.M. (1989). Plant growth-promoting rhizobacteria: Effects on growth and nitrogen fixation of lentil *(Lens esculenta Moench)* and pea (*Pisum sativum* L.). *Soil Biology and Biochemistry* 21: 511-517.

Chet I., Barak Z., and Oppenheim A. (1993). Genetic engineering of microorganisms for improved biocontrol activity. In I. Chet (ed.), *Biotechnology in plant disease control* (pp. 211-235). New York: John Wiley.

Chiarini L., Bevovino A., Tabacchioni S., and Dalmastri C. (1998). Inoculation of Burkholderia cepacia, Pseudomonas fluorescens and Enterobacter sp. on *Sorghum bicolor:* Root colonization and plant growth promotion of dual strain inocula. *Soil Biology and Biochemistry* 30: 81-87.

Cook R.J., Weller D.M., and Bassett E.N. (1988). Effect of bacterial seed treatments on growth and yield of recropped wheat in western Washington. *Biological and Cultural Tests for Control of Plant Disease* 3: 53.

Copping L.G. and Menn J.J. (2000). Biopesticides: A review of their action, applications and efficacy. *Pest Management Science* 56: 651-676.

Curl E.A. and Truelove B. (1986). *The rhizosphere*. New York: Springer-Verlag.

Dashti N., Zhang F., Hynes R., and Smith D.L. (1997). Application of plant growth-promoting rhizobacteria to soybean [*Glycine max* (L.) Merr.] increases protein and dry matter yield under short-season conditions. *Plant and Soil* 188: 33-41.

Dashti N., Zhang F., Hynes R., and Smith D.L. (1998). Plant growth promoting rhizobacteria accelerate nodulation and increase nitrogen fixation activity by field grown soybean [*Glycine max* (L.) Merr.] under short season conditions. *Plant and Soil* 200: 205-213.

Daza A., Santamaria C., Rodriguez-Navarro D.N., Camacho M., Orive R., and Temprano F. (2000). Perlite as a carrier for bacterial inoculants. *Soil Biology and Biochemistry* 32: 567-572.

de Brito-Alvarez M.A., Gagne S., and Antoun H. (1995). Effect of compost on rhizosphere microflora of the tomato and on the incidence of plant growth-promoting rhizobacteria. *Applied and Environmental Microbiology* 61: 194-199.

Defago G., Berling C., Burger U., Haas D., Kahr G., Keel C., Voisard C, Wirthner P.H., and Wuthrich B. (1990). Suppression of black root rot of tobacco by a *Pseudomonas* strain: Potential applications and mechanisms. In D. Hornby, R.J. Cook, and Y. Henis (eds.), *Biological control of soil-borne plant pathogens* (pp. 93-108). Wallingford, UK: CAB International.

de Freitas J.R., Banerjee M.R., and Germida J.J. (1997). Phosphate-solubilizing rhizobacteria enhance the growth and yield but not phosphorus uptake of canola (*Brassica napus* L.). *Biology and Fertility of Soils* 24: 358-364.

de Salamone I.E.G., Hynes R.K., and Nelson L.M. (2001). Cytokinin production by plant growth promoting rhizobacteria and selected mutants. *Canadian Journal of Microbiology* 47: 404-411.

Dobbelaere S., Croonenborghs A., Thys A., Ptacek D., Vanderleyden J., Dutto P., Labandera-Gonzalez C., Caballero-Mellado J., Aguirre J.F., Kapulnik Y., et al.

(2001). Responses of agronomically important crops to inoculation with *Azospirillum. Australian Journal of Plant Physiology* 28: 871-879.

Dong Z., McCully M.E., and Canny M.J. (1997). Does *Acetobacter diazotrophicus* live and move in the xylem of sugarcane stems? Anatomical and physiological data. *Annals of Botany* 80: 147-158.

Duquenne P., Chenu C., Richard G., and Catroux G. (1999). Effect of carbon source supply and its location on competition between inoculated and established bacterial strains in sterile soil microcosm. *FEMS Microbiology Ecology* 29: 331-339.

Elliott L.F., Gilmour C.M., Lynch J.M., and Tittemore D. (1984). Bacterial colonization of plant roots. In R.L. Todd and J.E. Giddens (eds.), *Microbial-plant interactions,* ASA Special Publication Number 47 (pp. 1-16). Soil Science Society of America—American Society of Agronomy—Crop Science Society of America.

Elo S., Maunuksela L., Salkinoja-Salonen M., Smolander A., and Haahtela K. (2000). Humus bacteria of Norway spruce stands: Plant growth promoting properties and birch, red fescue and alder colonizing capacity. *FEMS Microbiology Ecology* 31: 143-152.

Foldes T., Banhegyi I., Herpai Z., Varga L., and Szigeti J. (2000). Isolation of *Bacillus* strains from the rhizosphere of cereals and in vitro screening for antagonism against phytopathogenic, food-borne pathogenic and spoilage micro-oragnisms. *Journal of Applied Microbiology* 85: 840-846.

Founoune H., Duponnois R., Meyer J.M., Thioulouse J., Masse D., Chotte J.L., and Neyra M. (2002) Interactions between ectomycorrhizal symbiosis and on *Acacia holosericea:* Isolation of mycorrhiza helper bacteria (MHB) from a Soudano-Sahelian soil. *FEMS Microbiology Ecology* 41: 37-46.

Franke I.H, Fegan M., Hayward C., Leonard G., and Sly L.I. (2000). Molecular detection of *Gluconacetobacter sacchari* associated with the pink sugarcane mealybug *Saccharicoccus sacchari* (Cockerell) and the sugarcane leaf sheath microenvironment by FISH and PCR. *FEMS Microbiology Ecology* 31: 61-71.

Fred E.B., Baldwin I.L., and McCoy E. (1932). *Root nodule bacteria and leguminous plants.* Madison: University of Wisconsin.

Germida J.J. and Walley F.L. (1996). Plant growth-promoting rhizobacteria alter rooting patterns and arbuscular mycorrhizal fungi colonization of field-grown spring wheat. *Biology and Fertility of Soils* 23:113-120.

Ghosh S., Dickstein R., Knight T.J., and Langston-Unkefer P.J. (2001). Interactions between *Pseudomonas syringae* pv. *tabaci* and two rhizosphere hosts, *Medicago sativa* and *Avena sativa. Journal of Plant Biochemistry and Biotechnology* 10: 91-99.

Glass D.J. (1997). Commercialization of soil microbial technologies. In F.B. Metting Jr. (ed.), *Soil microbial ecology: Applications in agricultural and environmental management* (pp. 595-618). New York: Marcel Dekker, Inc.

Glick B.R. (1995). The enhancement of plant growth by free-living bacteria. *Canadian Journal of Microbiology* 41: 109-117.

Glick B.R., Penrose D.M., and Li J.P. (1998). A model for the lowering of plant ethylene concentrations by plant growth-promoting bacteria. *Journal of Theoretical Biology* 190: 63-68.

Goel A.K., Sindhu S.S., and Dadarwal K.R. (2002). Stimulation of nodulation and plant growth of chickpea (*Cicer arietinum* L.) by *Pseudomonas* spp. antagonistic to fungal pathogens. *Biology and Fertility of Soils* 36: 391-396.

Grayston S.J. and Germida J.J. (1991). Sulfur-oxidizing bacteria as plant growth promoting rhizobacteria for canola. *Canadian Journal of Microbiology,* 37: 521-529.

Grichko V.P. and Glick B.R. (2001). Amelioration of flooding stress by ACC deaminase-containing plant growth-promoting bacteria. *Plant Physiology and Biochemistry* 39: 11-17.

Gualtieri G. and Bisseling T. (2000). The evolution of nodulation. *Plant Molecular Biology* 42: 181-194.

Gutierrez-Manero F.J., Ramos-Solano B., Probanza A., Mehouachi J., Tadeo F.R., and Talon M. (2001). The plant-growth-promoting rhizobacteria *Bacillus pumilus* and *Bacillus licheniformis* produce high amounts of physiologically active gibberellins. *Physiologia Plantarum* 111: 206-211.

Hashem M.A. (2001). Problems and prospects of cyanobacterial biofertilizer for rice cultivation. *Australian Journal of Plant Physiology* 28: 881-888.

Hebber K.P., Atkinson D., Tucker W., and Dart P. J. (1992). Suppression of *Fusarium moniliforme* by corn root-associated *Pseudomonas cepacia. Soil Biology and Biochemistry* 24: 1009-1020.

Homma Y., Sata Z., Hirayama F., Konna K., Shirahama H., and Suzuki T. (1989). Production of antibiotics by *Pseudomonas cepacia* as an agent for biological control of soilborne plant pathogens. *Soil Biology and Biochemistry* 21: 723-728.

Howie W.J. and Suslow T.V. (1991). Role of antibiotic synthesis in the inhibition of *Pythium ultimum* in the cotton spermosphere and rhizosphere by *Pseudomonas fluorescens. Molecular Plant-Microbe Interactions* 4: 393-399.

Hurek T., Handley L.L., Reinhold-Hurek B., and Piche Y. (2002). *Azoarcus* grass endophytes contribute fixed nitrogen to the plant in an unculturable state. *Molecular Plant-Microbe Interactions* 15: 233-242.

Hurek T., Reinhold-Hurek B., van Montagu M., and Kellenberger E. (1994). Root colonization and systemic spreading of *Azoarcus* sp. strain BH72 in grasses. *Journal of Bacteriology* 176: 1913-1923.

Isopi R., Fabbri P., Del-Gallo M., and Puppi G. (1995). Dual inoculation of *Sorghum bicolor* (L.) Moench ssp. *bicolor* with vesicular arbuscular mycorrhizas and *Acetobacter diazotrophicus. Symbiosis* 18: 43-55.

James E.K. and Olivares F.L. (1998). Infection and colonization of sugar cane and other graminaceous plants by endophytic diazotrophs. *Critical Reviews in Plant Science* 17: 77-119.

James E.K., Olivares F.L., de Oliveira A.L.M., dos Reis F.B., da Silva L.G., and Reis V.M. (2001). Further observations on the interaction between sugar cane and *Gluconacetobacter diazotrophicus* under laboratory and greenhouse conditions. *Journal of Experimental Botany* 52: 747-760.

Keel C., Schnider U., Maurhofer M., Voisard C., Laville J., Burger U., Wirthner P., Haas D., and Defago G. (1992). Suppression of root diseases by *Pseudomonas fluorescens* CHAO: Importance of the bacterial secondary metabolite 2,4-diacetylphloroglucinol. *Molecular Plant-Microbe Interactactions* 5: 4-13.

Keel C., Voisard C., Berling C.H., Kahr G., and Defago G. (1989). Iron sufficiency, a prerequisite for suppression of tobacco black root rot by *P. fluorescens* strain CHAO under gnotobiotic conditions. *Phytopathology* 79: 584-589.

Keel C., Wirthner P., Oberansli T., Voisard C., Berger, U., Haas D., and Defago G. (1990). Pseudomonads as antagonists of plant pathogens in the rhizosphere: Role of the antibiotic 2,4-diacetylphloroglucinol in the suppression of black root rot of tobacco. *Symbiosis* 9: 327-341.

Kim K.Y., Jordan D., and McDonald G.A. (1998). Effect of phosphate-solubilizing bacteria and vesicular-arbuscular mycorrhizae on tomato growth and soil microbial activity. *Biology and Fertility of Soils* 26: 79-87.

Kloepper J.W. (1997). Plant growth-promoting rhizobacteria as biological control agents. In F.B. Metting Jr. (eds.), *Soil microbial ecology: Applications in agricultural and environmental management* (pp. 255-274). New York: Marcel Dekker, Inc.

Kloepper J.W., Lifshitz R., and Schroth M.N. (1988). *Pseudomonas* inoculants to benefit plant production. *ISI Atlas of Science, Animal and Plant Sciences* (pp. 60-64). Philadelphia: Institute for Public Information.

Kloepper J.W. and Schroth M.N. (1978). Plant growth-promoting rhizobacteria on radishes. In D.A. Angers (ed.), *Proceedings of the Fourth International Conference on Plant Pathogenic Bacteria,* Volume 2 (pp. 879-882). Station de Pathologie Vegetale et Phytobacteriologie, INRA, Gibert-Clarey, Tiurs. Paris: INRA.

Knight T.J. and Langston-Unkefer P.J. (1988). Enhancement of symbiotic dinitrogen fixation by a toxin-releasing plant pathogen. *Science* 241: 951-954.

Kremer R.J., Begonia M.F.T., Stanley L., and Lanham E.T. (1990). Characterization of rhizobacteria associated with weed seedlings. *Appied and Environmental Microbiology* 56: 1649-1655.

Kropp B.R., Thomas E., Pounder J.I., and Andrson A.J. (1996). Increased emergence of spring wheat after inoculation with Pseudomonas chlororaphis isolate 2E3 under field and laboratoryconditions. *Biology and Fertility of Soils* 23: 200-206.

Kumar V. and Narula N. (1999). Solubilization of inorganic phosphates and growth emergence of wheat as affected by *Azotobacter chroococcum* mutants. *Biology and Fertility of Soils* 28: 301-305.

Leeman M., van Pelt J.A., den Ouden F.M., Heinsbroek M., Bakker P.A.H.M., and Schippers B. (1995). Induction of systemic resistance against fusarium wilt of radish by lipopolysaccharides of *Pseudomonas fluorescens*. *Phytopathology* 85: 1021-1027.

Lemanceau P. and Alabouvette C. (1993). Suppression of fusarium wilts by fluorescent pseudomonads: Mechanisms and applications. *Biocontrol Science and Technology* 3: 219-234.

Lemanceau P., Bakker P.A.H. M., de Kogel W.J., Alabouvette C., and Schippers B. (1992). Effect of pseudobactin 358 production by *Pseudomonas putida* WCS358 on suppression of fusarium wilt of carnations by non pathogenic *Fusarium oxysporum* Fo47. *Appied and Environmental. Microbiology* 58: 2978-2982.

Li D.M. and Alexander M. (1988). Co-inoculation with antibiotic-producing bacteria to increase colonization and nodulation by rhizobia. *Plant and Soil* 108: 211-219.

Loper J.E. (1988). Role of fluorescent siderophore production in biological control of *Pythium ultimum* by a *Pseudomonas fluorescens* strain. *Phytopathology* 78: 166-172.

Ma W., Zalec K., and Glick B.R. (2001). Biological activity and colonization pattern of the bioluminesence-labeled plant growth-promoting bacterium *Kluyvera ascorbata* SUD165/26. *FEMS Microbiological Ecology* 35: 137-144.

Malik K.A., Bilal R., Mehnaz S., Rasul G., Mirza M.S., and Ali S. (1997). Association of nitrogen-fixing plant promoting rhizobacteria (PGPR) with kallar grass and rice. *Plant and Soil* 194: 37-44.

Manthey J.A., Crowley D.E., and Luster D.G. (1994). *Biochemistry of metal micronutrients in the rhizosphere*. Boca Raton, FL: Lewis Publishers.

Marchal K. and Vanderleyden, J. (2000). The "oxygen paradox" of dinitrogen-fixing bacteria. *Biology and Fertility of Soils* 30: 363-373.

Marek-Kozaczuk M. and Skorupska A.(2001). Production of B-group vitamins by plant growth-promoting Pseudomonas fluorescens strain 267 and the importance of vitamins in the colonization and nodulation of red clover. *Biology and Fertility of Soils* 33: 146-151.

Masalha J., Kosegarten H., Elmaci Ö., and Mengel K. (2000). The central role of microbial activity for iron acquisition in maize and sunflower. *Biology and Fertility of Soils* 30: 433-439.

Maurhofer M., Hase C., Meuwley P., Métraux J-P., and Défago G. (1994). Induction of systemic resistance of tobacco to tobacco necrosis virus by the root-colonizing *Pseudomonas fluorescens* strain CHA0: Influence of *gac*A and of pyoverdine production. *Phytopathology* 84: 139-146.

McCully M.E. (2001). Niches for bacterial endophytes in crop plants: A plant biologist's view. *Australian Journal of Plant Physiology* 28: 983-990.

Mehnaz S., Mirza M.S., Haurat J., Bally R., Normand P., Bano A., and Malik K.A. (2001). Isolation and 16S rRNA sequence analysis of the beneficial bacteria from the rhizosphere of rice. *Canadian Journal of Microbiology* 472: 110-117.

Molla A.H., Shamsuddin Z.H., Halimi M.S., Morziah M., and Puteh A.B. (2001). Potential for enhancement of root growth and nodulation of soybean co-inoculated with *Azospirillum* and *Bradyrhizobium* in laboratory systems. *Soil Biology and Biochemistry* 33: 457-463.

Mrkovacki N. and Milic V. (2001). Use of *Azotobacter chroococcum* as potentially useful in agricultural application. *Annals of Microbiology* 51: 145-158.

Nelson E.B., Chao W.-L., Norton J.M., Nash G.T., and Harman G.E. (1986). Attachment of *Enterobacter cloacae* to hyphae of *Pythium ultimum:* Possible role in the biological control of pythium pre-emergence damping-off. *Phytopathology* 76: 327-335.

Nelson E.B. and Maloney A.P. (1992). Molecular approaches for understanding biological control mechanisms in bacteria: studies of the interaction of *Enterobacter cloacae* with *Pythium ultimum*. *Canadian Journal of Plant Pathology* 14: 106-114.

Omar M.N.A., Mahrous N.M., and Hamouda A.M. (1996). Evaluating the efficiency of inoculating some diazotrophs on yield and protein content of 3 wheat cultivars under graded levels of nitrogen fertilization. *Annals of Agricultural Science* 41: 579-590.

Ownley B.H., Weller D.M., and Alldredge J.R. (1991). Relation of soil chemical and physical factors with suppression of take-all by *Pseudomonas fluorescens* 2-79. In C. Keel, B. Koller, and G. Defago (eds.), *Plant growth-promoting rhizobacteria—Progress and prospects.* IOBC/WPRS Bull. 14: 299-301.

Paau A.S. (1988). Formulations useful inapplying beneficial microorganisms to seed. *Tib. Tech.* 6: 276-279.

Pal S.S. (1998). Interactions of an acid tolerant strain of phosphate solubilizing bacteria with a few acid tolerant crops. *Plant and Soil* 198: 169-177.

Pan B., Vessey J.K., and Smith D.L. (2002). Response of field-grown soybean to co-inoculation with the plant growth promoting rhizobacteria *Serratia proteamaculans* or *Serratia liquefaciens,* and *Bradyrhizobium japonicum* pre-incubated with genistein. *European Journal of Agronomy* 17: 143-153.

Park C.S., Paulitz T.C., and Baker R. (1988). Biocontrol of *Fusarium* wilt of cucumber resulting from interactions between *Pseudomonas putida* and non-pathogenic isolates of *Fusarium oxysporum. Phytopathology* 78: 190-194.

Parr J.F., Hornick S.B., and Kaufman D.D. (1994). Use of microbial inoculants and organic fertilizers in agricultural production. Soil Microbial Systems Laboratory, Agricultural Research Service, U.S. Department of Agriculture. Available online at <www.agnet.org/library/article/eb394.htm>.

Patten C.L. and Glick B.R. (2002). Role of *Pseudomonas putida* indoleacetic acid in development of the host plant root system. *Applied and Environmental Microbiology* 68: 3795-3801.

Paulitz T.C. (1991). Effect of *Pseudomonas putida* on the stimulation of *Pythium ultimum* by seed volatiles of pea and soybean. *Phytopathology* 81: 1282-1287.

Pierson L.S. III and Thomashow L.S. (1993). Cloning and heterlogous expression of the phenazine biosynthetic locus from *Pseudomonas aureofaciens* 30-84. *Molecular Plant-Microbe Interactions* 5: 330-339.

Probanza A., Garcia J.A.L., Palomino M.R., Ramos B., and Manero F.J.G. (2002). *Pinus pinea* L. seedling growth and bacterial rhizosphere structure after inoculation with PGPR *Bacillus* (*B. licheniformis* CECT 5106 and *B. pumilus* CECT 5105). *Applied Soil Ecology* 20: 75-84.

Ratti N., Kumar S., Verma H.N., and Gautam S.P. (2001). Improvement in bioavailability of tricalcium phosphate to *Cymbopogon martinii* var. *motia* by rhizobacteria, AMF and *Azospirillum inoculation. Microbiological Research* 156: 145-149.

Raupach G.S., Liu L., Murphy J.F., Tuzun S., and Kloepper J.W. (1996). Induced systemic resistance in cucumber and tomato against cucumber mosaic cucumovirus using plant growth-promoting rhizobacteria (PGPR). *Plant Disease* 80: 891-894.

Raverkar K.P. and Konde B.K. (1988). Effect of *Rhizobium* and *Azospirillum lipoferum* inoculation on the nodulation, yield and nitrogen uptake of peanut cultivars. *Plant and Soil* 106: 249-252.

Richardson A.E. (2001). Prospects for using soil microorganisms to improve the acquisition of phosphorus by plants. *Australian Journal of Plant Physiology* 28: 897-906.

Riggs P.J., Chelius M.K., Iniguez A.L., Kaeppler S.M., and Triplett E.W. (2001). Enhanced maize productivity by inoculation with diazotrophic bacteria. *Australian Journal of Plant Physiology* 28: 829-836.

Rodgers P.B. (1993). Potential of biopesticides in agriculture. *Pesticide Science* 39: 117-129.

Salisbury F.B. (1994). The role of plant hormones. In R.E. Wilkinson (ed.) *Plant-environment interactions* (pp. 39-81). New York: Dekker.

Schippers B., Lugtenberg B.J.J., and Weisbeek P.J. (1987). Plant growth control by fluorescent pseudomonads. In I. Chet (ed.), *Innovative approaches to plant disease control* (pp. 19-39). New York: Wiley.

Schippers B., Scheffer R.J., Lugtenberg B.J.J., and Weisbeek P.J. (1995). Biocoating of seeds with plant growth-promoting rhizobacteria to improve plant establishment. *Outlook on Agriculture* 24: 179-185.

Schroth M.N. and Becker J.O. (1990). Concepts of ecological and physiological activities of rhizobacteria related to biological control and plant growth promotion. In D. Hornby and R.J. Cook (eds.), *Biological control of soil-borne plant pathogens* (pp. 389-414). Wallingford, UK: CAB International.

Schultze M. and Kondorosi A. (1998). Regulation of symbiotic root nodule development. *Annual Review of Genetics* 32: 33-57.

Seligy V.L. and Rancourt J.M. (1999). Antibiotic MIC/MBC analysis of Bacillus-based commercial insecticides: Use of bioreduction and DNA-based assays. *Journal of Indian Microbiology and Biotechnology* 22: 565-574.

Shabayev V.P., Smolin, V.Y., and Mudrik V.A. (1996). Nitrogen fixation and CO_2 exchange in soybeans *(Glycine max L.)* inoculated with mixed cultures of different microorganisms. *Biology and Fertility of Soils* 23: 425-430.

Shapira R., Ordentlich A., Chet I., and Oppenheim A. B. (1989). Control of plant diseases by chitinase expressed from cloned DNA in *Escherichia coli*. *Phytopathology* 79: 1246-1249.

Shishido M., Breuil C., and Chanway C.P. (1999). Endophytic colonization of spruce by plant growth-promoting rhizobacteria. *FEMS Microbiology Ecology* 29: 191-196.

Singh C.S. and Subba Rao N.S. (1979). Associative effect of *Azospirillum brasilense* with *Rhizobium japonicum* on nodulation and yield of soybean *(Glycine max)*. *Plant and Soil* 53: 387-392.

Singh S. and Kapoor K.K. (1999). Inoculation with phosphate-solubilizing microorganisms and a vesicular arbuscular mycorrhizal fungus improves dry matter yield and nutrient uptake by wheat grown in a sandy soil. *Biology and Fertility of Soils* 28: 139-144.

Slininger P.J. and Jackson M.A. (1992). Nutritional factors regulating growth and accumulation of phenazine-1-carboxylic acid by *Pseudomonas fluorescens* 2-79. *Applied Microbiology and Biotechnology* 37: 388-392.

Srinath J., Bagyaraj D.J., and Satyanarayana B.N. (2003). Enhanced growth and nutrition of micropropagated *Ficus benjamina* to *Glomus mosseae* co-inoculated with *Trichoderma harzianum* and *Bacillus coagulans*. *World Journal of Microbiology and Biotechnology* 19: 69-72.

Srinivasan M., Petersen D.J., and Holl F.B. (1996). Influence of indoleacetic-acid-producing *Bacillus* isolates on the nodulation of *Phaseolus vulgaris* by *Rhizobium etli* under gnotobiotic conditions. *Canadian Journal of Microbiology* 42: 1006-1014.

Steenhoudt O. and Vanderleyden J. (2000). *Azospirillum,* a free-living nitrogen-fixing bacterium closely associated with grasses: Genetic, biochemical and ecological aspects. *FEMS Microbiology Reviews* 24: 487-506.

Subba Rao N.S. (1981). *Biofertilizers in agriculture and forestry*. Enfield, NH: Science Publishers, Inc.

Taiz L. and Zeiger E. (1998). *Plant physiology*. Sunderland, MA: Sinauer Associates, Inc.

Timmusk S., Nicander B., Granhall U., and Tillberg E. (1999). Cytokinin production by *Paenibacillus polymyxa*. *Soil Biology and Biochemistry* 31: 1847-1852.

Van Driesche R.G. and Bellows T.S. Jr. (1996). *Biological control.* New York: Chapman & Hall, An International Thomson Publishing Company.

Van Peer R., Niemann G.J., and Schippers B. (1991). Induced resistance and phytoalexine accumulation in biological control of fusarium wilt of carnation by *Pseudomonas* sp. strain WCS417r. *Phytopathology* 81: 728-734.

Verma S.C., Ladha J.K., and Tripathi A.K. (2001). Evaluation of plant growth promoting and colonization ability of endophytic diazotrophs from deep water rice. *Journal of Biotechnology* 91: 127-141.

Vessey, J.K. (2003). Plant growth promoting rhizobacteria as biofertilizers. *Plant and Soil* 255: 571-588.

Vessey J.K. and Buss T.J. (2002). *Bacillus cereus* UW85 inoculation effects on growth, nodulation, and N accumulation in grain legumes. Controlled-environment studies. *Canadian Journal of Plant Science* 82: 282-290.

Voisard C., Bull C.T., Keel C., Laville J., Maurhofer M., Schnider U., Défago G., and Haas D. (1994). Biocontrol of root diseases by *Pseudomonas fluorescens* CHA0: Current concepts and experimental approaches. In F. Ògara, D.N. Dowling, and B. Boeston (eds.), *Molecular ecology of rhizosphere microorganisms: Biotechnology and the release of GMOs* (pp. 67-89). Weinheim, Germany: VCH Verlagsgesellschaft mbH.

Voisard C., Keel C., Haas D., and Defago G. (1989). Cyanide production by *Pseudomonas fluorescens* helps suppress black root rot of tobacco under gnotobiotic conditions. *EMBO Journal* 8: 351-358.

von Wiren N., Khodr H., and Hider R.C. (2000). Hydroxylated phytosiderophore species possess an enhanced chelate stability and affinity for iron(III). *Plant Physiology* 124: 1149-1157.

Walley F.L. and Germida J.J. (1997). Response of spring wheat *(Triticum aestivum)* to interactions between *Pseudomonas* species and *Glomus clarum* NT4. *Biology and Fertility of Soils* 24: 365-371.

Walter J.F. and Paau A.S. (1997). Microbial inoculant production and formulation. In F.B Metting Jr. (ed.), *Soil microbial ecology: Applications in agricultural and environmental management* (pp. 579-594.) New York: Marcel Dekker, Inc.

Wei G., Kloepper J.W., and Tuzun S. (1991). Induction of systemic resistance of cucumber to *Colletotrichum orbiculare* by select strains of plant-growth promoting rhizobacteria. *Phytopathology* 81: 1508-1512.

Wei G., Kloepper J.W., and Tuzun S. (1996). Induced systemic resistance to cucumber diseases and increased plant growth by plant growth-promoting rhizobacteria under field conditions. *Phytopathology* 86: 221-224.

Weller D.M. (1988). Biological control of soilborne plant pathogens in the rhizosphere with bacteria. *Annual Review of Plant Pathology* 26: 379-407.

Weller D.M. and Thomashow L.S. (1993). Microbial metabolites with biological activity against plant pathogens. In R.D. Lumsden and J.L. Vaughn (eds.), *Pest management: Biologically based strategies* (pp. 172-180). Washington, DC: American Chemical Society.

Wiwat C., Siwayaprahm P., and Bhumiratna A. (1999). Purification and characterization of chitinase from *Bacillus circulans* No.4.1. *Current Microbiology* 39: 134-140.

Yesmin L. and Banerjee M.R. (2001). Bacterial viability and biological seed treatment of canola. In *Proceedings of Soils & Crops 2001 Workshop,* February 22 & 23, Saskatoon, Saskatchewan, Canada, pp. 274-282.

Zehnder G., Kloepper J.W., Yao C., and Wei G. (1997). Induction of systemic resistance in cucumber against cucumber beetles (Coleoptera: Chrysomelidae) by plant growth-promoting rhizobacteria. *Journal of Economic Entomology* 90: 391-396.

Zhang F., Dashti N., Hynes R.K., and Smith D.L. (1996). Plant growth promoting rhizobacteria and soybean [*Glycine max* (L.) Merr.] nodulation and nitrogen fixation at suboptimal root zone temperatures. *Annals of Botany* 77: 453-459.

Zhang F., Dashti N., Hynes R.K., and Smith D.L. (1997). Plant growth promoting rhizobacteria and soybean [*Glycine max* (L.) Merr.] growth and physiology at suboptimal root zone temperatures. *Annals of Botany* 79: 243-249.

Chapter 7

Sustainable Agriculture and the Rhizobia/Legumes Symbiosis

João Ruy Jardim Freire
Enilson Luiz Saccol de Sá

INTRODUCTION

All living organisms are composed of carbon, hydrogen, oxygen, nitrogen, and the so-called minor elements sulfur, potassium, molybdenum, boron, etc. Carbon, hydrogen, and oxygen are easily available in nature. However, nitrogen, available at high level in the atmosphere (78 percent), is blocked for direct utilization by the living organisms. Only a very small fraction of global nitrogen is available for plant utilization.

Nitrogen is the essential element for deoxyribonucleic acid (DNA) and ribonucleic acid (RNA), the base of the genetic code, for the constitution of amino acids, proteins, enzymes, receptors, catalyzers, and the structural components of all plants and animals (Smil, 1997).

However, plants and animals are not able to synthesize the inorganic nitrogen composts necessary to produce organic composts with nitrogen needed for their metabolism. The nitrogen molecule (N_2), as atmospheric gas, is formed by two atoms of N connected by three links and is highly stable. In the atmosphere, only electric discharges can "break" this molecule.

However, some bacteria, known as diazotrophic bacteria, are able to fix the N_2 and produce NH_3 utilizing nitrogenase enzyme. From NH_3, all nitrogen-containing molecules that can be utilized by superior organisms are formed. All nitrogen-fixing microorganisms are prokaryotes, and the list has been enlarged in recent years due to the growing importance of the nitrogen fixation process to agriculture systems production. Besides those diazotrophic microorganisms associated with crops there are indications that other diazotrophic microbes can play an important role in the planetary nitrogen cycle. However, adequate methodologies are lacking for the evaluation of diazotrophic population dynamics and their ecological importance.

There is great diversity in the diazotrophic group. Some are free-living bacteria in the soil, such as *Beijerinckia, Derxia,* and some species of *Enterobacter* and *Escherichia;* others are plant-associative bacteria such as *Azotobacter, Azotococcus, Azospirillum,* and *Herbaspirillum*. Also, some diazotrophics belonging to the heteromorphic group of the genera *Pseudomonas* have higher versatility in the use of several energy sources compared to other diazotrophic microorganisms. Some diazotrophic species of this group predominate in the rhizosphere of certain plant species (Chen et al., 1994).

The most important group of diazotrophic microorganisms is able to establish symbiosis with plants of the family Leguminosae where they elicit the formation of specialized organs, nodules, in which they reduce atmospheric nitrogen to ammonia. Usually the nodules are formed on the roots of plants but in some plants, for example, *Sesbania,* they are formed on the stem. Bacterial symbiosis has a high economic and ecological importance in agriculture mainly due to its economy in the use of nitrogen fertilizers and its reduction of environmental pollution.

DIVERSITY OF RHIZOBIA

The rhizobial species described so far are very diverse and do not form an evolutionary homogenous clade. They belong to the α-2 subclass of Proteobacteria and are phylogenetically intertwined with nonsymbiotic bacteria. These microorganisms are distributed in the order Rhizobiales (Garrity et al., 2001):

Family—Rhizobiaceae
Genus—*Rhizobium*
Genus—*Sinorhizobium*

Family—Phyllobacteriaceae
Genus—*Allorhizobium*
Genus—*Mesorhizobium*

Family—Bradyrhizobiaceae
Genus—*Bradyrhizobium*

Family—Hyphomicrobiaceae
Genus—*Azorhizobium*

Family—Methylobacteriaceae
Genus—*Methylobacterium*

Some organisms of the order Rhizobiales can produce nodules on the plant stem, such as the genus *Allorhizobium,* which can fix nitrogen in nodules on the stem of *Neptunia natans,* a legume of flooded areas of Senegal (De Lajudie et al., 1998) and *Azorhizobium caulinodans,* which produces stem nodules in *Sesbania rostrata* (Boivin et al., 1997).

Methylobacterium nodulans presents a unique property among rhizobia, i.e., its facultative growth on methanol. These rhizobia were isolated from *Crotalaria* legumes and were assigned to a new species within the *Methylobacterium* genus (Sy et al., 2001).

Symbionts of legumes exhibiting ecological and agronomic potential should be characterized prior to their use in sustainable agriculture and environmental management.

Rhizobium-*Legume Symbiosis*

The plants of the family Leguminosae include more than 13,000 species. However, only about 200 are cultivated for economic exploitation for human food and animal feed, fibers, soil improvement, soil recuperation, gums, and other uses.

The rural economy of developing countries is largely based on cultivation of graminaceous crops such as wheat, maize, oats, barley, and also roots and tubercles. On a smaller scale, leguminous plants are cultivated as subsistence crops, and as cattle pastures. An important exception is soybean. In recent years large areas are also under cultivation with peas, *Cicer,* and field beans *(Phaseolus).* There is, however, a great potential for increasing the use of leguminous plants for human food and utilization in production systems of sustainable agriculture (Freire et al., 1988).

Soybean is the best example of the benefits of symbiotic nitrogen fixation. This crop has made it possible in some countries, such as Brazil and Argentina, to extraordinarily increase the availability of proteins for human consumption and animal feed for poultry and swine.

Intensive agriculture, which is largely based on the use of nitrogen chemical fertilizers, is the opposite of sustainable agriculture, which is based on the repositioning of nitrogen used by plant growth through the supply of massive application of organic residues and such other crop systems as crop rotation, allied crops, intercrops, consortium with pastures, and succession of legumes crops/pastures.

After the oil crisis of the 1970s interest in biological nitrogen fixation increased extraordinarily. This interest was evident in the large number of meetings and congresses sponsored by international organizations, such as Food and Agriculture Organization (FAO), United Nations Educational,

the advantages of the seeding of leguminous cultures in winter before a summer crop, such as maize. When legumes are used as cover crops there is an increase in nitrogen availability for the first crop in succession (immediate effect). The winter crops, such as *Lathyrus sativus, Lupinus angustifolius, Vicia sativa,* and *Pisum arvense,* allowed significant gains in maize summer crops, extract nitrogen from the soil equal to 74 $kg \cdot ha^{-1}$ (Aita et al., 1994). In 5 years of nontillage experimental work, maize yield reached 2,380 $kg \cdot ha^{-1}$ when seeded after *Vicia villosa*. However, higher yields were obtained when, after the winter legume crop, nitrogen fertilizer was applied to the maize crop (Frye et al., 1985).

Besides the immediate effect, the long-term use of legumes as cover crops increases soil nitrogen availability, which is called the residual effect. A long-term experiment carried out over ten years with three crop productions and three tillage systems showed that conservation tillage plus the crop production system of oat and *Vicia sativa* before maize and *Vigna unguiculata* increased maize N uptake by 26 percent and maize yield by 19 percent (Amado et al., 1999).

An alternative nitrogen source, especially for small farms on which maize is manually harvested, is the use of such intercropped summer species such as *Canavalia ensiformis, Cajanus cajan, Stizolobium niveum,* and *Glycine* sp. In comparison with the control treatment under reduced tillage, without cover crops and mineral nitrogen, the legumes increased maize yield in a magnitude that ranged from 12 to 43 percent (281 to 1,031 $kg \cdot ha^{-1}$), equivalent to those promoted by amounts of mineral nitrogen varying from 50 to 86 $kg \cdot ha^{-1}$ (Spagnollo et al., 2002).

Besides the nitrogen supplied to the crops, the use of conservation cropping systems, in substitution for traditional fallow/maize systems used by some farmers in the south of Brazil, was an efficient tool to store soil organic matter and increase agricultural carbon sequestration and, therefore, mitigate the greenhouse effect. The consortium maize plus velvet bean *(Stizolobium cinereum),* in eight years, acted as a sink of 15.5 $Mg \cdot ha^{-1}$ of CO_2, while the treatment of fallow/maize had a net emission of 4.32 $Mg \cdot ha^{-1}$ of CO_2 (Amado et al., 2001) and the net carbon dioxide removal by the crop system oat plus common vetch *(Vicia sativa)*/maize plus cowpea *(Vigna unguiculata)* in 9 years was about 44 $Mg \cdot ha^{-1}$ (Bayer et al., 2000). The higher carbon and nitrogen pools in that crop system were associated with greater recalcitrance of mineral-associated soil organic matter to biological decomposition, resulting from its interaction with variable charge minerals (Bayer et al., 2001).

FUTURE PERSPECTIVES

The world consumption of nitrogen fertilizer has been increasing in recent years. In 1999 alone 88 million tons were utilized. Apart from the consumption of nonrenewable energy sources, environmental pollution from nitrogen fertilizers escaping the root zone is high (Peoples et al., 1994).

Besides the fact that biological nitrogen fixation is an important and integral component of sustainable agricultural systems, the use of legumes offers more flexible management than nitrogen fertilizer because the pool of organic nitrogen also slowly becomes available to nonlegume species. In addition, the use of legumes in rotations offers control of crop diseases and pests (Graham and Vance, 2000).

The use of legumes in sustainable agriculture tends to increase with the adoption of nontillage systems and with more research's data on crop rotation systems and allied cropping using leguminous plants. In this scenario, the inoculation of legumes with rhizobial strains selected for high nitrogen fixation capacity should therefore increase.

New advantages from the use of rhizobial strains in agricultural systems have been reported recently. Rhizobia have been found as endophytic bacteria not only of legumes (Sturz et al., 1997; O'Callaghan et al., 1999) but also of nonlegumes. In Egyptian soils cultivated with rice in rotation with clover, indigenous *Rhizobium leguminosarum* bv. *trifolii* was found invading rice roots, reaching an internal population density of about 1.1×10^6 endophytes per gram fresh weight of rice (Yanni et al., 1997). *Rhizobium etli,* which nodulates the common bean *(Phaseolus vulgaris),* is a natural maize endophyte in the traditional fields in Mexico (Gutiérrez-Zamora and Martinez-Romero, 2001). Nevertheless, the relative contribution of rhizosphere and endophytic rhizobia to plant growth promotion is not well-known.

Another important role is attributed to rhizobia as plant-growth-promoting rhizobacteria (PGPR) (Höflich et al., 1995; Yanni et al., 2001). The beneficial effect of PGPR may be due to phosphate solubilization (Chabot et al., 1996), production of the plant growth regulators indole-3-acetic acid (IAA) and cytokinin (Noel et al., 1996), induction of systemic resistance in potato roots against infection by the cyst nematode (Reitz et al., 2000), and other traits.

Yield increases caused by inoculation of nonlegumes with PGPR rhizobia have been reported in field trials with maize, wheat, and barley (Höflich et al., 1994); rice under field conditions (Yanni et al., 1997), and maize and lettuce under field conditions (Chabot et al., 1996).

Several reports indicated rhizobia as endophytic and plant-growth-promoting rhizobacteria in legumes and nonleguminous plants. However, there is a need for further research to elucidate the rhizobial diversity, the mechanisms of plant interaction, the mechanisms of rhizospheric microbial interactions, contribution to nitrogen fixation, and plant growth promotion. More research efforts should be applied to studies of rhizobial diversity and to investigate its contribution to global nitrogen fixation and to determine if it is possible to produce inoculants with those new rhizobial strains.

CONCLUSIONS

In developed countries the pressure for higher food production in subsidized agriculture has led to the abandonment of the sound principles of sustainable agriculture. The United States, for example, has obtained higher yields of maize and wheat by heavy applications of fertilizers, especially nitrogen. In Europe all crops and grass pastures are also dependent on nitrogen fertilizer. Damage to the environment is apparent. In some developing countries that model has also been followed. However, the principles and systems of sustainable agriculture can be applied to any type of agriculture, extensive or intensive, with reduction in fertilizer use, especially nitrogen, resulting in beneficial effects on the environment.

The increasing interest in sustainable agriculture systems is enhancing the use of legumes inoculated with more efficient strains of rhizobia in agricultural systems such as intercropping, crop rotation, allied crops, and no-tillage systems. Nitrogen fixation is the best-known benefit of legume inoculation with rhizobia. Other benefits can arise from the increased sequestering of carbon dioxide from the atmosphere and its incorporation into soil organic matter, which may be a useful way of reducing the greenhouse effect.

It is important to consider that we are just starting to understand the role of rhizobia in soil, in the rhizosphere, and within the roots of leguminous and nonleguminous plants. It is also necessary to increase our knowledge of rhizobial diversity and the evolution of nitrogen-fixing ability in prokaryotes.

The increase in the human population requires food supplies must be increased; sustainable agriculture systems appear to be the best choice to achieve this goal. Biological nitrogen fixation is an important part of this scenario.

REFERENCES

Aita, C.; Ceretta, C.A.; Thomas, A.L.; Pavinato, A.; and Bayer, C. (1994). Espécies de inverno como fonte de nitrogênio para milho no sistema de cultivo mínimo e feijão em plantio direto. *Revista Brasileira de Ciência do Solo* 18:101-108.

Allison, J.R. (1992). The economics of using legumes in crop systems. In M. Mulongoy, M. Gueye, and D.S.C. Spencer (eds.), *Biological nitrogen fixation and sustainability of tropical agriculture* (pp. 323-332). Chichester, England: John Wiley and Sons.

Amado, T.J.C.; Bayer, C.; Eltz, F.L.F.; and Brum, A.C.R. (2001). Potencial de culturas de cobertura em acumular carbono e nitrogênio no solo no plantio direto e a melhoria da qualidade ambiental. *Revista Brasileira de Ciência do Solo* 25:189-197.

Amado, T.J.C.; Mielniczuk, J.; Fernandes, S.B.V.; and Bayer, C. (1999). Culturas de cobertura, acúmulo de nitrogênio total no solo e produtividade de milho. *Revista Brasileira de Ciência do Solo* 23:679-686.

Andrew, C.S. (1977). Nutritional restraints on legume-symbiosis. In J.M. Vincent, A.S. Whitney, and Bose Whitney (eds.), *Exploiting the legume/Rhizobium symbiosis in tropical agriculture* (pp. 253-274). Honolulu: University of Hawaii.

Barni, N.A.; Gomes, J.E.S.; and Gonçalves, N.A. (1978). Resposta da soja (*Glycine max* (L.) Merril) à adubação nitrogenada no florescimento. *Agronomia Sulriograndense* 14:243-250.

Bayer, C.; Martin-Neto, L.; Mielniczuk, J.; Pillon, C.N.; and Sangoi, L. (2001). Changes in soil organic matter fractions under subtropical no-till cropping systems. *Soil Science Society of America Journal* 65:1473-1478.

Bayer, C.; Mielniczuk, J.; Amado, T.J.C.; Martin-Neto, L.; and Fernandes, S.V. (2000). Organic matter storage in a sandy clay loam Acrisol affected by tillage and cropping systems in southern Brazil. *Soil and Tillage Research* 54:101-109.

Bliss, F.A. (1993). Breeding common bean for improved biological nitrogen fixation. *Plant and Soil* 152:71-79.

Boivin, C.; Ndoye, I.; Lortet, G.; Ndiaye, A.; De Lajudie, P.; and Dreyfus, B. (1997). The *Sesbanis* root symbionts *Sinorhizobium saheli* and *S. teranga* bv. *sesbaniae* can form stem nodules on *Sesbania rostrata*, although they are less adapted to stem nodulation then *Azorhizobium caulinodans. Applied and Environmental Microbiology,* 63:1040-1047.

Brockwell, J. (1980). Experiments with crop and pasture legumes—Principles and practice. In F.J. Bergersen (ed.), *Method for evaluating biological nitrogen fixation* (pp. 417-488). Chichester, England: John Wiley and Sons.

Brockwell, J.; Diatloff, A.; Roughley, R.J.; and Date, R.A. (1982). Selection of rhizobia for inoculants. In J. Vincent (ed.), *Nitrogen fixation in legumes* (pp. 173-191). Sidney: Academic Press.

Castellanos-Ramos, J.; Pe a-Cabriales, J.J.; and Rojas-Martinez, I. (1993). Análisis retrospectivo del uso de inoculantes com cepas elite en frijol (*Phaseolus vulgaris* L.). *Turrialba* 43: 89-99.

Chabot, R.; Antoun, H.; Kloepper, J.W.; and Beauchamp, C.J.(1996). Effect of phosphorus on root colonization and growth promotion of maize by bioluminescent mutants of phosphate-solubilizing *Rhizobium leguminosarum* bv. *phaseoli. Soil Biology and Biochemistry* 30:1615-1618.

Chen, Y.K.; Barraquio, M.W.L.; and Knowles, R. (1994). N-fixing pseudomonads and related soil bacteria. *FEMS Microbiology Reviews* 13:95-117.

Date, R. (1996). Selection of strains for inoculant production. In A.P. Balatti and J.R. Freire (eds.), *Legume inoculants. Selection and characterization of strains. Production, use and management* (pp. 1-14). Buenos Aires: Editorial Kingraf.

De Lajudie, P.; Laurent-Fulele, E.; Willems, A.; Torck, U.; Coopman, R.; Collins, M.D.; Kersters, K.; Dreyfus, B.; and Gillis, M. (1998). *Allorhizobium undicola* gen. nov., sp. nov., nitrogen-fixing bacteria that efficiently nodulate *Neptunia natans* in Senegal. *International Journal of Systematic Bacteriology* 48:1277-1290.

Domit, L.A.; Costa, J.A.; Vidor, C.; and Pereira, J.S. (1990). Inoculação de sementes de cereais de estação fria com *Bradyrhizobium japonicum* e seu efeito na soja cultivada em sucessão. *Revista Brasileira de Ciência do Solo* 14:313-320.

Duhan, J.S. and Dudeja, S.S. (1999). Competitiveness does not correlate with siderophore production in *Rhizobium-Cajanus* cajan symbiosis. *Symbiosis* 26: 79-87.

Fabiano, E.; Gualtieri, G.; Pritsch, C.; Polla, G.; and Arias, A. (1994). Extent of high-affinity iron transport systems in field isolates of rhizobia. *Plant and Soil* 164: 177-185.

Fox, R.L. and Kang, B.T. (1977). Some major fertility problems of tropical soils. In J.M. Vincent, and A.S. Whitney, and J. Bose (eds.), *Exploiting the legume/Rhizobium symbiosis in tropical agriculture* (pp. 183-210). Honolulu: University of Hawaii.

Freire, J.R.J. (1977). Inoculation of soybeans. In J.M. Vincent, A.S. Whitney, and J. Bose (eds.), *Exploiting the legume/Rhizobium symbiosis in tropical agriculture* (pp. 335-379). Honolulu: University of Hawaii.

Freire, J.R.J. (1982). Important limiting factors in soil for the rhizobium-legume symbiosis. In M. Alexander (ed.), *Biological nitrogen fixation* (pp. 51-74). New York: Plenum Publishers.

Freire, J.R.J.; Kolling, J.; and Scholles, D. (1988). Legume nitrogen fixation potential for maximizing agriculture production in developing countries. In S.T. Chang, K.Y. Chan, and N.Y.S. Woo (eds.), *Recent advances in biotechnology and applied biology* (pp. 667-681). Hong Kong: The Chinese University Press.

Freire, J.R.J. and Scholles, D. (1996). The use and management of rhizobia/legume inoculants. In A.P. Balatti and J.R. Freire (eds.), *Legume inoculants: Selection and characterization of strains. Production, use and management* (pp. 131-148). Buenos Aires: Editorial Kingraf.

Frye, W.W.; Smith, W.G.; and Williams, R.J. (1985). Economics of winter crops as a source of nitrogen for no-till corn. *Journal of Soil and Water Conservation* 40: 246-249.

Garrity, G.M.; Winters, M.; and Searles, D.B. (2001). *Taxonomic outline of the procariotic genera—Bergey's manual of systematic bacteriology,* Second edition. New York: Springer-Verlag.

Gibson, A.H. (1980). Methods for legumes in glasshouses and controlled environment cabinets. In F.J. Bergersen (ed.), *Method for evaluating biological nitrogen fixation* (pp. 139-184). Chichester, England: John Wiley and Sons.

Gill, P.R. Jr.; Barton, L.L.; Scoble, M.D.; and Neilands, J.B. (1991). A high affinity iron transport system of *Rhizobium meliloti* may be required for efficient nitrogen fixation in planta. *Plant and Soil* 130:211-217.

Graham, P.H. and Vance, C.P. (2000). Nitrogen fixation in perspective: An overview of research and extension needs. *Field Crops Research* 65:93-106.

Gutiérrez-Zamora, M.L. and Martinez-Romero, E. (2001). Natural endophytic association between *Rhizobium etli* and maize (*Zea mays* L.). *Journal of Biotechnology* 91:117-126.

Hardarson, G.; Bliss, F.A.; Cigales-Rivero, M.R.; Henson, R.A.; Kipe-Nolt, J.A.; Longueri, L.; Manrique, A.; Peña-Cabriales, J.J.; Pereira, P.A.A.; Sanabria, C.A.; and Tsai, S.M. (1993). Genotypic variation in biological nitrogen fixation by common bean. *Plant and Soil* 152:59-70.

Höflich, G.; Wiehe, W.; and Hecht-Buchholz, C. (1995). Rhizosphere colonization of different crops with growth promoting *Pseudomonas* and *Rhizobium* bacteria. *Microbiology Research* 150:139-147.

Höflich, G.; Wiehe, W.; and Kühn, G. (1994). Plant growth stimulation by inoculation with the symbiotic and associative rhizosphere microorganisms. *Experientia* 50:897-905.

Kang, B.T.; Spikens, L.; Wilson, G.F.; and Nangju, D. (1981). Leucaena (*Leucaena leucocephala* (Lam) de Wit) prunings as nitrogen source for maize (*Zea mays* L.). *Fertilizer Research* 2:279-287.

Kang, B.T.; Wilson, G.F.; and Spikens, L. (1981). Alley cropping maize (*Zea mays* L.) and Leucaena (*Leucaena leucocephala* (Lam) de Wit) in southern Nigeria. *Plant and Soil* 63:165-179.

Keyser, H.H. and Munns, D.N. (1979). Tolerance of Rhizobia to acidity, aluminum, and phosphate. *Soil Science Society of America Journal* 43:519-523.

Kipe-Nolt, J.A.; Montealegre, C.; and Tohme, J. (1992). Restriction of nodulation by a broad host range *Rhizobium tropici* strain CIAT 899 in wild accessions of *Phaseolus vulgaris* L. *New Phythologist* 120:489-494.

Kolling, J.; Scholles, D.; and Selbach, P.A. (1988). Peletização e inoculação de alfafa em diferentes doses de calcário. II—Experimento a campo. *Agronomia Sulriograndense* 24:287-295.

Kuan, C.Y. (1982). The importance of legume cover crop establishment for cultivation of rubber *(Hevea brasiliensis)* in Malasia. In P.H. Grahan and S.C. Harris (eds.), *Biological nitrogen fixation technology for tropical agriculture* (pp. 369-378). CIAT, Cali.

Langsdale, G.W.; Clark, R.L.; and Bruce, R.R. (1992). The role of legumes in sustaining soil productivity and controlling soil erosion. In M. Mulongoy, M. Gueye, and D.S.C. Spencer (eds.), *Biological nitrogen fixation and sustainability of tropical agriculture* (pp. 361-375). Chichester, England: John Wiley and Sons.

Montealegre, C.; Graham, P.H.; and Kipe-Nolt, J.A. (1995). Preference in the nodulation of *Phaseolus vulgaris* cultivar RAB 39. *Canadian Journal of Microbiology* 4:992-998.

Munns, D.N. (1986). Acid soil tolerance in legumes and rhizobia. *Advances in Plant Nutrition* 2:63-91.

Neilands, J.B. (1981). Iron absorption and transport in microorganisms. *Annual Review of Nutrition* 1:27-46.

Noel, T.C.; Sheng, C.; Yost, C.K.; Pharis, R.P.; and Hynes, M.F. (1996). *Rhizobium leguminosarum* as a plant growth promoting rhizobacterium: Direct growth promotion of canola and lettuce. *Canadian Journal of Microbiology* 42:279-283.

O'Callaghan, K.J.; Davey, M.R.; and Cocking, E.C. (1999). Xylem colonization of *Sesbania rostrata* by *Azorhizobium caulinodans* ORS571. In E. Martínez and G. Hérnandez (eds.), *Highlights of nitrogen fixation research* (pp. 145-147). New York: Kluwer Academic/Plenum Publishers.

Peña-Cabriales, J.J.; Grageda-Cabrera, O.A.; Kola V.; and Hardarson, G. (1993). Time course and N_2 fixation in common bean (*Phaseolus vulgaris* L.). *Plant and Soil* 152:115-121.

Peoples, M.B.; Mosier, A.R.; and Freney, J.R. (1994). Minimizing gaseous losses on nitrogen. In P.E. Bacon (ed.), *Nitrogen fertilization and the environment* (pp. 562-602). New York: Marcel Dekker.

Reitz, M.; Rudolph, K.; Schröder, J.; Hoffmann-Hergarten, S.; Hallmann, J.; and Sikora, R.A. (2000). Lipopolysacharides of *Rhizobium etli* strain G12 act in potato roots as an inducing agent of systemic resistance to infection by the cyst nematode *Globodera pallida. Applied and Environment Microbiology* 66:3515-3518.

Ronchi, A.L. and Balatti, A. (1996). Production of inoculants. In A.P. Balatti and J.R. Freire (eds.), *Legume inoculants. Selection and characterization of strains. Production, use and management* (pp. 39-54). Buenos Aires: Editorial Kingraf.

Roughley, R.J. and Pulsford, D.J. (1982). Production and control of legume inoculants. In J. Vincent (ed.), *Nitrogen fixation in legumes* (pp. 193-209). Sidney: Academic Press.

Sarrantonio, M. (1992). A protocol for screening legumes as soil-improving crops. In M. Mulongoy, M. Gueye, and D.S.C. Spencer (eds.), *Biological nitrogen fixation and sustainability of tropical agriculture* (pp. 41-47). Chichester: John Wiley and Sons.

Scholles, D.; Kolling, J.; and Selbach, P.A. (1983). Resposta da soja ao nitrogenio mineral na presença e na ausência de calcário e de resteva de trigo. *Revista Brasileira de Ciência do Solo* 7:305-309.

Smil, V. (1997). Global population and the nitrogen cycle. *Scientific American,* July, 76-81.

Somasegaran, P. and Hoben, H.J. (1994). *Handbook for rhizobia.* New York: Springer-Verlag.

Spagnollo, E.; Bayer, C.; Wildner, I.P.; Ernani, P.R.; Albuquerque, J.A.; and Proença, M.M. (2002). Leguminosas estivais intercalares como fonte de nitrogênio para o milho no sul do Brasil. *Revista Brasileira de Ciência do Solo* 26: 417-423.

Sturz, A.W.; Christie, B.R.; Matheson, B.G.; and Nowak, J. (1997). Biodiversity of endophytic bacteria wich colonize red clover nodules, roots, stems and foliage and their influence on host growth. *Biology and Fertility of Soils* 25:13-19.

Sy, A.; Giraud, E.; Jourand, P.; Garcia, N.; Willems, A.; Lajudie, P.; Prin, Y.; Neyra, M.; Gillis, M.; Boivin-Masson, C.; Dreyfus, B.; and de-Lajudie, P.

(2001). Methylotrophic *Methylobacterium* bacteria nodulate and fix nitrogen in symbiosis with legumes. *Journal of Bacteriology* 183(1):214-220.

Triplett, E.W. and Sadowsky, M.J. (1992). Genetic of competition for nodulation of legumes. *Annual Review of Microbiology* 46:399-428.

Vargas, A.T. and Graham, P.H. (1989). Cultivar and pH effects on competition for nodule sites between isolates of *Rhizobium* in beans. *Plant and Soil* 117:195-200.

Vincent, J.M. (1970). *A manual for the practical study of the root-nodule bacteria.* Oxford: Blackwell Sci. Publ.

Von Uexküll, H.R. and Mutert, E. (1995). Global extent, development and economic impact of acid soils. *Plant and Soil* 171:1-16.

Yanni, Y.G.; Rizk, R.Y.; Corich, V.; Squartini, A.; Ninke, K.; Philip-Hollingsworth, S.; Orgambide, G.; de Bruijn, F.; Stoltzfus, J.; Buckley, D.; et al. (1997). Natural endophytic association between *Rhizobium leguminosarum* bv. *trifolii* and rice roots and assessment of its potential to promote rice growth. *Plant and Soil* 194:99-114.

Yanni, Y.G.; Rizk, R.Y.; Abd El-Fattah, F.K.; Squartini, A.; Corich, V.; Giacomini, A.; de Bruijn, F.; Rademaker, J.; Maya-Flores, J.; Ostrom, P.; et al. (2001). The beneficial plant growth-promoting association of *Rhizobium leguminosarum* bv. *trifolii* with rice roots. *Australian Journal of Plant Physiology* 28:1-26.

Chapter 8

Wild-Legume Rhizobia: Biodiversity and Potential As Biofertilizer

Hamdi Hussein Zahran

INTRODUCTION

Nitrogen fixation is the reduction of atmospheric nitrogen to ammonia. It is the basic phenomenon that makes nitrogen available for the biosphere. Biological N_2 fixation (BNF) is the most important way for N to enter soil. It is an efficient source of fixed N_2, which plays an important role in land remediation. Deficiency in mineral N often limits plant growth, therefore symbiotic relationships have evolved between plants and a variety of microorganisms (Freiberg et al., 1997). Nitrogen is a key input to most crops and pastures that do not belong to the leguminous family. Because nitrogen in its various forms is usually costly, it is an important decision for farmers to provide it to their crops in the most economic way and to choose between chemical fertilizers and legumes to provide the nitrogen (Schilizzi and Pannell, 2001). Similarly, because there are various types of fertilizers and different species of legume crops, the primary question farmers may ask is, Which fertilizers and legumes provide nitrogen in the most economic way? The N_2-fixing *Rhizobium*-legume symbioses are important sources of BNF, and they are more effective as a biofertilizer than other N_2-fixing systems (e.g., heterotrophs), associative bacteria, and actinorhizal plants (e.g., *Frankia-Casuarina* and *Frankia-Atriplex* symbioses) (Abdel-Ghaffar, 1989; Wullstein, 1989; Caucas and Abril, 1996; Sayed et al., 1997; Zahran, 1999). Values estimated for various legume crops and pasture species commonly fall in the range of 200 to 300 kg $N \cdot ha^{-1}$ per year (Peoples et al., 1995). The nodulated wild legumes have potential for N_2 fixation, reforestration, and control soil erosion, and are also useful in providing a vegetational cover (biofertilizer) for degraded lands and arid ecosystems (Ahmad et al., 1984; Jha et al., 1995; Requena et al., 2001).

The present chapter highlights the significance of a new group of root-nodule bacteria (rhizobia from wild legumes) for soil fertility and their potential as biofetilizers, and also reviews the present taxonomic position.

TAXONOMY AND BIODIVERSITY OF WILD-LEGUME RHIZOBIA

The taxonomy of rhizobia is an expanding field because more rhizobia are being isolated and characterized every day, especially from the tropical and Mediterranean zones, where diversity is still poorly documented (Zakhia and De Lajudie, 2001). Progress in the taxonomy of rhizobia is due to an increasing number of available effective techniques in bacterial characterization. Classification of rhizobia has undergone great changes during recent years due to new phylogenetic and polyphasic data leading to the description of new taxa (Young and Haukka, 1996; Perret et al., 2000; Silva et al., 2003). However, only 23 percent of the total number of legume species (that is between 16,500 and 19,500) have been characterized for their microsymbionts so far (Zakhia and De Lajudie, 2001). This predicts that other new groups of legume symbiotic bacteria (genera and species) may emerge in the future.

The rhizobia of wild legumes are diverse group of root-nodule bacteria isolated from various naturally growing trees and herbs on cultivated and noncultivated lands, including legumes native to arid regions of developing countries. The rhizobia of wild legumes are genetically diverse. They are intimately associated with different hosts (and are thus promiscuous), while few are more selective and specific, and have a narrow host range (Tan et al., 1999; Doignon-Bourcier et al., 2000; Perret et al., 2000). These observations emphasize the ability of various wild legumes to host in their nodules bacterial strains belonging to totally different genera and species. The diversity of rhizobial populations that nodulate wild legumes are also known for some rhizobia from crop legumes. Common beans *(Phaseolus vulgaris)* harbor diverse strains of rhizobia belonging to different *Rhizobium* species, e.g., *R. leguminosarum, R. etli,* and *R. tropici* (Van Berkum et al., 1996).

The rhizobia of wild leguminous trees have been characterized and mainly affiliated with certain genera of rhizobia—*Bradyrhizobium, Mesorhizobium, Rhizobium,* and *Sinorhizobium*—after extensive analysis of the phenotypic, genotypic, and molecular characteristics (Zhang et al., 1991; Lindström and Zahran, 1993; De Lajudie et al., 1994; Haukka et al., 1996; Swelim et al., 1997; Zahran, 1997a; Lafay and Burdon, 1998, 2001; McInroy et al., 1999; Tan et al., 1999; Wang et al., 1999; Nick, De Lajudie, et al., 1999; Tighe et al., 2000; Frioni et al., 2001; Zahran, 2001; Odee et al.,

2002; Zahran et al., 2003). The four rhizobia-nodulating *Acacia* species (*A. cyanophylla, A. gummifera, A. horrida,* and *A. raddiana*) were affiliated with *Sinorhizobium* (*S. meliloti* and *S. fredii*) and *R. galegae-Agrobacterium tumefaciens* species, after analyzing both the 16SrRNA gene (rDNA) and the 16S-23SrRNA spacer by PCR with RFLP analysis (Khbaya et al., 1998). *Bradyrhizobium* species were the most abundant, and *R. tropici* dominated among the *Rhizobium* and *Mesorhizobium* isolates of these shrubby legumes. Nick, De Lajudie, et al. (1999) characterized and identified 25 Sudanese and 5 Kenyan strains, isolated from the root nodules of *Acacia senegal* and *Prosopis chilensis,* based on several molecular and genetic methods of analysis. Two strains were classified with genus *Sinorhizobium saheli* and seven with *S. terangae;* the rest did not cluster with any of the established species. The same strains of rhizobia were further characterized using DNA-DNA-blot hybridization and rep-PCR genomic fingerprinting (Nick, Jussila, et al., 1999), which are rapid and dependable genomic screening methods used to classify new rhizobial isolates of unknown taxonomic status. About 50 rhizobial isolates from root nodules of *Mimosa affinis,* a small leguminous plant native to Mexico, were identified as *Rhizobium etli* on the basis of the results of PCR-RFLP and RFLP analyses of small subunit rRNA genes, multilocus enzyme electrophoresis, and DNA-DNA homology (Wang et al., 1999). Strains of *R. etli* bv. *mimosae* formed nitrogen-fixing nodules on another legume tree, *L. leucocephala.* A novel bacterium isolated from root nodules of *L. leucocephala* was described (Wang et al., 2002). This species is distinguished from other *Sinorhizobium* species by DNA-DNA hybridization, 16SrRNA gene restriction fragments and sequence, and some distinctive phenotypic features. It showed 98 percent 16SrRNA gene sequence similarity to some *Sinorhizobium* species. The name *Sinorhizobium morelense* sp. *nov.* was proposed.

About 58 root-nodule isolates from *Lespedeza* spp. (annual or perennial herb or shrub plants in China and the United States) were characterized using numerical taxonomic analysis of phenotypic features, SDS-PAGE analysis of whole cell proteins, DNA-DNA hybridization, 16SrRNA gene sequence analysis, and cross nodulation with selected legume species (Yao et al. 2002). *Lespedeza* spp. were promiscuous hosts for rhizobia; the majority of rhizobial strains were affiliated with three species—*Sinorhizobium saheli, Bradyrhizobium japonicum,* and *B. elkanii*—and a few strains were assigned to a novel species of the genus, *Bradyrhizobium yuanmingense.* The genetic diversity among 95 isolates of *Astragalus adsurgens* was investigated using molecular methods (Gao et al., 2001). These isolates were differentiated by AFLP, REP, ERIC, and BOX-PCR fingerprinting analysis. The results suggested that *A. adsurgens* rhizobia belong to each of the genera *Agrobacterium, Mesorhizobium, Rhizobium,* and *Sinorhizobium.* Forty-eight rhizobial isolates from root nodules of *Indigofera* spp. and *Kummerowia stipulacea* (annual and perennial wild legumes grown in China) were characterized by a polyphasic approach (Wei et al., 2002). The isolates of *Indigofera* were identified as *Rhizobium,* whereas those from *K. stipulacea* were affiliated with *Sinorhizobium.* Considering the distinctive features among these two groups and related species within the genera *Rhizobium* and *Sinorhizobium,* two novel species were proposed (*R. indigoferae* sp. *nov.* and *S. kummerowiae* sp. *nov.*). Rhizobia from the wild herb legumes *Phaseolus, Desmodium,* and *Macroptilium* species in Mexico were identified (Parker, 2002) and rRNA genetic markers were analyzed. The bacteria were related to genus *Bradyrhizobium.*

In general, recent taxonomic studies of rhizobia have led to a total of 6 genera and 28 recognized species, compared to 1 genus and 6 species in 1982 (Zakhia and De Lajudie, 2001). However, the rhizobia of the wild legumes examined so far are currently affiliated with four of these genera. The majority were identified as members of the genera *Rhizobium* and *Sinorhizobium,* but few strains were classified into the genera *Mesorhizobium* and *Bradyrhizobium.* It appears that some legumes may be nodulated by root-nodule bacteria other than rhizobia with its broad meaning. Most of the bacteria that establish symbioses with legumes belong to the α subclass of Proteobacteria, although other species from the β subclass, such as *Burkholderia* and *Ralstonia,* can also nodulate legumes (Chen et al., 2001; Moulin et al., 2001). The bacteria isolated from *Crotalaria* in Senegal were phylogenetically separated from all recognized rhizobia (Sy et al., 2001). They form a new species in the *Methylobacterium* genus, for which the name *M. nodulans* has been proposed. This is a very interesting find which may change the traditional concepts of rhizobiology and broaden the borders of root-nodule bacteria nodulating various forms of legumes.

SYMBIOTIC NITROGEN-FIXING SYSTEMS ESTABLISHED WITH WILD RHIZOBIA

The naturally growing (wild) legumes are represented by some herbs, shrubs, and trees. Leguminous trees are widely distributed; they are valuable for fixing atmospheric nitrogen. For example, *Acacia* and *Prosopis* are widespread in arid regions of Africa and the Middle East (Jenkins et al., 1987; Mitchelsen and Rosendahl, 1990; El-Amin et al., 1999; Räsänen et al., 2001). Plants of the genus *Acacia* are pioneer plants, which play an important role for preservation and fertility of poor and eroded soils in Africa. These legumes produce an extensive, deep root system, in addition to their potential to fix atmospheric nitrogen. A high percentage of the examined leguminous trees has been proved to form effective (N_2-fixing) nodules (Kadiata et al., 1996; Masutha et al., 1997; Tissue et al., 1997). The N_2 fixed (measured as a percentage of the total nitrogen) in some leguminous trees, e.g., *Albizzia, Glyricidia,* and *Leucaena,* ranged between 28 to 72, 44 to 84, and 20 to 74 percent, respectively (Kadiata et al., 1996). Some leguminous trees fix about 43 to 581 kg $N \cdot ha^{-1}$, compared to about 15 to 210 kg $N \cdot ha^{-1}$ for grain legumes (Dakora and Keya, 1997). Leaf pruning of some tree legumes with high N_2-fixing potential is a very important component of sustainability in agroforestry and soil fertility. The pruning of a shrub such as *Sesbania sesban,* intercropped with one hectare of cereals, provided about 448 kg of N, 31 kg of P, 125 kg of K, 114 kg of Ca, and 27.3 kg of Mg (Dakora and Keya, 1997). The foliage of this legume, with its high content of N and other essential elements, is in fact an ideal biofertilizer. Wild (naturally growing) herb legumes form effective nitrogen-fixing systems with some strains of rhizobia (Jha et al., 1995; Zahran, 1998), an association which is significant to soil fertility in arid regions. For example, five nodulated wild herb legumes *(Medicago intertexta, Melilotus indicus, Trifolium resupinatum, Trigonella hamosa,* and *Alhagi murarum),* in cultivated lands in Egypt, formed N_2-fixing symbioses with *Rhizobium* strains that were as effective as those established between the cultivated legume crops and their compatible rhizobia (Zahran, 1998). Therefore, wild herb legumes can be cultivated in reclaimed desert soil under arid conditions (Zahran, 1999; Zahran et al., 1999). Spontaneous forage legumes, for example, *Hedysarum corronarium, Medicago truncatula, Trifolium squarrosum, Scorpiurus vermiculatus,* grown in arid regions of North Africa and the Mediterranean also formed effective nitrogen-fixing symbioses (Ouzzane and Abdel-Guerfi, 2000), as did *Lablab purpureus* (Abdel-Wahab et al., 2002). Therefore, the wild herb legumes appear to play a significant role in soil fertility.

Effective (nodulating) rhizobia from wild legumes may be used as an inoculum for wild and crop legumes. Some rhizobia are more efficient in nitrogen-fixation activity with other hosts than their compatible hosts (Tantawy et al., 2002). Wange (1989) obtained effective symbiosis between *Acacia* rhizobia and each of peanut and cowpea. Rhizobia isolated from *Acacia* and *Prosopis,* grown in the arid regions of Sudan, formed effective symbiosis with legumes, for example, *Phaseolus vulgaris, Vicia faba,* and *Medicago sativa* (Zhang et al., 1991). Rhizobia from the wild tree legumes *(A. nilotica* and *S. sesban)* and the herb legumes *(A. murarum, M. indicus,* and *T. resupinatum)* formed successful symbiosis with some grain-legume crops, for example, *Vicia faba, Vigna sinensis, Pisum sativum,* and *Medicago sativa* (Zahran et al., 1999), and those isolated from the tree *Pericopsis mooniana* formed effective (nitrogen-fixing) symbioses on the original host, as well as on the herb legume siratro *(Macroptilium atropurpureum),* which is a broad spectrum host (Lalani Wijesundara et al., 2000). Inoculation with these isolates improved nodulation and nitrogen fixation of *P. mooniana,* and a 50 percent increase in plant dry matter under a low-level N fertilizer during a 12-month growth period was obtained (Lalani Wijesundara et al., 2000). Sinorhizobia, which had been used as inocula for *Acacia* and *Prosopis* seedlings, increased plant yield significantly compared with uninoculated seedlings (Räsänen et al., 2001). Legume growth may be improved if inoculated by a genetically modified rhizobia; these strains may be able to establish functional symbioses even in the presence of unfavorable environmental conditions. A 10 kb DNA fragment, constructed from a wild-type strain of *Sinorhizobium* (able to nodulate *Acacia* and showing multiple resistance to several antibiotics, 4 percent NaCl, low and high pH, heavy metals such as copper and mercury, and temperatures as high as 43°C), has been transferred to a sensitive strain of *Rhizobium etli* (Defez et al., 2000).

A few wild legumes grow in pure stand, but most grow and serve better in mixed plantings with grass. Some of the shrub (tree) and herb nitrogen-fixing legumes can be intercropped with some economic crops, such as cereals, for improving crop yield. The performance of some trees such as *Leucaena* and *Sesbania,* when intercropped with some annual grasses, e.g., barley, pearl millet, as well as Rhodes-, rye- and Sudan grasses, was successful under Egyptian semiarid conditions (Abbas et al., 2001). Intercropping with *Sesbania,* for example, increased biomass yield of barley by about 180 percent. Similarly, wild legumes were used for intercropping with nonlegumes. Dinitrogen inputs of some herbaceous legumes, for example, *Mucuna pruriens* and *Lablab purpureus,* into tropical cover-crop systems, for example, *Zea mays* and *Imperata cylindrica,* were significant (Ibewiro et al., 2000). The proportion of N_2 derived from fixation in these two legumes at 12 weeks varied from 52 to 90 percent, depending on whether the treatments were N fertilized, inoculated or uninoculated, and on cover-crop systems. Legumes are important in mixed pastures because they contain large amounts of protein and thus enrich a diet of grass when the two are combined. All legumes have the ability to enrich the soil with nitrogen, although this probably does not occur until the plant or part of its roots dies. Mixed pastures of legumes and grass reduces the need for nitrogenous fertilizer. Trannin et al. (2000) found that the main method of belowground N transfer from a legume to the associated crop occurred through decomposing roots rather than via root exudates or direct mycorrhizal hyphae transfer.

BEHAVIOR OF WILD-LEGUME RHIZOBIA UNDER EXTREME CONDITIONS

As the agricultural soils in various habitats of the world becomes more saline, the need for wild plants adapted to salinity will increase. Plants that are able to make productive use of saline land as well as plants that prevent further salinization will be required. The latter may assist in the rehabilitation of salty land. Several environmental conditions (e.g., salinity and drought) inhibit growth and activity of nitrogen-fixing legumes and threaten agriculture in arid and semiarid climates in several countries, including Egypt. Nearly 40 percent of the world's land surface can be categorized as having potential salinity problems (Zahran, 1997b), most areas of which are confined to the tropics and Mediterranean region. Salinity is a widespread problem caused, in part, by an imbalance between rainfall and transpiration in dryland cropping systems. Soil infertility is often due to the low amount of nitrogenous material in soil and the large quantities of salt in soil and water. The introduction of plants capable of surviving under these conditions (salt-tolerant plants) is a recommended practical solution. Salt-tolerant and nitrogen-fixing varieties of legumes exist (Zahran, 1991, 1999). *Medicago sativa* and *Lotus maroccanus* are salt-tolerant perennial legumes, the latter of which has better growth in 120 mM NaCl (Schachtman and Kelman, 1991). Another legume, *T. fragiferum,* has been reported as a promising salt-tolerant perennial species (Rogers and West, 1993; Rumbaugh et al., 1993). Lentil *(Lens culinaris)* formed effective symbioses with a salt-tolerant strain of *Rhizobium* spp. isolated from lentil (Rai and Singh, 2000). The wild herb *Lablab purpureus* exhibited mild resistance to conditions prevailing in arid climates (Abdel-Wahab et al., 2002); nitrogenase activity, leghemoglobin, and protein of nodules were moderately affected by mild levels of salinity and drought. Free-living rhizobia are often more tolerant to salt stress than the symbiotic process itself. Nevertheless, rhizobia vary in their response to salt stress.

Growth of a number of rhizobia was usually reduced by exposure to 100 mM NaCl, while others such as *R. meliloti* (Mohammad et al., 1991; Embalomatis et al., 1994) and rhizobial strains from *Acacia, Prosopis,* and *Leucaena* (Zhang et al., 1991; Zahran et al., 1994; Lal and Khanna, 1995; Kulkarni and Nautiyal, 2000; Lippi et al., 2000) were tolerant to 300 to 850 mM NaCl. *Rhizobium* sp. *(Sesbania)* grew in yeast-extract mannitol broth (YEB) containing 28 percent salt (NaCl wt/vol) for up to 18 h of incubation at 30°C (Kulkarni et al., 2000). This strain survived at 65°C in YEB medium at pH 7.0 for up to 4 min. These results suggest that possessing the trait for high salt tolerance might be of some evolutionary significance for the survival of rhizobia in alkaline soils at high pH and temperature. Variation in stress conditions should be taken into account when selecting salt-tolerant (osmotolerant) strains for inoculation of legumes in arid regions. Osmotolerant rhizobia efficiently use osmoprotecting substances when exposed to high salinity conditions, probably depending on their availability in the environment (Botsford and Lewis, 1990; Smith et al., 1994; Zahran et al., 1997; Welsh, 2000). In the presence of high levels of salt (up to 300 to 400 mM NaCl), the level of the intracellular free glutamate increases in cells of some species of *L. leucocephala* rhizobia (Yap and Lim, 1983) and *R. meliloti* (Le Rudulier and Bernard, 1986). Changing of the surface antigenic lipopolysaccharides (LPS) and polysaccharides by salt stress might impair rhizobium-legume interaction (Zahran 1992a; Zahran et al., 1994). Salt stress may repress the gene of LPS synthesis. Genes (chromosomal region) are involved in LPS biosynthesis of *Sinorhizobium meliloti* (Lagares et al., 2001). LPS mutant of *S. meliloti* resulted in a reduced ability to elicit the formation of plant nodules and altered the infection thread structures that aborted prematurely (Keating et al., 2002). LPS are essential for the development of root nodules of legumes (Bhat and Carlson, 1992; Tao et al., 1992). To select or construct rhizobia strains with superior symbiotic performance in saline environments, the mechanisms of rhizobial tolerance to salinity should be identified. The occurrence of genes required for salt tolerance under free-living conditions are important for the establishment of effective symbioses with the legume host. Therefore, it should be possible to improve symbiotic efficiency by manipulating the expression of bacterial genes required for salt tolerance.

Inoculation of legumes by salt-tolerant strains of rhizobia may enhance their nodulation and nitrogen fixation under salt stress. Salt-tolerant strains of rhizobia isolated from *Acacia redolens,* growing in saline areas of Australia, produced effective nodules on both *A. redolens* and *A. cyclops,* grown in sand at salinity levels up to 80 mM NaCl (Craig et al., 1991). The growth, nodulation, and nitrogen fixation (N content) of *Acacia ampliceps,* inoculated with salt-tolerant *Rhizobium* strains in sand culture, were resistant to salt levels up to 200 mM NaCl (Zou et al., 1995). The rhizobia isolated from *Acacia nilotica* in different agroclimatic zones and tolerant to 850 mM NaCl formed effective nitrogen-fixing nodules on *Acacia* trees grown at 150 mM NaCl (Lal and Khanna, 1995). The reports reviewed previously underscore the fact that some of the highly salt-tolerant rhizobia form effective (nitrogen-fixing) root nodules on some tree legumes in saline soils with moderate salinity (less than 100 mM NaCl). This treatment may improve fertility of saline soils in arid lands. In fact, the salt-tolerant rhizobia isolated from wild legumes have specific characteristics that offer them ecological significance. These associations between wild legumes and rhizobia may be used as biofertilizers in salty soils.

Many perennial legumes grow wild in the Mediterranean Basin, North and South America, and east Africa. Perennials that persist in climates such as that of northwest Egypt are of special interest. Here the climate comprises mild winters, short growing seasons, and hot summers. Few perennial legumes, for example, *Lotus corniculatus,* occur along the Egyptian coasts. *Trifolium fragiferum* has been reported in Egypt, and *Onobrychis ptolemaica* is found in the Sinai Peninsula and was originally collected near Cairo (Kernick, 1978). The key to the survival of perennials is their ability to resist drought in semiarid and arid environments and depend on individual plant longevity and vegetative reproduction. Perennial legumes, for example, leucerne *(M. sativa),* produced roots that have access to water deep in the soil profile (Collino et al., 1997). Drought-tolerant rhizobia had been used as inoculants for legume production in arid and semiarid environments (Zahran, 1999; Athar, 2000). Rhizobia from *Prosopis* (Soushtari and Pepper, 1985) and *Cicer* (Athar, 2000), exhibited greater survival under dry conditions. The nitrogen-fixing legume-*Rhizobium* symbioses that would be selected for arid climates are those which grow rapidly when temperature and moisture conditions are favorable (Johnson et al., 1981). Naturally occurring forage legumes (annuals or perennials) in semiarid or arid climates are nodulated by nitrogen-fixing root nodules (Abdel-Ghaffar, 1989; Zahran, 1998). In arid regions, high soil temperature affects both the free-living and symbiotic life of rhizobia (Zahran, 1999). For most rhizobia, the optimum temperature range for growth in culture is 28 to 31°C, and many are unable to grow at 38°C (Graham, 1992). However, some strains of rhizobia, for example, *R. leguminosarum* bv *phaseoli,* were reported to be heat tolerant (35 to 40°C) and to form effective symbioses with their host legumes (Hungria et al., 1993; Michiels et al., 1994). These associations will be relevant for cultivation in arid climates. Similarly, some strains of rhizobia from the woody legumes *Acacia* and *Prosopis* (Zahran et al., 1994) grew at 40 to 44°C. However, some strains of the rhizobia surviving under salt or heat stress may lose their infectivity, due to plasmid curing or to alterations in cellular polysaccharides necessary for infection (Zahran, 1992a,b, 1999; Zahran et al., 1994). The rhizobia that survive under unfavorable conditions may modify their genetic characteristics, which are reflected in their symbiotic activities. It has recently been reported that an antibiotic-resistant rhizobial strain from *L. leucocephala* was nonnodulating (Wang et al., 2002).

FUTURE PERSPECTIVES

Symbiotic and free-living nitrogen-fixing organisms have the potential to reduce our overall dependence on man-made forms of N fertilizer. There is a considerable economic incentive to explore ways to increase the efficiency of biological nitrogen fixation as a biofertilizer resource. Although usually the microsymbiont will be modified by recombinant or classic genetic manipulation, the host plant can also be manipulated for increased nitrogen fixation capacity. Identification and analysis of host plant genes involved in nodule morphogenesis and functioning has lagged behind that of *Rhizobium* (Fischer, 1994; Morris et al., 1999). New molecular biology and genetic manipulation techniques (tools) are now available that give plant breeders new opportunities to create and manipulate genetic variation to produce improved plants more easily (Morris et al., 1999). The wild herb legumes *Lotus corniculatus* and *L. japonicus* can be easily transformed with *Agrobacterium.* These plants therefore are attractive experimental samples for manipulating root nodule symbiosis. These techniques enable potentially beneficial traits to be introduced into root nodules. Leghemoglobin is a protein that has similarities to mammalian hemoglobin and is believed to facilitate O_2 diffusion to the respiring bacteroids without irreversibly inactivating the O_2-sensitive nitrogenase enzyme complex (De Bruijn et al., 1994). Leghemoglobin genes from several legumes have been cloned, including *Medicago truncatula* (Baker et al., 1988; Gallusci et al., 1991) and *Vicia faba* (Frühling et al., 1997). The promoter region of the leghemoglobin gene from *Sesbania rostrata (Srglb3)* has been characterized in *L. corniculatus* (De Bruijn et al., 1994). Biotechnologies (molecular genetics, genetic engineering, etc.) have been used to improve the symbiotic characteristics and nitrogen fixation on the basis of established knowledge of microbial genetics (Arber, 2000). Rhizobial genes required for symbiotic nitrogen fixation include those involved in *Nod* factor synthesis, nodule development, synthesis of the nitrogen-fixing apparatus, and bacteroid metabolism. McCardell et al. (1999) suggested that either structure or regulatory genes can be altered to enhance and improve symbiotic nitrogen fixation. An approach involved the modification of the expression of the *nifA* gene; some *R. meliloti nifA*-enhanced strains had a 7 to 15 percent increase in biomass compared with the wild-type parent (McCardell et al., 1999). On the other hand, a number of plant genes (nodulin genes whose expression is specifically induced in root tissue as a consequence of the interaction with rhizobia) are also known. Hence, the formation of effective (nitrogen-fixing) root nodules requires a coordinated temporal and spatial expression of both plant and bacterial genes. The root-nodule bacteria from wild legumes are good candidates for future research dealing with the regulation of nitrogen-fixation genes. Wild legumes in arid climates have specific traits that may be transferred to crop legumes to improve their tolerance to unfavorable conditions. Rhizobia that have adapted to drastic environmental conditions in arid climate may harbor specific genes which are expressed under these conditions. Transfer of chromosomal DNA from the salt-tolerant (grow at 30 percent NaCl) *Bacillus* species into a strain of *R. leguminosarum* was successful (El-Saidi and Ali, 1993). This strain became salt tolerant and grew at about 10 to 15 percent NaCl. Plant yield of lentil, cultivated in reclaimed soils, was improved when inoculated with this genetically modified strain of *Rhizobium*.

One of the most significant and interesting future perspectives in research dealing with salt tolerance in leguminous plants is the identification and cloning of some genes regulated by salt stress. Some of these genes had been cloned in the model legume *Arabidopsis thaliana* (Shi et al., 2000; Shi and Zhu, 2002; Venema et al., 2002). These genes encode putative Na^+/H^+ antiporters. In saline environments, plants accumulate Na^+ in vacuoles through the activity of tonoplast Na^+/H^+ antiporters. The first gene for a putative plant vacuolar Na^+/H^+ antiporter, AtNHX1, was isolated from *Arabidopsis* and shown to increase plant tolerance to NaCl (Venema et al., 2002). Salt stress and abscisic acid regulated the expression of the vacuolar Na^+/H^+ antiporter gene AtNHX1 in *A. thaliana* (Shi and Zhu, 2002). Na^+ efflux by cells is mediated by a plasma-membrane-type Na^+/H^+ antiporter in *A. thaliana* (Shi et al., 2000). Also, it has been demonstrated (Venema et al., 2002) that the purified and reconstituted AtNHX1 (the vacuolar Na^+/H^+ antiporter of *A. thaliana*) not only transport Na^+ but also K^+ with similar affinity. They also reported a role for intracellular antiporters in organelle pH control and osmoregulation. Na^+ toxicity in the plant cells cytosol can be prevented by compartmentation of the ion, an operation depending on tonoplast Na^+ transporters responsible for Na^+ influx into the vacuoles. The search for genes coding for salt-tolerant wild legume Na^+/H^+ antiporters, which have not been identified, is the matter for future research to take on to understand the mechanism of salt tolerance in legumes. Some species of rhizobia are very versatile, that is, very flexible with respect to the strain of *Rhizobium* that is suitable. Other species are highly specific in their requirements. If nodules do not occur, an appropriate strain of *Rhizobium* inoculum is frequently added to the seed, sometimes as a fine cap or pelleting. However, for successful symbiosis and nitrogen fixation, the conditions of rhizobial inoculants (e.g., peat based) must be optimized. Changes in the physiological and agricultural characteristics of peat-based rhizobium inoculants after long-term storage was reported (Revellin et al., 2000). The manufacture of rhizobial inoculants (carriers) is a new biotechnology which needs further research work to improve the conditions of storage. Rhizobial inoculants are of prime importance in reclaimed desert soil where wild legumes are cultivated. Inoculations of legumes are a common practice in these habitats. Requena et al. (2001) reported that the introduction of target indigenous species associated with a managed community of microsymbionts is a successful biotechnological tool to aid the recovery of desertified ecosystems.

CONCLUSIONS

The root-nodule bacteria (rhizobia) from wild legumes are multipurpose bacteria with very interesting characteristics. They have a wide host range, a characteristic that offers these legumes an ecological advantages. These rhizobia can form nodules with other wild or crop legumes, and also can be a source for genetic information to improve symbiotic characteristics of other rhizobia. The taxonomy of these bacteria were recently studied using different modern molecular and genetic tools. Several strains were related to or classified into the existing species and genera of rhizobia; other strains could not be identified, being different in characteristics from the existing ones. Some root-nodule bacteria were related to new genera, for example, *Burkholderia, Methylobacterium,* and *Ralstonia*. This finding points out that several new species and genera of root-nodule bacteria may emerge in the future. Some rhizobia from naturally growing legumes in semiarid climates are more tolerant to certain ecological stresses (e.g., salt, severe drought, elevated temperatures) than rhizobia from cultivated legumes and these rhizobia do successfully establish effective symbioses under stress conditions. The significance of rhizobia from wild legumes is their symbiotic nitrogen-fixation activity, which improves soil fertility and plant productivity. The symbiotic bacteria (rhizobia) provide an ideal biofertilizer in terms of soil fertility and avoidance of health hazards.

REFERENCES

Abbas M., Monib M., Rammah A., Fayez M., and Hegazy N.A. (2001). Intercropping of sesbania *(Sesbania sesban)* and leucaena *(Leucaena leucocephala)* with five annual grasses under semiarid conditions as affected by inoculation with specific rhizobia and associative diazotrophs. *Agronomie* 21: 517-525.

Abdel-Ghaffar A.S. (1989). Aspects of microbial activities and nitrogen fixation in Egyptian desert soils. *Arid Soil Research and Rehabilitation* 3: 281-294.

Abdel-Wahab A.M., Shabeb M.S.A., and Younis M.A.M. (2002). Studies on the effect of salinity, drought stress and soil type on nodule activities of *Lablab*

purpureus (L.) sweet (Kashrangeeg). *Journal of Arid Environments* 51: 587-5602.
Abdel-Wahab S.M., El-Bakry A.A., Tantawy H., and El-Batanony N.H. (2002). Phenotypic characterization of some wild rhizobial isolates isolated from some wild legumes in Egypt. *Al-Azhar Journal of Microbiology* 58: 87-102.
Ahmad M.H., Rafique Uddin M., and Mc Laughlin W. (1984). Characterization of indigenous rhizobia from wild legumes. *FEMS Microbiology Letters* 24: 197-203.
Arber W. (2000). Genetic variation: Molecular mechanisms and impact on microbial evolution. *FEMS Microbiology Reviews* 24: 1-7.
Athar M. (2000). Drought tolerance by lentil rhizobia *(Rhizobium leguminosarum)* from arid and smiarid areas of Pakistan. *Letters in Applied Microbiology* 26: 38-42.
Baker D.G., Gallusci P., Lullien V., Khan H., Gherardi M., and Huguet T. (1988). Identification of two groups of leghemoglobin genes in alfalfa *(Medicago sativa)* and a study of their expression during root nodule development. *Plant Molecular Biology* 11: 761-772.
Bhat U.R. and Carlson R.W. (1992). Chemical characterization of pH-dependent structure epitopes of lipopolysaccharide from *Rhizobium leguminosarum* bv *phaseoli*. *Journal of Bacteriology* 174: 2230-2235.
Botsford J.L. and Lewis T.A. (1990). Osmoregulation in *Rhizobium meliloti:* Production of glutamic acid in response to osmotic stress. *Applied and Environmental Microbiology* 56: 488-494.
Caucas V. and Abril A. (1996). *Frankia* sp. infects *Atriplex cordobensis*-cross-inoculation assay and symbiotic efficiency. *Phyton* 59: 103-110.
Chen W.M., Laevens S., Lee T.M., Coenye T., De Vos P., Mergeay M., and Vandamme P. (2001). *Ralstonia taiwanensis* sp. *Nov.* isolated from root nodules of *Mimosa* species and spuitum of a cystic fibrosis patient. *International Journal of Systematic and Evolutionary Microbiology* 51: 1729-1735.
Collino D.J., Lopez A.E., Dardanelli J., Sereno R., and Racca R.W. (1997). Effect of soil water availability on water use strategies and dry matter production by two alfalfa cultivars differing in winter dormancy. *Phyton* 61: 45-53.
Craig G.F, Atkins C.A., and Bell D.T. (1991). Effect of salinity on growth of *Rhizobium* and their infectivity and effectiveness on two species of *Acacia*. *Plant and Soil* 133: 253-262.
Dakora F.D. and Keya O. (1997). Contribution of legume nitrogen fixation to sustainable agriculture in sub-Saharan Africa. *Soil Biology and Biochemistry* 29: 809-817.
De Bruijn F.J., Chen R., Fujimoto S.Y., Pinaev A., Silver D., and Szczyglowski K. (1994). Regulation of nodulin gene expression. *Plant and Soil* 161: 59-68.
Defez R., Senatore B., and Camerini D. (2000). Genetically modified rhizobia as a tool to improve legume growth in semiarid conditions. In *(Abstracts) Mediterranean Conference of Rhizobiology, Workshop on symbiotic nitrogen fixation for Mediterranean areas,* Montpellier, France, July 9-13.
De Lajudie P., Willems A., Pot B., Dewettinck D., Maestrojuan G., Neyra M., Collins M.D., Dreyfus B., Kersters K., and Gillis M. (1994). Polyphasic taxonomy

of rhizobia: Enmendation of the genus *Sinorhizobium* and description of *Sinorhizobium meliloti* comb. nov., *Sinorhizobium saheli* sp. nov., and *Sinorhizobium teranga* sp. nov. *International Journal of Systematic Bacteriology* 44: 715-733.

Doignon-Bourcier F., Willems A., Coopman R., Laguerre G., Gillis M., and De Lajudie P. (2000). Genotypic characterization of *Bradyrhizobium* strains nodulating small Senegalense legumes by 16S-23SrRNA intergeneric gene spacers and amplified fragment length polymorphism fingerprint analyses. *Applied and Environmental Microbiology* 66: 3987-3997.

El-Amin H.B., Mahdi A.A., and Mekki I.I. (1999). The role of symbiotic biological nitrogen fixation in Sudan's agriculture. In *Proceedings of the International Symposium on Biological Nitrogen Fixation and Crop Production,* Cairo, Egypt, May 11-13, pp. 60-72.

El-Saidi M.T. and Ali A.M.M. (1993). Growing different field crops under high salinity levels and utilization of genetically engineered rhizobia and *Azotobacter* salt drought tolerant strains. In H. Leigh and A. Al-Masoom (eds.), *Towards the rational use of high salinity tolerant plants,* Volume 2 (pp. 59-65). Dorcrecht, the Netherlands: Kluwer Academic Publishers.

Embalomatis A., Papacosta D.K., and Katinakis P. (1994). Evaluation of *Rhizobium meliloti* strain isolated from indigenous populations in northern Greece. *Journal of Agriculture and Crop Science* 172: 37-80.

Fischer H.-M. (1994). Genetic regulation of nitrogen fixation in rhizobia. *Microbiology Reviews* 58: 352-386.

Freiberg C., Fellay R., Bairoch A., Broughton W.J., Rosenthal A., and Perret X. (1997). Molecular basis of symbiosis between *Rhizobium* and legumes. *Nature* 387: 394-401.

Frioni L., Rodríguez A., and Meerhoff M. (2001). Differentiation of rhizobia isolated from native legume trees in Uruguay. *Applied Soil Ecology* 16: 275-282.

Frühling M., Roussel H., Gianinazzi-Pearson V., Pühler A., and Perlick A.M. (1997). The *Vicia faba* leghemoglobin gene VfLb29 is induced in root nodules and in roots colonized by the arbuscular mycorrhizal fungus *Glomus fasciculatum. Molecular Plant Microbe Interactions* 10: 124-131.

Gallusci P., Dedieu A., Journet E.P., Huguet T., and Baker D.G. (1991). Synchronous expression of leghemoglobin genes in *Medicago truncatula* during nitrogen-fixing root nodule development and response to exogeneously applied nitrate. *Plant Molecular Biology* 17: 335-349.

Gao J., Terefework Z., Chen W., and Lindström K. (2001). Genetic diversity of rhizobia isolated from *Astragalus adsurgens* growing in different geographical regions of China. *Journal of Bacteriology* 91: 155-168.

Graham P.H. (1992). Stress tolerance in *Rhizobium* and *Bradyrhizobium* and nodulation under adverse soil conditions. *Canadian Journal of Microbiology* 38: 475-484.

Haukka K., Lindström K., and Young P.W. (1996). Diversity of partial 16S rRNA sequences among and within strains of African rhizobia isolated from *Acacia* and *Prosopis. Systematic and Applied Microbiology* 19: 352-359.

Hungria M., Franco A.A., and Sprent J.I. (1993). New sources of high temperature tolerant rhizobia for *Phaseolus vulgaris* L. *Plant and Soil* 149: 103-109.

Ibewiro B., Sanginga N., Vanlauwe B., and Merckx R. (2000). Evaluation of symbiotic dinitrogen inputs of herbaceous legumes into tropical cover-crop systems. *Biology and Fertility of Soils* 32: 234-242.

Jenkins M.B., Virginia R.A., and Jarrel W.M. (1987). Rhizobial ecology of the woody legume mosquite *(Prosopis glandulosa)* in the Sonoran desert. *Applied and Environmental Microbiology* 33: 36-40.

Jha P.K., Nair S., Gopinathan M.C., and Babu C.R. (1995). Suitability of rhizobia-inoculated wild legumes *Argyrolobium flaccidum, Astragalus graveolens, Indigofera gangetica* and *Lespedeza stenocarpa* in providing a vegetational cover in an unreclaimed lime stone quarry. *Plant and Soil* 177: 139-149.

Johnson D.A., Rumbaugh M.D., and Asay K.H. (1981). Plant improvement for semi-arid range lands: Possibilities for drought resistance and nitrogen fixation. *Plant and Soil* 58: 279-303.

Kadiata B.D., Mulongoy K., and Isirimash N.O. (1996). Time course of biological nitrogen fixation, nitrogen absorption and biomass accumulation in three woody legumes. *Biology, Agriculture and Horticulture* 13: 253-266.

Keating D.H., Willits M.G., and Long S.R. (2002). A *Sinorhizobium meliloti* lipo-polysaccharide mutant altered in cell surface sulfation. *Journal of Bacteriology* 184: 6681-6689.

Kernick M.D. (1978). *Ecological management of arid and semiarid rangelands in Africa and the Near and Middle East,* Volume 4. FAO: Rome.

Khbaya B., Neyra M., Normand P., Zerhari K., and Filali-Maltouf A. (1998). Genetic diversity and phylogeny of rhizobia that nodulate *Acacia* spp. in Morocco assessed by analysis of rRNA genes. *Applied and Environmental Microbiology* 64: 4912-4917.

Kulkarrni S. and Nautiyal C.S. (2000). Effect of salt and pH stress on temperature-tolerant *Rhizobium* sp. NBRI330 nodulating *Prosopis juliflora. Current Microbiology* 40: 221-226.

Kulkarrni S., Surange S., and Nautiyal C.S. (2000). Crossing the limit of *Rhizobium* existence in extreme conditions. *Current Microbiology* 40: 402-409.

Lafay B. and Burdon J.J. (1998). Molecular diversity of rhizobia occurring in native shrubby legumes in Southeastern Australia. *Applied and Environmental Microbiology* 64: 3989-3997.

Lafay B. and Burdon J.J. (2001). Small-subunit rRNA genotyping of rhizobia nodulating Australian *Acacia* spp. *Applied and Environmental Microbiology* 67: 396-402.

Lagares A., Hozbor D.F., Niehaus K., Pich Otero A.J.L., Lorenzen J., Arnold W., and Pühler A. (2001). Genetic characterization of a *Sinorhizobium meliloti* chromosomal region involved in lipo-polysaccharide biosynthesis. *Journal of Bacteriology* 183: 1248-1258.

Lal B. and Khanna S. (1995). Selection of salt tolerant *Rhizobium* isolates of *Acacia nilotica. World Journal of Microbiology and Biotechnology* 10: 637-639.

Lalani Wijesundara, T.I., Van Holm, L.H.J., and Kulasooriya, S.A. (2000). Rhizobiology and nitrogen fixation of some tree legumes native to Sri Lanka. *Biology and Fertility of Soils* 30: 535-543.

Le Rudulier D. and Bernard T. (1986). Salt tolerance in *Rhizobium:* A possible role of betaines. *FEMS Microbiology Reviews* 39: 67-72.

Lindström K. and Zahran H.H. (1993). Lipopolysaccharide pattern in SDS-PAGE of rhizobia that nodulate leguminous trees. *FEMS Microbiological Letters* 107: 327-330.

Lippi D., De Paolis M.R., Di Mattia E., Grego S., Pietrosanti T., and Cacciari I. (2000). Effect of salinity on growth and starvation-survival of a tropical *Rhizobium* strain. *Biology and Fertility of Soils* 30: 276-280.

Masutha T.H., Muofhe M.L., and Dakora F.D. (1997). Evaluation of N_2 fixation and agroforestry potential in selected tree legumes for sustainable use in South Africa. *Soil Biology and Biochemistry* 29: 993-998.

McCardell A., Sadowsky M.J., and Cregan P.B. (1999). Genetics and improvement of biological nitrogen fixation. In F. Blaine Metting Jr. (ed.), *Soil microbial ecology, applications in agricultural and environmental management* (pp. 151-175). New York: Marcel Dekker, Inc.

McInroy S.G., Campbell C.D., Haukka K.E., Odee D.W., Sprent J.I., Wang W-J., Young J.P.W., and Sutherland J.M. (1999). Characterization of rhizobia from African acacias and other tropical woody legumes using Biolog and partial 16S rRNA sequencing. *FEMS Microbiology Letters* 170: 111-117.

Michiels J., Verreth C., and Vanderleyden J. (1994). Effects of temperature stress on bean-nodulating *Rhizobium* strains. *Applied and Environmental Microbiology* 60: 1206-1212.

Mitchelsen A. and Rosendahl S. (1990). The effect of VA mycorrhizal fungi, phosphorous and drought stress on the growth of *Acacia nilotica* and *Lacuna leucocephala*. *Plant and Soil* 124: 7-13.

Mohammad R.M., Akhavan-Kharazian M., Campbel W.F., and Rumbaugh M.D. (1991). Identification of salt-and drought-tolerant *Rhizobium meliloti* strains. *Plant and Soil* 134: 271-276.

Morris P., Judith Webb K., Robbins M.P., Skøt L., and Jorgensen B. (1999). Application of biotechnology to *Lotus* breeding. Trefoil: *The Science and Technology of* Lotus. CSSA Special publication no. 28.

Moulin L., Munive A., Dreyfus B., and Boivin-Masson C. (2001). Nodulaion of legumes by members of -subclass of Proteobacteria. *Nature* 411: 948-950.

Nick G., De Lajudie P., Eardly B.D., Suomalainen S., Paulin L., Zhang X., Gillis M., and Lindström K. (1999). *Sinorhizobium arboris* sp. nov. and *Sinorhizobium kostiense* sp. nov., isolated from leguminous trees in Sudan and Kenya. *International Journal of Systematic Bacteriology* 49: 1359-1368.

Nick G., Jussila M., Hoste B., Nemi M., Kaijalainen S., De Lajudie P., Gillis M., De Bruijn F.J., and Lindström K. (1999). Rhizobia isolated from root nodules of tropical leguminous trees characterized using DNA-DNA dot-blot hybridisation and rep-PCR genomic fingerprinting. *Systematic and Applied Microbiology* 22: 287-299.

Odee D., Haukka W.K., McInroy S.G., Sprent J.I., Sutherland J.M., and Young J.P.W. (2002). Genetic and symbiotic characterization of rhizobia isolated from tree and herbaceous legumes grown in soils from ecologically diverse sites in Kenya. *Soil Biology and Biochemistry* 34: 801-811.

Ouzzane A. and Abdel-Guerfi A. (2000). Nodulation in four species of spontaneous forage legumes. In (Abstracts) Mediterranean Conference of Rhizobiology, Workshop on symbiotic nitrogen fixation for Mediterranean areas, Montpellier, France, July 9-13.

Parker M.A. (2002). Bradyrhizobia from wild *Phaseolus, Desmodium,* and *Macroptilium* species in Northern Mexico. *Applied and Environmental Microbiology* 68: 2044-2048.

Peoples M.B., Herridge D.F., and Ladha J.K. (1995). Biological dinitrogen fixation: An efficient source of nitrogen for sustainable agriculture production. *Plant and Soil* 174: 3-28.

Perret X., Staehelin C., and Broughton W.J. (2000). Molecular basis of symbiotic promiscuity. *Microbiology and Molecular Biology Reviews* 64: 180-201.

Rai R. and Singh R.P. (2000). Effect of salt stress on interaction between lentil *(Lens culinaris)* genotypes and *Rhizobium* spp. strains: Symbiotic N_2 fixation in normal and sodic soils. *Biology and Fertility of Soils* 29: 187-190.

Räsänen L.A., Sprent J.I., and Lindström K. (2001). Symbiotic properties of sinorhizobia isolated from *Acacia* and *Prosopis* nodules in Sudan and Senegal. *Plant and Soil* 235: 193-210.

Requena N., Perez-Solis E., Azcon-Aguilar C., Jeffries P., and Barea, J-M. (2001). Management of indigenous plant-microbe symbioses aids restoration of desertified ecosystems. *Applied and Environmental Microbiology* 67: 495-498.

Revellin C., Meunier G., Giraud J.J., Sommer G., Wadoux P., and Catroux G. (2000). Changes in the physiological and agricultural characteristics of peat-based *Bradyrhizobium japonicum* inoculants after long-term storage. *Applied Microbiology and Biotechnology* 54: 206-211.

Rogers M.E. and West D.W. (1993). The effects of root zone salinity and hypoxia on shoot and root growth in *Trifolium* species. *Annals of Botany* (London) 72: 503-509.

Rumbaugh M.D., Pendery B.M., and James D.W. (1993). Variatrion in the salinity tolerance of strawberry clover (*Trifolium fragiferum* L.). *Plant and Soil* 153: 265-267.

Sayed W.F., Wheeler C.T., Zahran H.H., and Shoreit A.A.M. (1997). Effect of temperature and moisture on the survival and symbiotic effectivness of *Frankia* spp. *Biology and Fertility of Soils* 25: 349-353.

Schachtman D.O. and Kelman W.M. (1991). Potential of *Lotus* germplasm for the development of salt, aluminium and manganese tolerant pasture plants. *Australian Journal of Agricultural Research* 42: 139-150.

Schilizzi S. and Pannell D.J. (2001). The economics of nitrogen fixation. *Agronomie* 21, 527-537.

Shi H., Ishitani M., Kim C., and Zhu J.-K. (2000). The *Arabidopsis thaliana* salt tolerance gene SOS1 encodes a putative Na^+/H^+ antiporter. *Proceedings of the National Academy of Sciences USA* 97: 6896-6901.

Shi H. and Zhu J.-K. (2002). Regulation of expression of the vacuolar Na^+/H^+ antiporter gene AtNHX1 by salt stress and abscisic acid. *Plant Molecular Biology* 550: 543-550.

Shoushtari N.H. and Pepper I.L. (1985). Mesquite rhizobia isolated from the Sonoran desert: Competitiveness and survival in soil. *Soil Biology and Biochemistry* 17: 803-806.

Silva C., Vinuesa P., Eguiarte L.E., Martinez-Romero E., and Souza V. (2003). *Rhizobium etli* and *Rhizobium gallicum* nodulate common bean *(Phaseolus vulgaris)* in a traditionally managed milpa plot in Mexico: population genetics and biogeographic implications. *Applied and Environmental Microbiology* 69: 884-893.

Smith L.T., Smith G.M., Desouza M.R., Pocard J.M., LeRudulier D., and Madkour M.A. (1994). Osmoregulation in *Rhizobium meliloti:* Mechanism and control by other environmental signals. *Journal of Experimental Zoology* 268: 162-165.

Swelim D.M., Hashem F.M., Kuykendall L.D., Hegazi N.I., and Abdel-Wahab S.M. (1997). Host specificity and phenotypic diversity of *Rhizobium* strains nodulating *Leucaena, Acacia* and *Sesbania* in Egypt. *Biology and Fertility of Soils* 25: 224-232.

Sy A., Giraud E., Jouran P., Garcia N., Willems A., DeLajudie P., Prin Y., Neyra M., Gillis M., Boivin-Masson C., and Dreyfus B. (2001). Methylotrophic *Methylobacterium* bacteria nodulate and fix nitrogen in symbiosis with legumes. *Journal of Bacteriology* 183: 214-220.

Tan Z.Y., Wang E.T., Peng G.X., Zhu M.E., Martinez-Romero E., and Chen W.X. (1999). Characterization of bacteria isolated from wild legumes in the northwestern regions of China. *International Journal of Systematic Bacteriology* 49: 1457-1469.

Tantawy H, El-Bakry A.A., El-Batanony N.H., and Abdel-Wahab S.M. (2002). Ecological and symbiotic characterization of wild rhizobia nodulating some wild legumes in Egypt. *Al-Azhar Journal of Microbiology* 58 (in press).

Tao H., Brewin N.J., and Noel K.D. (1992). *Rhizobium leguminosarum* CFN42 lipopolysaccharide antigenic changes induced by environmental conditions. *Journal of Bacteriology* 174: 2222-2229.

Tighe S.W., De Lajudie P., Dipietro K., Lindström K., Nick G., and Jarvis B.D.W. (2000). Analysis of cellular fatty acids and phenotypic relationships of *Agrobacterium, Bradyrhizobium, Mesorhizobium, Rhizobium* and *Sinorhizobium* species using the Sherlock microbial identification system. *International Journal of Systematic and Evolutionary Microbiology* 50: 787-801.

Tissue D.T., Magonigal J.P., and Thomas R.B. (1997). Nitrogenase activity and N_2 fixation are stimulated by elevated CO_2 in a tropical N_2 fixing tree. *Oecologia* 109: 28-33.

Trannin W.S., Urquiaga S., Guerra G., Ibijbijen J., and Cadisch G. (2000). Interspecies competition and N transfer in a tropical grass-legume mixture. *Biology and Fertility of Soils* 32: 441-448.

Van Berkum P., Beyene D., and Eardly B.D. (1996). Phylogenetic relationships among *Rhizobium* species nodulating the common bean (*Phaseolus vulgaris* L.). *International Journal of Systematic Bacteriology* 46: 240-244.

Venema H., Quintero F.J., Pardo J.M., and Donaire J.P. (2002). The *Arabidopsis* Na^+/H^+ exchanger ATNHX1 catalyzes low affinity Na^+ and K^+ transport in reconstituted liposomes. *The Journal of Biological Chemistry* 277: 2413-2418.

Wang E.T., Rogel M.A., Garcia-de los Santos A., Martinez-Romero J., Cevallos M.A., and Martinez-Romero E. (1999). *Rhizobium etli* bv. *Mimosae,* a novel biovar isolated from *Mimosa affinis*. *International Journal of Systematic Bacteriology* 49: 1479-1491.

Wang E.T., Tan Z.Y., Willems, A.Y., Fernandez-Lopez M., Reinhold-Hurek B., and Martinez-Romero E. (2002). *Sinorhizobium morelnse* sp. nov., a *Leucaena leucocephala*-associated bacterium that is highly resistant to multiple antibiotics. *International Journal of Systematic and Evolutionary Microbiology* 52: 1687-1693.

Wange S.S. (1989). Response of groundnut (*Arachis hypogaea* L.) to inoculation with *Rhizobium* strains isolated from wild arboreal legumes. *Journal of Applied Microbiology and Biotechnology* 5: 135-141.

Wei G.H., Wang, E.T., Tan Z.Y., Zhu M.E., and Chen W.X. (2002). *Rhizobium indigoferae* sp. nov. and *Sinorhizobium kummerowia* sp. nov., respectively isolated from *Indigofera* spp. and *Kummerowia stipulacea. International Journal of Systematic and Evolutionary Microbiology* 52: 2231-2239.

Welsh D.T. (2000). Ecological significance of compatible solute accumulation by micro-organisms: From single cells to global climate. *FEMS Microbiology Reviews* 24: 263-290.

Wullstein L.H. (1989). Evaluation and significance of associative dinitrogen fixation for arid soil rehabilitation. *Arid Soil Research and Rehabilitation* 3: 259-265.

Yao Z.Y., Kan F.L., Wang E.T., Wei G.H., and Chen W.X. (2002). Characterization of rhizobia that nodulate legume species of the genus *Lespedeza* and description of *Bradyrhizobium yuanmingense* sp. nov. *International Journal of Systematic and Evolutionary Microbiology* 52: 2219-2230.

Yap S.F. and Lim S.T. (1983). Response of *Rhizobium* sp. UMKL 20 to sodium chloride stress. *Archives of Microbiology* 135: 224-228.

Young J.P.W. and Haukka K.E. (1996). Diversity and phylogeny of rhizobia. *New Phytologist* 133: 87-94.

Zahran H.H. (1991). Conditions for successful *Rhizobium*-legume symbiosis in saline environments. *Biology and Fertility of Soils* 12: 73-80.

Zahran H.H. (1992a). Characterization of root-nodule bacteria indigenous in the salt-affected soils of Egypt by lipopolysaccharide, protein and plasmid profiles. *Journal of Basic Microbiology* 32: 279-287.

Zahran H.H. (1992b). DNA-DNA hybridization of some root-nodule bacteria indigenous in the salt-affected soils of Egypt. *Folia Microbiologica* 37: 295-298.

Zahran H.H. (1997a). Chemotaxonomic characterization of some fast-growing rhizobia nodulating leguminous trees. *Folia Microbiologica* 42: 367-380.

Zahran H.H. (1997b). Diversity, adaptation and activity of the bacterial flora in saline environments. *Biology and Fertility of Soils* 25: 211-223.

Zahran H.H. (1998). Structure of root nodules and nitrogen fixation in Egyptian wild herb legumes. *Biologia Plantarum* 41: 575-585.

Zahran H.H. (1999). *Rhizobium*-legume symbiosis and nitrogen fixation under severe conditions and in an arid climate. *Microbiology and Molecular Biology Reviews* 63: 968-989.

Zahran H.H. (2001). Rhizobia from wild legumes: Diversity, taxonomy, ecology, nitrogen fixation and biotechnology. *Journal of Biotechnology* 91: 143-153.

Zahran H.H., Abdel-Fattah M., Ahmad M.S., and Zaki A.Y. (2003). Polyphasic taxonomy of symbiotic rhizobia from wild leguminous plants growing in Egypt. *Folia Microbiologia* 48: 510-520.

Zahran H.H., Ahmad M.S., Abdel-Fattah M., and Zaki A.Y. (1999). Phenotypic characteristics, cross nodulation and nitrogen fixation of root-nodule bacteria isolated from wild leguminous plants in Egypt. In *Proceedings of the International Symposium on Biological Nitrogen Fixation and Crop Production,* Cairo May 11-13, pp. 77-90.

Zahran H.H., Mohammad E.M., Emam M.M. and Ismael S.S. (1997). The chemical composition, structure, and ultrastructure of a halotolerant rhizobia isolated from Egypt. In *Proceedings of the 9th Microbiology Conference,* Cairo, Egypt, March 25-27, pp. 121-148.

Zahran H.H., Rasanen L.A., Karsisto M., and Lindström K. (1994). Alteration of lipopolysaccharide and protein profiles in SDS-PAGE of rhizobia by osmotic and heat stress. *World Journal of Microbiology and Biotechnology* 10: 100-105.

Zakhia F. and De Lajudie P. (2001). Taxonomy of rhizobia. *Agronomie* 21: 569-576.

Zhang X.P., Karsisto M., Harper R., and Lindström K. (1991). Diversity of *Rhizobium* bacteria isolated from the root nodules of leguminous trees. *International Journal Systematic Bacteriology* 41: 104-113.

Zou N., Dort P.J., and Marcar N.E. (1995). Interaction of salinity and rhizobial strains on growth and nitrogen fixation by *Acacia ampliceps*. *Soil Biology and Biochemistry* 27: 409-413.

Chapter 9

Potential of Arbuscular Mycorrhizae in Organic Farming Systems

Tanu
Anil Prakash
Alok Adholeya

INTRODUCTION

The conversion from intensive/conventional agriculture to alternative forms of agriculture, including organic agriculture, shows the change in human attitudes toward the excessive use of chemical fertilizers, pesticides, herbicides, etc. The increased use of synthetic fertilizers in cultivation has helped feed the world's growing population, but their excessive use has also polluted the environment and caused a decline in soil productivity (Higa, 1991; Parr et al., 1994). There is evidence that many of these synthetic chemicals alter biological ecosystems, initiate changes in nontarget organisms, and exert influence on soil microorganisms. Nitrous oxide (N_2O), a greenhouse gas, evolved from soils due to the excessive use of fertilizers and now causes damage to the ozone layer. All these factors have resulted in a change in the attitude toward the conventional system of agriculture and a shift toward exploring alternative means. Traditional practices such as organic farming are now being considered as remedies for the hazardous effects that synthetic fertilizers have had on soil biology (Thampan, 1995; Das and Biswas, 2002). Organic agriculture emphasizes the functional dynamic interactions among soil, plants, animals, humans, ecosystems, and environment. Organic farming means farming in the spirit of organic relationships. Organic agriculture involves the judicious use of resources and ecological processes that are otherwise naturally present in an agriculture ecosystem. It is a production system that avoids or largely excludes the use of synthetically produced fertilizers, pesticides, growth regulators, and livestock feed additives. To the maximum extent feasible, organic farming systems rely on crop rotation, crop residues, animal manures, green manures, off-farm organic waste, and aspects of biological pest control to maintain soil productivity and tilth, to supply plant nutrients, and to control pests. Organic farming also considers the soil a living system that harbors and develops the activities of beneficial organisms. The overall aim of organic agricultural management is to reduce off-farm inputs and minimize waste and environmental degradation, while encouraging resilience to disturbances such as disease outbreaks and adverse climate conditions.

The results of many field trials, including those of the long-term manurial experiments conducted at different centers in India, revealed that the combined use of mineral fertilizers and organic manures is the most appropriate method to achieve a higher crop yield. Such a system also provides stability to farming systems where high intensity cropping models are practiced (Singh and Dwivedi, 1996).

MYCORRHIZAE IN ORGANIC FARMING

The use of beneficial microorganisms in combination with organic manures and inorganic fertilizers in balanced proportions is an upcoming alternative means of agriculture for sustaining plant productivity (Reganold et al., 1990; Parr et al., 1994; Das and Biswas, 2002). A sustainable system can be perceived as one with an ever-positive change in the measured output of a system, which can be sustainable over a period, and one where outputs do not diminish, even when fresh inputs are not available. Liebhart et al. (1989) defined sustainable agriculture as "agriculture that can evolve indefinitely towards greater human utility, greater efficiency of resource use and a balance with the environment that is favorable to human beings and also to most other species." The Consultative Group on International Agricultural Research in 1988 stated, "Sustainable agriculture is the successful management of resources for agriculture to satisfy the changing human needs, while maintaining the quality of environment and conserving the natural resources." A sustainable system uses its inputs most efficiently and judiciously to maximize the productivity and profitability with least adverse impact on ecological balance.

The goals of conventional, production-oriented agriculture continue to be valid, but there are now additional priorities (Cook, 1982). These priorities stem from the realization that production-based healthy soil that can support high-yielding plants is a fragile resource in limited supply which must be protected and conserved to sustain agricultural production (Reganold et al., 1990). In recent years, there has been increasing evidence that the microbial communities of soil and plants have an important role in sustainable agriculture. Among the microorganisms living in the rhizosphere of plants, arbuscular mycorrhizal (AM) fungi have been found to be essential components of sustainable soil-plant systems (Bethlenfalvay and Linderman, 1992; Hooker and Black, 1995; van der Heijden et al., 1998). Mycorrhizal fungi are the most common fungal associations formed in nearly all cultivated plants—agricultural, horticultural, or forestry plant species. AM fungi are emerging as important colonizers not only of roots and the rhizosphere but also of the bulk soil (Barea et al., 1993; Lovato et al., 1995). The association appears to play as great a role in conserving soil as in enhancing plant productivity (Bethlenfalvay and Barea, 1994). They mediate the movement of mineral nutrients from the soil to the plant and of carbon compounds from plant to the soil. Although this exchange is known to benefit plants, only recently have its effects on soil aggregation been recognized and fully appreciated (Tisdall and Oades, 1979; Tisdall, 1991).

The most important mycorrhizal response is one that results in increased efficiency of mineral uptake, especially of poorly mobile ions, and there are numerous reports dealing with the nutritional benefits that plants derive from mycorrhizal associations (Gianinazzi et al., 1990; Barea, 1991; Gaur et al., 1998; Sharma and Adholeya, 2000). In all types of mycorrhizae, fungal hyphae permeate soil and reach beyond the depletion zones developed around the nonmycorrhizal roots (Bolan, 1991; Li et al., 1991).

Thus, mycorrhizal roots explore a larger soil volume and have greater absorptive areas than nonmycorrhizal roots (Bethlenfalvay and Linderman, 1992). Indian soils are usually deficient in phosphorus (P), an element that is required in relatively large concentrations as compared to the other major nutrients. When applied to soils, phosphorus quickly gets fixed and is immobilized (Mosse, 1973), rendering it unavailable to the plant. Mycorrhizal fungi help in its mobilization/solubilization and increase the uptake of phosphorus in plants (Harley and Smith, 1983; Bolan, 1991; Li et al., 1991). The role of organic manures in increasing P bioavailability has been well established. The availability is further increased with AM fungi and other phosphate-solubilizing microorganisms in the presence of slow-release P fertilizers (Das and Biswas, 2002). The inoculation of plants with mycorrhizal fungi during the seedling stage and subsequent transplantation in well-manured fields can certainly substitute for chemical fertilizers, particularly P, to a large extent (Bansal and Mukerji, 1994). This way the production can be sustained without affecting soil productivity and fertility. The conservation and augmentation of mycorrhizal fungi is one of the most important approaches in sustaining plant productivity. Mäder et al. (2000) emphasized the role of AM symbiosis in agroecosystems in which synthetic mineral fertilizers are replaced by organic ones, used at low doses. They concluded that AM fungi (under organic systems) may have substantially contributed to plant nutrition in the biologically cultivated grass-clover system which was highly colonized by AM fungi—it was reduced by only 10 to 12 percent compared to the conventional system, the yield of winter wheat, which was colonized only to a minor extent and was reduced by 19 to 26 percent. The greatest yield decrease (34 to 45 percent) was observed for potatoes, where no colonization by AM fungi was detected. However, large-scale application of AM fungi in plant production is currently limited by a lack of sufficient amount of AM inocula (Miller et al., 1986).

The aim of this chapter is to discuss recent developments concerning the potential of AM fungi in organic agriculture. Given the effects of AM fungi symbiosis on plant growth and health, this symbiosis, if appropriately managed, would permit a satisfactory reduction of chemical fertilizers and pesticides inputs, the key aspects of sustainable plant production. In fact, the potential of AM fungi as biofertilizers and bioprotectors which enhance crop productivity is well recognized, although it has not yet been well exploited (Lovato et al., 1995). To provide an understanding of the concepts and background of the benefits of mycorrhizae, some aspects of cultural and agronomic practices and their overall management in organically managed ecosystems are reviewed here.

FACTORS AFFECTING MYCORRHIZAL INOCULUM POTENTIAL

In soils, AM fungi are found as spores, hyphae, or infected root pieces. All these are sources of inoculum, with the extraradical mycelium as the main source (Sylvia and Jarstfer, 1992). The inoculum potential is a product of the abundance and vigor of the propagules in the soil and can be quantified by determining the rate of colonization of a susceptible host under a set of standard conditions (Moorman and Reeves, 1979; Hayman, 1982; Liu and Luo, 1994). Various methods are in use to quantify the infectious propagules (Franson and Bethlenfalvay, 1989; Plenchette et al., 1989; Sharma et al., 1996; Gaur et al., 1998).

Agricultural systems are disturbed biological systems in which several practices may be detrimental to AM fungi populations (Thompson, 1994). Discussed are some practices that influence the AM fungi inoculum potential.

Crop Rotation and Fallow Systems

Fallowing is an integral part of agricultural systems. Usually, the longer the fallow period the better crops grow, but clean fallowing results in a decline in the AM fungi inoculum and in AM colonization of the next crop (Black and Tinker, 1979; Thompson, 1987, 1990). Selection of crop species is an important management factor. By growing crop plants that provide a high density of AM in the soil, natural AM populations can be effectively managed to achieve results that might be considered the objective of artificial inoculation of AM fungi. Crop species that are nonhosts of AM fungi are equivalent to fallows in their effects on AM populations (Ocampo et al., 1980; Thompson, 1991). Long-term studies were monitored by Thompson (1987) at Queensland, Australia, and found that fallowing resulted in the decline in propagules of AM fungi in the soil and reduced colonization of the crop in the field. Furthermore, he conducted inoculation trials and found that increasing inoculum abundance in the soil overcame the deleterious impact of fallowing. He recommended that farmers avoid planting mycorrhizal-dependent crops, such as linseed, sunflower, and soybean, following periods of fallow or after a nonhost plant such as canola that leads to a reduced number of AM propagules. Continuous monocultures can both decrease populations of AM fungi spores (Rao et al., 1995) and shift the AM fungi species composition of the community toward species which may not be beneficial to the crop (Johnson and Pflegler, 1992).

Variations in AM inoculum levels can also be accompanied by variations in populations of plant pathogens (Thompson, 1994). Reid and Bowen (1979) observed a reduction in AM colonization rate and inoculum density affected due to flooding. The flooding of rice fields affects the inoculum potential throughout the growing season (Nopamornbodi et al., 1988).

Tillage Practices

Soil disturbance through tilling can dramatically affect the functioning of mycorrhizae in a sustainable agricultural system (Douds and Millner, 1999; Brito et al., 2001). Experiments have shown that the efficacy of AM mycelia in the soil as propagules and organs for P uptake decreases with increasing soil disturbance (Jasper et al., 1989; Fairchild and Miller, 1990; McGonigle and Miller, 1996; Goss and Varennes, 2001). Greater AM colonization in maize was recorded early in the growing season in soils under zero/no tillage than in moldboard plowing (McGonigle and Miller, 1993). Disturbance may not affect AM colonization in soils with high spore populations (Jasper et al., 1991). The effect of soil disturbance seems to happen because AM hyphae are killed in the soil (Evans and Miller, 1990).

Miller et al. (1995) documented an interesting case where disturbance of an arable, no-till soil resulted in reduced AM development and subsequently less absorption of P by seedlings of maize in a field. They hypothesized that soil disturbance reduced the effectiveness of mycorrhizal symbiosis and later experimentally concluded that, under nutrient-deficient conditions, the ability of mycorrhizal seedlings to associate with intact hyphal networks in soil may be highly advantageous for crop establishment.

Spore populations may increase with cultivation (Kruckelmann, 1975). Tillage also affects the distribution of spores within the soil. Untilled soil had more spores in the 8 to 15 cm depth (Abbott and Robson, 1991). Anderson et al. (1987) found that after seven years of continuous maize cultivation, untilled soils contained more residues of phosphate fertilizer and the plants showed greater root length with slightly higher AM colonization than did moldboard-plowed soils. Disturbing the macrostructure of long-term, zero-tilled soils reduced AM colonization and P and Zn uptake of the mycorrhizal hosts, maize, and wheat plants, but not of nonmycorrhizal hosts such as spinach and rapeseed (Evans and Miller, 1988). Injection of the fungicide Benomyl decreased AM colonization of maize and P concentration of maize shoots in undisturbed soil (Evans and Miller, 1988).

Chemicals and Fertilizers

Apart from crop rotation and tillage practices, AM symbiosis is influenced by the degree and type of fertilization and plant protection practices (Bethlenfalvay, 1992; Johnson and Pflegler, 1992). Soils under low-input (organic/biological farming) agriculture have a higher rate of AM infection and propagules than do soils under conventional (fertilizer inputs were given) farming (Sattelmacher et al., 1991; Douds et al., 1993, 1995). The authors suggested that these differences were due to high inputs of fertilizers and pesticides in the conventional fields as well as the influence of a different crop rotation. In winter wheat a two- to threefold increase in colonization by AM fungi at a biological farm site was observed as compared with a neighboring conventional farm (Ryan et al., 1994). It was concluded that the amount of soluble P was the main cause for these differences. Replicated field studies comparing AM symbiosis under biological and conventional farming practices has been applied hitherto only to short-term trials. In a pot trial with strawberries, Werner et al. (1990) found an increase in the degree of AM colonization upon transition from conventional to biological management already in the second year after conversion. Similarly, in field experiments in a rotation including potato and wheat, a change from conventional, high-input to a low-input system resulted an increase in AM root colonization (Limonard and Ruissen, 1989).

Mäder et al. (2000) reported that soils from low-input farming systems under wheat-rye-grass rotation had a greatly enhanced capacity to initiate AM symbiosis. The relative differences in this capacity remained similar when propagules of the AM fungus *Glomus mosseae* were experimentally inoculated, although overall root colonization by AM fungi was 2.8 times higher. Among fertilizers, levels of phosphatic fertilizer were found to be critical. Increased rates of P fertilizer application result in decreased AM colonization percentage but increased length of root colonized by AM fungi. Further additions of P fertilizer decreased colonization and root length (Amijee et al., 1989; Bolan, 1991). High rates of application of N fertilizers also inhibited AM fungi (Sylvia and Neal, 1990). However, AM colonization was found to be positively correlated with N and negatively correlated with P in one of the studies conducted by Thompson (1986) in maize and wheat.

There have been several reports on the effect of pesticides on AM fungi. Generally, broad-spectrum biocidal fumigants are very toxic to AM fungi (Menge, 1982; Johnson and Pflegler, 1992). In general, insecticides and nematicides at recommended levels do not inhibit AM fungi (Thompson, 1994). Another aspect of AM-herbicide interactions is the possibility that AM fungi may improve uptake of residual herbicides. Mycorrhizal soybeans suffered more damage than did nonmycorrhizal soybeans in soil with low concentrations of residual atrazine. Hyphae of AM fungi were shown to remove ^{14}C residues of atrazine from soil and transfer them to maize (Nelson and Khan, 1992).

Organic Amendments and AM Fungi

Organic amendments are known to influence AM fungi both positively and negatively. Many authors have reported the impact of organic amendments on mycorrhizal development (Sheikh et al., 1975; Harinikumar and Bagyaraj, 1988, 1989; Calvet et al., 1993; Gaur and Adholeya, 2000; Gryndler et al., 2001). Douds et al. (1997) reported enhanced spore populations of AM fungi *Glomus* spp. in soils amended with leaf and chicken litter applied over a period of three years in a maize-vegetable crop rotation. Gavito and Varela (1993) reported frequent presence of spores in organic debris that may provide protection and thereby increase spore survivability. The inhibitory effects of farmyard manure and sewage sludge have been reported by Guidi et al. (1988) and Christie and Kilpatrick (1992) respectively. The inhibitory effect of sewage sludge amendments on AM was attributed to the large quantities of heavy metals. Heavy metals led to the decrease in AM spore population (del Val et al., 1999b) and the reduction/delay of the root-colonizing ability (Gildon and Tinker, 1981; Koomen et al., 1990; Griffioen et al., 1994; Leyval et al., 1994; Leyval and Weissenhorn, 1996; Vidal et al., 1996). However, del Val et al. (1999a) reported that there is a limit to the tolerance and adaptation of different AM ecotypes to heavy metals in sewage sludge. Most of the studies indicate that the response of AM to organic amendments depends upon the type of organic amendment and on the type and quantity of mycorrhizal propagules.

SELECTION OF AM FUNGI

The selection of inoculant AM fungi requires knowledge of the intrinsic characters of fungi in the context of the range of soils and environment conditions under which the inoculant fungi will be used. Effective fungi for large-scale agriculture may be those that enhance plant growth where phosphate is limiting. Soil and environmental conditions and the quantity of ineffective propagules of the fungus will have an overriding influence on whether or not the mycorrhizae benefits the plant (Haas and Krikun, 1985; Jagpal and Mukerji, 1991a,b). The criteria for selecting AM fungi will depend on details of the local environment, soil conditions, and host plant to be used. The criteria for selecting inoculants to benefit the plant directly by alleviating phosphorus deficiency are the ability to

1. colonize roots rapidly following inoculation;
2. absorb phosphate from soil;
3. transfer phosphorus to the plant;
4. increase plant growth;
5. persist in soil as required;
6. reestablish mycorrhizae in the following seasons (if required); and
7. form propagules that withstand manipulation during and after inoculum production.

INTEGRATION OF AM FUNGI AND COMPOSTS FOR DISEASE CONTROL

The management of crop diseases caused by root pathogens has become a challenging research area in plant pathology. Increasing knowledge and concern about the environmental consequences of pesticide application have aroused interest in alternative methods of plant protection. The roots of most plant species live in symbiosis with certain soil fungi for establishing mycorrhizae. The fungus biotrophically colonizes the root cortex and develops an extramatrical mycelium that helps the plant acquire mineral nutrients from soil (Harley and Smith, 1983). It has been recognized that mycorrhizal symbiosis plays a key role in nutrient cycling in the ecosystem and protects plants against environmental and cultural stress (Barea and Jeffries, 1995).

Generally, AM fungi cause few changes in root morphology, but the physiology of the host plant changes significantly. The improved potential for mineral uptake from the soil accounts for changes in the nutritional status of the host tissues, which in turn changes the structural and biochemical aspects of root cells. This can alter membrane permeability and thus the quality and quantity of root exudation. Altered exudation induces changes in the composition of microorganisms in the rhizospheric soil, which could be termed the mycorrhizosphere. The net effect of these alterations is a tolerant plant, better able to withstand environmental stress and plant diseases (Linderman, 1988). To evaluate the influence of AM fungi on the incidence and development of diseases, the variables, namely the plant pathogen, the symbiotic fungus, and environmental conditions, should be considered. Mostly, the host mediates the interaction between the pathogen and the symbiont. The characterization of these interactions should therefore include information on the mechanisms involved, because interactions vary with the host-symbiont-pathogen combination; it is difficult to make generalizations about the effect of AM fungi on diseases.

ORGANIC MANURE AND AMF INOCULUM PRODUCTION

Since AM fungi are obligate symbionts, they must be produced on living roots. This is problematic when mass production is needed and runs the risk of contamination with plant pathogens. AM fungi depend for reproduction on photosynthates of the plant (Harley and Smith, 1983) or on root components when the root is manipulated to grow meristematically on artificial growth media under axenic conditions (Mugnier and Mosse, 1987).

The sources of AM inocula are defined by the biology of AM fungi. All infective structures of the fungi can be used, including chlamydospores or pieces of the mycelium produced inside or outside the host roots. Infected roots and the infected substrates (carriers containing infected roots or/and AM mycelium and spores) are common sources of AM inoculum.

APPLICATION OF AM FUNGI

There are two types of situations where inoculation is feasible: (1) when a low population of effective propagules is present and (2) when indigenous fungi are absent or in low levels in the rooting medium. It is obvious that the second alternative is the more promising and thus the corresponding biotechnology can be applied (Azcon-Aguilar and Barea, 1997). Application of AM fungi in agriculture requires an integrated approach where the appropriate dose and type of inoculum applied at the right stage are important (Lovato et al., 1995). A special case is that of micropropagated plants, as produced through tissue culture where AM fungi are totally absent (Vidal et al., 1992; Vestberg and Estaun, 1994). The majority of nursery-produced plants have a subsequent stage in the field. Therefore, bioassays to determine the natural potential and the field situations where AM fungi inoculation is advisable must first be carried out (Azcon-Aguilar and Barea, 1997).

FUTURE PERSPECTIVES

Biofertilizers and organic manures, if properly incorporated in a particular agroecosystem, can definitely help compensate for the detrimental effects being imposed due to conventional agricultural (high input of chemicals) practices and can help increase the productivity of many economically important plant species, such as clover (Mäder et al., 2000). Taking into consideration the potential damages rendered by the use of chemical fertilizers, the integrated use of AM biotechnology along with the organic farming can be a feasible and rewarding solution. However, for maximum benefits, inoculation with efficient AM fungi and careful selection of host-fungus-organic amendment combinations are necessary. This integrated system (organo-biofertilizer) may help to minimize the dependence upon high inputs of energy in the form of synthetic chemicals and will substitute cultural and biological techniques for these inputs. As the labile and available nutrients within an ecosystem are largely contained in the organic matter, there is a strong link between availability of organic matter and both biodiversity and nutrient cycling. Organic amendments provide the soil with vast numbers of beneficial microbes and the habitat the microbes need to proliferate. The interactions between AM fungi and these beneficial microbes could also be a biotechnological tool for benefiting the growth and health of plants under organically managed agricultural practices.

Thus, for the appropriate management of AM biotechnology in organically managed crop production systems, it is important to understand functional mycorrhizal symbiosis under the right organic farming practices (crop rotation, crop residue management, tillage, and degree of fertilization) for achieving long-term sustainability in crop production and soil health.

CONCLUSIONS

Arbuscular mycorrhizal fungi inhabit plant roots and the surrounding bulk soil. Mycorrhizae benefit the plant in many ways, the most important of which is improving the uptake of immobile nutrients from soil, especially phosphorus. Mycorrhizal symbiosis is influenced by management practices such as the degree and type of fertilization and farming systems. Proper understanding of the impacts of agronomic and farming practices upon AM fungi symbiosis to aid and ensure sustainability in terms of soil health vis-à-vis productivity is needed. By integrating AM fungi with low-input or organic farming, the benefit may be more through an increased ability to take up mineralized soil nutrients present at low concentrations. Cultural practices such as high doses of inorganic fertilization in soil may affect the indigenous population of AM fungi, which may in turn influence the mineral acquisition by plants in the rhizosphere. Crop profitability can be optimized by taking AM fungi into consideration and selecting optimal doses of organic manures that should also match the native AM inoculum level that exists in the soil.

REFERENCES

Abbott L.K. and Robson A.D. (1991). Factors influencing the occurrence of VA mycorrhizal fungi. *Agric Ecos Environ* 35: 121-150.

Amijee F., Tinker P.B., and Stribley D.P. (1989). The development of endomycorrhizal root systems. VII. A detailed study of effects of soil phosphorus on colonization. *New Phytol* 111: 435-446.

Anderson E.L., Miller P.D., and Kunishi H.M. (1987). Maize rot length density and mycorrhizal infection as influenced by tillage and soil phosphorus. *J Plant Nutri* 10: 1346-1349.

Azcon,-Aguilar C. and Barea J.M. (1997). Applying mycorrhiza biotechnology to horticulture: Significance and potentials. *Scientia Hort* 68: 1-24.

Bansal M. and Mukerji K.G. (1994). Efficacy of root litter as a biofertilizer. *Biology and Fertility of Soils* 18: 228-230.

Barea J.M. (1991). Vesicular arbuscular mycorrhizae as modifiers of soil fertility. *Advances in Soil Science* 15: 1-40.

Barea J.M., Azcon R., and Azcon-Aguilar C. (1993). Mycorrhiza and crops. In I. Tommerup (ed.), *Advances in plant pathology, mycorrhiza: A synthesis* (pp. 167-189). London: Academic Press.

Barea J.M. and Jeffries P. (1995). Arbuscular mycorrhizas in sustainable soil plant systems. In A. Varma and B. Hock (eds.), *Mycorrhiza: Structure, function, molecular biology and biotechnology* (pp. 521-560). Berlin: Springer-Verlag.

Bethlenfalvay G.J. (1992). Mycorrhizae in the agricultural plant soil system. *Symbiosis* 14: 413-425.

Bethlenfalvay G.J. and Barea J.M. (1994). Mycorrhizae in sustainable agriculture. I. Effects on seed yield and soil aggregation. *Amer J Alternative Agric* 9 (4): 157-161.

Bethlenfalvay G.J. and Linderman R.G. (1992). Mycorrhizae in sustainable agriculture. Special publication. Madison, WI: ASA.

Black R. and Tinker P.B. (1979). The development of endomycorrhizal root systems. II. Effect of agronomic factors and soil conditions on the development of vesicular-arbuscular mycorrhizal infection in barley and on the endophyte spore density. *New Phytol* 83: 401-413.

Bolan N.S. (1991). A critical review on the role of mycorrhizal fungi in the uptake of phosphorus by plants. *Plant Soil* 134: 189-207.

Brito I., Antunes P., and Carvalho M. (2001). Effect of conventional tillage vs no till on indigenous arbuscular mycorrhizal fungi of *Triticale* and wheat under field conditions, *Diversity and Integration in Mycorrhizas* Proceedings of third international conference of Mycorrhiza, July 8-13, Adelaide Convention Centre, Southern Australia.

Calvet C., Estaun V., and Camprubi A. (1993). Germination, early mycelial growth and infectivity of a vesicular-arbuscular mycorrhizal fungus in organic substrates. *Symbiosis* 14: 405-411.

Christie P. and Kilpatrick D.J. (1992). Vesicular-arbuscular mycorrhizal infection in cut grassland following long-term slurry application. *Soil Biol Biochem* 24: 325-330.

Cook G.W. (1982). *Fertilizing for maximum yield,* Third edition. New York: Macmillan Publishing Co.

Das S. and Biswas B.C. (2002). Organic farming—prospects and problems. *Fertilizer News* 47(12): 105-112.

del Val C., Barea J.M., and Azcon-Aguilar C. (1999a). Assessing the tolerance to heavy metals of arbuscular mycorrhizal fungi isolated from sewage sludge-contaminated soils. *Appl Soil Ecol* 11: 261-269.

del Val C., Barea J.M., and Azcon-Aguilar C. (1999b). Diversity of arbuscular mycorrhizal fungus populations in heavy-metal contaminated soils. *Applied and Environmental Microbiology* 65(2): 718-723.

Douds D.D. Jr., Galvez L., Frank-Snyder M., Reider C., and Drinkwater L.E. (1997). Effect of compost addition and crop rotation point upon VAM fungi. *Agri Eco Environ* 65: 257-266.

Douds D.D. Jr., Galvez L., Janke R.R., and Wagoner P. (1995). Effect of tillage and farming system upon populations and distribution of vesicular-arbuscular mycorrhizal fungi. *Agric Ecos Environ* 52: 111-118.

Douds D.D. Jr., Janke R.R., and Peters S.E. (1993). VAM fungus spore populations and colonization of roots of maize and soyabean under conventional and low-input sustainable agriculture. *Agri Ecos Environ* 43: 325-335.

Douds D.D. Jr. and Millner P. (1999). Biodiversity of arbuscular mycorrhizal fungi in agro-ecosystems. *Agric Ecosyst Environ* 74: 77-93.

Evans D.G. and Miller M.H. (1988). Vesicular arbuscular mycorrhizas and the soil disturbance induced reduction of nutrient absorption in maize. I. Casual relationships. *New Phytol* 110: 67-74.

Evans D.G. and Miller M.H. (1990). The role of the external mycelial net work in the effect of soil disturbance upon vesicular arbuscular mycorrhizas colonization of maize. *New Phytol* 114: 65-71.

Fairchild G.L. and Miller M.H. (1990). Vesicular arbuscular mycorrhizas and the soil-disturbance-induced reduction of nutrient absorption in maize III Influence of phosphorus amendments to soil. *New Phytol* 114: 641-650.

Franson R.L. and Bethlenfalvay G.J. (1989). Infection unit method of vesicular-arbuscular mycorrhizal propagule determination. *Soil Sci Soc Am J* 53: 754-756.

Gaur A. and Adholeya A. (2000). On-farm production of VAM inoculum and vegetable crops in marginal soil amended with organic matter. *Tropical Agriculture* 77: 21-26.

Gaur A., Adholeya A., and Mukerji K.G. (1998). A comparison of AM fungi inoculants using *Capsicum* and *Polianthes* in marginal soil amended with organic matter. *Mycorrhiza* 7: 307-312.

Gavito M.E. and Varela L. (1993). Seasonal dynamics of mycorrhizal associations in maize fields under low input agriculture. *Agric Ecos Environ* 45: 275-282.

Gianinazzi S., Trouvelot A., and Gianinazzi-Pearson V. (1990). Role and use of mycorrhizas in horticultural crop production. *XXIII International Horticulture Congress,* August 27-September 1, Florence, pp. 25-30.

Gildon A. and Tinker P.B. (1981). A heavy metal-tolerant strain of mycorrhizal fungus. *Trans Br Mycol Soc* 77: 648-649.

Goss M.J. and Varennes A.D. (2001). Effect of soil disturbance on the efficacy of indigenous mycorrhizal fungi in enhancing nitrogen by soybeans. *Diversity and Integration in Mycorrhizas* Proceedings of Third International Conference of Mycorrhiza, July 8-13, Adelaide Convention Centre, Southern Australia.

Griffioen W.A.J., Iestwaart J.H., and Ernst W.H.O (1994). Mycorrhizal infection of *Argrostis capillaris* population on a copper contaminated soil. *Plant Soil* 158: 83-89.

Gryndler M., Hrselova H., Vosatka M., Votruba J., and Klir J. (2001). Organic fertilization changes the response of mycelium of arbuscular mycorrhizal fungi and their sporulation to mineral NPK supply. *Folia Microbiol* 46(6): 540-542.

Guidi G., Pera A., Giovannetti M., Poggio G., and Bertoldi M. (1988) Variations of soil structure and microbial population in compost amended soil. *Plant Soil* 106: 113-119.

Haas J.H. and Krikun J. (1985). Efficacy of endomycorrhizal fungus isolates and inoculum quantities required for growth response. *New Phytol* 100: 613-621.

Harinikumar K.M. and Bagyaraj D.J. (1988). *Effect of intercropping and organic soil amendments on native VAM.* Ist Asian Conference on Mycorrhizae. University of Madras, India, pp. 207-208.

Harinikumar K.M. and Bagyaraj D.J. (1989). Effect of cropping sequence, fertilizers and farmyard manure on vesicular arbuscular mycorrhizal fungi in different crops over three consecutive seasons. *Biol Fertil Soils* 7: 173-175.

Harley J.L. and Smith S.E. (1983). *Mycorrhizal symbiosis.* New York: Academic Press.

Hayman D.S. (1982). The physiology of vesicular-arbuscular endo-mycorrhizal symbiosis. *Can J Bot* 6: 944-963.

Higa T. (1991). Effective microorganisms: A biotechnology for mankind. In J.F. Parr, S.B. Hornick, and C.E. Whitman (eds.), *Proceedings of the First International Conference on Kyusei Nature Farming* (pp. 8-14). Washington, DC: U.S. Department of Agriculture.

Hooker J.E and Black K.E (1995). Arbuscular mycorrhizal fungi as components of sustainable soil-plant systems. *Critical Reviews in Biotechnology* 15(3/4): 201-212.

Jagpal R. and Mukerji K.G. (1991a). Reforestation in waste lands using vesicular arbuscular mycorrhizae. In S.A. Abod, P.M. Tahir, L.M. Tsai, N.A. Shukar, A.S. Sajap, and D. Manikam (eds.), *Recent development in tree plantations of humid/subhumid tropics of Asia* (pp. 488-494). Selonger, Malaysia: University Pertanian Malaysis.

Jagpal R. and Mukerji K.G. (1991b). VAM fungi in reforestation. In B.L. McMichael and H. Persson (eds.), *Plant roots and their environment* (pp. 309-313). Amsterdam, the Netherlands: Elsevier.

Jasper D.A., Abbot L.K., and Robson A.D. (1989). *Acacias* respond to additions of phosphorus and to inoculation with VA-mycorrhizal fungi in soils stockpiled during mineral sand mining. *Plant Soil* 115: 99-108.

Jasper D.A., Abbott L.K., and Robson A.D. (1991). The effect of soil disturbance on VAM fungi in soils from different vegetation types. *New Phytol* 118: 471-476.

Johnson N.C. and Pflegler F.L. (1992). VA-mycorrhizae and cultural stresses. In G.J. Bethlenfalvay (ed.), *Mycorhiza in sustainable agriculture* (pp. 71-100). Madison, WI: Agronomy Society of America.

Koomen I., McGrath S.P., and Giller K. (1990). Mycorrhizal infection of clover is delayed in soils contaminated with heavy metals from past sewage sludge applications. *Soil Biol Biochem* 22: 871-873.

Kruckelmann H.W. (1975). Effects of fertilizers, soils, soil tillage, and plant species on the frequency of *Endogone* chlamydospores and mycorrhizal infection in arable soils. In F.E. Sanders, B. Mosse, and P.B. Tinker (eds.), *Endomycorrhizas* (pp. 511-525). London: Academic Press.

Leyval C. and Weissenhorn I. (1996). Tolerance to metals of arbuscular mycorrhizal fungi from heavy metal polluted soils. A summary of results. In C. Azcon-Aguilar, J.M. Barea (eds.), *Mycorrhizae in integrated systems, from genes to plant development* (pp. 452-454). Luxemburg: European Commission.

Leyval C., Weissenhorn I., Glashoff A., and Berthelin J. (1994). Influence of heavy metals on germination of arbuscular mycorrhizal fungal spores in soils. *Acta Bot Gallica* 4: 523-538.

Li X.L., Marschner H., and George E. (1991). Acquisition of phosphorus and copper by VA mycorrhizal hyphae and root to shoot transport in white clover, *Plant Soil* 136: 49-57.

Liebhart W.C., Andrews R.W., Culik M.N., Hardwood R.R., Janke R.R., Radke J.K., and Rieger-Schwartz S.L. (1989). Crop production during conversion from conventional to low-input methods. *Agronomy J* 81:150-159.

Limonard T. and Ruissen M.A. (1989). The significance of VA-mycorrhiza to future arable farming in the Netherlands. *Netherlands J Plant Pathol* 951: 129-136.

Linderman R.G. (1988). Mycorrhizal interactions with the rhizosphere effect. *Phytopathol.* 78: 366-371.

Liu R.J. and Luo X.S. (1994). A new method to quantify the inoculum potential of arbuscular-mycorrhizal fungi. *New Phytol* 128: 89-92.

Lovato P.E., Schuepp H., and Giganinazzi S. (1995). Application of arbuscular mycorrhizal fungi in orchard and ornamental plants. In A.K Varma and B. Hock (eds.), *Mycorrhiza structure, function, molecular biology and biotechnology* (pp. 521-559). Heidelberg: Springer.

Mäder P., Edenhofer S., Boller T., Wiemken A., and Niggli U. (2000). Arbuscular mycorrhizae in a long term field trial comparing low-input (organic,biological) and high-input (conventional) farming systems in a crop rotation. *Biology and fertility of soils* 31: 150-156.

McGonigle T.P. and Miller M.H. (1993). *Does the more effective mycorrhizal symbiosis found with reduced tillage increase corn yield?* Annual report, Land Resource Science, University of Guelph, Canada, pp. 69-70.

McGonigle T.P. and Miller M.H. (1996). Mycorrhizae, phosphorus absorption and yield of maize in response to tillage. *Soil Sci Soc Am J* 60: 1856-1861.

Menge J.A. (1982). Effect of soil fumigants and fungicides on vesicular arbuscular fungi. *Phytopathol* 72: 1125-1132.

Miller J.C. Jr., Rajzpakse S., and Garber R.K. (1986). Vesicular arbuscular mycorrhizae in vegetable crops. *Hortic Sci* 21: 974-984.

Miller M.H., McGonigle T.P., and Addy H.D. (1995). Functional ecology of arbuscular mycorrhizas as influenced by phosphate fertilization and tillage in an agricultural ecosystem. *Crit Rev Biotechnol* 15: 241-255.

Moorman T.B. and Reeves F.B. (1979). The role of endomycorrhizae in revegetation practices in the semi-arid West. II. A bioassay to determine the effect of land disturbance on endomycorrhizal population. *Amer J Bot* 66: 14-18.

Mosse B. (1973). Advances in study of vesicular-arbuscular mycorrhiza. *Ann Rev Phytopathol* 11: 171-196.

Mugnier J. and Mosse B. (1987). Vesicular-arbuscular endomycorrhizal infection in transformed root-inducing T-DNA roots grown axenically. *Phytopathol* 77: 1045-1050.

Nelson D.S. and Khan S.U. (1992). Uptake of atrazine by hyphae of glomus vesicular-arbuscualr mycorrhizae and root systems of corn (*Zea mays* L.). *Weed Science* 40(1): 161-170.

Nopamornbodi O., Thamsurakul S., Vasuvat Y., and Charoensook S. (1988). Survival of VA mycorrhixal fungi after paddy. In A. Mahadevan, N. Raman, and K. Natarajan (eds.), *Mycorrhizae for green Asia* (pp. 241-242). Madras, India: University of Madras.

Ocampo J.A., Martin J., and Hayman D.S. (1980). Influence of plant interactions on vesicular-arbuscular mycorrhizal infections. I. Host and non-hosts plant grown together. *New Phytol* 84: 27-35.

Parr J.F., Hornick S.B., and Kaufman D.D. (1994). Use of microbial inoculants and organic fertilizers in agricultural production. *Proceedings of the International Seminar on the Use of Microbial and Organic Fertilizers in Agricultural Production.* Taipei, Taiwan: Food and Fertilizer Technology Center.

Plenchette C., Perrin R., and Duvert P. (1989). The concept of soil infectivity and the method for its determination as applied to endomycorrhizas. *Can J Bot* 67: 112-115.

Rao A.V., Tarafdar J.C., Sharma S.K., and Agrawal R.K. (1995). Influence of cropping systems on soil biochemical properties in an arid rain-fed environment. *J Arid Environ* 31: 237-244.

Reganold J.P., Papendick R.I., and Parr J.F. (1990). Sustainable agriculture. *Scientific American* 262: 112-120.

Reid C.P.P. and Bowen G.D. (1979). Effects of moisture on VA mycorrhiza formation and root development in Medicago. In J.L. Harley and R.S. Russel (eds.), *The soil-root interface* (pp. 211-215). London: Academic Press.

Ryan M.H., Chilvers G.A., and Dumaresq D.C. (1994). Colonization of wheat by VA-mycorrhizal fungi was found to be higher on a farm managed in an organic manner than on a conventional neighbour. *Plant Soil* 160: 33-40.

Sattelmacher B., Reinhard S., and Pomikalko A. (1991). Differences in mycorrhizal colonization of rye grown in conventional or organic (biological dynamic) farming systems. *J Agron Crop Sci* 167: 350-355.

Sharma M.P. and Adholeya A. (2000). Enhanced growth and productivity following inoculation with indigenous AM fungi in four varieties of onion in an alfisol. *Biol Agric Hort* 18 : 1-14.

Sharma M.P., Gaur A., Bhatia N.P., and Adholeya A. (1996). Growth responses and dependence of *Acacia nilotica* var. *cupriciformis* on the indigenous arbuscular mycorrhizal consortium of a marginal wasteland soil. *Mycorrhiza* 6: 441-446.

Sheikh N.A., Saif S.R., and Khan A.G. (1975). Ecology of endogone: II Relationship of endogone spore population with chemical soil factors. *Islamabad J Sci* 2: 6-9.

Singh G.B. and Dwivedi B.B. (1996). Integrated nutrient management for sustainability. *Indian Farming* 46(8): 9-15.

Sylvia D.M. and Jarstfer A.G. (1992). Sheared root inocula of vesicular-arbuscular mycorrhizal fungi. *Appl Environ Microbiol* 58: 229-232.

Sylvia D.M. and Neal L.H. (1990). Nitrogen affects the phosphorus response of VA mycorrhiza. *New Phytol* 115: 303-310.

Thampan P.K. (ed.) (1995). *Organic agriculture.* Gandhi Nagar, Cochin, Kerala: Peekay Tree Crops Development Foundation.

Thompson J.P. (1986). Soilless cultures of vesicular-arbuscular mycorrhizae of cereals: Effects of nutrient concentration and nitrogen source. *Can J Bot* 64: 2282-2294.

Thompson J.P. (1987). Decline of vesicular-arbuscular mycorrhizae in long fallow disorder of field crops and its expression in phosphorus deficiency of sunflower. *Aust J Agric Res* 38: 847-867.

Thompson J.P. (1990). Soil sterilization methods to show VA-mycorrhizae aid P and Zn nutrition of wheat in vertisols. *Soil Biol and Biochem* 22: 229-240.

Thompson J.P. (1991). Improving the mycorrhizal conditions of the through cultural practices and effects of growth and phosphorus uptake by plants. In G. Johansen, K.K. Lee, and K.L. Sahrawat (eds.), *Phosphorus nutrition of grain legumes in the semi-arid tropics* (pp. 117-137). Patancheru, Andhra Pradesh, India: ICRISAT.

Thompson J.P. (1994). What is potential for the management of mycorrhizas in agriculture? In A.D. Robson, L.K. Abbot, and N. Malajczuk (eds.), *Management of mycorrhizas in agriculture, horticulture and forestry* (pp. 191-200). Dordrecht, the Netherlands: Kluwer Academic Publishers.

Tisdall J.M. (1991). Fungal hyphae and structural ability of soil. *Aust J Soil Res* 29: 729-743.

Tisdall J.M. and Oades J.M. (1979). Stabilization of soil aggregates by the root systems of rye grass. *Aust J Soil Res* 17: 429-441.

van der Heijden M.G.A., Klironomos J.N., Ursic M., Moutoglis P., Streitwolf E.R., Boller T., Wiemken A., and Sanders I.R. (1998). Mycorrhizal fungal diversity determines plant diversity, ecosystem variability and productivity. *Nature* 396: 69-72.

Vestberg M. and Estaun V. (1994). Micropropagated plants, an opportunity to positively manage mycorrhizal activities. In S. Gianinazzi and H. Schuepp (eds.), *Impact of mycorrhizas on sustainable agriculture and natural ecosystems* (pp. 217-226). Basel, Switzerland: Birkhauser.

Vidal M.T., Azcon-Aguilar C., and Barea J.M. (1996). Effect of heavy metals (Cu, Cd and Cu) on arbuscular mycorrhizal formation. In C. Azcon-Aguilar and J.M. Barea (eds.), *Mycorrhizae in integrated systems: From genes to plant development* (pp. 487-490). Luxemburg: European Commission.

Vidal M.T., Azcon-Aguilar C., Barea J.M., and Pliego-Alfaro F. (1992). Mycorrhizal inoculation enhances growth and development of micropropagated plants of avocado. *Hort Science* 27: 785-787.

Werner M.R., Robert A.K., and Gliessman S.R. (1990). Colonization of strawberry roots by VA mycorrhizal fungi in agroecosystems under conventional and transitional organic management. *Biol Agric Hort* 7: 139-151.

Chapter 10

Role of Mycorrhizae in Forestry

Hanna Dahm

INTRODUCTION

Mycorrhiza is the result of a plant-microbial interaction in the soil of great ecological significance. Mycorrhizae are formed by association between a plant root and a fungus, and by far the majority of vascular plants are involved in these associations. There are a number of different forms of mycorrhizal association, but in many cases the fungus involved (the mycobiont) enhances plant uptake of nutrients from the soil, increases rootlet size and longevity, protects the root system from many pathogens, and can also absorb and translocate water to the plant host (Killham, 1994).

The mechanism of enhanced nutrient uptake by mycorrhizal plants is the increased soil volume exploited by the extramatrical mycelium of the mycorrhizal fungus. Nutrients can therefore be transported from beyond the narrow nutrient depletion zone that surrounds the nonmycorrhizal roots. In addition to the mycorrhizal fungus exploiting the greater soil volume, the fungal mycelium can often absorb nutrients at lower solution concentrations than can the uninfected plant root.

The mycorrhizal association can also enable the plant host to access nutrients (particularly nitrogen) in an organic form that would otherwise be unavailable. In addition, some mycorrhizal plants can be connected via hyphal strands and this link provides a means of nutrient transfer from plant to plant (Killham,1994).

ECOLOGICAL SIGNIFICANCE OF MYCORRHIZAE

Most land plants form mycorrhizae. Many species of fungi are involved in ectomycorrhizal associations: of a forest, a single tree species, an individual tree, or even a small segment of lateral root. Even a single tree species can have numerous species of fungi capable of forming ectomycorrhizae on its roots; a single fungus can enter into ectomycorrhizal association with numerous tree species. Some fungi are apparently host specific; others have broad host ranges and form ectomycorrhizae with members of numerous tree genera in diverse families (Marx, 1980). Evaluation of the occurrence and function of linkages between plants via shared mycorrhizal fungi requires a full understanding of compatibility and specificity phenomena. Specificity phenomena in mycorrhizal associations directly affect plant community development in a variety of ways. Mycorrhizal dependence, fidelity to a mycorrhizal class, and mycobiont propagule availability are prime determinants of the early stages of community development and affect long-term plant dominance. During community succession, shared compatibility for mycorrhizal fungal associates determines the linkage potential for those plants and influences plant interactions. Ecological linkages of plants via mycorrhizal interactions may contribute to the formation of functional guilds of plants and fungi. Shared compatibility for mycorrhizal symbionts also influences the ability of plants to migrate to new locations and so may affect plant migrations during periods of rapid global climate change (Molina et al., 1992).

Microorganisms are present in great numbers near the feeder roots of trees, and they play vital roles in numerous physiological processes (Marx, 1980). These dynamic microbial processes include saprophytism and symbiosis (parasitic and mutualistic) (Smith and Smith, 1990).

The most widespread mutualistic symbiosis of trees is the mycorrhizal association between root-inhabiting fungi and the feeder roots of trees. Mycorrhizal associations are so common under natural soil conditions that a nonmycorrhizal tree is the exception.

Mycorrhizal symbioses are part of a great array of symbioses between heterotrophic and autotrophic organisms. In this symbiotic system, one of the partners is carbon autotrophic and provides fixed carbon compounds for the system. That is, the heterotrophic partners are adapted to receive a supply of carbon directly from the photosynthetic products of the autotrophic partners, rather than primarily or solely from humus or dead tissues indirectly derived from photosynthesis after the death of the autotroph. In autotrophic mycorrhizal systems, carbon compounds synthesized in the green tissues not only nourish the host itself but also pass into the fungus and so into the external mycelium as its source of carbon. Soil-derived nutrients absorbed by the mycelium in the soil pass into the fungal phase and so into the host tissue (Varma, 1998).

TYPES OF MYCORRHIZAE ON FOREST TREES

Ectomycorrhizae

This type occurs naturally on many forest trees (see Photos 10.1 and 10.2). All members of the gymnosperm family Pinaceae (pine, spruce, larch, hemlock, etc.) as well as certain angiosperms (willow, poplar, aspen, hickory, pecan, oak, birch, beech, eucalyptus, etc.) form ectomycorrhizae (Marx, 1980).

Some of these trees can be either ectomycorrhizal or endomycorrhizal, depending on soil conditions. Ectomycorrhizal infection is initiated from spores or hyphae of the fungal symbionts inhabiting the rhizosphere of the feeder roots. Mycelia of the mycorrhizal fungi are stimulated by root exudates, growing over the feeder root surface and forming a fungus mantle. Following mantle development, hyphae develop around root cortical cells and form the Hartig net. The Hartig net is the main distinquishing feature of ectomycorrhizae.

Most fungi that form ectomycorrhizae with forest trees are members of Basidiomycetes. Certain Ascomycetes (e.g., truffles) also form mycorrhizae.

PHOTO 10.1. Seedlings of pine with well-developed ectomycorrhizae in the forest nursery.

PHOTO 10.2. Seedlings of pine with poorly developed ectomycorrhizae in the forest nursery.

Ectomycorrhizal fungi aid the growth and development of trees. For some trees, such as *Pinus,* they are indispensable for growth under natural conditions. Trees with ectomycorrhizae have a much larger, physiologically active, root-fungus area for nutrient and water absorption than do trees with few or no ectomycorrhizae.

Ectomycorrhizae are able to absorb and accumalate nitrogen, phosphorus, potassium, and calcium in the fungus mantles more rapidly and for longer periods of time than nonmycorrhizal feeder roots. Ectomycorrhizal fungi help break down certain complex minerals and organic substances in the soil and transmit nutrients from these materials to the tree. Ectomycorrhizae also appear to increase the tolerance of trees to drought, high soil temperatures, soil toxins, and extremes of soil pH. Ectomycorrhizae deter infection of feeder roots by root pathogens (Marx, 1980) (see Photos 10.3, 10.4, and 10.5).

Endomycorrhizae

This type is formed on most economically important agronomic and forage crops, as well as on fruit and nut trees.

Important forest trees such as maple, elm, gum, ash, sweetgum, sycamore, and black walnut form endomycorrhizae. Some trees, such as alder can form both endo- and ectomycorrhizae.

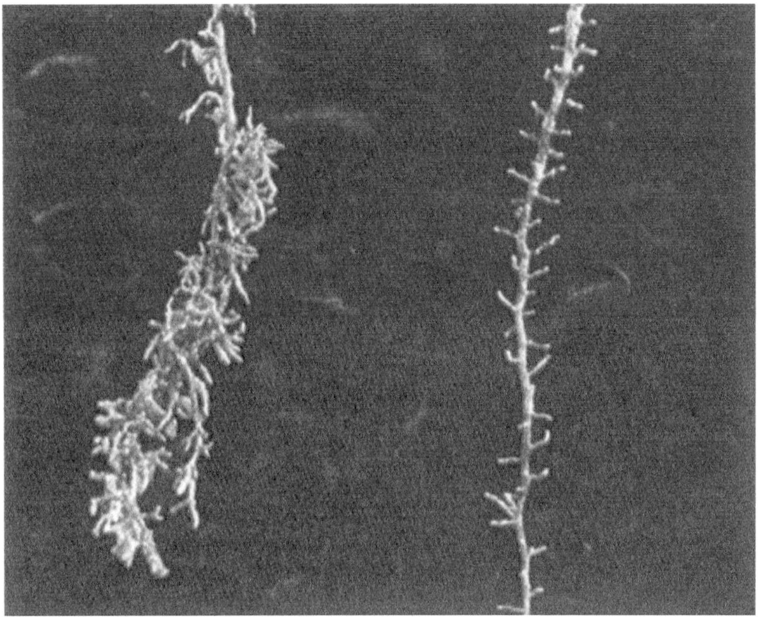

PHOTO 10.3. Seedling of pine with (left) and without (right) ectomycorrhiza.

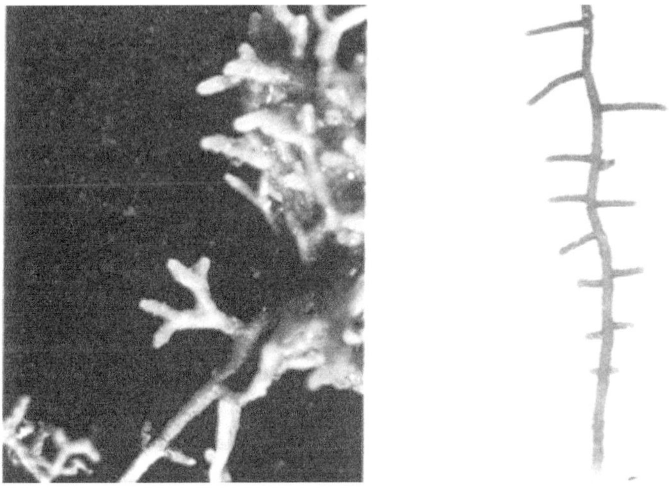

PHOTO 10.4. Ectomycorrhizal feeder root (mantle) (left); nonmycorrhizal feeder root (no mantle) (right).

PHOTO 10.5. Seedlings having ectomycorrhizae on more than 75 percent of the roots are pictured at left. Seedlings having ectomycorrhizae on about 5 percent of the roots are at right.

Endomycorrhizal fungi form a loose network of hyphae on the feeder root surfaces and do not develop the dense fungus mantle. Hyphae of endomycorrhizal fungi penetrate the cell walls and progress into the cortical cells of the root (Marx, 1980).

These infective hyphae may develop specialized absorbing or nutrient-exchanging structures (haustoria) called arbuscules in the cytoplasmic matrix of the cortical cells. The term *vesicular-arbuscular* mycorrhizae has been coined to denote this type of endomycorrhizae.

Root-containing vesicular-arbuscular associations look superficially similar to nonmycorrhizal roots, unless they have been processed to determine the internal features.

The fungi that form endomycorrhizae with trees are mainly Phycomycetes. Many endomycorrhizal fungi of trees belong to the family Endogonaceae. Spores of these fungi are able to survive in soil for years without a host. Most species appear to have a very broad host range.

It was found that certain hardwood trees depend on endomycorrhizae for normal growth, for example, the *Pinus* species depends on ectomycorrhizae (Mc Gonigle and Fitter, 1990).

Ectendomycorrhizae

Mycorrhizae with many of the characteristics of ectomycorrhizae but also exhibiting a high degree of intracellular penetration have been described as ectendomycorrhizae. Ectendomycorrhizae have a limited distribution in forest soils and are found primarily on the roots of normally ectomycorrhizal trees. In ectendomycorrhizal associations, a few cells are penetrated by the fungus. The senescent cortex is fully colonized by hyphae in the late Hartig net zone (Harley and Smith, 1983). Mikola (1965) found ectendomycorrhizae to be common on pine seedlings in forest nurseries, particularly in areas that were formerly agricultural soils. Ectendomycorrhizal fungi belong to *Ascomycetes: Wilcoxina mikolae, Phialophora finlandia, Chloridium paucisporum* (Wang and Wilcox, 1985). *Wilcoxina mikolae* is the first colonizer of young seedlings. In the second or third season in the nursery, sheathed ectomycorrhizae may also appear and their proportion gradually increase. However, pine seedlings are often still completely ectendomycorrhizal at the time of transplant to the field. Ectendomycorrhizae have also been found in seedlings of natural regeneration of pine forest and occasionally even in root systems of old trees (Mikola, 1988).

Some of the genera of trees have ecto- as well as endomycorrhizae, for instance, *Juniperus, Cupressus, Salix, Malus, Pyrus, Tilia, Eucalyptus, Arbutus.* Ectomycorrhizae are most common in the families Pinaceae, Salicaceae, Betulaceae, and Fagaceae; in other families they exist in only some genera (Meyer, 1973). This presence in distinct systematic categories might have various reasons:

1. Certain chemical compounds are involved in the mechanism regulating the formation of the mycorrhizae; these substances apparently are synthesized in some taxa and not in others, or they might be formed in definite amounts.
2. The woody plants with ectomycorrhizae succeed only in regions or in soils where the symbiotic fungus is present or where it is capable of growing.

Under natural conditions some trees are incapable of developing without ectomycorrhizae. This group of trees could be named obligate ectomycorrhizae. The reason for their debilitation in the absence of ectomycorrhiza might be due to the following: The truly ectomycorrhizal fungus acts primarily as a parasite and takes advantage of the tree. On infection the tree develops defensive mechanisms. Only when a state of equilibrium has been reached between the pathogenic capacity of the fungus and the defensive mechanisms of the host is the balance maintained and can both partners live together mutualistically for a longer period, provided that the association is advantageous to both partners (Meyer, 1973).

In the absence of truly mycorrhizal fungi, other fungi may penetrate into the root. The tree responds to this invasion by forming toxic compounds such as the tannins that fill the outer cells of the root cortex. Furthermore, the cell walls of the root cortex become suberized and, as in true ectomycorrhizae, the formation of root hairs is suppressed. Under such conditions the normal function of the root as an organ for absorption might be greatly diminished and consequently the trees decline (Meyer, 1973).

The absence of a fungal sheath or a thin, weakly developed mantle and intracellular infection of the cortex in trees that normally have ectomycorrhizae are signs of a pathogenic relationship between the fungus and the tree.

Obligately ectomycorrhizal trees are, for example, members of the genera *Abies, Larix, Picea, Pinus, Carpinus, Fagus,* and *Quercus*. Plantations of these trees may fail to grow well in the absence of truly ectomycorrhizal fungi. Facultative ectomycorrhizal genera are *Cupressus, Juniperus, Salix, Betula, Corylus, Alnus, Pyrus, Acer,* and *Eucalyptus* (Meyer, 1973). These trees are capable of thriving in the absence of true ectomycorrhizal fungi. Most of them are the first invaders of a wasteland and act as pioneers in forest succession. Facultative ectomycorrhizal trees improve the site for the trees that follow, and under their influence true ectomycorrhizal fungi may colonize the area.

OCCURRENCE OF MYCORRRHIZAL FUNGI IN SOILS OF NATIVE FORESTS

Fungi associated with forest trees are mostly basidiomycetes such as *Amanita* sp., *Lactarius* sp., *Pisolithus* sp., and *Rhizopogon* sp., although some ascomycetes form mycorrhizae for example, *Cenococcum* sp., *Elaphomyces* sp., and *Tuber* sp.

In native forests many fungi form ectomycorrhizae and are best adapted to the conditions prevailing on that particular site.

In the competition between different tree species and fungi, an equilibrium is reached which enables the best adapted species to form ectomycorrhizal forests of a relatively great stability. Many mycorrhizal fungi are capable of living in symbiosis with several tree species (Meyer, 1973). On the other hand, a given tree species can form mycorrhizae with a series of fungi, so that in many cases tree and fungus species are interchangeable. Some fungal species form a mycorrhiza with only one particular host tree, for example, *Suillus grevillei* and *S. tridentinus* with *Larix.* Conventional wisdom generalizes that most ectomycorrhizal fungi are broad-host-ranging and express little host specificity. However, ectomycorrhizal fungi vary from narrow to broad in host range. Similarly, ectomycorrhizal hosts may associate with relatively few to very high numbers of fungus species; several plant taxa show strong coevolution with particular fungal taxa. Consequently, not all host-fungus combinations are compatible. The overlap between hosts for shared compatibility with common fungi will determine their linkage potential (Molina et al., 1992).

In plant pathology, specificity typically implies a highly restricted association between species of host and species or strain of a pathogen. Mycorrhizal associations do not express strong specificity. Six phenomena that encompass the spectrum of specificity in mycorrhizal associations can be recognized (Molina et al., 1992):

1. *Dependency vs. independency*—Defines whether plants form or do not form mycorrhizae;
2. *Facultative vs. obligate symbionts*—Defines the ability or inability of symbionts to complete life cycles in the absence of mycorrhiza formation;
3. *Fidelity to a class of mycorrhiza*—Recognizes that most symbionts typically form one class of mycorrhiza (vesicular-arbuscular or ectomycorhiza) but that some can form two or more classes;
4. *Host range of mycorrhizal fungi*—Describes a spectrum that varies from a narrow range (typically restricted to a host genus) to an intermediate range (often restricted to a host family) to a broad range (typically extending across diverse host-families and orders);
5. *Host receptivity*—Defines the number and diversity of mycorrhizal fungi accepted by a particular host, also ranging from narrow (low number) to broad (high number);
6. *Ecological specificity*—The influence of biotic and abiotic factors on ability of plants to form functional mycorrhizas with particular fungi in natural soils

SOIL TYPE AND HUMUS PROPERTIES

Differences in mycorrhizal frequency in forests arise from differences in soil properties and types of tree associations. According to Melin (1925) development of mycorrhizae is influenced by the activity of the fungus infecting the roots. The activity of the fungus, however, is related to a variety of soil properties. A pH value of about 5 is optimal for many mycorrhizal fungi. A deficiency of soil nitrogen, phosphorus, potassium, or calcium can stimulate development of mycorrhizae (Meyer, 1973).

ECOLOGICAL SPECIFICITY

Harley and Smith (1983) coined the term *ecological specificity* to emphasize that actual mycorrhizal development and therefore host-fungus associations in nature may differ strongly from those produced experimentally, particularly in pure culture syntheses. Ecological specificity include the influence of biotic and environmental interactions on the development and function of mycorrhizae in natural soils. Interactions between plants (e.g., companion or neighboring plant influences), between fungi (competitive to facilitative), or between plant and fungi (e.g., stimulation of spore germination) exemplify components of the biotic complex that determine whether certain plants and fungi enter into mycorrhizal symbioses.

Ectomycorrhizal fungi vary in their physiological response to environmental conditions. Danielson and Visser (1989) used the term *soil specificity* to indicate the importance of soil factors in influencing mycorrhizal development by certain fungi. Several *Rhizopogon* species form ectomycorrhiza only on one host species (e.g., *Pseudotsuga* or *Pinus*). When inoculated into dual host cultures, these fungi fail to form ectomycorrhiza with companion plants even though hyphae grow vigorously from ectomycorrhiza of the specific host and make abundant contact with companion plants.

However, some *Rhizopogon* species that were restricted to *Pinus* in monoculture were able to spread in dual culture on to the companion tree species. These examples showed that not only restricted host specificity but also companion plants can influence the mycorrhizal potential of ectomycorrhizal fungi. Mycelia of an ectomycorrhizal fungus actively growing in the soil and drawing energy from a photosynthetic host can have a different colonizing potential and thus host range than spores of the fungus residing in the soil (Molina, 1980).

COMPETITION AMONG ECTOMYCORRHIZAL FUNGI IN FOREST SOIL

In nature, a single plant is generally associated with several species of mycorrhizal fungi. The number of fungal partners is particularly high in the case of large, perennial plants such as trees and more particularly in trees with ectomycorrhizae. Many factors detrmine the structure of these mycorrhizal populations: microheterogenity of the soil, age and activity of the roots, etc. However, a key factor is the differential ability of the different fungi to infect the roots under a given range of conditions and in the presence of other symbionts. New roots are continuously formed, and the different fungi compete to infect them.

Most information about these competitive interactions is purely descriptive, and little is known about the mechanisms involved. However, it is clear that the dispersal strategy of each fungal species plays a major role (Garbaye, 1991). Chemical interactions through antibiotic substances are also probably involved.

Numerous species of mycorrhizal fungi are known to grow in forest soils. In beechwood soil, within a square of 10 cm × 10 cm, populations of up to 15 different species of ectomycorrhizae have been found in the organic soil layers (Brand, 1991). It is likely that competition exists among the different fungi. Mycorrhizal fungi of forest trees must compete with other species for their place on or in the tree roots, which are their sources of energy.

Patterns of competition can often be observed when examining the different ectomycorrhizal species that colonize the laterals of a long root. Black mycelium of *Cenococcum geophilum,* for example, often can be seen growing along roots and on the mantle surface of other ectomycorrhizae. Microscopic examination of long roots and mycorrhizal mantles shows that various mycelia can grow along the root (Brand and Agerer, 1986). Especially in spring, and after periods of rain, root tips can burst through the mantle of their fungal partners but are rapidly reinfected, sometimes by the same fungus but often by a different species.

Competition between fungi occurs within as well as on the surface of mycorrhizal root. Intracellular infections in root cortex cells of ectomycorrhizae have frequently been observed and reported (Brand, 1991). Usually they have been interpreted to be signs of increasing pathogenicity of an ectomycorrhizal fungus, or to be typical of aging mycorrhizae. They also occur in typical ectomycorrhizae of conifers in nurseries.

Endomycorrhizal infections in ectomycorrhizae may arise as a result of ingrowths of the Hartig net of an ectomycorrhizal fungus, but mixed associations also exist in which two fungi share one root. One of the associates forms the typical structures of an ectomycorrhiza (the fungal mantle and Hartig net), while a second fungus forms haustoria in the cortical cells of the root.

FAVORABLE EFFECTS OF MYCORRHIZAE ON PLANT GROWTH

Increase of Absorbing Surface

The main role of mycorrhizal association is to acquire nutrients by exploring the soil volume with hyphae that are both more responsive and more extensive than the roots themselves. The fungi associated with the roots produce external hyphae which allow the host to draw on nutrients far away from the roots, within a radius of 20 cm or more.

Phosphate Nutrition

Phosphorus plays a major role in the transfer of genetic information, being part of the nucleic acids (DNA and RNA) as well as in energetic processes being part of ATP. Phosphate assimilation by mycorrhizal roots has been divided into three major steps (Gianinazzi-Pearson and Gianinazzi, 1985):

1. *Mobilization and absorption*—Mycorrhizal trees can utilize some sources of phosphate unavailable to nonmycorrhizal ones. Organic or mineral-insoluble forms of phosphorus seem to be mobilized by ectomycorrrhizae. Williamson and Alexander (1975) reported that there was two to eight times more phosphatase activity on the surface of mycorrhizal beech roots than on nonmycorrhizal roots. Such activity might hydrolyze organic phosphorus. Several mycorrhizal fungi are able to dissolve Ca phytate due to phytase activity rather than pH changes in the medium. *Pinus sylvestris* seedlings grown with ^{32}P-labeled Fe phosphate, apatite, and humus, absorbed more labeled P when mycorrizal with *Suillus luteus* than when nonmycorrhizal (Ritter and Lyr, 1965).
2. *Accumulation*—Polyphosphate compounds are a major reserve in ectomycorrhizal fungi and represent up to 30 percent of total phosphate. P accumulation in mycorrhizae as polyphosphate granules in the fungal sheath was detected (Lapeyrie et al., 1984). Polyphosphate is compartmentalized in the fungal vacuole and forms of metachromatic inclusions that can be visualized by transmission electronic microscopy. The ability of the fungal sheath in ectomycorrhiza to accumulate large amounts of phosphate as polyphosphate and rapidly mobilize them when required makes them a potentially important storage organ for the host plant.

3. *Transfer to the host tissues*—In vesicular-arbuscular mycorrhizae, phosphorus transfer to the host is presumed to be coupled with carbohydrate transfer to the fungus. Recent investigations have shown that the Hartig net is organized in a way that bidirectional transfer of ions and molecules is promoted (Kottke and Oberwinkler, 1987). This organization consists in lack of septation of the hyphae and their intimate juxtaposition, resulting in a coenocytic structure, functionally equivalent to the transfer cells in vascular plants.

Nitrogen Nutrition

Nitrogen is one of the most important nutrients limiting growth in forest systems and is required by trees for normal growth and development. It also appears to be the most important nutrient in increasing forest productivity. Although shown that mycorrhizae stimulate nitrate and ammonium uptake by woody plants, the translocation within the fungus and between the fungus and the host is still not explained.

Ectomycorrhizal fungi can be involved in different stages of nitrogen metabolism. Absorbed ammonium is assimilated into organic compounds via two pathways (Plassard et al., 1985). An NADP-dependent glutamate dehydrogenase (GDH) which catalyzes the amination of oxoglutarate giving glutamate; the GS/GOGAT pathway with the sequential enzymes: glutamine synthetase (GS) which catalyzes the amination of glutamic acid, giving glutamine and glutamate synthase (GOGAT) which catalyzes the formation of glutamate from glutamine. The GDH pathway predominates in fungi. The GS/GOGAT pathway is the major route of NH_4 assimilation in higher plant roots (Oaks and Hirel, 1985). Even if most investigations of the nitrogen nutrition of ectomycorrhizal plants take into consideration the simple form of nitrogen, recent results propose that colonization might provide access to organic nitrogen compounds which constitute the major source of this nutrient in organic forest soils.

The ability to use proteins as a source of nitrogen requires the intervention of proteases. Such enzymes were detected in mycelial extracts and culture filtrates of a mycorrhizal fungi (Dahm and Strzelczyk, 1995).

Water Uptake

Although it has been shown that endomycorrhizal fungal associates can modulate the drought resistance of host plants through mechanisms as diverse as enhanced water uptake, osmotic adjustment, altered elasticity of the cell walls, or altered symplastic water content, information on the role of ectomycorrhizal associations in plant water relations remains scarce (Dosskey et al., 1991). The most specific feature of ectomycorrhizae is the fungal sheath, which completely covers the absorbing short roots. Most of the flux of water between the soil and the plant has to pass through the fungus. Therefore, ectomycorrhizal infection can lead not only to enhancement of plant nutrient content but also to additional uptake of water. The capacity of fungal mycelium to supply water in sufficient quantities is important to sustain transpiration and photosynthesis. Ectomycorrhizal plants are more resistant to drought stress.

Sulfur

Enhanced growth and biomass accumulation is largely dependent on an increased rate of protein synthesis. This is of particular significance in young trees where additional biomass is mainly invested into living, protein-containing tissue. But also in mature trees that preferentially accumulate wood, enhanced protein synthesis is required to adapt the cellular metabolism to an increased need for precursors of structural components of the cell wall such as cellulose and lignin. Since reduced sulfur in the sulfur-containing amino acids cystine, cysteine, and methionine is an essential constituent of protein, enhanced protein synthesis not only relies on an improved availability of organic nitrogen but also of sulfur in the oxidation state (Rennenberg, 1999). As a consequence, enhanced acquisition and assimilation of nitrogen in ectomycorrhizal symbiosis alone is not sufficient to improve growth.

Association of the root of trees with ectomycorrhizal fungi seems to interact with the processes of sulfur nutrition in different ways, but the results of these interactions are dependent on both the plant and the fungal partners (Rennenberg, 1999). Mycorrhizal colonization of the roots can alter the response of sulfate uptake to sulfate availability in the soil and enhances xylem loading and, hence xylem transport of sulfate to the leaves. Simultaneously, sulfate reduction in the roots seems to be stimulated by ectomycorrhizal association. Increase of sulfate reduction in the leaves of mycorrhizal trees can result in enhanced phloem transport of reduced sulfur from the leaves to the roots (Kreuzwieser and Rennenberg, 1998).

Stimulating Effect on Metabolite Production

Mycorrhizal fungi supply vitamins and growth-stimulating substances such as auxins, gibberellins, and cytokinins. Investigators demonstrated the production of vitamins (such as biotin, thiamine, and panthotenic acid) as well as hormones, cytokinins (e.g., zeatin), and indole compounds, particularly IAA (indole-3-acetic acid), by numerous ectomycorrhizal fungi (Chalot et al., 1988; Strzelczyk et al., 1991; Strzelczyk et al., 1997). According to Allen and Mac Mahon (1985) the formation of mycorrhizae represents a special adaptation to surviving in unsuitable conditions and under stress in the habitat. In general, adaptations of plants to changes in the external environment results in changes in plant growth and development, and are controlled by plant hormones (Gogala, 1991).

Slankis (1973) proposed a hormonal theory of the regulation of mycorrhizae formation:

1. Root structures resembling ectomycorrhizae can be induced in pine roots either by fungal exudates or by supraoptimal concentrations of synthetic auxins.
2. Ectomycorrhizae undergo the same type of morphogenesis when subjected to high nitrogen concentration, which arrest the formation of new mycorrhizae and cause termination of the symbiosis.
3. IAA and other indole compounds abound in mycorrhizal roots, while only traces are present in nonmycorrhizal ones.
4. Production of IAA in vitro by mycorrhizal fungi was suppressed in a medium rich in nitrogen.
5. Auxins influence carbohydrate transport.

Slankis (1973) developed his theory on the role of auxins in the development and ramification of mycorrhizal short roots on the basis of experiments with synthetic auxins or with crude fungal exudates.

Rudawska (1980) has shown that some amino acids diminished the content of auxins in fungal filtrates in the presence of tryptophan in the medium. Pokojska and Strzelczyk (1988) found that organic acids decreased the production of auxinlike substances through mycorrhizal fungi growing in L-tryptophan-supplemented media. Among the plant growth regulators, cytokinins are probably the most important for mycorrhizae because of their influence on the growth and development not only of the higher plant themselves but also of the mycorrhizal fungi (Gogala, 1991). Cytokinins influenced the content of K, Ca, P, and N in the mycelium of some mycorrhizal fungi. Cytokinins increase the fluidity of hyphal membranes. This change could influence ions and water transport processes through the fungal membrane (Pohleven, 1988, 1989).

Miller (1967) isolated zeatin and riboside zeatin from a culture of ectomycorrhizal fungus *Rhizopogon luteolus*. Crafts and Miller (1974) detected production of cytokinins by several mycorrhizal fungi, and they identified them as *trans*-zeatin and *trans*-ribosylzeatin. Strzelczyk and Kampert (1987) found that almost all fungi isolated from mycorrhizae of *Pinus sylvestris* produced cytokinins. Some vitamins of the B-group influence the production of cytokinins in these fungi. Little is known about gibberellins in mycorrhizal fungi. Gibberellin-like substances were detected in fungi isolated from mycorrhizae of *Pinus sylvestris* (Strzelczyk et al., 1976).

Protective Effect Against Pathogens

Ectomycorrhizal fungi have long been known to provide protection to forest trees against antagonistic fungi. *Laccaria laccata* has been reported to be very effective in protecting roots of Douglas fir and spruce seedlings against *Fusarium oxysporum* (Mortier et al., 1988). Zak (1964) and Marx (1972) suggested a number of mechanisms by which mycorrhizal fungi might protect roots against soil pathogens: physical protection by the mantle, competition with the pathogen for available nutrients, and stimulation of an antagonistic microflora associated with the mantle.The fungus mantle may create a unique and totally different type of mechanical barrier against pathogenic infection. In mature ectomycorrhizae the fungus mantle is a tightly interwoven hyphal network that completely covers the root tip and cortex.

Ectomycorrhizae may also afford protection by secreting antibiotics originating from the fungal symbiont or by stimulating host root cells to secrete antimicrobial metabolites which may inhibit or kill potential pathogens. Inoculating fumigated nursery soil with spores of *Laccaria laccata,* a common ectomycorrhizal symbiont in nursery and forest soils, has been shown to suppress *Fusarium* sp. disease (Schisler and Linderman, 1989). This symbiont produces diffusible extracellular metabolites and induces primary host roots to produce phenolic compounds, which may account for the protective effect of *L. laccata* against *Fusarium* sp.

The production of antifungal, antibacterial, or antiviral compounds by more than 100 species of ectomycorrhizal fungi suggests that ectomycorrhizal fungi other than *L. laccata* could also influence disease (Sylvia and Sinclair, 1983a,b). Ectomycorhizae also produce volatile compounds that can inhibit the growth of several root pathogens. Ectomycorrhizae could release significant amounts of ethylene into the mycorrhizosphere, which may play a pivotal role in soil fungistasis (Gogala, 1991).

Mycorrhizosphere

Mycorrhizosphere is the rhizosphere of a root infected by mycorrhizal fungus. As the fungus uses some of the root exudates and modifies root functions, microbial communities in the mycorrhizosphere differ from those in the rhizosphere and in the soil. A wide variety of organisms lives in close vicinity of roots or mycorrhizae, taking advantage of the various organic compounds released by the plant. These organisms include members of most taxonomic groups of aerobic and anaerobic heterotrophic organisms, from bacteria through fungi and protozoa to animals (Garbaye, 1991).

Specialized mycorrhizosphere organisms depend completely on the simple organic molecules (sugars, amino acids, organic acids) released by the plant and its fungal symbionts. They cannot multiply significantly in the nonrhizospheric soil because of the lack of these short-lived substances. As a consequence, they compete strongly with one another as well as with the symbionts and pathogens in their root-surface-colonizing preinfection stage. This group includes some fungi, but the most common microorganisms in this category are the so-called rhizobacteria, such as fluorescent *Pseudomonas, Bacillus,* and *Azospirillum*. Some of them are able to fix atmospheric nitrogen and significantly contribute to the maintenance of the fertility of ecosystems; some others are the extensively studied PGPR (plant-growth-promoting rhizobacteria), which are used for improving crop yields (Dahm et al., 1993).

Rhizosphere microflora is often favorable to mycorrhizal development and stability. These mycorrhizal helper bacteria may produce a substrate used by mycorrhizal fungus, growth factors, and vitamins. Free-living nitrogen-fixing bacteria may favor mycorrhizal development, contributing ammonium and amino acids to the fungus (Garbaye, 1991).

Like most microorganisms, mycorrhizal fungi accumulate secondary metabolites, which are toxic to the fungus itself and limit its growth. Some mycorrhizospheric bacteria are able to break down these molecules, and are thus favourable to the development of the fungus. A given microorganism could alter root exudates by stimulating the root to produce more or different substances, or by metabolizing part of the exudates. Infection of roots by mycorrhizal fungi might be stimulated by helper bacteria which release cellulolytic and pectolytic enzymes, making it easier to separate the outer cells of the root cortex. Phytohormones could also be involved, which are produced by wide range of rhizosphere microorganisms and control the morphology of rootlets.

A key factor determining whether a plant can survive in a nutrient-deficient soil is its ability to mobilize some elements, such as phosphorus, from insoluble minerals or organic molecules. In this respect, mycorrhizal fungi are not necessarily more efficient than the root itself, and it has been shown that mycorrhizosphere bacteria play a major role in mobilizing elements for the benefit of the plant-fungus association (Garbaye, 1991).

FACTORS AFFECTING ECTOMYCORRHIZAL REDUCTION

Type and Severity of Disturbance

Fewer ectomycorrhizae on Douglas fir seedlings grow in burned clear-cuts compared with unburned clear-cuts. Amaranthus et al. (1987) found 90 percent less ectomycorrhiza and 44 percent less basal area growth on Douglas fir seedlings grown in clear-cut and severely burned soils compared with undisturbed forest soil. Because ectomycorrhizae predominate in the organic layers of the soil, the degree of organic matter lost from a site can influence mycorrhizae populations. Harwey et al. (1976) found that up to 95 percent of the active ectomycorrhizae were located in humus and decaying wood in a mature Douglas fir-larch *(Larix)* forest.

The importance of decaying wood to support ectomycorrhizal populations may be most critical following disturbances, such as wildfires. Decaying wood also acts as habitat for small mammals, which are important in distributing fungal spores of several belowground mycorrhizal fungi (Maser et al., 1978).

Forest practices that increase soil compaction and erosion adversely affect mycorrhizal formation and seedling establishment (Amaranthus, 1992). Increasing soil density can decrease fungal growth. In addition, decreasing soil structure restricts the movement of oxygen and water into soil and flushing out carbon dioxide. Populations of mycorrhizal fungi can be altered when levels of these basic elements become extreme. The diversity and abundance of these organisms is also reduced when topsoil is severely disturbed and eroded.

Litter and soil organic matter are important in maintaining soil physical properties. Slash burning can remove up 75 percent of soil organic matter and leads to nitrogen volatilization (Parke et al., 1983b). Resulting ash can cause large increases in available phosphorus, potassium, calcium, and magnesium that are subject to rapid loss by leaching. Once organic matter is removed, it may take several years to reaccumulate. Although an apparent allelopathic response of seedling ectomycorrhizae to forest litter has been reported, ectomycorrhizae of older trees develop better in humus and litter layers than in mineral soil (Parke et al., 1983a). Mycorrhizal fungi present in litter layers have been implicated in nutrient cycling. However, most of these fungi are considered incapable of saprobic growth and lack the enzymes necessary for utilization of complex carbohydrates such as cellulose and lignin. Ectomycorrhizal-fungus-containing litter has been exploited as an inoculum for more than a century and is probably the most widely used source of ectomycorrhizal inoculum for forest nurseries worldwide (Marx, 1980). Advantages of using litter include reliability and the relative ease of obtaining large amounts of viable inoculum of fungi presumably adapted to the forest sites from which they were taken; disadvantages include the potential for introducing pathogens into a nursery and the inability to select particular fungus species with specific desirable attributes such as tolerance to drought, high temperatures, or low pH soils (Parke et al., 1983a).

Nitrogen Fertilizer Treatment

Nitrogen fertilization reduces soil microbial biomass and activity (Arnebrant and Söderström, 1992).

Forest fertilization may result in decreased production of fruit bodies of ectomycorrhizal fungi as well as a decreased number of ectomycorrhizal species (Wästerlund, 1982). However, the fruit body production of *Paxillus involutus* and *Lactarius rufus* has been reported to be unaffected or even increased as a result of forest fertilization (Ohenoja, 1988). The number of root tips colonized by ectomycorrhizal fungi may also decrease after fertilization (Ahlström et al., 1988), but Laiho et al. (1987) and Shubin (1988) found no effect on total number of ectomycorrhizal root tips after fertilization. The observed effects of fertilization on the ectomycorrhizal fungi may either be direct or indirect, mediated via effect on the host tree. A decreased fine root production has been reported as a result of forest fertilization (Ahlström et al., 1988). Plants have also been shown to increase their shoot: root ratio at higher N levels (Ingestad et al., 1986), and this has been interpreted in terms of a changed carbon allocation pattern resulting in relatively less carbon being available to the roots (Rufty et al., 1990).

Fungicides

In forest nurseries, pine seedlings are adapted to rather high fertilization levels and have mainly ectendomycorrhizae, which may not be typical of planting sites (Väre, 1989). Fungicide application in forest nurseries against fungal diseases may retard or even prevent the formation of proper ectomycorrhizae in root systems. Active ingradients of fungicides can be heavy metals (e.g., copper oxychloride), but many fungicides (e.g., propiconazole) function by inhibiting the reactions in the pathway of sterol biosynthesis. Fungal sterols (ergosterols) are primarily involved in the cell membranes modulating fluid properties of membrane lipids (Weete, 1989). Fungicides may cause increased transparency and gradual granulation and degeneration of cytoplasm in the fungal symbiont of mycorrhizal roots (Manninen et al., 1998).

Climatic Conditions

Climate influences seedling growth and ectomycorrhizae formation (Pilz and Perry, 1984). Dry climates may limit the activity of mycorrhizal fungi by decreasing the length of time for spore production, germination, and mycelial growth (Amaranthus and Perry, 1987). Seedlings in moist climates may be able to survive longer without mycorrhizae than those in dry climates, increasing their chances of becoming colonized. Moisture content also affects uptake of certain nutrients by mycorrhizae (Gadgil, 1972). Seedlings growing in cold climates may also require rapid, early mycorrhizal colonization to take advantage of the short growing season and obtain the necessary nutrients and water to survive the long cold season and early frosts. According to Parke et al. (1983b), maximum formation of mycorrhizae in disturbed and undisturbed forest soils of Oregon occured at 18.5 to 24°C, and there were no significant qualitative or quantitative differences between disturbed (clear-cut) or undisturbed (forest) soils. Mycorrhizae formation was moderate even at the lowest temperature tested (7.5°C) but was greatly reduced or prevented at or above 29.5°C. Treatment of soil at 35°C for 1 week did not appear to adversely affect viability of ectomycorrhizal fungus propagules, but young mycorrhizae subjected to the same treatment appeared to be severely injured. Thus, the ability of native mycorrhizal fungi to grow at low soil temperatures is especially important as they may contribute to the survival of seedlings outplanted into climatic zones characterized by warm, dry summers following cool, wet winters and springs.

Biotic Conditions

Little is known of the persistence and distribution of ectomycorrhizae in the absence of living hosts. It has been suggested by Hacskaylo (1973) that ectomycorrhizal fungi do not persist long in the absence of host-supplied substrate. Logged and burned sites are rapidly invaded by woody shrubs. These shrubs (e.g., members of the Ericaceae and Fagaceae) form mycorrhizae with many of the same fungi as do members of the Pinaceae (Molina and Trappe, 1982). These shrubs help preserve mycorrhizae fungus diversity during periods of rapidly changing aboveground community structure. Some woody shrub species may act as biological reservoirs of not only mycorrhizal frungi but other microflora as well. Survival of propagules and input of spores is insufficient to maintain ectomycorrhizal inocula potential in the absence of living hosts (Amaranthus and Perry, 1987). Invasion of sites by nonectomycorrhizal plants over several years can seriously affect reforestation, particularly in the case of ectomycorrhizal tree species growing on difficult sites where seedlings must establish ectomycorrhizae rapidly to survive. Ectomycorrhizae formation and seedling survival generally decreases as the length of time between disturbance and reforestation increases (Perry et al., 1989).

RECEPTIVENESS OF FOREST SOILS TO ECTOMYCORRHIZAL ASSOCIATION

Soil fumigation, fertilization, and other nursery practices can affect population levels of mycorrhizal fungi and select for strains adapted to nursery conditions. Mycorrhizae are sometimes deficient in modern container nurseries (Perrin et al., 1996). A few dominant aggressive fungi, well adapted to the nursery, commonly colonize seedlings before outplanting. Aggressive symbionts such as *Thelephora terrestris* may quickly dominate root systems, reducing the ectomycorrhizal diversity within the first years of seedling development. Mycorrhizal fungi on nursery seedlings often disappear after outplanting into forest sites and are replaced by the indigenous mycorrhizal fungus. The ability of ectomycorrhizal fungi to persist or reinvade soil after clear-cutting varies with geographical location and is affected by preparation of sites (Stenström et al., 1990). Dramatic decreases in or total elimination of active mycorrhizal fungi sometimes persists for years after clear-cutting (Perrin et al., 1996). Failure in field plantation or poor field performance of seedlings is likely related to the lack or low level of native mycorrhizal fungi and the ineffectiveness of mycorrhizae developed at the nursery. For these reasons, positive effects can be expected on outplanted seedling performance after inoculation with selected mycorrhizal fungi (Stenström and Ek, 1990). However, neutral or negative effects of mycorrhizal inoculation on seedling growth and survival have also been observed (Stenström et al., 1990). These results may be related to the ability of the selected fungus to persist on the root after outplanting. Perrin et al. (1996) claimed that native mycobionts colonize short roots in the region of new root growth, whereas inoculated symbionts tend to extend directly from mycorrhizae developed prior to transplanting.

Most investigators have reported a decline or loss of native ectomycorrhizal fungi in soils after clear-cutting. Soil infectivity decreases rapidly with time, and the proportion of ectomycorrhizal fungal types shifts after disturbance. It is very likely that ectomycorrhizal populations will be low or nonexistent in many sites to be planted, especially in fields set aside from agriculture and devoid of an indigenous ectomycorrhizal inoculum. The receptiveness of the soil will then depend more on its physical and chemical properties than on the microbial component, and inoculation with a selected mycosymbiont could be advantageous. The success of controlled inoculation of ectomycorrhizal species will depend on disturbances in mycorrhizal soil infectivity in the absence of host plants and on the ecological competence of the species or strains selected for inoculation at the nursery level (Perrin et al., 1996).

Artificial regeneration using direct sowing or planting of young plants from nurseries plays a major role in reforestation practices. Even if vigorous nonmycorrrhizal seedlings are produced on fertilized soil, the use of frequent applications of concentrated inorganic fertilizers leads to a progressive independence of forest plants from mycorrhizal fungi. As Le Tacon (1982) pointed out, the higher the fertilizer level, the fewer the symbionts that remain and the more fertilizers needed. A favorable development of mycorrhizae requires a soil low in such nutrients as nitrogen, phosphorus, potassium, or calcium (Chalot et al., 1988). High levels of nitrogen and phosphorus in soil decrease the sucrose content of short roots, which in turn decreases their susceptibility to ectomycorrhizal development.

PRACTICAL ASPECTS OF MYCORRHIZAE: PERSPECTIVES

The use of specific fungi to form ectomycorrhizae on forest tree seedlings in the nursery can improve performance after outplanting. The primary purpose of inoculation is to provide seedlings with adequate ectomycorrhiza for planting in man-made forests (Marx et al., 1991).

In selection of ectomycorrhizal fungi for use in inoculating soils in tree nurseries, several criteria are important. Among these are optimal pH and pH tolerance of the candidate isolates. An isolate should show promise for adapting to the pH of the nursery soil as well as the soil at the intended planting site. The widespread problem of acid rain creates a need for fungi to be tolerant to H_2SO_4 as well. The fungi selected for inoculation should be highly effective in mycorrhizae with the host tree and well adapted to the stand conditions, including stress factors (drought, temperature, etc.).

The interrelations between the accompanying microorganisms should also be known. It was demonstrated that enhanced mycorrhizae formation on *Pinus radiata* seedlings occurred when a mixed soil microflora was used in combination with the ectomycorrhizal fungus.

Four types of inocula have been used for mycorrhizal inoculation in nurseries:

1. Soil from natural forests, plantations, or nurseries
2. Mycorrhizal seedlings
3. Pure cultures of mycorrhizal fungi
4. Inoculation with fungal spores

Soil inocula have been most widely used in forestry practice. Application of mycorrhizal soil is the easiest and most simple method of nursery inoculation. Nevertheless, it has several drawbacks. Soil inoculum contains a mixture of both mycorrhizal fungi and other soil organisms and perhaps also root pathogens.

Soils from natural pine forests and healthy plantations contain a large variety of mycorrhizal fungi, among them some species that are effective and well suited for nursery conditions (Perrin et al., 1996). On the other hand, soil inoculum from an old nursery may contain damping-off fungi, nematodes, and other nursery pests.

Soil for inoculation should not be taken from old nurseries or even natural forests where dangerous pathogens (*Armillaria mellea, Phytophthora cinnamomi,* etc.) are known to exist.

The application of living seedlings for inoculation is quite a reliable method. The species of fungi are usually not known, but if the mother trees are vigorous and healthy and if inspection reveals good mycorrhizal structure, the conclusion can be drawn that efficient fungi are present. If unfavorable soil properties such as excessively high pH are indicated, this can be corrected by appropriate soil treatment, for example, acidification.

Theoretically, pure cultures and spores seem to be the best inocula; therefore their use is most often recommended. Pure cultures are easy to handle and there is no risk of introduction of pests. Unfortunately, mycorrhizal fungi in pure cultures are slow growing. The inoculum must be produced in a laboratory and shipped to nurseries for use. The procedure is rather expensive. If it happens that the fungus dies immediately after inoculation, probably because of the competition with other soil fungi and other microorganisms (mycolysis), mycorrhizal infection does not take place.

Inoculum by spores would have many advantages. They need little space and the risk of disease transmission is eliminated. Spores can be collected only during a brief period of the year, which may be a less favorable season for inoculation, and therefore spores have to be stored. Here a sufficient knowledge about spore viability and symbiotic efficiency of different fungal species is required.

In a recent review of mycorrhizal research, some authors state that we may know less than we think about mycorrhizae, since we have consistently based broad hypotheses and conclusions on studies of a small number of taxa (Kilronomos and Kendrick, 1993). Populations of mycorrhizal fungi are thought to have occupied the same soil habitats for millions of years, slowly adapting to changes in site conditions (Trappe and Molina, 1986).

Some mycorrhizal fungi appear to have worldwide distribution patterns and have apparently adapted to a wide range of habitats, but it is known that soil factors such as pH restrict the distribution of many taxa. Unfortunately, information about the soil and climatic conditions where isolates used in experiments were originally obtained is usually not provided. These data are urgently needed to allow isolates to be selected for use in inoculation programs and to establish correlations between their taxonomy and physiology.

The functional diversity of mycorrhizal fungi includes variations between individual species and isolates of fungi in the following capacities: obtaining limiting soil nutrients (inorganic forms of phosphorus, nitrogen, organic forms of nitrogen, trace elements); amelioration of adverse soil conditions (toxic concentrations of metal ions, extremes in pH, high conductivity, nutrient imbalances such as high Mg/Ca ratios); responses to severe climatic conditions (limited or excessive water supply, temperature extremes); compatibility with different hosts; tolerance of adverse soil conditions (disturbance, microbial competition); survival and spread in soil by spores, mycelia; capacity for inoculum production (spore production, growth in sterile culture).

The fact that most mycorrhizal research has been concerned with plant growth responses with little consideration for the fungi involved has also helped create the false impression that most of these fungi are functionally equivalent (Abbott and Robson, 1991). Fungal properties that would help determine the effectiveness of mycorrhizal associations include the amount of soil hyphae produced relative to root colonization, the rate of hyphal growth and root colony initiation, and physiological characteristics which regulate nutrient absorption or nutrient translocation by hyphae and exchange with the host.

CONCLUSIONS

Many papers have been published on mycorrhizae, most of which relate to forest trees (Parke et al., 1983a,b; Mikola, 1988; Danielson and Visser, 1989; Baar and de Vries, 1995). Mycorrhizal associations of higher plants are among the most widely known microbial processes in nature. In natural conditions, the symbiosis forms spontaneously and contributes to plant growth. Many forest trees cannot grow without ectomycorrhizae. This point has practical significance to afforestation programs with normally ectomycorrhizal trees in areas of the world where their symbiotic fungi do not occur naturally.

Mycorrhizal associations function as dynamic biological linkages. Understanding the extent, diversity, and function of these linkages is vital to predict future productivity. Although inoculation may be necessary in certain reforestation or afforestation situations (e.g., severe, long-term degradation), maintaining the biological diversity of mycorrhizal fungi and other soil microbes through wise resource management may be the best way to apply mycorrhizal technology. Diversity in both plant and microbial communities stabilizes the plant-soil system during environmental fluctuations, particularly following adverse disturbance (Perry et al., 1987). Specificity phenomena in mycorrhizal associations act as primary sources of functional biodiversity in terrestial ecosystems.

REFERENCES

Abbott L.K. and Robson A.D. (1991). Factors influencing the occurrence of vesicular arbuscular mycorrhizas. *Agriculture, Ecosystems and Environment* 35:121-150.

Ahlström K., Persson H., and Börjesson I. (1988). Fertilization in a mature Scots pine (*Pinus sylvestris* L.) stands-effects on fine roots. *Plant Soil* 106: 179-190.

Allen M.F. and Mac Mahon J.A. (1985). Impact of disturbance on cold desert fungi: Comparative microscale dispersion patterns. *Pedobiologia* 28: 215-224.

Amaranthus M.P. (1992). Mycorrhizas, forest disturbance and regeneration in the Pacific Northwestern United States. In D.J. Read, D.H. Lewis, A.H. Fitter, and I.J. Alexander (eds.), *Mycorrhizas in ecosystems* (pp. 202-207). Wallingford, Oxon, UK: C.A.B. International.

Amaranthus M.P. and Perry D.A. (1987). Effect of soil transfer on ectomycorrhiza formation and the survival and growth of conifer seedlings in disturbed forest sites. *Can. J. For. Res.* 17: 944-950.

Amaranthus M.P., Perry D.A., and Borchers S.L. (1987). Reduction of native mycorrhizae reduces growth of Douglas-fir seedlings. In D.M. Sylvia and J.H.

Graham (eds.), *Mycorrhizae in the next decade* (pp. 80-126). Gainesville, FL: University of Flordia.

Arnebrant K. and Söderström B. (1992). Effect of different fertilizer treatments on ectomycorrhizal colonization potential in two Scots pine forests in Sweden. *For. Ecol. Manag.* 53: 77-89.

Baar J. and de Vries F.W. (1995). Effect of manipulation of litter and humus layers on ectomycorrhizal colonization potential in Scots pine stands of different age. *Mycorrhiza* 5: 267-272.

Brand F. (1991). Ektomykorrhizen ouf *Fagus sylvatica*. Charakterisierung und Identifizierung, ökologische Kennzeichnung und unsterile Kultivierung. *Libri Botanici* 2: 1-228.

Brand F. and Agerer R. (1986). Studies on ectomycorrhizae. VIII—Mycorrhizae formed by *Lactarius subdulcis, L. vellereus* and *Laccaria amethystina* on beech. *Zeitschrift für Mykologie* 52 (2): 287-320.

Chalot M., Battut P.M., Botton B., Le Tacon F., and Garbaye J. (1988). Recent advances in physiological aspects of ectomycorrhizal effects on tree development. *Acta Ecologica* 9 (4): 333-351.

Crafts C.B. and Miller C.O. (1974). Detection and identification of cytokinins produced by mycorrhizal fungi. *Plant Physiol.* 54: 586-588.

Dahm H., Różycki H., Strzelczyk E., and Li C.Y. (1993). Production of B-group vitamins by *Azospirillum* spp. grown in media of different pH at different temperatures. *Zentrbl. Mikrobiol.* 148: 195-203.

Dahm H. and Strzelczyk E. (1995). Impact of vitamins on cellulolytic, pectolytic and proteolytic activity of mycorrhizal fungi. *Symbiosis* 18: 233-250.

Danielson R.M. and Visser S. (1989). Host response to inoculation and behaviour of introduced and indigenous ectomycorrhizal fungi of jack pine grown on oilsands tailings. *Can. J. For. Res.* 19: 1412-1421.

Dosskey M.G., Boersma L., and Linderman R.G. (1991). Role of the photosynthate demand of ectomycorrhizas in the response of Douglas fir seedlings to drying soil. *New Phytologist* 117: 327-334.

Gadgil P.D. (1972). Effect of waterlogging on mycorrhizas of radiata pine and Douglas-fir. *New Zealand J. For. Sci.* 2: 222-226.

Garbaye J. (1991). Biological interactions in the mycorrhizosphere. *Experientia* 47: 370-375.

Gianinazzi-Pearson V. and Gianinazzi S. (1985). The physiology of improved phosphate nutrition in mycorrhizal plants. In *Physiological and genetical aspects of mycorrhizae* (pp. 101-120). Proceedings of the First European Symposium on Mycorrhizae, Dijon. Paris: I.N.R.A.

Gogala N. (1991). Regulation of mycorrhizal infection by hormonal factors produced by hosts and fungi. *Experientia* 47: 331-340.

Hacskaylo E. (1973). Carbohydrate physiology of ectomycorrhizae. In G.C. Marks and T.T. Kozlowski (eds.), *Ectomycorrhizae: Their ecology and physiology* (pp. 207-230). London: Academic Press.

Harley J.L. and Smith E.E. (1983). *Mycorrhizal symbiosis.* London: Academic Press.

Harwey A.E., Larsen M.J., and Jurgensen M.F. (1976). Distribution of ectomycorrhizae in a mature douglas-fir/larch forest soil in western Montana. *Forest Sci.* 22: 393-398.

Ingestad T., Arveby A.S., and Kähr M. (1986). The influence of ectomycorrhiza on nitrogen nutrition and growth of *Pinus sylvestris* seedlings. *Physiol. Plant.* 68: 575-582.

Killham K. (1994). The mycorrhizal association—A plant/root fungus interaction. In *Soil Ecology* (pp. 66-75). Cambridge, UK: Cambridge University Press.

Kilronomos J.N. and Kendrick W.B. (1993). Research on mycorrhizas: Trends in the past 40 years as expressed in the "Mycolit" database. *New Phytol.* 125: 595-600.

Kottke I. and Oberwinkler F. (1987). The cellular structure of the Hartig net: Coenocytic and transfer cell-like organization. *Nord. J. Bot.* 7: 85-95.

Kreuzwieser J. and Rennenberg H. (1998). Sulphate uptake and xylem loading of mycorrhizal beech roots. *New Phytol.* 140: 319-329.

Laiho O., Sarjala T., Hyvärinen R., and Rautiainen L. (1987). Effect of fertilization on mycorrhiza in pine stands. *Folia Forest,* 699.

Lapeyrie F.F., Chilvers G.A., and Douglas P.A. (1984). Formation of metachromatic granules following phosphate uptake by mycelial hyphae of an ectomycorrhizal fungus. *New Phytol.* 98: 345-360.

Le Tacon F. (1982). Perspectives de la maitrise de la mycorhization en sylviculture. In *Les Mycorhizes: Biologie et Utilisation* (pp. 273-285). Dijon: I.N.R.A.

Manninen A.M., Laatikainen T., and Holopainen T. (1998). Condition of Scots pine fine roots and mycorrhiza after fungicide application and low-level ozone exposure in a 2-year field experiment. *Trees* 12: 347-355.

Marx D.H. (1972). Ectomycorrhizae as biological deterrents to pathogenic root infection. *Ann. Rev. Phytopathol.* 10: 426-434.

Marx D.H. (1980). Ectomycorrhizal fungus inoculations: A tool for improving forestation practices. In P. Mikola (ed.), *Tropical mycorrhiza research* (pp. 13-71). Oxford: Oxford University Press.

Marx D.H., Ruehle J.L., and Cordell C.E. (1991). Methods for studying nursery and field response of trees to specific ectomycorrhiza. In J.R. Norris, D.J. Read, and A.K. Varma (eds.), *Methods in microbiology* (pp. 383-411). London: Academic Press.

Maser C., Trappe J.M., and Nussbaum R.A. (1978). Fungal-small mammal interrelationships with emphasis on Oregon coniferous forests. *Ecology* 59: 799-809.

Mc Gonigle T.P. and Fitter A.H. (1990). Ecological specificity of vesicular-arbuscular mycorrhizal associations. *Mycol. Res.* 94: 120-122.

Melin E. (1925). *Untersuchungen über die Bodeutung der Baummykorrhiza. Eine ökologisch—physiologische Studie,* as cited in Molina R., Massicotte H., and Trappe J. (1992). Specificity phenomena ion mycorrhizal symbioses: Community-ecological consequences and practical implications. In M.J. Allen (ed.), *Mycorrhizal functioning: An integrative plant-fungal process* (pp. 357-423). New York: Chapman and Hall.

Meyer F.H. (1973). Distribution of ectomycorrhizae in native and mane-made forests. In G.C Marks and T.T. Kozlowski (eds.), *Ectomycorrhizae—Their ecology and physiology* (pp. 79-105). New York: Academic Press.

Mikola P. (1965). Studies on the ectotrophic mycorrhiza of pine. *Acta Forestalia Fennica* 79: 1-56.

Mikola P. (1988). Ectendomycorrhiza of conifers. *Silva Fennica* 22: 19-27.

Miller C.O. (1967). Zeatin and zeatin-riboside from a mycorrhizal fungus. *Science* 157: 1055-1057.

Molina R. (1980). Patterns of ectomycorrhizal host-fungus specificity in the Pacific Northwest. Doctoral thesis, Oregon State University, Corvallis.

Molina R., Massicotte H., and Trappe J.M. (1992). Specificity phenomena in mycorrhizal symbioses: Community ecological consequences and practical implications. In M.F. Allen (ed.), *Mycorrhizal functioning: An intergrative plant-fungal process* (pp. 357-423). New York: Chapman and Hall.

Molina R. and Trappe J.M. (1982). Lack of mycorrhizal specificity by the ericaceous hosts *Arbutus menziesii* and *Arctostaphylos uva-ursi. New Phytol.* 90: 495-509.

Mortier F., Le Tacon F., and Garbaye J. (1988). Effect of inoculum type and inoculation dose on ectomycorrhizal development, root necrosis and growth of Douglas fir seedlings inoculated with *Laccaria laccata* in a nursery. *Ann. Sci. For.* 45: 301-310.

Oaks A. and Hirel B. (1985). Nitrogen metabolism in roots. *Ann. Rev. Plant Physiol.* 36: 345-365.

Ohenoja E. (1988). Behaviour of mycorrhizal fungi in fertilized forests. *Karstenia* 28: 27-30.

Parke J.L., Linderman R.G., and Trappe J.M. (1983a). Effects of forest litter on mycorrhiza development and growth of Douglas-fir and western red cedar seedlings. *Can. J. For. Res.* 13 (4): 666-671.

Parke J.L., Linderman R.G., and Trappe J.M. (1983b). Effect of root zone temperature on ectomycorrhiza and vesicular-arbuscular mycorrhiza formation in disturbed and undisturbed forest soils of southwest Oregon. *Can. J. For. Res.* 13 (4): 657-665.

Perrin E., Parlade X., and Pera J. (1996). Receptiveness of forest soils to ectomycorrhizal association: 1. Concept and method as applied to the symbiosis between *Laccaria bicolor* (Maire) Orton and *Pinus pinaster* Art. or *Pseudotsuga nmenziesii* (Mirb.) Franco. *Mycorrhiza* 6: 469-476.

Perry D.A., Amaranthus M.P., Borchers J., Borchers S.L., and Brainerd R. (1989). Bootstrapping in ecosystems. *Bioscience* 39: 230-237.

Perry D.A., Molina R., and Amaranthus M.P. (1987). Mycorrhizae, mycorrhizospheres and reforestation: Current knowledge and research needs. *Can. J. For. Res.* 17: 929-940.

Pilz D.P. and Perry D.A. (1984). Impact of clearcutting and slash burning on ectomycorrhizal associations of Douglas-fir. *Can. J. For. Res.* 14: 94-100.

Plassard C., Martin F., Mousain D., and Salsac L. (1985). Physiology of nitrogen assimilation by mycorrhiza. In *Physiological and Genetical Aspects of Mycor-*

rhizae. Proceedings of the first European Symposium on Mycorrhizae (pp. 111-120). Dijon, Paris: I.N.R.A.

Pohleven F. (1988). *The role of cytokinins on ion transport in the mycelium of ectomycorrhizal fungus Suillus variegatus* (Sow.ex Fr.) O. Kuntze. Doctoral thesis, University of Ljubljana, Yugoslavia.

Pohleven F. (1989). The influence of zeatin on the ion absorption and transport in mycorrhizal fungus *Suillus variegatus. Biol. Vestn.* 37: 67-78.

Pokojska A. and Strzelczyk E. (1988). Effect of organic acids on production of auxin-like substances by ectomycorrhizal fungi. *Symbiosis* 6: 211-224.

Rennenberg H. (1999). The significance of ectomycorrhizal fungi for sulfur nutrition of trees. *Plant and Soil* 215: 115-122.

Ritter G. And Lyr H. (1965). The significance of ectomycorrhizal fungi for the utilization of different sources of phosphate by *Pinus sylvestris* L. and fruiting—bodies of *Hebeloma mesophaeum. Zetschr. für Mikrobiol.* 5, (4): 274-283.

Rudawska M. (1980). Effect of pine root extractives on growth regulators and IAA-oxidase activity in pure cultures of mycorrhizal fungi. *Acta Physiol. Plant.* 2: 133-144.

Rufty T.W. Jr., Mac Kown C.T., and Volk R.J. (1990). Alterations in nitrogen assimilation and partitioning in nitrogen stressed plants. *Physiol. Plant.* 79: 85-95.

Schisler D.A. and Linderman R.G. (1989). Selective influence of volatiles purged from coniferous forest and nursery soils on microbes of a nursery soil. *Soil Biol. Biochem.* 21: 389-396.

Shubin V.I. (1988). Influence of fertilizers on the fruiting of forest mushrooms. *Acta Bot. Fenn.* 135: 85-87.

Slankis V. (1973). Hormonal relationships in mycorrhizal development. In G.C. Marks (ed.), *Ectomycorrhizae* (pp. 232-298). New York: Academic Press.

Smith S.E. and Smith F.A. (1990). Structure and function of the interfaces in biotrophic symbioses as they relate to nutrient transport. *New Phytol.* 114: 1-38.

Stenström E. and Ek M. (1990). Field growth of *Pinus sylvestris* following nursery inoculation with mycorrhizal fungi. *Can. J. For. Res.* 20: 914-918.

Stenström E., Ek M., and Unestam T. (1990). Variation in field response of *Pinus sylvestris* to nursery inoculation with four different ectomycorrhizal fungi. *Can. J. For. Res.* 20: 1796-1803.

Strzelczyk E., Dahm H., and Pachlewski R. (1991). B-group vitamins production by mycorrhizal fungi in response to pH. *Plant and Soil* 137: 237-241.

Strzelczyk E., Dahm H., Pachlewski R., and Różycki H. (1997). Production of indole compounds by the ectomycorrhizal fungus *Cantharellus cibarius* Fr. *Pedobiologia* 41: 402-411.

Strzelczyk E. and Kampert M. (1987). Effect of B-group vitamins on cytokinin-like substances production by ectomycorrhizal fungi of pine (*Pinus sylvestris* L.). *Symbiosis* 3: 135-146.

Strzelczyk E., Sitek J., and Kowalski S. (1976). Production of gibberellin-like substances by fungi isolated from mycorrhizae of pine (*Pinus sylvestris* L.). *Acta Microbiol. Polon.* 2 (7): 145-153.

Sylvia D.M. and Sinclair W.A. (1983a). Phenolic compounds and resistance to fungal pathogens induced in primary roots of Douglas-fir seedlings by the ectomycorrhizal fungus *Laccaria laccata*. *Phytopathology* 73: 390-397.

Sylvia D.M. and Sinclair W.A. (1983b). Suppresive influence of *Laccaria laccata* on *Fusarium oxysporum* and Douglas-fir seedlings. *Phytopathology* 73: 384-389.

Trappe J.M. and Molina R. (1986). Taxonomy and genetics of mycorrhizal fungi: their interactions and relevance. In V. Gianinazzi-Pearson and S. Gianinazzi (eds.), *Mycorrhizae: Physiology and genetics* (pp. 133-146). Paris: I.N.R.A.

Väre H. (1989). The mycorrhizal condition of weakened Scots pine saplings grown on ploughed sites in northern Finland. *Can. J. For. Res.* 19: 341-346.

Varma A. (1998). Mycorrhizae—The friendly fungi: What we know, what should we know, and how do we know? In A. Varma (ed.), *Mycorrhiza mannual* (pp. 1-24). Berlin: Springer-Verlag.

Wang C.J.K. and Wilcox H.E. (1985). New species of ectendomycorrhizae and pseudomycorrhizal fungi: *Phialophora finlandia, Chloridium paucisporum* and *Phialocephala fortinii*. *Mycologia* 77: 955-958.

Wästerlund I. (1982). Försvinner tallens mykorrhizasvampar vid gödsling? [Do pine mycorrhizae disappear following fertilization?] *Sven. Bot. Tidskr.*, 76: 411-417.

Weete J.D. (1989). Structure and function of sterols in fungi. Adv. *Lipid Res.* 23: 115-167.

Williamson B. and Alexander I. (1975). Acid phosphatases localized in the sheath of beech mycorrhiza. *Soil Biol. Biochem.* 7: 195-198.

Zak B. (1964). Role of mycorrhizae in root disease. *Ann. Rev. Phytopathol.* 2: 377-382.

Chapter 11

Physiological and Molecular Aspects of Osmotic Stress Alleviation in Arbuscular Mycorrhizal Plants

Juan Manuel Ruiz-Lozano

INTRODUCTION

In many terrestrial ecosystems plants are naturally exposed to adverse environmental factors that have a negative effect on plant survival, development, and productivity. Drought and salinity are considered the most important abiotic factors limiting plant growth and yield in these areas (Sánchez-Díaz and Aguirreolea, 1993). Drought and salinity share a common osmotic component (by water limitation for plant) as the main factor (although not the only one in the case of salinity) responsible of their negative effects on plant development. Thus, osmotic stress is broadly used to refer to situations where insufficient water availability limits plant growth and development (Zhu et al., 1997).

Although the different plant species can vary in their sensitivity and response to the decreased water potential caused by water deficit, it may be assumed that all plants have encoded capability for stress perception, signaling, and response (Bohnert et al., 1995). Plants can respond to osmotic stress at morphological, anatomical, and cellular levels with modifications that allow the plant to avoid the stress or to increase its tolerance (Bray, 1997). The morphological and anatomical adaptations can be of vital importance for some plant species, but they are not a general response of all plant species. In contrast, the cellular responses to osmotic stress seem to be conserved in the plant kingdom. Some of the cellular responses, such as osmotic adjustment by synthesizing compatible osmolytes, are common to all cellular organisms (Csonka and Hanson, 1991). Osmotic adjustment is achieved by accumulation of compatible solutes in order to keep a gradient of water potential favorable to water entrance in the plant and maintenance of cell turgor.

If water deficit persists and the osmotic adjustment is insufficient to avoid plant damage, there are other mechanisms by means of which the plants cope with drought stress. First, plants can increase the water permeability in their tissues. Aquaporins are water channel proteins that, when opened, facilitate the passive movement of water molecules down a water potential gradient. These proteins belong to the large major intrinsic protein (MIP) family of transmembrane channels that are represented in all kingdoms (Chrispeels and Agre, 1994). Two classes of plant aquaporins, located in the plasma membrane and tonoplast, respectively, have been identified so far (Johnson et al., 1990; Kammerloher et al., 1994). Transcripts of *Mip* (major intrinsic protein) genes whose abundance changes under osmotic stress have been isolated from root cDNA libraries of ice plants (Bohnert et al., 1995). The encoded MIPs are homologous to plant and animal aquaporins (Chrispeels and Agre, 1994). It has been suggested that vacuolar and plasma membrane aquaporins, acting in concert, are responsible for the cytosolic osmoregulation that is necessary for maintaining normal metabolic processes. However, inhibition studies of aquaporins in vivo and antisense mutant studies have also suggested that, in addition to cytosolic osmoregulation, aquaporins are important for the bulk of water in the whole plant (Kjelbom et al., 1999). Second, there is an increasing interest in the study of a group of proteins called late embryogenesis abundant (LEA) proteins, which accumulate in the seeds during their maturation phase, when tolerance to desiccation is needed. A number of studies have demonstrated that these proteins also accumulate in the vegetative plant organs during periods of water deficit, and it has been proposed that LEA proteins play an important role in the maintenance of the structure of other proteins and membranes, in the sequestration of ions, binding of water, and functioning as molecular chaperones (Close, 1996). Nowadays, it is known that overexpression of LEA proteins in plants and yeast confers tolerance to osmotic stresses (Xu et al., 1996; Imai, 1996).

The aim of this chapter is to summarize the current knowledge about the osmotic stress alleviation in arbuscular mycorrhizal (AM) plants and to discuss new perspectives for molecular studies that can contribute to gain a global understanding of the different mechanisms by which the AM symbiosis protects host plants against water deficit.

Oxidative Stress

Water deficit in plants also induces an increase of free radicals in cells, thus generating oxidative stress (Moran et al., 1994; Sgherri and Navari-Izzo, 1995; Iturbe-Ormaetxe et al., 1998). Many of the degenerative reactions associated with several abiotic stresses are mediated by reduced oxygen species such as superoxide radicals or hydrogen peroxide. By themselves they are relatively unreactive, but they can form species that are damaging to essential cellular components. In presence of metal ions (e.g., iron), superoxide and hydrogen peroxide can react in a Haber-Weiss reaction to form hydroxyl radicals. These radicals (and their derivatives) are among the most reactive species known to chemistry, capable of reacting indiscriminately to cause lipid peroxidation, the denaturation of proteins and the mutation of DNA (Halliwell and Gutteridge, 1989; Bowler et al., 1992). However, plants possess a number of antioxidant mechanisms that protect them from the production of these kinds of radicals. Gamble and Burke (1984) appear to have been the first to consider the relationship between antioxidant systems and water deficit, and showed that glutathione reductase (GR) activity in wheat leaves was higher in droughted than in well-watered plants. Since then, plenty of reports have shown experimental evidences of a positive correlation between tolerance to water deficit and increased antioxidant activities (for a review, see Smirnoff, 1993; Noctor and Foyer, 1998).

In addition to the naturally protecting systems of plants against stress, in nature, plants grow associated to a number of soil microorganisms that can be beneficial. Among them, mycorrhizal fungi are widespread microorganisms able to establish a symbiotic association with the roots of most terrestrial plants. Plants improve their ability for nutrient uptake and their tolerance to biotic and abiotic stresses while the fungus gets a protected ecological niche and plant photosynthates (Smith and Read, 1997). There are several types of mycorrhizal associations, but the most abundant is the so-called arbuscular mycorrhizae, which involves about 90 percent of the plant species on earth. AM symbiosis is present in all natural ecosystems, even in those affected by adverse environmental conditions (Barea and Jeffries, 1995). The AM symbiosis can be defined as a specialized system for nutrients uptake and transfer, more efficient than the roots themselves. That is especially true for nutrients with low mobility in the edaphic solution such as the P (Varma and Hock, 1998). Nevertheless, the physiological role of the AM symbiosis is not limited to the uptake and transfer of nutrients to the host plant. Many other beneficial effects for the host plant and for the ecosystems have been described. It is well accepted that the AM symbiosis improves the rooting and establishment of plants, they facilitate the nutrient cycling, protect the host plants against biotic and abiotic stresses and ameliorate soil properties, which favours the soil diversity as well as the natural plant succession (Sánchez-Díaz and Honrubia, 1994; Ruiz-Lozano, Azcón, and Gómez, 1995; Azcón-Aguilar and Barea, 1996).

AM Symbiosis and Osmotic Stress Alleviation

Several ecophysiological studies investigating the role of AM symbiosis against osmotic stresses have demonstrated that the symbiosis often results in altered rates of water movement into, through, and out of the host plants, with consequent effects on tissue hydration and plant physiology (Augé, 2001). In the earliest work on the subject, Safir et al. (1971, 1972) concluded that AM symbiosis probably affected the water relations of plants indirectly through improved P nutrition. The notion that AM effects on water relations were mainly nutritional in nature was prevalent for several years. Nevertheless, considering further reports during the 1990s, many presenting more than one-time measurements of stomatal conductance or water potential, it now seems incontrovertible that AM fungi can modify host water relations in a way entirely unrelated to improved P acquisition (Augé, 2001). Moreover, it is being accepted that the contribution of AM symbiosis to plant drought tolerance is the result of accumulative physical, nutritional, physiological, and cellular effects.

The studies carried out so far have suggested possible mechanisms by which AM symbiosis can alleviate drought stress in the host plants. Among these mechanisms the most important are listed in the following sections.

UPTAKE AND TRANSFER OF WATER THROUGH THE FUNGAL HYPHAE TO THE HOST PLANT

The pioneer studies carried out by Allen (1982) and Hardie (1985) suggested a possible role of AM hyphae in the water uptake and transfer to the host plant. AM fungal hyphae, with a diameter of 2 to 5 μm, can penetrate soil pores inaccessible to root hairs (10 to 20 μm diameter) and take water that is not available for nonmycorrhizal plants. Allen (1991) estimated that the rate of water transport from extraradical hyphae to the root was 0.28 $ng \cdot s^{-1}$ per entry point, a level sufficient to maintain normal water relations. Faber et al. (1991) measured rates of water transport in hyphae crossing air gaps between compartments ranging from 375 to 760 nL $H_2O \cdot h^{-1}$. On the contrary, other authors (Fitter, 1985; George et al., 1992; Koide, 1993) have predicted rates of water uptake by hyphae on the basis of hyphal entry points per unit of root length, hyphal cross-sectional areas, and water potential gradients. Their predictions suggested that hyphal water transport rates were negligible. Since no clear conclusion could be drawn on that topic, new studies were developed. Ruiz-Lozano and Azcón (1995) designed an experiment with lettuce plants grown in containers that had three compartments arranged vertically (see Figure 11.1).

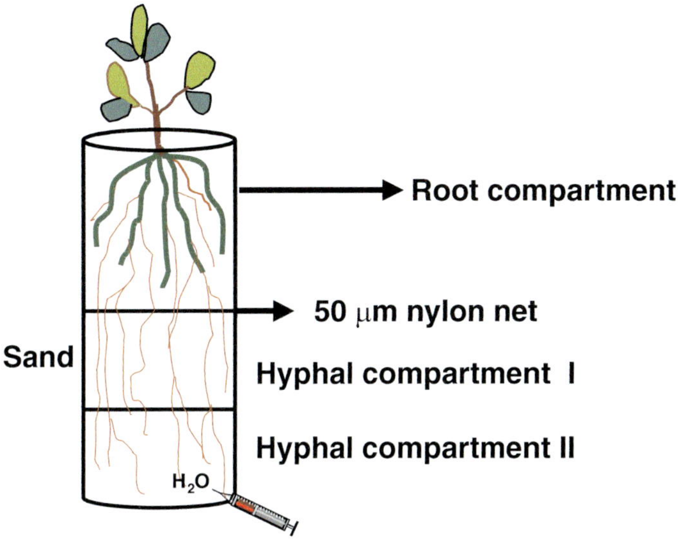

FIGURE 11.1. Experimental system used. Reprinted from *Physiologia Plantarum,* 95, figure 1, by J. M. Ruiz-Lozano and R. Azcón, Hyphal contribution to water uptake in mycorrhizal plants as affected by the fungal species and water status, 427-478, Copyright 1995, with permission from *Physiologia Plantarum.*

The upper compartment (root compartment) contained the mycorrhizal plants and was separated from the next compartment by a 50 μm nylon screen that allowed penetration by AM hyphae but not by roots. In the lower compartment (located at 10 cm from the root compartment) water was applied by injection to reach –0.05 MPa. A control treatment without water injected in this compartment was also used. Results obtained showed that addition of water to the lowermost compartment increased plant fresh weight by 150 percent in *Glomus fasciculatum*-infected plants and by 215 percent in those of *Glomus deserticola* as compared to the uninoculated P-fertilized treatment (see Figure 11.2).

Similarly, leaf water content and gas exchange in mycorrhizal plants also increased with water application to the second hyphal compartment. Despite the fact that the two AM species used in the experiment differed in their efficiency for hyphal water uptake and transport, results indicated that the positive effect of mycorrhizal fungi was enhanced by water addition to the hyphal compartment and that the water uptake by the host plant increased due to the presence of the AM fungi (Ruiz-Lozano and Azcón, 1995).

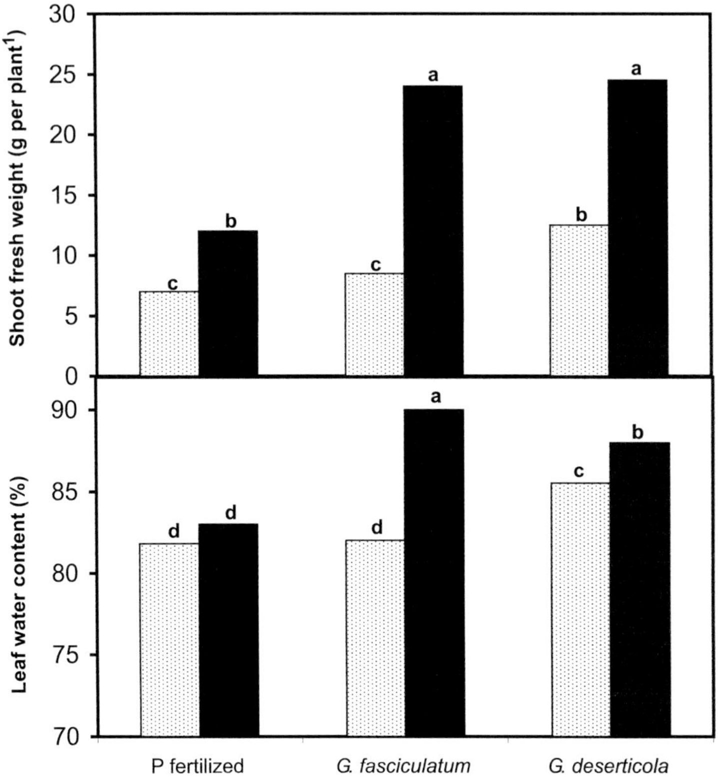

FIGURE 11.2. Shoot fresh weight and leaf water content in P-fertilized or mycorrhizal *(G. fasciculatum* and *G. deserticola)* lettuce plants grown with two levels of soil water content in the hyphal compartment II: no water (light columns) or water added to reach –0.05 MPa (dark columns); different letters in each column refer to significant differences according to Duncan's multiple range test. Reprinted from *Physiologia Plantarum,* 95, figure 2, by J. M. Ruiz-Lozano and R. Azcón, Hyphal contribution to water uptake in mycorrhizal plants as affected by the fungal species and water status, 427-478, Copyright 1995, with permission from *Physiologia Plantarum.*

AMELIORATION OF PLANT GAS EXCHANGE

A number of studies have demonstrated that, during soil drying, mycorrhizal plants often maintain higher gas exchange rates than nonmycorrhizal plants of similar size and nutrient status (Augé et al., 1987; Bethlenfalvay et al., 1987; Augé, 1989; Augé, Stodola, et al., 1992; Sánchez-Díaz and

Honrubia, 1994; Ruiz-Lozano, Azcón, and Gómez, 1995; Ruiz-Lozano, Gómez, and Azcón, 1995; Goicoechea et al., 1997). It has also been found that mycorrhizal and nonmycorrhizal plants often show different critical points or thresholds of stomatal behavior during drought episodes. In fact, in wheat plants leaf water potential was about 0.2 MPa lower in *Glomus fasciculatum*-colonized plants than in similar-sized nonmycorrhizal plants when stomata began to close (Allen and Boosalis, 1983). In the same way, leaf water potential at stomatal closure was about 0.7 MPa lower in roses colonized by *Glomus deserticola* or *Glomus intraradices* than in similar-sized nonmycorrhizal roses (Augé et al., 1986). Duan et al. (1996) found that mycorrhizal plants maintained higher stomatal conductance, transpiration rate, and shoot water than nonmycorrhizal plants. These higher foliar water status characteristics were associated with lower xylem-sap abscisic acid (ABA) concentrations and lower ABA fluxes to leaves in mycorrhizal plants at low soil water content. The mechanism by which AM symbiosis achieved such effect was not clear, but these authors suggested that mycorrhizal fungi probably increased the capability of the root system to scavenge water in dried soil, resulting in less strain to foliage and hence higher stomatal conductance and shoot water content at a particular soil water potential. In contrast, other studies have suggested that AM influence on transpiration resided in their own leaves, even after they had been separated from their root systems (Green et al., 1998).

CHANGES IN THE WATER RETENTION PROPERTIES OF THE SOIL

Not finding a plant-based explanation for mycorrhizal influences on stomatal conductance and other leaf water relations, Augé et al. (2001) suggested that perhaps AM and non-AM plants behave differently during drought because the symbiosis affects the soil water retention properties. The physical, chemical, and biological actions of AM hyphae and hyphal exudates on soils typically affect soil structure (Jastrow and Miller, 1991; Oades and Waters, 1991). Soil structure affects soil moisture retention properties (Hamblin, 1985). Therefore, AM symbiosis may influence soil moisture retention properties (Augé et al., 2001). Thus, they compared the water relations of mycorrhizal and nonmycorrhizal soils and found that seven months of mycorrhization of a Sequatchie loam by *Glomus intraradices* altered the soil moisture characteristic curve relative to nonmycorrhizal soils with similar rooting densities. Once soil matrix potential began to decline, changes in soil matrix potential per unit change in soil water content were smaller in mycorrhizal than in the nonmycorrhizal soil. Within the range of about –1 to –5 MPa, the mycorrhizal soil had to dry more than the nonmycorrhizal soil to reach the same soil matrix potential. The study revealed that the mycorrhizal soil had significantly more water-stable aggregates and substantially higher extraradical hyphal densities than the nonmycorrhizal soil and that correlated well with the improved moisture retention properties of the mycorrhizal soil.

STIMULATION OF ASSIMILATIVE ACTIVITIES ESSENTIAL FOR PLANT GROWTH

Among these activities, nitrate reductase (NR) plays a crucial role. Nitrate ions are the main N source for many of the higher plants (Azcón et al., 2001). In fact, in most soils the major part of the inorganic N pool consists of nitrate because of rapid nitrification of ammonium mineralized from organic material or applied to the soil as fertilizer (Schmidt, 1982). NR is the first enzyme in the nitrate assimilation pathway and probably represents the rate-limiting step in this process (Campbell, 1988). However, NR levels are drastically decreased by adverse environmental conditions such as drought stress (Sánchez-Díaz and Aguirreolea, 1993). This is mainly due by a lower flux of nitrate from the soil to the roots since NR is a typical example of a plant enzyme induced by its substrate (nitrate) (Ruiz-Lozano and Azcón, 1996). The studies carried out so far have demonstrated that AM fungi have the gene set for assimilatory nitrate reduction (Kaldorf et al., 1994) and that NR activity is located in arbuscule-containing cells (Kaldorf et al., 1998). In addition, it has been shown that AM symbiosis allows higher levels of NR activity in plants subjected to drought stress than in the nonmycorrhizal ones (see Table 11.1), and that such enhanced NR activity correlates with a higher tolerance of mycorrhizal plants to drought in terms of plant biomass production (Tobar et al., 1994; Ruiz-Lozano and Azcón, 1996).

OSMOTIC ADJUSTMENT

As soil dries and the soil water potential acquires more negative values, plants must decrease their water potential so that a favorable gradient for water uptake is maintained between the soil and the roots. The most important mechanism to achieve such effect is to decrease the plant osmotic potential by active accumulation of organic ions or solutes, known as osmoregulation (Morgan 1984; Sánchez-Díaz and Aguirreolea, 1993; Hoekstra et al., 2001). The osmotic adjustment allows the cell to maintain the

TABLE 11.1. Total NR activity (μmol NO^{2-} g·FW^{-1}·h^{-1}) and specific NR activity (μmol NO^{2-}·mg $protein^{-1}$) in mycorrhizal *(Glomus mosseae, G. fasciculatum, G. deserticola)* or uninoculated (unfertilized and P-fertilized) lettuce plants grown under well-watered or drought-stressed conditions.

Treatments	Total NR	Specific NR
Well watered		
Control	18.2d	10.1cd
P fertilized	19.2d	11.3c
G. mosseae	36.8b	21.6b
G. fasciculatum	25.6c	15.1c
G. deserticola	42.2a	24.8a
Drought stressed		
Control	7.8f	4.6d
P fertilized	11.6e	7.3d
G. mosseae	21.4d	12.6c
G. fasciculatum	21.0d	12.4c
G. deserticola	36.6b	21.5b

turgence and the processes that depend on it, such as cellular expansion and growth, stomatal aperture, or photosynthesis. The solutes that participate in osmotic adjustment are inorganic ions (mainly K^+ and Cl^-) or uncharged organic compounds such as proline or glycine betaine, as well as such sugars as sacarose, pinitol, or mannitol.

Proline is a nonprotein amino acid that forms in most tissues as they are subjected to water stress, and together with sugar is readily metabolized upon recovery from drought (Kameli and Losel, 1993; Singh et al., 2000). In plants the proline biosynthesis route is thought to resemble the pathway in bacteria (Kishor et al., 1995). In *Escherichia coli,* the reaction starts with the phosphorylation of glutamate by γ-glutamil kinase (γ-GK) to form γ-glutamil phosphate, which is reduced to glutamic-γ-semialdehyde (GSA) by GSA dehydrogenase. GSA spontaneously cyclizes to Δ^1-pyrroline-5-carboxylate (P5C), which is reduced by P5C reductase (P5CR) to proline. The accumulation of proline in plant tissues during periods of drought is due primarily to de novo synthesis, although a reduced rate of catabolism has been also observed. In addition to acting as an osmoprotectant, proline also serves as a sink for energy to regulate redox potentials, as a hydroxy radical scavenger, as a solute that protects macromolecules against denaturation, and as a means of reducing the acidity in the cell (Kishor et al., 1995).

To date, the studies carried out in relation to osmoregulation in AM symbiosis are scarce, although an increase in the accumulation of proline in mycorrhizal plants subjected to drought in comparison to the nonmycorrhizal ones has been observed (Ruiz-Lozano, Azcón, and Gómez, 1995; Ruiz-Lozano and Azcón, 1997; Goicoechea et al., 1998). Also, it has been shown that mycorrhizal colonization and drought interact in modifying free amino acid and sugar pools in roots of rose plants (Augé, Foster, et al., 1992). Finally, a greater osmotic adjustment has been recently found in the leaves of mycorrhizal *Ocimum basilicum* plants than in nonmycorrhizal ones during a lethal drought episode (Kubikova et al., 2001).

PROTECTION AGAINST THE OXIDATIVE DAMAGE GENERATED BY DROUGHT

It is well-known that drought causes oxidative stress in plants and that many of the degenerative reactions associated with drought stress are mediated by reduced oxygen species (Moran et al., 1994; Sgherri and Navari-Izzo, 1995; Iturbe-Ormaetxe et al., 1998). The information available so far about the activity of antioxidant enzymes in AM symbiosis is scarce and has focused mainly in the role of superoxide dismutases (SODs). Little attention has been paid to the role of other important antioxidant enzymes such as catalase, ascorbate peroxidase, or glutathione reductase. The pioneer studies on this subject showed that the AM fungus *Glomus mosseae* possesses Cu Zn-SOD activity and that mycorrhizal clover roots, apart from the constitutively expressed isoforms, exhibited two new SOD isoforms as compared to nonmycorrhizal roots: a mycCu Zn-SOD and a Mn-SOD (Palma et al., 1993). Later studies showed that mycorrhizal lettuce plants subjected to drought had increased SOD activity as compared to the corresponding nonmycorrhizal controls (Ruiz-Lozano et al., 1996). These results were later confirmed by means of molecular techniques (Ruiz-Lozano et al., 2001b). Three cDNAs putatively encoding two Mn-SODs and one Fe-SOD from lettuce were cloned using degenerate primers designed on conserved regions of Mn-SODs. The most interesting results were obtained in relation to the *Mn-sod II* gene (see Figure 11.3).

The expression of this gene in P-fertilized plants was unaffected by drought stress. In contrast, important changes in transcript accumulation occurred in mycorrhizal plants both as a consequence of fungal presence and of drought stress. Under well-watered conditions, fungal presence greatly decreased *Mn-sod II* gene expression (52 percent in the case of

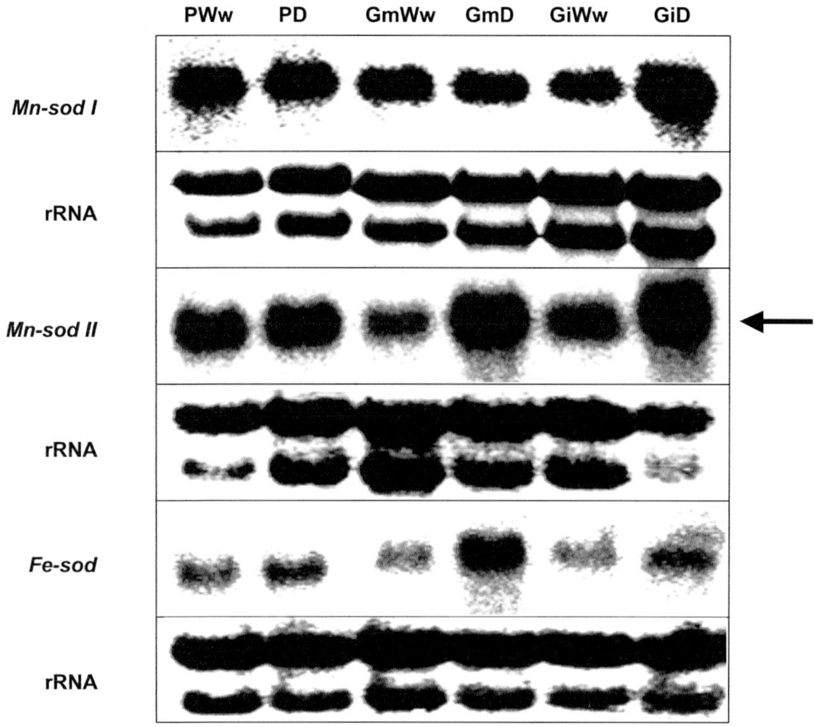

FIGURE 11.3. Northern blot or root total RNA (20 mG) from P-fertilized (P) or mycorrhizal (*G. mosseae,* Gm or *G. intraradices,* Gi) lettuce plants cultivated under well-watered (Ww) or drought-stressed (D) conditions. Blots were hybridized with probes for *Mn-sod I, Mn-sod II,* or *Fe-sod* genes. The panel under each Northern analysis shows the amount of rRNA loaded for each treatment. Reprinted from *Journal of Experimental Botany,* 52, figure 1, by J. M. Ruiz-Lozano, C. Collados, J. M. Barea, and R. Azcón, Cloning of cDNAs encoding SODs from lettuce plants which show differential regulation by arbuscular mycorrhizal symbiosis and drought stress, 2241-2242, Copyright 2001, with permission from *Journal of Experimental Botany.*

G. mosseae and 29 percent in the case of *G. intraradices*) relative to nonmycorrhizal plants. In contrast, when the plants were subjected to drought stress, both mycorrhizal treatments induced a significant increase of *Mn-sod II* transcript accumulation. This increase was over 50 percent in *G. mosseae*–colonized plants and over 138 percent in *G. intraradices*–colonized plants relative to nonmycorrhizal plants. For both, the increase in SOD activity (Ruiz-Lozano et al., 1996) and the increase in *Mn-sod II* gene expression (Ruiz-Lozano et al., 2001b) were well correlated with enhanced tolerance to drought, in terms of plant growth maintenance, by both mycorrhizal treatments (decrease in plant growth as consequence of drought by only 10 to 16 percent in mycorrhizal plants as compared to 30 percent in P-fertilized plants). In addition, recent studies have also shown that mycorrhizal symbiosis can alleviate drought-induced nodule senescence in legume plants (Ruiz-Lozano et al., 2001a). The most interesting results found in such studies was the important reduction of the oxidative damage to lipid proteins in nodules of mycorrhizal plants subjected to drought as compared to the nodules of nonmycorrhizal plants (see Figure 11.4).

Such reduction of oxidative damage to biomolecules seemed to be the main mechanism by which the AM symbiosis protected the root nodules in legume plants against the premature nodule senescence induced by drought stress. In a recent study, it was found that mycorrhizal symbiosis considerably increased the glutathione reductase activity both in roots and nodules

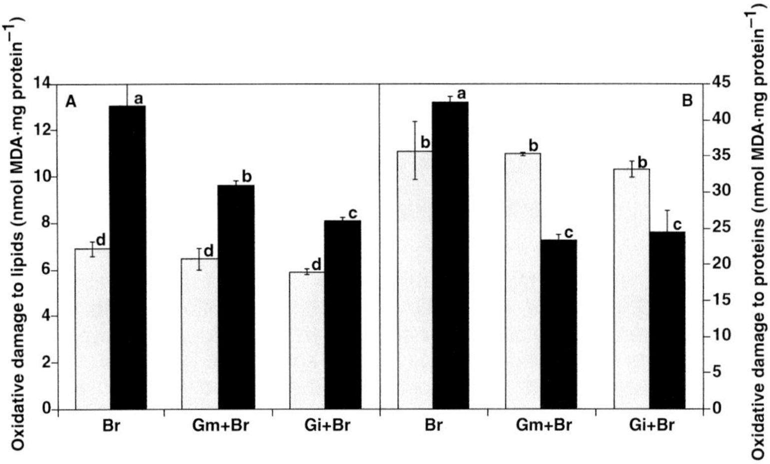

FIGURE 11.4. Effect of drought stress and microbial treatment on oxidative damage to lipids (a) or proteins (b) in nodules of soybean plants. Treatments are designed as Br, *Bradyrhizobium japonicum;* Gm+Br, *Glomus mosseae* plus *Bradyrhizobium japonicum;* Gi+Br, *Glomus intraradices* plus *Bradyrhizobium japonicum.* Plants were either well watered (light columns) or drought stressed for ten days (dark columns). Different letters in each column correspond to significant differences according to Duncan's multiple range test. Reprinted from *New Phytologist,* 151, figure 4, by J. M. Ruiz-Lozano, C. Collados, J. M. Barea, and R. Azcón, Arbuscular mycorrhizal symbiosis can alleviate drought-induced nodule senscence in soybean plants, 493-502, Copyright 2001, with permission from *New Phytologist.*

of soybean plants subjected to drought stress (Porcel et al., 2003). Glutathione reductase is an important component of the ascorbate-glutathione cycle since it is the enzyme which regenerates the oxidized glutathione into its reduced form (Noctor and Foyer, 1998).

All these results suggest that mycorrhizal protection against oxidative stress caused by drought is perhaps one of the most important mechanisms by which AM symbiosis increases the tolerance of host plants against drought. These observations agree with recent findings by Bartels (2001) who proposed that both the prevention of oxidative stress and the elimination of reactive oxygen species are the most interesting approaches to gain plant tolerance against several abiotic stresses, including drought.

FUTURE PERSPECTIVES AND CONCLUSIONS

Although in recent years there has been an increase in the understanding of water relations in mycorrhizal plants and the physiological processes involved in the enhanced tolerance of mycorrhizal plants to water limitation, there are still many unknown aspects which must be elucidated. Molecular techniques now allow a better understanding of the processes involved in a given phenomenon. Hence, it is likely that the use of molecular techniques will allow shedding further light on the different mechanisms which play a role in the protection of the host plants against water deficit.

There are a number of aspects related to plant drought tolerance that have not yet been investigated in mycorrhizal symbiosis. In that sense, aquaporins are water channels that can facilitate the passage of water across biological membranes. Recent studies have demonstrated that AM symbiosis induces the expression of genes encoding aquaporins (Roussel et al., 1997; Krajinski et al., 2000). Nevertheless, these studies were carried out under well-watered conditions. The high expression of genes encoding aquaporins in tissues involved in water transport suggests a role in transcellular water flow through living cells (Barrieu et al., 1998). Since other studies have suggested that aquaporins contribute significantly to the hydraulic conductivity of cells (Tazawa et al., 1996) and that they have a role in cellular osmoregulation (Kjelbom et al., 1999), it could be of interest to study at a molecular level whether such overexpression of MIP proteins induced by AM symbiosis increases the water permeability of the host plant, and whether this has an effect on the water relations of mycorrhizal plants under drought stress conditions.

Another important mechanism involved in drought tolerance is the induction of LEA proteins (Close, 1996). No studies have been carried out so far in relation to LEA proteins in mycorrhizal plants subjected to drought.

Hence, it seems of interest to investigate also whether these proteins have any function in the protection of host plants against drought by AM symbiosis. In the same way, we should increase our knowledge about the molecular mechanisms by which AM symbiosis protects the host plants against the oxidative damage induced by drought. For that, the possible existence of genes encoding for antioxidant enzymes others than SODs (e.g., catalase, ascorbate peroxidase, glutathione reductase) in AM fungi should be examined, as well as, their expression patterns in AM symbiosis under osmotic stress.

REFERENCES

Allen M.F. (1982). Influence of vesicular-arbuscular mycorrhizae on water movement through *Bouteloua gracilis* Lag ex Steud. *New Phytologist* 91: 191-196.

Allen M.F. (1991). *The ecology of mycorrhizae*. Cambridge: Cambridge University Press.

Allen M.F. and Boosalis M.G. (1983). Effects of two species of VA mycorrhizal fungi on drought tolerance of winter wheat. *New Phytologist* 93: 67-76.

Augé R.M. (1989). Do VA mycorrhiza enhance transpiration by influencing host phosphorus status? *Journal of Plant Nutrition* 12: 743-753.

Augé R.M. (2001). Water relations, drought and vesicular-arbuscular mycorrhizal symbiosis. *Mycorrhiza* 11: 3-42.

Augé R.M., Foster J.G., Loescher W.H., and Stodola A.W. (1992). Symplastic sugar and free amino acid molality of *Rosa* roots with regard to mycorrhizal colonization and drought. *Symbiosis* 12: 1-17.

Augé R.M., Scheckel K.A., and Wample R.L. (1986). Greater leaf conductance of well-watered VA mycorrhizal rose plants is not related to phosphorus nutrition. *New Phytologist* 103: 107-116.

Augé R.M., Scheckel K.A., and Wample R.L. (1987). Leaf water and carohydrate status of VA mycorrhizal rose exposed to water deficit stress. *Plant and Soil* 99: 291-302.

Augé R.M., Stodola A.J.W., Brown M.S., and Bethlenfalvay, G.J. (1992). Stomatal responses of mycorrhizal cowpea and soybean to short-term osmotic stress. *New Phytologist* 120: 117-125.

Augé R.M., Stodola A.J.W., Tims J.E., and Saxton A.M. (2001). Moisture retention properties of a mycorrhizal soil. *Plant and Soil* 230: 87-97.

Azcón-Aguilar C. and Barea J.M. (1996). Arbuscular mycorrhizas and biological control of soil-borne plant pathogens—An overview of the mechanisms involved. *Mycorrhiza* 6: 457-464.

Azcón R., Ruiz-Lozano J.M., and Rodríguez R. (2001). Differential contribution of arbuscular mycorrhizal fungi to plant nitrate uptake (^{15}N) under increasing N supply to the soil. *Canadian Journal of Botany* 79: 1175-1180.

Barea J.M. and Jeffries P. (1995). Arbuscular mycorrhizas in sustainable soil-plant systems. In A. Varma and B. Hock (eds.), *Mycorrhiza: structure, function, molecular biology and biotechnology* (pp. 521-561). Berlin: Springer-Verlag.

Barrieu F., Chaumont F., and Chrispeels, M.J. (1998). High expression of the tonoplast aquaporin ZmTIP1 in epidermal and conducting tissues of maize. *Plant Physiology* 117: 1153-1163.

Bartels D. (2001). Targeting detoxification pathways: An efficient approach to obtain plants with multiple stress tolerance. *Trends in Plant Sciences* 6: 284-286.

Bethlenfalvay G.J., Brown M.S., Mihara K.L., and Stafford A.E. (1987). The *Glycine-Glomus-Bradyrhizobium* symbiosis V. Effects of mycorrhizae on nodule activity and transpiration in soybean under drought stress. *Plant Physiology* 85: 115-119.

Bohnert H.J., Nelson D.E., and Jensen R.G. (1995). Adaptations to environmental stresses. *Plant Cell* 7: 1099-1111.

Bowler C., Van Montagu M., and Inzé D. (1992). Superoxide dismutase and stress tolerance. *Annual Review of Plant Physiology and Plant Molecular Biology* 43: 83-116.

Bray E.A. (1997). Plant responses to water deficit. *Trends Plant Sci.* 2: 48-54.

Campbell, W.J. (1988). Nitrate reductase and its role in nitrate assimilation in plants. *Physiologia Plantarum* 74: 214-219.

Chrispeels M.J. and Agre P. (1994). Aquaporins: Water channel proteins of plant and animal cells. *Trends in Biochemical Sciences* 19: 421-425.

Close T. (1996). Dehydrins: Emergence of a biochemical role of a family of plant dehydration proteins. *Physiologia Plantarum* 97: 795-803.

Csonka L.N. and Hanson A.D. (1991). Prokaryotic osmoregulation: Genetics and physiology. *Annual Review of Microbiology* 45: 569-606.

Duan X., Newman D.S., Reiber J.M., Green C.D., Saxton A.M., and Augé R.M. (1996). Mycorrhizal influence on hydraulic and hormonal factors implicated in the control of stomatal conductance during drought. *Journal of Experimental Botany* 47: 1541-1550.

Faber B.A., Zasoski R.J., and Munns D.N. (1991). A method for measuring hyphal nutrient and water uptake in mycorrhizal plants. *Canadian Journal of Botany* 69: 87-94.

Fitter A.H. (1985). Functioning of vesicular-arbuscular mycorrhizas under field conditions. *New Phytologist* 99: 257-265.

Gamble P.E. and Burke J.J. (1984). Effect of water deficit on the chloroplast antioxidant system. I. Alterations in glutathione reductase activity. *Plant Physiology* 76: 615-621.

George E., Häuser K.U., Vetterlein D., Gorgus E., and Marschner H. (1992). Water and nutrient translocation by hyphae of *Glomus mosseae*. *Canadian Journal of Botany* 70: 2130-2137.

Goicoechea N., Antolin M.C., and Sánchez-Díaz M. (1997). Gas exchange is related to the harmone balance in mycorrhizal or nitrogen-fixing alfalfa subjected to drought. *Physiologia Plantarum* 100: 989-997.

Goicoechea N., Szalai G., Antolín M.C., Sánchez-Díaz M., and Paldi E. (1998). Influence of arbuscular mycorrhizae and *Rhizobium* on free polyamines and proline levels in water-stressed alfalfa. *Journal of Plant Physiology* 153: 706-711.

Green C.D., Stodola A., and Augé R.M. (1998). Transpiration of detached leaves from mycorrhizal and nonmycorrhizal cowpea and rose plants given varying abscisic acid, pH, calcium and phosphorus. *Mycorrhiza* 8: 93-99.

Halliwell B. and Gutteridge J.M.C. (1989). *Free radicals in biology and medicine.* Oxford: Clarendon Press.

Hamblin A.P. (1985). The influence of soil structure on water movement, crop root growth, and water uptake. *Advances in Agronomy* 38: 95-158.

Hardie K. (1985). The effect of removal of extraradical hyphae on water uptake by vesicular-arbuscular mycorrhizal plants. *New Phytologist* 101: 677-684.

Hoekstra F.A., Golovina E.A., and Buitink J. (2001). Mechanisms of plant desiccation tolerance. *Trends in Plant Science* 6: 431-438.

Imai R. (1996). A *lea*-class gene of tomato confers salt and freezing tolerance when overexpressed in *Saccharomyces cerevisiae. Gene* 170: 243-248.

Iturbe-Ormaetxe I., Escuredo P.R., Arrese-Igor C., and Becana M. (1998). Oxidative damage in pea plants exposed to water deficit or paraquat. *Plant Physiology* 116: 173-181.

Jastrow J.D. and Miller R.M. (1991). Methods for assessing the effects of biota on soil structure. *Agriculture Ecosystems and Environment* 34: 279-303.

Johnson K.D., Höfte H., and Chrispeels M.J. (1990). An intrinsic tonoplast protein of proteins storage vacuoles in seeds is structurally related to a bacterial solute transporter (GlpF). *Plant Cell* 2: 525-532.

Kaldorf M., Schemelzer E., and Bothe H. (1998). Expression of maize and fungal nitrate reductase in arbuscular mycorhiza. *Molecular Plant-Microbe Interactions* 11: 439-448.

Kaldorf M., Zimmer W., and Bothe H. (1994). Genetic evidence for the occurrence of assimilatory nitrate reductase in arbuscular mycorrhizal fungi. *Mycorrhiza* 5: 23-28.

Kameli A. and Losel D.M. (1993). Carbohydrates and water status in wheat plants under water stress. *New Phytologist* 125: 609-614.

Kammerloher W., Fischer U., Pienchottka G.P., and Schäffner A.R. (1994). Water channels in the plant plasma membrane cloned by immunoselection from a mammalian expression system. *Plant Journal* 6: 187-199.

Kishor P.B., Hong Z., Miao G.H., Hu C.A., and Verma, D.P.S. (1995). Overexpression of Δ^1-pyrroline-5-carboxilate synthetase increases proline production and confers osmotolerance in transgenic plants. *Plant Physiology* 108: 1387-1394.

Kjelbom P., Larsson C., Johansson I., Karlsson M., and Johanson U. (1999). Aquaporins and water homeostasis in plants. *Trends in Plant Sciences* 4: 308-314.

Koide R. (1993). Physiology of the mycorrhizal plant. *Advances in Plant Pathology* 9: 33-54.

Krajinski F., Biela A., Schubert D, Gianinazzi-Pearson V., Kaldenhoff R., and Franken P. (2000). Arbuscular mycorrhiza development regulates the mRNA abundance of *Mtaqp1* encoding a mercury-insensitive aquaporin of *Medicago truncatula*. *Planta* 211: 85-90.

Kubikova E., Moore J.L., Ownlew B.H., Mullen M.D., and Augé R.M. (2001). Mycorrhizal impact on osmotic adjustment in *Ocimum basilicum* during a lethal drying episode. *Journal of Plant Physiology* 158: 1227-1230.

Moran J.F., Becana M., Iturbe-Ormaetxe I., Frechilla S., Klucas R.V., and Aparicio-Tejo P. (1994). Drought induces oxidative stress in pea plants. *Planta* 194: 346-352.

Morgan J.M. (1984). Osmoregulation and water stress in higher plants. *Annual Review of Plant Physiology* 33: 299-319.

Noctor G. and Foyer C.H. (1998). Ascorbate and glutathione: keeping active oxygen under control. *Annual Review of Plant Physiology and Plant Molecular Biology* 49: 249-279.

Oades J.M. and Waters A.G. (1991). Aggregate hierarchy in soils. *Australian Journal of Soil Research* 29: 815-828.

Palma J.M., Longa M.A., del Rio L.A., and Arines J. (1993). Superoxide dismutase in vesicular-arbuscular red clover plants. *Physiologia Plantarum* 87: 77-83.

Roussel H., Bruns S., Gianinazzi-Pearson V., Hahlbrock K., and Franken P. (1997). Induction of a membrane intrinsic protein-encoding mRNA in arbuscular mycorrhiza and elicitor-stimulated cell suspension cultures of parsley. *Plant Science* 126: 203-210.

Ruiz-Lozano J.M. and Azcón R. (1995). Hyphal contribution to water uptake in mycorrhizal plants as affected by the fungal species and water status. *Physiologia Plantarum* 95: 427-478.

Ruiz-Lozano J.M. and Azcón R. (1996). Mycorrhizal colonisation and drought stress exposition as factors affecting nitrate reductase activity in lettuce plants. *Agriculture Ecosystems and Environment* 60: 175-181.

Ruiz-Lozano J.M. and Azcón R. (1997). Effect of Ca application on the tolerance of mycorrhizal lettuce plants to polyethylene glycol-induced water stress. *Symbiosis* 23: 9-22.

Ruiz-Lozano J.M., Azcón R., and Gómez M. (1995). Effects of arbuscular mycorrhizal *Glomus* species on drought tolerance: Physiological and nutritional plant responses. *Applied and Environmental Microbiology* 61:456-460.

Ruiz-Lozano J.M., Azcón R., and Palma J.M. (1996). Superoxide dismutase activity in arbuscular-mycorrhizal *Lactuca sativa* L. plants subjected to drought stress. *New Phytologist* 134: 327-333.

Ruiz-Lozano J.M., Collados C., Barea J.M., and Azcón R. (2001a). Arbuscular mycorrhizal symbiosis can alleviate drought-induced nodule senescence in soybean plants. *New Phytologist* 151: 493-502.

Ruiz-Lozano J.M., Collados C., Barea J.M., and Azcón R. (2001b). Cloning of cDNAs encoding SODs from lettuce plants which show differential regulation by arbuscular mycorrhizal symbiosis and by drought stress. *Journal of Experimental Botany* 52: 2241-2242.

Ruiz-Lozano J.M., Gómez M., and Azcón R. (1995). Influence of different *Glomus* species on the time-course of physiological plant responses of lettuce to progressive drought stress periods. *Plant Science* 110: 37-44.

Safir G.R., Boyer J.S., and Gerdemann J.W. (1971). Mycorrhizal enhancement of water transport in soybean. *Science* 172: 581-583.

Safir G.R., Boyer J.S., and Gerdemann J.W. (1972). Nutrient status and mycorrhizal enhancement of water transport in soybean. *Plant Physiology* 49: 700-703.

Sánchez-Díaz M. and Aguirreolea J. (1993). Efectos fisiológicos que causa la falta persistente de agua en los cultivos. *Phytoma* 51: 26-36.

Sánchez-Díaz M. and Honrubia M. (1994). Water relations and alleviation of drought stress in mycorrhizal plants. In S. Gianinazzi and H. Schüepp (eds.) *Impact of arbuscular mycorrhizas on sustainable agriculture and natural ecosystems* (pp. 167-178). Basel, Switzerland: Birkäuser Verlag.

Schmidt E. (1982). Nitrification in soil. In F.J. Stevenson (ed.), *Nitrogen in agricultural soils* (pp. 253-287). Madison, WI: American Society of Agronomy.

Sgherri C.L.M. and Navari-Izzo F. (1995). Sunflower seedlings subjected to increasing water deficit stress: Oxidative stress and defence mechanisms. *Physiologia Plantarum* 93: 25-30.

Singh D.K., Sale P.W.G., Pallaghy C.K., and Singh V. (2000). Role of proline and leaf expansion rate in the recovery of stressed white clover leaves with increased phosphorus concentration. *New Phytologist* 146: 261-269.

Smirnoff N. 1993. The role of active oxygen in the response of plants to water deficit and desiccation. *New Phytologist* 125: 27-58.

Smith S.E. and Read D.J. (1997). *Mycorrhizal symbiosis*. San Diego: Academic Press.

Tazawa M., Asai K., and Wasaki N. (1996). Characteristics of Hg- and Zn-sensitive water channels in the plasma membrane of *Chara* cells. *Botanical Acta* 109: 388-396.

Tobar R.M., Azcón R., and Barea J.M. (1994). Improved nitrogen uptake and transport from ^{15}N-labeled nitrate by external hyphae of arbuscular mycorrhiza under water-stressed conditions. *New Phytologist* 126: 119-122.

Varma A. and Hock B. (1998). *Mycorrhiza: Structure, function, molecular biology and biotechnology*. Berlin: Springer-Verlag.

Xu D.P. Duan X., Wang B., Hong B., Ho T.D., and Wu R. (1996). Expression of a late embryogenesis abundant protein gene, *HVA1*, from barley confers tolerance to water deficit and salt stress in transgenic rice. *Plant Physiology* 110: 249-257.

Zhu J.K., Hasegawa P.M., and Bressan R. (1997). Molecular aspects of osmotic stress in plants. *Critical Reviews in Plant Sciences* 16: 253-277.

Chapter 12

Arbuscular Mycorrhizal Inoculation in Nursery Practice

Miguel Marin

INTRODUCTION

The concept of sustainable agriculture can be defined as the maintenance of soil fertility and structure over a long period of time such that the economic yields from crop plants can be achieved through minimal inputs of fertilizer that are necessary to reach such yields (Bethlenfalvay and Linderman, 1992). Modern systems of agriculture, particularly in nursery production, have been developed that rely on high inputs of artificial readily soluble fertilizers to sustain them. As an alternative, in sustainable systems we should be seeking a modification of existing strategies such that fertilizer inputs are reduced but not eliminated, and maximum use is made of the added soil microbiota in efficient nutrient capture and in cycling nutrients to the plant root system (Barea and Jeffries, 1995).

The ability of the root systems to establish symbiotic relationships with some of the soil microorganisms (bacteria, actinomycetes, and fungi) of the rhizosphere represents one of the most successful strategies that land plants have developed to survive, with abiotic and biotic stresses imposed during the colonization of terrestrial ecosystems (Allen, 1996). This phenomenon can be approached by using sustainable agriculture to produce adequate high-quality plant production. One of these beneficial components of soil biota include mycorrhizae, which are mutualistic associations occurring between the roots of most plant species and certain groups of fungi (Harley, 1989). The term *mycorrhiza* was proposed by Frank (1885) and refers to an association between fungi ("myco") and root ("rhiza"), as a mutualistic symbiosis where the host plant receives mineral nutrients via fungal myce-

The author wishes to thank Dr. José Lorenzo García Férriz, director of R and D Department in the company Comercial Técnica y Viveros, SA (COTEVISA), for discussions and suggestions about the nurseries and the mycorrhizae.

lium while heterotropic fungus obtains carbon compounds produced after photosynthesis of the host (Sylvia, 1998). In 1973, Rambelli suggested the term *mycorrhizosphere* to substitute for *rhizosphere* because of the significant and dynamic microbial interactions found in the soil of mycorrhizal roots. Therefore, mycorrhizal fungi are often considered biofertilizers, even though they do not add inorganic nutrients to the soil. Rather, they improve the exploitation of soil resources for poorly mobile nutrients.

The purpose of this chapter is to (1) present fundamental information on arbuscular mycorrhizae in horticultural plants, (2) observe the possible application of mycorrhizal fungi as biofertilizers, and (3) analyze the possibilities of a rational management of AM activity with respect to how it might improve nursery plant production.

TYPES OF MYCORRHIZAE

There are six types of mycorrhiza based in the structural features. Two basic categories are designated, one in which the root surface is sheathed in a fungal mantle (sheathing) and one lacking a mantle but in which hyphae proliferate internally (endo). The defining structures of each type are fungal pegs (monotropoid), Hartig net and intracellular penetration (arbutoid—also the subtype "ectendo"), mantle, external mycelial network (ecto), peloton (orchid), hyphal complexes in hair roots (ericoid), and arbuscules or hyphal coils (arbuscular). All of them build a living bridge between plant roots and bulk soil in most ecosystems (Peterson and Farquhar, 1994). Also, the morphologically and functionally different groups of mycorrhizal fungi are distinguished by having the plant exchange compartment either outside (ectomycorrhiza) or inside (endomycorrhiza) the root epidermis cells (Harley and Smith, 1983).

Arbuscular Mycorrhizal Fungi

Arbuscular mycorrhizae (AM), or vesicular-arbuscular mycorrhizae (VAM), are endomycorrhizae which got their name from the distinct fungal tree-shaped, short-lived structure that develops in plant root cells (arbus = tree). The AM is the most widespread group, found from the arctic to the tropics, and is the dominating mycorrhizae in deserts, grasslands, agroecosystems, and broad-leaved forests in temperate to tropical biomes (Smith and Read, 1997). The AM is the most intensively studied type of mycorrhiza because it is present in most agricultural and natural ecosystems and plays an important role in plant growth, health, and productivity (Harley and Smith, 1983; Gianinazzi, Gianinazzi-Pearson, and Troubelot, 1990;

Lovato et al., 1995). Only a few genera belonging to families Cruciferae, Chenopodiaceae, and Cyperaceae are not found, due to the excretion of substances, in and around the roots, which are toxic to the growth of fungi (Glenn et al., 1988). Arbuscular mycorrhizal fungi (AMF) are the least diverse group of mycorrhizal fungi on the species level, containing only 154 species in the order Glomales of the division Zygomycetes (Walker, 1992; Morton, 1995). AMF lack plant host specificity and are obligate symbionts, germinating but not growing in the absence of responding plant roots (Smith and Read, 1997). The wide range of hosts has only marginally been screened, but a compilation by Trappe (1987), based on 6,500 plant species (e.g., 3 percent of the unknown but estimated number of angiosperms), suggested that the range covers 50 percent of the dicots and 50 percent of the monocots. Many bryophytes and pteridophytes also develop AM symbiosis (Harley, 1989). There are reports of AM plants with such different adaptations as being aquatic (Cooke and Lefor, 1998), epiphytic (Koske and Gemma, 1995), halophytic or desert cacti, rainforest trees (Smith and Read, 1997), C3 and C4 grasses (Hetrick et al., 1991), and agricultural crops (Smith and Read, 1997). One of the first terrestrial plants, *Aglaophyton,* in a 400-million-year-old fossil, shows arbuscules identical to the type found in roots today (Pirozynski and Dalpe, 1989). The long co-evolutionary history of plants and AMF probably explains the lack of host specificity and the global distribution of AMF, an important characteristic in their agricultural management.

Endomycorrhizal Colonization

The AMF form morphologically distinct resting spores in the soil and can be multiplied in the presence of a host plant. Some of these spores can be surface sterilized and used to produce new spores in axenic seedlings or root organ culture (Mosse and Hepper, 1975). The close relationship of AMF with their host plants is mirrored by their obligatory biotrophic status. In the absence of a host, their growth is limited to a relatively short time (20 to 30 days) after which many modifications in fungal morphology point to a cessation of hyphal growth. Presence of the root allows development of vegetative mycelium, which, under favorable conditions, can colonize 60 to 90 percent of the length of the root system (Becard and Piche, 1989).

Mycorrhizal colonization begins with the hyphae that arise from soil-borne propagules, large resting spores of the AMF or mycorrhizal root fragments. The fungal hyphae penetrate the root between the epidermal cells and form an appressorium in the first cell layers. This stage marks the autotrophic growth of the fungus. The colonizing hyphae pass through the intercellular spaces and then enter the root tissues, spreading between and through cells of the cortical root layers. Once the hyphae have reached the inner cortex they grow into the cells and form treelike structures called arbuscules. These branched hyphae are closely surrounded by the intact host plasmalemma and represent a large surface of cellular contact between both symbionts. These facilitate the exchange of metabolites between hosts and fungus. The arbuscules are probably the main transfer site of mineral nutrients, mainly phosphorus, from the fungus to the plant and of carbon compounds to the fungus (Smith and Gianinazzi-Pearson, 1990). As internal colonization spreads, the extraradical hyphae ramify and grow along the root surface forming more penetration points. They also grow outward into the surrounding soil, thus developing an extensive three-dimensional network of mycelium, which interfaces with soil particles.

MYCORRHIZAE EFFECTS ON PLANT PHYSIOLOGY

Nutrient Uptake

The hyphal network of AMF within the soil is a vital component of the soil ecosystem. This mycelium is the functional organ for the uptake and translocation of nutrients to and from the mycorrhizae. Mycorrhiza increases the absorbing surface area of root plants by the narrow-diameter structures (hyphae) that allow more nutrients to be removed from the soil and produced enzymes that enhance mineralization of organic matter. Smith and Gianinazzi-Pearson (1988) indicate that the length of the external hyphae growing in soil associated with mycorrhizal roots reaches an average of 1 $m \cdot cm^{-1}$ root, but values of up to 10 to 14 $m \cdot cm^{-1}$ root have also been recorded. These mycelial networks can extend several centimeters outward from the root surface; bridging over the zone of nutrient return, the plant provides the fungus with sugars, amino acids, and vitamins essential for its growth (Harley and Smith, 1983). The activities of mycorrhizal fungi result in the accumulation of nutrients in the root region, for they absorb them at a distance and translocate them to the roots. Mycorrhizae, like all plant structures, have a life span; they may grow apically, producing a new active region while their proximal region ages and senesces, or the whole organ senesces together. During senescence the hyphae may remain active after the cells begin to degenerate (Heap and Newman, 1980).

The increases of the total absorption surface of mycorrhized plants improve their access of immobile elements, mainly phosphorus due to the capacity of the AMF to absorb phosphate from soil (Asimi et al., 1980). In addition, mycorrhizal infection results in an increase in the uptake of copper, zinc (Lambert et al., 1979; George et al., 1994; Ortas et al., 1996), nickel (Killham and Firestone, 1983), cadmium (Guo et al., 1996), and chloride and sulfate (Buwalda et al., 1983) in areas beyond the root's depletion zone. Furthermore, mycorrhizal fungi may permit some plant-to-plant transport of nitrogen (Bethlenfalvay et al., 1991). Nitrogen released from senescing roots of one plant may be taken up by the mycorrhizal hyphae and transported to other plants (Hamel et al. 1991). It has also been shown that external AM hyphae are able to take up N both as NH_4^+ (Ames et al., 1983; Johansen et al., 1992, 1994; Frey and Schüepp, 1993) and NO_3^- (George, Häussler, Kothari, et al., 1992; Johansen et al., 1993; Tobar et al., 1994). Some evidence also suggests that mycorrhizal fungi can modify host water relations, at least on some occasions and to some extent, in a way entirely unrelated to improved phosphorus acquisition. It seems also evident that AM symbiosis, host phenology, phosphorus, and carbon nutrition are usually so inextricably linked in nature that most field effects of AMF on host water balance are probably at least partially related to modified plant size or developmental rates (Auge, 2001).

Carbon Allocation

AMF drain up to 20 percent of photosynthetic carbon assimilated by plants (Jakobsen and Rosendahl, 1990). This transfer of carbon to the fungus has sometimes been considered a drain on the host. However, the host plant may increase photosynthetic activity following mycorrhizal colonization, thereby compensating for carbon lost to the soil (Dixon et al., 1994). Occasionally, plant growth suppression has been attributed to mycorrhizal colonization, but usually this occurs only under low light (photosynthetic limiting) or high phosphorus conditions (very common in commercial nursery productions). The AMF provide plants in return large amounts of nutrients (P, N, K, Zn, and Cu) (Cooper and Tinker, 1978; Ames et al., 1983; Francis and Read, 1984) and water from the soil (George et al., 1994; Marschner and Dell, 1994).

In an ecosystem, the flow of carbon to soil attracts a unique rhizosphere microbial community, the mycorrhizosphere, altogether producing a functionally diverse and dynamic soil biota (Schreiner et al., 1997), which is fundamental for plant nutrition in natural systems and sustainable agriculture.

Mycorrhizal Protection

In general, AMF plants are better nourished and better adapted to their environment. The effect of mycorrhizae on plant growth can be direct (such as when hyphae take up nutrients and translocate them to the host) or indirect (such as when an infection of the root system by the fungi causes morphological and physiological changes in the host plant) (George, Häussler, Vetterlein, et al., 1992). Thus, in colonized plants, increases in rates of photosynthesis (Dixon et al., 1994) and protection against environmental stresses including cold (Charest et al., 1993; Paradis et al., 1995), drought (Osonubi et al., 1992), salinity (Al-Karaki, 2000) and pollution (Leyval et al., 1994) have been reported.

AM associations, in general, can reduce or even suppress damage caused by soilborne plant pathogens (Linderman, 1994; Hooker, Jaizme-Vega, and Atkinson, 1994). Mycorrhizae have been suggested as biocontrol agents; however, it is important to point out that the enhancement of root resistance/tolerance to pathogen attack in AM plants is not exerted with the same effectiveness by all AMF, is not applicable to all pathogens, and is not expressed in all substrates or in all environment conditions (Ryan and Graham, 2001). Resistance to root pathogens can be due to AMF occupying the root niche and to increased plant vigor, and the establishment of seedlings is the transport of photo derivatives from an unshaded to a shaded plant via the AMF mycelium (Eissenstat and Newman, 1990). Actually, worldwide phaseout of methyl bromide by 2005 could conceivably increase reliance on AMF to maintain host tolerance to soil pathogens, and new studies must be carried out to clarify this question.

Mycorrhizae may increase the action of biological control agents (*Pseudomonas, Trichoderma,* etc.) on root diseases (Linderman, 1994; Edwards et al., 1998), but the opposite effect has been described too (Green et al., 1999). There are examples demonstrating that prior colonization of selected AMF protects plants against fungi, such as *Phytophthora* (Vestberg et al., 1994), *Fusarium* (St-Arnaud et al., 1997), *Pythium* (Thomas et al., 1994), *Verticillium* (Karagiannidis et al., 2002), and *Sclerotium* (Prados-Ligero et al., 2002), or nematodes such as *Pratylenchus* (Pinochet et al., 1998) and *Meloidogyne* (Al-Raddad, 1995), etc.

Plant Hormone Modification

There is evidence that plant growth regulators are directly related to plant-AMF interactions (Allen et al., 1980). The invasion of a mycorrhizal fungus into root tissue is accompanied by accumulation of plant hormones such as cytokinins (Allen et al., 1980; Baas and Kuiper, 1989; Druge and Schonbeck, 1992) or abscisic acid (Danneberg et al., 1992). Other plant hormones, such as jasmonic acid, are said to play a signaling role during response of plants to pathogens (Gundlach et al., 1992). This compound was reported to affect AMF in pot culture (Regvar et al., 1996), perhaps by modulating the plant defense mechanisms. Also, the application of indolebutyric acid (IBA) with AM inoculation in citrus rootstock seedlings increases their vegetative growth (Dutra et al., 1996).

INTERACTION OF ARBUSCULAR MYCORRHIZAL FUNGI WITH SOIL MICROBIOTA

AMF interactions with a wide diversity of microorganisms in the rhizosphere appear to play a key role in ecological and biotechnological approaches to improving sustainability of soil-plant ecosystems. The AM formation can be a positive, neutral, or negative effect on the rhizosphere microorganisms (Azcon-Aguilar and Barea, 1992). AM symbiosis can modulate the activity and number of the microbial components in the soil surrounding the mycorrhizal system, giving way to the so-called mycorrhizosphere (Linderman, 1992).

Mycorrhizal fungi are able to establish relationships with specific types of organisms (Azcon-Aguilar and Barea, 1997):

1. *N_2-fixing bacteria.* AMF and *Rhizobium* act synergistically under low fertility conditions, resulting in greater nitrogen and phosphorus content in combination than when each is inoculated onto legume alone. However, appropriate *Rhizobium*-AMF combination must be selected (Azcon et al., 1991).
2. *Phosphate-solubilizing microorganisms (PSM).* The interaction between these bacteria and AMF permit the use of rock and organic phosphates as an alternative phosphorus source (Barea et al., 1993).
3. *Plant-growth-promoting rhizobacteria (PGPR).* These bacteria are characterized by their ability to colonize the rhizosphere environment and improve plant growth. Direct effects have been described as promotion of seedling emergence, root formation, and nodulation (Kloepper et al., 1991). There is experimental evidence that some PGPR, mainly *Pseudomonas* spp. and *Bacillus* spp., interact with AMF by developing synergistic activities which lead to improved plant growth (Linderman, 1992).
4. *Soil-structure-building microorganisms.* Free-living microrganisms and AMF improve soil quality through the formation of stable aggregates. AMF also produce glycoprotein extracellularly on the mycelia in the bulk soil, which together with the physical network of hyphae helps to aggregate soil (Wright and Upadhyaya, 1998), thus improving aeration and water percolation.

5. *Plant pathogens.* AMF colonization interacts with soilborne pathogens in root systems and may be a protection by competition in the same tissue.

MYCORRHIZAE AS SUBSTITUTE FOR FERTILIZERS

Why Do We Need to Change Current Fertilizer Management?

The world's population is expected to grow from 6 billion to around 10 billion by 2050 (Byrnes and Bumb, 1998). To feed a world with huge increases in population and to sustain the well-being of humans, a large increase in food production must be achieved. The projected need for increase in food production must be accomplished on existing cultivated areas because the expansion into new land is limited due to environmental concerns, urbanization, and increasing water scarcity. However, recent trends indicate that the growth rate in crop production has begun to decline in the past ten years and possibly cannot keep pace with the projected increase in global food demand in the upcoming decades (for references, see Cakmak, 2002). Soil productivity is decreasing globally due to soil degradation in the form of erosion, nutrient depletion, water scarcity, acidity, salinization, depletion of organic matter, and poor drainage.

To meet the challenge of food security, agricultural production must increase on existing land, and therefore crop production must be intensified per unit of agricultural land. Mineral nutrients are the major contributors to enhancing crop production and maintaining soil productivity and preventing soil degradation. In general, improving the nutritional status of plants by applying fertilizers and maintaining soil fertility has been the critical step in the doubling of food production for both developed and developing countries since the beginning of the green revolution (Loneragen, 1997). Nearly all increases in projected food requirements in the next decades will be the result of enhancements in yield per unit area and intensive use of agricultural land. Byrnes and Bumb (1998) estimated that fertilizer consumption should increase up to 300 million tons to match required demands of food production and nutrient removal from soils by 2020. In view of this estimate, in the future developing countries should pursue new policies and take new measures to provide more support to farmers regarding the supply of fertilizers. Enhanced use of fertilizers raises concerns due to adverse effects on the environment in terms of eutrophication of surface waters, pollution of drinking water, and gaseous emissions causing global warming.

Therefore, nutrient use efficiency and improved soil management become an important challenge, particularly for nitrogen and phosphorus fertilizers.

AMF As Biofertilizers

Biofertilizers can be defined as living fertilizer compounds of microbial inoculants or groups of microorganisms that are able to fix atmospheric nitrogen or solubilize phosphorus, decompose organic material, or oxidize sulfur in the soil. One of these fertilizers is AMF, which are probably the most abundant fungi in agricultural soils, accounting for somewhere between 5 and 50 percent of the biomass of soil microorganisms (Olsson et al., 1999). AMF can increase plant growth by enhancing uptake of mineral nutrients from soil, particularly diffusion-limited ions such as phosphorus (Bryla and Koide, 1990).

Availability of phosphorus to plant roots is limited both in acidic and alkaline soils, mainly due to formation of sparingly soluble phosphate compounds with Al and Fe in acidic and Ca in alkaline soils (Marschner, 1995). Every year, large amounts of phosphorus fertilizers are applied to soils for crop production globally, and only 10 to 20 percent of the applied phosphorus fertilizers can be absorbed by plants (Holford, 1997). The remaining is rapidly transformed into unavailable phosphorus forms, which are not readily absorbed by plants roots. Some plant species have evolved several adaptive mechanisms to improve their ability to cope with soils having insufficient levels available. Enhancements of root systems to better explore the soil for phosphorus are a typical response of plants to phosphorus deficiency, particularly by increasing the production and elongation of root hairs (Gahoonia and Nielsen, 1998), by acidifying the rhizosphere in order to increase the solubility of phosphorus sources, or by producing phosphatases, which function to make available organically bound phosphorus (Grinsted et al., 1982; Bekele et al., 1983), and also by the formation of mycorrhizal association (Marschner, 1995). Crops that responded negatively to a reduction in AM colonization in low phosphorus fertility or fumigation soils generally had a relatively low specific root length and high mycorrhizal dependency (Ryan and Graham, 2001). However, Thompson (1987, 1991) reported that wheat *(Triticum aestivum),* barley *(Hordeum vulgare),* and oats *(Avena sativa),* with relatively high specific root lengths (length/dry mass), were mostly unaffected by reduced AM colonization. Also, the establishment of AMF in root systems produced changes related to enzymatic activity of either acid or alkaline phosphatases (Fries et al., 1998; Joner and Jakobsen, 1995; Saito, 1995) that contribute to greater soil nutrimental phosphatases. Phosphorus-regulated gene expression in AMF includes phosphatases (Tisserant et al., 1998), phosphorus transporters (Versaw, 2001), and proteins related to polyphosphate metabolism (Dexheimer et al., 1996). A high-affinity phosphorus transporter is expressed in the extraradical mycelium of the *G. versiforme* (Harrison and Van Buuren, 1995). In *G. intraradices,* a similar phosphorus transporter is regulated by phosphorus availability in the external medium and possibly also by the phosphorus status of the host root (Maldonado-Mendoza et al., 2001). The phosphorus-related processes of polyphosphate accumulation and intracellular phosphatase activities are mainly localized to the vacuolar compartment (Gianinazzi and Gianinazzi-Pearson, 1978; Smith and Read, 1997).

Other advantages of AMF for phosphorus absorption include increasing the absorptive surface areas in contact with the soil solution (Koide, 1998). AMF hyphae are much longer than most root hairs, which allow them to explore pockets of soil that are unavailable to nonmycorrhizal roots. In the nursery system, this characteristic of mycorrhizal symbiosis is not very common because plants are usually cultivated in limited substrates.

Finally, AMF enhancement of plant growth and development at low and moderate phosphorus content may also be attributed to increased uptake of Fe, Mn, Zn, and Cu (Jones et al., 1991). Thus, in some crop cultures, AMF can be substituted for copper fertilizer in the culture of citrus seedlings (Timmer and Leyden, 1980), for zinc fertilizer in the greenhouse culture of rice (Sharma et al., 1988), etc. However, the application of AMF as biofertilizers must be joined by bioprotector and biostimulator functions to readily become a reality with their employment for the production of orchard and ornamental crops (Lovato et al., 1995).

MYCORRHIZAL SYMBIOSIS IN HORTICULTURAL SYSTEMS

The effects of AMF on the growth and development of horticultural plants have been studied and described in many research papers (Lovato et al., 1995). Horticulture is one of the target systems in which AM biotechnology can express its potential. This is due to the characteristics of the components and plant production strategies in horticultural practices which make it necessary and easy to inoculate plants with AMF. The increased uptake of soil minerals by colonized plants means that it is possible to consider substantially reducing the application of fertilizers and biocides and at the same time obtain equivalent or even higher crop yields.

Principally, plant propagation in horticultural systems usually starts from seedlings (from seeds or micropropagation), cuttings, or grafting produced or developed in soil or in substrates that have been treated to lower the level of pathogenic organisms. In the case of micropropagation, sterility is obviously a key component. Therefore, the most common propagation techniques include treatments whose aim is to eliminate or reduce the microbial population in the rooting medium, although this obviously affects the beneficial ones, such as the AMF. However, these practices provide the most rewarding situation for the application of mycorrhizal biotechnology by reintroducing appropriately selected AMF at suitable stages of plant production schedules.

It is accepted that the first evidence of the positive influence of AM symbiosis on horticultural production was provided by Menge et al. (1977). He demonstrated that AM inoculation was a prerequisite to get the establishment of citrus plants in biocide-treated nursery beds. Since that publication a number of experiments have been carried out, as reviewed by Nemec (1986), Gianinazzi, Gianinazzi-Pearson, and Trouvelot (1990), Gianinazzi, Trouvelot, and Gianinazzi-Pearson (1990), Barea et al. (1993), Chang (1994), Lovato et al. (1995, 1996), and Singh (1999a,b, 2002). The AMF have been applied to several horticultural plants and much information has been published about them, including:

1. vegetable crops and spices, such as lettuce, onion, leek, celery, asparagus, pepper, cucumber, beans, cotton, tomato, strawberry, potato, muskmelon, wild cardoon, and watermelon;
2. temperate fruit crops, including citrus, apple, almond, peach, peach-almond hybrid, grapevine, blackberry, pear, kiwifruit, raspberry, cherry, and plum;
3. tropical plantation crops, among them coffee, rubber, cacao, papaya, bananas, oil palm, cocoa, avocado, pineapple, and passion fruit, cherimoya; and
4. floricultural crops (pot plants, foliage plants, and cut flowers), such as rose lilac, verbena primula, cyclamen, chrysanthemum, geranium, begonia, hydrangea, gerbera, liquidambar, ampelopsis, and marigold.

Therefore, in horticultural crops the main effects of AM inoculation include these practical production benefits (Barea et al., 1993; Chang, 1994):

- Enhanced seedling growth
- Reduced phosphate requirements, fertilizers in general
- Increased survival rate and development of micropropagated plantlets
- Increases resistance to soilborne pathogens

- Increased resistance to abiotic stresses (transplant shock, water stress, etc.)
- Earlier flowering and fruiting
- Increased crop uniformity
- Improved rooting of cuttings
- Increased soil aeration and drainage
- Increase resistance to invasive weeds
- Increased fruit production

ENDOMYCORRHIZAL TECHNOLOGY IN DIFFERENT PROPAGATION SYSTEMS

Fertilizer inputs in horticultural crops can be reduced with mycorrhizae because they act as a promoter of plant growth and can obtain plants with high vigor. But it is essential to know the proper moment of mycorrhizae application for fungi inocula to have the most effect on plant growth. Therefore, discussed here are the different systems in nursery of plant propagation in relation to mycorrhizal applications.

Plants from Seeds

The propagation system through seeds is related to recollection, conservation, treatment, and germination. The seeds keep the genetic variability of the germplasm, which is necessary before new problems (pedoclimatic conditions, genetic compatibility, and disease tolerances) occur in the crop. But genetic variability is just a disadvantage in commercial nurseries. For these reasons, seed propagation in commercial nurseries has been replaced for several crops by other more efficient propagation methods (cutting, micropropagation). In relation to the mycorrhization of seeds in seedbeds, (1) it is not practiced by the slow germination of some seeds, mainly in fruit tree nurseries; (2) the production in the germination process of some species of inhibited substances for mycorrhizal colonization (Alarcon and Ferrera-Cerrato, 1999); and (3) in most of the agricultural crops the plantlets depend on their nutrients in the seed for initial growth and not on mycorrhizae (Allsopp and Stock, 1995). AMF inoculation should be done before transplantation to develop direct contact between fungus and roots. However, mycorrhization in seedbeds has the advantage of low quantity of inoculum and can occur in horticultural crops with initial high growth rates after seed germination (Marin, Ybarra, et al., 2002).

Cuttings

Another propagation method of plants is cutting. The use of AMF with this propagation system has seldom been employed (Linderman and Call, 1977; Kormanik et al., 1982), and its effects on root cutting are not well-known. AMF have the capacity to promote lateral roots from a cutting because of the physiological activity of these fungi to synthesize growth regulators (auxins and cytokinins) (Gianinazzi, 1991), as discussed earlier. In cutting propagation, different factors must be considered: type of cutting and its physical distance from the tree, and the concentrations of plant growth regulators employed to take new roots. From the studies, it has been observed that the inoculation of AMF in olive cutting substrate favored lateral root formation (Citernesi et al., 1998).

AMF inoculation has been used to evaluate activity in root formation in different horticulture plants (Davies et al., 1987; Chulan and Martin, 1992; Habte and Byappanahalli, 1994; Davies et al., 2000) in fruit cuttings of apple (Gnekow and Marschner, 1989), olive (Citernesi et al., 1998), and kiwi (Powell and Santhanakrishnan, 1986; Calvet et al., 1989).

Micropropagated Plants

Micropropagation is an excellent tool for rapid production of homogeneous, genetically improved, important cultivars (George and Sherrington, 1984). It has become a fundamental practice for the commercial nursery propagation of many kinds of plants. The uses and advantages of micropropagation for commercial nurseries have been described by different authors (Hartmann et al., 1990; Hooker, Gianinazzi, et al., 1994; Vestberg and Estaun, 1994). However, this technique reduces or even eliminates the population of beneficial microorganisms in the soil, such as mycorrhizal fungi, at least during the early stages after in vitro acclimatization. This phase is a critical step in the micropropagation cycle (Debergh and Read, 1991; Subhan et al., 1997), and the lack of beneficial microorganisms can adversely affect survival and growth of in vitro produced plantlets (Pons et al., 1983). The horticultural plants multiplied by micropropagation techniques frequently develop AMF colonization and have high degree of dependence on this symbiosis for normal development (Nemec, 1986).

Although AM symbiosis has been synthesized during the in vitro propagation of various horticultural species (Chavez and Ferrera-Cerrato, 1990) and woody plants (Pons et al., 1983; Ravolanirina et al., 1987), this technique presents a number of difficulties to a commercial nursery. Ex vitro inoculation is obviously easier than in vitro. Mycorrhizal inoculation ex vitro can be performed either after the rooting phase in vitro, at the very beginning of the acclimatization stage, or after the acclimatization, before the beginning of the hardening phase under greenhouse conditions. Several studies have been published describing experiments of this kind. For example, the AM colonization can be beneficial for slower-growing woody species (Fortuna et al., 1996; Azcon-Aguilar and Barea, 1997). The micropropagated crop plants include grapevines (Ravolanirina et al., 1989; Schubert et al., 1990; Schellenbaum et al., 1991), oil palm (Blal et al., 1990), raspberries (Varma and Schuepp, 1994), banana (Jaizme-Vega and Azcon, 1995), pineapple (Jaizme-Vega and Azcon, 1991, 1995; Guillemin et al., 1992; Varma and Schuepp, 1994), strawberry (Niemi and Vestberg, 1992; Williams et al., 1992; Vestberg et al., 1994), pear and peach clonal rootstocks (Rapparini et al., 1994), kiwifruit (Schubert et al., 1992), apple, peach and plum rootstocks (Sbrana et al., 1994), *Prunus* rootstocks (Estaun et al., 1994; Berta et al., 1995) avocado (Vidal et al., 1992; Azcon-Aguilar et al., 1992), olive (Marin, García-Férriz, et al., 2002), and hydrangea (Varma and Schuepp, 1994).

Differences in root growth and in the development rate of target plants are important for the choice of inoculation time. At an early ex vitro stage, plantlets still possess a certain degree of heterotrophy, and the effectiveness of AM inoculation is dependent on the development of autotrophy. Thus, an inoculation protocol should be designed for each case (Vestberg and Estaun, 1994). Similarly, the selection of the right fungal partner is critical for efficient AM development in micropropagated plantlets (Guillemin et al., 1992; Vestberg, 1992; Lovato et al., 1995). Moreover, the influence of the fungal strain on symbiosis efficiency also varies with the growing media and with the fertilization regime.

MANAGEMENT OF ARBUSCULAR MYCORRHIZA IN NURSERY PLANT PRODUCTION

When considering inoculation with mycorrhizal fungi, growers should think about what they hope to gain. The benefits of mycorrhizae can be separated into nursery effects and outplanting effects. The experience at many nurseries has been that high-quality crops can be grown without mycorrhizal inoculation because the nursery environment supplies all the growth requirements of a seedling. However, because of the need to reduce the impact of nursery practices of the biocide and fertilizer usage and due to restrictive environmental legislations, new production systems are being developed that emphasize the use of slow-release fertilizers, minimize the use of biocides, and more efficiently utilize water. Also, one of the most important benefits is from a marketing standpoint because seedlings with well-developed mycorrhizae are widely considered to be high-quality nursery stock. So AMF biotechnology can potentially impact nursery production systems in many ways.

Nursery Practices

The beneficial effects of mycorrhizal symbiosis are less evident in the favorable conditions of a nursery than in the harsh conditions of a final field site. Mycorrhizal inoculation is less likely to accelerate the nursery crop than it is to improve the customer's planting success. Thus, inoculating nursery stock with mycorrhizal fungi is not enough. Various cultural practices in conventional nurseries may reduce the distribution and effectiveness of the mycorrhizal symbiosis and must be consistent with its needs. Some considerations one should consider when assessing the use of mycorrhizae in nursery conditions are discussed in the next sections.

Substrates

Artificial potting mixes and other substrates are commonly used in nurseries where the soil is rarely incorporated. Thus, AMF are not normally present in these rooting media. Because the receptivity of such substrates to AMF is variable, we selected those potting mix materials that favor AM formation and functioning. Soil properties, especially pH, could be taken into consideration, especially as to how AMF vary in their responses. Tests are needed to define appropriate substrate-crop-AMF combinations. Successful colonization has been reported to take place in substrates containing sand, gravel, peat, expanded clay, pumice, perlite, bark, cocoa fiber, sawdust, vermiculite, or a mixture of these (Gianinazzi, Trouvelot, and Gianinazzi-Pearson, 1990; Nemec, 1992).

Inorganic Fertilizers

The main problem of AMF as biofertilizer is the adverse effect of high soil phosphorus levels on AM formation (Mosse, 1973; Menge et al., 1978; Jasper et al., 1979; Abbott et al., 1984; Baath and Spokes, 1989), which is principally caused by concentrations being higher in the roots (Sanders, 1975). In addition, prolonged nitrogen deposition or fertilization may eventually reverse any initial increases in mycorrhizal abundance (Rillig et al., 2002). This nitrogen savings effect is probably due to an increased efficiency of nitrogen use resulting from improved phosphorus nutrition of the plant. In horticultural systems, host growers are unwilling to risk low production through reduced fertilizer and biocide inputs. Moreover, AM colonization is generally severely limited by high phosphorus inputs in vegetable systems. Considering the high production costs of horticultural and ornamental crops, the incentive to inoculate AMF is weaker than in agricultural crops, as phosphorus fertilizer is a very low cost component of vegetable production. An economy analysis by Olsen et al. (1999) of pepper production found that reducing phosphorus inputs to favor AM fungal activity resulted in only a 2.1 percent savings in production costs. But the application of AM inocula may represent only a small contribution to the input costs too, particularly if AM management is associated with other biotechnologies such as micropropagation.

In nursery practice, phosphorus fertilizers may have a positive effect on AM activity when substrate fertility is low. In the same way, mycorrhizal plants reach an acceptable level of productivity at a dose of phosphorus lower than those needed for nonmycorrhizal controls to reach the same productivity, but there is a level of phosphorus application at which mycorrhizal plants do not derive additional benefits of plant growth over the nonmycorrhizal counterparts. As a rule, the benefits of mycorrhizae are greatest when soil phosphorus levels are at or below 50 ppm (50 $mg \cdot kg^{-1}$). Mycorrhizal infection of roots declines above this level with little if any occurring above 100 ppm phosphorus even when soil is inoculated with a mycorrhizae mix.

The development and application of slow-release fertilizers (e.g., Osmocote) in nursery production can be a good method of fertilization. Therefore, appropriate fertilization regimes must be carefully established with regard to quantities, nutrient composition, and application schedules (Lovato et al., 1995).

Biocides

Intensive nursery production depends on high inputs of biocides (fungicides, herbicides, and insecticides). Actually, the application of these biocides is necessary because biocontrol agents have not yet showed satisfactory results in commercial nurseries. Several works showed that fungicides may appreciably decrease mycorrhizal formation (Trappe et al., 1984; Abbott and Robson, 1991). For example, species of *Gigaspora* may be particularly vulnerable to agricultural practices (chemical and mechanical) on nontarget organisms (Schreiner and Bethlenfalvay, 1996). In the same sense, the biocide rates used are an important factor at the moment of their application, for example, captan reduced colonization of *Glomus fasicu-*

latum when applied at a recomemmded rate (Sreenivasa and Bagyaraj, 1989), but stimulated colonization and spore formation when applied at lesser rates (Guillemin and Gianinazzi, 1992). Different groups of fungicides show differential effects on the AM symbiosis. Benzimidazole fungicides (e.g., Benomyl) are consistently detrimental, as are the substituted aromatic hydrocarbons (Johnson and Pfleger, 1992). Anyway, it has been demonstrated that certain biocide treatments are compatible with AMF. Fosetyl-Al, folpet, propamocarb, metalaxyl, and others appear to have no significant effect on AM formation and thus are not recommended for AM inoculum production (Seymour et al., 1994; Marin, Ybarra, et al., 2002). In the same sense, the effect of insecticides on AMF colonization varies as does the AM species studied (Marin, Ybarra, et al., 2002). Finally, most of the biocide treatments do not prevent a normal infectiveness of AMF inocula, but their use must be controlled in each nursery AMF inoculations. Therefore, it must be expected that when the plants are transplanted into the field, and the remains disappear from the biocides, the beneficial effects of mycorrhizal symbiosis will be showed in mycorrhized plants.

Choice of Fungi

Although AMF are not host specific, initial studies have shown that mycorrhizal fungi vary widely in their effectiveness (Abbott and Robson, 1981; Graham et al., 1982). The AMF should be chosen for the specific crop production, not for convenience in the nursery. Therefore, the first step in any inoculation program is to obtain an isolate that is both infective (able to penetrate and spread in the root) and effective (able to enhance growth or stress tolerance in the host). Native endomycorrhizal fungi from nursery or close fields are preferred (Stahl et al., 1988; Camprubi and Calvet, 1996), but isolation and management are not always possible and are expensive in most small nurseries. However, several companies can make this work for the nursery. The nursery manager can select a large list of available commercial inoculum with different mycorrhizal species that these companies are selling with instructions of use. You can find them easily on the World Wide Web with a good search engine.

Although there is no specificity between fungus and host plant, preferences can be expressed in field or nursery experiments. That is, some fungal species work better with particular host plants. *Glomus intraradices* has provided good growth responses in a wide range of host plants (Camprubi and Calvet, 1996), and *G. etunicatum* has been most effective in moderately acidic soils (Clark and Zeto, 1996). Another option is inoculum composed of a mixture of different AMF, made by companies for a broad range of nursery crops (Daft, 1983; Schreiner and Bethlenfalvay, 1996).

Mycorrhizal Inoculum

The material that carries propagules of mycorrhizal fungi is called inoculum. Several kinds of fungal structures can serve as propagules. In mycorrhization with AMF, the propagules are usually spores, mycelium, infected roots, and soil representing the complete life cycle of the fungus (Camprubi and Calvet, 1996). Spores are considered the most resistant to adverse environmental conditions but are slower than other propagules to colonize new roots. The spores are in the range of 0.1 mm in diameter, the largest of all fungal spores. As a result, they quickly settle out of suspension and do not readily pass through apertures of small diameter. Thus, AMF inoculum does not suit itself to distribution through a liquid handling system. Material applied to the surface of soil or even a very open container mix is likely to remain on the surface and out of reach of the roots.

AMF inoculation is usually accomplished with different types of inocula. A disuse method, it incorporates the organic soil from areas with mycorrhized plants as inocula into the nursery soil at a rate of 5 to 10 percent per volume. This method is simple and appropriate for most farm-level or community nurseries. However, it entails moving large amounts of soil and may transfer pathogens from forest soils to the nursery. It is not possible to sterilize forest soils because the process will also kill the AMF. For larger nurseries, a similar method is to construct beds of infected AMF soils. The nursery bed is completely filled with infected soil. Next, seed of the appropriate host plant is sown at a close spacing. Once well established, the roots of the infected plants and AMF will permeate the soil in the nursery bed. The soil and roots will be removed and used as inoculum. Mix them into the nursery soil (5 to 10 percent per volume). This method, while more expensive and management intensive than collecting soil beneath a healthy host plant, is appropriate if many seedlings are to be produced over a number of years. Healthy host plants should be maintained in the AM production bed to assure continued supply of AMF inocula.

Pot culture in pasteurized soils has been the most widely used method for producing AMF inocula, but it is time-consuming, bulky, and often not pathogen free. To overcome these problems, an alternative inoculum will be using various soil-free methods, namely the soilless growth media (for references, see Sylvia and Jarstfer, 1994). Here, the mycorrhizal propagules are around or inside a carrier that may be sand, vermiculite, soil, peat, clay, or other solid substrate. There are different methods of production; aeroponic systems (Hung et al., 1991; Jarstfer and Sylvia, 1994), hydroponics (Elmes et al., 1984; Mosse and Thompson, 1984), and axenic culture of AMF with transformed or nontransformed living roots of various hosts (Chabot et al., 1992; Diop et al., 1994) have been used successfully to produce AMF-colonized root inocula. Colonized roots produced without substrates can be sheared (Sylvia and Jarstfer, 1992). Jarstfer and Sylvia (1994) tested three types of aeroponic systems and chambers, for example, an atomizing disc, pressurized spray through microirrigation nozzle, or ultrasonically generated fog of nutrient solution with a droplet size 3 to 10 mm in diameter, and concluded that the pump and spray nozzle systems were the most versatile and reliable for aeroponic production of AM fungi. The authors found that the ultrasonic fog system producing relatively large size (>3 mm in diameter) droplets failed to provide enough free moisture on the surface of the roots, resulting in poor root growth and failure of AMF infection to spread and less spore production.

There are several companies that you can find on the World Wide Web which offer these inoculum with their own recommendations for the application of the product in each nursery condition. Here, the amount of inoculum required for a particular application is an important question, because it directly influences both the cost of the operation and the chances for success. For container use, the recommended number of propagules per plant has drifted downward, from several hundred a few years ago to much lower numbers now. A good recommendation is based on your experience for your conditions, to have spatial dispersion of propagules in the medium, and on empirical tests of dosage rates.

Timing of Mycorrhizal Inoculation

The benefit of an AM symbiosis depends on when in a plant's life stage its roots are colonized by AM fungi. The outcome is also affected by the growth rate of both the fungi and plant. Moreover, the AM symbiosis is also influenced by the composition of plant and fungal species (Schreiner et al., 1997), by the hierarchical structure of AMF species in the root niche, and by their inherited genetic and functional diversity (Smith and Gianinazzi-Pearson, 1988). In addition, the result of AM symbiosis is affected by the soil properties, soil and plant treatments, and the presence and amount of soil microbes.

A guiding principle in mycorrhizal work is that the inoculum must go into the root zone (Ferguson and Menge 1986; Koide et al., 1999). The roots of new seedlings must be able to grow through the inoculum, which is unlikely if the inoculum is placed on the soil surface. For plants grown from seed in their final containers, the most cost-effective option may be to mix the inoculum into the container medium. This is likely to use more inoculum than might be necessary with other inoculation methods but requires less labor. Nurseries that do not prepare their own container mixes may be able to persuade their medium suppliers to incorporate the inoculum. In bare-root horticultural plants, some new commercial inocula are also applied as a root dip, drench, or mix with the transplant water/fertilizer solutions.

Inoculation may be of greatest benefit to the plant when done at the earliest possible stage. However, germination and rooting stages can be difficult times to inoculate because facilities commonly have low light intensity, heavy use of chemicals, and very wet medium, making mycorrhizal colonization difficult. For plants that are moved one or more times during the production cycle, the first transplant may be a more practical time to inoculate.

Verification of Mycorrhization

No mycorrhizal program in a commercial nursery can be successful without a means of verifying results, since inoculation can and will fail for numerous reasons. Cold temperatures, excess fertility, incompatible biocides, incompatible medium ingredients, contaminated water, and wet medium may prevent colonization either indoors or outdoors. It is critical that environmental conditions and procedural changes be checked at every step.

A problem in mycorrhization control is that mycorrhizal roots look just like ordinary roots, and sometimes the observation of AM under light microscope is necessary. Exceptions are some species of the genera *Gigaspora* whose large spores can be observed on the roots of nursery plants. To distinguish AMF from the host tissues, staining procedures are utilized from the advantage of the fungi's chitinous cell walls (Grandmaison et al., 1988). Thus mycorrhizal studies often require procedures for estimating the proportion of roots in a sample that contains mycorrhizal structures, after clearing and staining them. The principal handling techniques of harvesting, cleaning, staining, and destaining AM roots are described in Brundrett et al. (1996). A mycorrhizal laboratory requires the use of toxic chemicals. The usual dyes are trypan blue (the most common), chlorazol black E, acid fuchsin—all of which are carcinogens; in addition, HCl is applied for the acidification of roots after clearing with KOH. An alternative method with ink and acetic acid has been described (Vierheilig et al., 1998).

The most frequently used root measuring procedure is a modification of the grid line intersects method (Newman, 1966; Giovannetti and Mosse, 1980), in which roots are randomly dispersed in a 9 cm diameter petri plate with grid lines. The observer scans along these grid lines with a dissecting microscope to quantify intersection between grid lines and roots that are designated as either colonized or nonmycorrhizal. The proportion of root length that is mycorrhizal and total root length can then be calculated from a conversion factor derived from the total length of grid lines and the area of the dish. Giovannetti and Mosse (1980) recommended that a minimum of 100 intersections should be used to assess a sample, and found that the accuracy was improved if samples were rerandomized and counted several times. It is also possible to use this method with larger circular or rectangular dishes. However, with rectangular dishes care must be taken to ensure that the number of vertical and horizontal lines are fully representative of the area of the dish. It is best to test new dishes with a sample of precisely known lengths.

A much simpler procedure, in which an observer simply provides a visual estimate of the degree of mycorrhizal colonization (within 5 or 10 percent), can also be reliable (Giovannetti and Mosse, 1980). Although this method is subjective and prone to operator bias, it still can provide sufficient information when precise values are not required (e.g., for pot culture quality control or when looking at samples from the field).

When mycorrhizal colonization is being assessed using a dissecting microscope, it is always a good idea to make slides from randomly selected subsamples of roots for observations with a compound microscope. This will allow fungi that do not stain well (such as many *Acaulospora* and some *Glomus* species) to be seen, and the contribution of saprobic or parasitic fungi to root colonization to be determined. The contribution can also be estimated by counting different AMF (recognized by root morphology) separately.

FUTURE PERSPECTIVES

The long-term goal of AMF research is to continue the evaluation of fungal species (or strains) in diverse culture conditions to different woody and herbaceous nursery crop species from propagation to production to transplanting and establishment in the landscape. The steps in predicting benefits of nursery management practices on mycorrhizae should include an understanding of the capacity of the fungal species present to form mycorrhizae and to function effectively. Also, if AMF are to play a positive role in the areas of disease suppression or tolerance of stress after transplants of greenhouse transplants, P inputs must be lowered substantially in horticultural systems.

A more in-depth evaluation of the available commercial inocula is clearly necessary. A future management strategy will be to obtain transgenic mycorrhizal fungi. New strains with improved fertilizer uptake, resistance for biotic and abiotic stresses, and/or effectiveness on a specific host would help realize the goal of high productivity with low environmental cost. It is likely that inoculation with AMF will become an integral part of most horticultural systems in the future.

Commercial inoculation production of AMF has increased during the past decade, although its future prospect is still uncertain. The obligate biotrophic nature of AMF inevitably raises the cost for inoculum production. Experiences to develop soilless culture systems seem to be promising because this enables the production of clean spores (Jarstfer and Sylvia, 1994). Also, in the nursery application of any microbial inoculum, it is essential to verify that the inoculated microorganisms caused the plant response in nursery or field. Various molecular techniques have been developed to distinguish the inoculated strain from other indigenous strains (Saito and Marumoto, 2002). However, these techniques to identify AMF are still not routinely used because of genetic heterogeneity in AMF and high cost to a commercial nursery.

Finally, there is still a gap in the research and development framework in this area that concerns technology transfer. Undoubtedly, as soon as the potential of AMF inoculation has been clearly demonstrated to the growers, commercial nurseries will begin to include AMF inoculation as a standard procedure in their protocols.

CONCLUSIONS

For improving the biological basis for long-term agricultural sustainability, we need to place greater emphasis on management of renewable resources within the crop production. Mycorrhizal fungi are real alternatives for promoting plant health and productivity in agriculture systems. Nursery production is a target system in which AM biotechnology has been demonstrated. Also, commercial inocula and the technical know how to apply these commercial AMF inocula for horticulture crops are now available. AMF inoculations increase crop growth and yet reduce within commercial propagation system time, energy, and chemical inputs (fertilizers and biocides) and enhance crop vigor and productivity with recycled water. It will make horticultural production systems not only more economically competitive but also more sustainable. The nursery industry wants to reduce chemical costs and usage, improve plant quality, and command a higher price for their product. The value-added benefits in producing superior plants need to be demonstrated so that the nursery industry can market and command a higher market price for mycorrhizal nursery crops.

Maximum benefits will be obtained only from inoculation with efficient AMF and a careful selection of compatible host-fungus-substrate combinations. In general, an early establishment of the AM status, after sowing or at outplanting, is a key factor to improve plant performance in horticultural practices. Soil analyses of phosphorus, interpreted in the context of adequacy for plant growth in a nursery, will assist in identifying sites that discourage or encourage the presence and activity of AMF. Furthermore, the inclusion of AMF in nursery soils can be incorporated into decision-making processes for selection phosphate fertilizer application rates.

It is important to rethink that the main benefits of AMF are solely producing larger plants and increasing phosphorus uptake. In a commercial nursery where fertility is generally not a limiting factor, AMF growth differences are frequently more minimal, as studies show. However, during commercial production, if there is less plant stress that occurs because of AMF colonization, which subsequently helps reduce the usage of biocides and water, then producers can gain in environmental and economic savings.

REFERENCES

Abbott L.K. and Robson A.D. (1981). Infectivity and effectiveness of five endomycorrhizal fungi: Competition with indigenous fungi in field soils. *Australian Journal of Agricultural Research* 32 (6): 631-639.

Abbott L.K. and Robson A.D. (1991). Factors influencing the occurrence of vesicular-arbuscular mycorrhizas. *Agriculture, Ecosystems and Environment* 35 (2-3): 121-150.

Abbott L.K., Robson A.D., and De Boer G. (1984). The effect of phosphorus on the formation of hyphae in soil by the vesicular-arbuscular mycorrhizal fungus, *Glomus fasciculatum*. *New Phytologist* 97 (3): 437-446.

Alarcon A. and Ferrera-Cerrato R. (1999). Arbuscular mycorrhizae management on fruit plant propagation systems. *Terra* 17 (3): 179-191.

Al-Karaki G.N. (2000). Growth of mycorrhizal tomato and mineral acquisition under salt stress. *Mycorrhiza* 10 (2): 51-54.

Allen M.F. (1996). The ecology of arbuscular mycorrhizas: A look back into the 20th century and a peek into the 21st. *Mycological Research* 100 (7): 769-782.

Allen M.F., Moore T.S., and Christensen M. (1980). Phytohormone changes in *Bouteloua gracilis* infected by vesicular-arbuscular mycorrhizae. I. Cytokinin increases in the plant hosts. *Canadian Journal of Botany* 58 (1): 371-374.

Allsopp N. and Stock W.D. (1995). Relationships between seed reserves, seedling growth and mycorrhizal responses in 14 related shrubs (Rosidae) from a low-nutrient environment. *Functional Ecology* 9 (2): 248-254.

Al-Raddad A.M. (1995). Interaction of *Glomus mosseae* and *Paecilomyces lilacinus* on *Meloidogyne javanica* of tomato. *Mycorrhiza* 5 (3): 233-236.

Ames R.N., Reid C.P.P., Porter L.K., and Cambardella C. (1983). Hyphal uptake and transport of nitrogen from two ^{15}N labelled sources by *Glomus mosseae*, a vesicular-arbuscular mycorrhizal fungus. *New Phytologist* 95 (3): 381-396.

Asimi S., Gianinazzi-Pearson V., and Gianinazzi S. (1980). Influence of increasing soil phosphorus levels on interactions between vesicular-arbuscular mycorrhizae and *Rhizobium* in soybeans. *Canadian Journal of Botany* 58 (12): 2200-2205.

Auge R.M. (2001). Water relations, drought and vesicular-arbuscular mycorrhizal symbiosis. *Mycorrhiza* 11 (1): 3-42.

Azcon R., Rubio R., and Barea J.M. (1991). Selective interactions between different species of mycorrhizal fungi and *Rhizobium meliloti* strains, and their effects on growth, N_2-fixation (^{15}N) and nutrition of *Medicago sativa* L. *New Phytologist* 117 (3): 399-404.

Azcon-Aguilar C., Barcelo A., Vidal M.T., and de la Viña G. (1992). Further studies on the influence of mycorrhizae on growth and development of micropropagated avocado plants. *Agronomie* 12 (10): 837-840.

Azcon-Aguilar C. and Barea J.M. (1992). Interactions between mycorrhizal fungi and other rhizospher microorganisms. In M.F. Allen (ed.), *Mycorrhizal functioning* (pp. 163-198). New York: Chapman and Hall.

Azcon-Aguilar C. and Barea J.M. (1997). Applying mycorrhiza biotechnology to horticulturae: Significance and potentials. *Scientia Horticulturae* 68 (1-4): 1-24.

Baas R. and Kuiper D. (1989). Effects of VAM infection and phosphate on *Plantago major* sp. *pleiosperma* in relation to internal cytokinin concentrations. *Physiological Plantarum* 76 (2): 211-215.

Baath E. and Spokes J. (1989). The effect of added nitrogen and phosphorus on mycorrhizal growth response and infection in *Allium schoenoprasum. Canadian Journal of Botany* 67 (11): 3227-3232.

Barea J.M., Azcón R., and Azcón-Aguilar C. (1993). Mycorrhiza and crops. In I. Tommerup (ed.), *Advances in plant pathology,* Volume 9: *Mycorrhiza: A synthesis* (pp. 167-189). London: Academic Press.

Barea J.M. and Jeffries P. (1995). Arbuscular mycorrhizas in sustainable soil-plant systems. In A.K. Varma and D. Hock (eds.), *Mycorrhiza: Structure, function, molecular biology and biotechnology* (pp. 521-560). Berlin: Springer-Verlag.

Becard G. and Piche Y. (1989). New aspects on the acquisition of biotrophic status by a vesicular-arbuscular mycorrhizal fungus, *Gigaspora margarita. New Phytologist* 112 (1): 77-83.

Bekele T., Cino B.J., Ehlert P.A.I., Van der Mass A.A., and Van Diest A. (1983). An evolution of plant-borne factors promoting the solubilization of alkaline rock phosphates. *Plant and Soil* 75 (3): 361-378.

Berta G., Trotta A., Fusconi A., Hooker J.E., Munro M., Atkinson D., Giovannetti M., Morini S., Fortuna P., Tisserant B., et al. (1995). Arbuscular mycorrhizal induced changes to plant growth and root system morphology in *Prunus cerasifera. Tree Physiology* 15 (5): 281-293.

Bethlenfalvay G.J. and Linderman R.G. (1992). Preface. In G.J. Bethlenfalvay and R.G. Linderman (eds.), *Mycorrhizae in sustainable agriculture* (pp. viii-xiii). Madison, WI: American Society of Agronomy, Inc.; Crop Science Society of America, Inc.; Soil Science Society of America, Inc.

Bethlenfalvay G.J., Reyes-Solis M.G., Camel S.C., and Ferrera-Cerrato R. (1991). Nutrient transfer between the root zones of soybean and maize plants connected by a common mycorrhizal mycelium. *Physiology Plantarum* 82 (3): 423-432.

Blal B., Morel C., Gianinazzi-Pearson V., Fardeau J.C., and Gianinazzi S. (1990). Influence of vesicular-arbuscular mycorrhizae on phosphate fertiliser efficiency in two tropic acid soils planted with micropropagated oil palm (*Elaeis guineensis* Jacq.). *Biology and Fertility of Soils* 9 (1): 43-48.

Brundrett M., Bougher N., Dell B., Grove T., and Malajczuk N. (1996). *Working with mycorrhizas in forestry and agriculture* (pp. 179-183). Canberra: Australian Centre for International Agricultural Research Monograph 32.

Bryla D.R. and Koide R.T. (1990). Regulation of reproduction in wild and cultivated *Lycopersicon esculentum* Mill. by vesicular-arbuscular mycorrhizal infection. *Oecologia* 84 (1): 74-81.

Buwalda J.G., Stribley D.P., and Tinker P.B. (1983). Increase uptake of anions by plants with vesicular-arbuscular mycorrhizas. *Plant and Soil* 71 (3): 463-467.

Byrnes B.H. and Bumb B.L. (1998). Population growth, food production and nutrient requeriments. *Journal of Crop Production,* 1-27.

Cakmak I. (2002). Plant nutrition research: Priorities to meet human needs for food in sustainable ways. *Plant and Soil* 247 (1): 3-24.

Calvet C., Pera J., Estuan V., and Camprubi A. (1989). Vesicular-arbuscular mycorrhizae of kiwifruit in an agricultural soil: Inoculation of seedlings and hardwood cuttings with *Glomus mosseae. Agronomie* 9 (2): 181-185.

Camprubi A. and Calvet C. (1996). Isolation and screnning of mycorrhizal fungi from citrus nurseries and orchads and inoculation studies. *Hort Science* 31 (3): 366-369.

Chabot S., Becard G., and Piche Y. (1992). Life cycle of *Glomus intraradices* in root organ culture. *Mycologia* 84 (3): 315-321.

Chang D.C.N. (1994). What is the potential for management of vesicular-arbuscular mycorrhizae in horticulture? In A.D. Robson, L.K. Abbott, and N. Malajczuk (eds.), *Management of mycorrhizas in agriculture, horticulture and forestry* (pp. 187-190). Dordrecht, the Netherlands: Kluwer Academic Press.

Charest C., Dalpe Y., and Brown A. (1993). The effect of vesicular-arbuscular mycorrhizae and chilling on two hybrids of *Zea mays* L. *Mycorrhiza* 4 (2): 89-92.

Chavez Mc.G. and Ferrera-Cerrato R. (1990). Effect of vesicular mycorrhizae on tisuue culture derived plantlets of strawberry. *HortScience* 25 (8): 903-905.

Chulan H.A. and Martin K. (1992). The vesicular-arbuscular (VA) mycorrhiza and its effects on growth of vegetatively propagated *Theobroma cacao* L. *Plant and Soil* 144 (2): 227-233.

Citernesi A.S., Vitagliano C., and Giovannetti, M. (1998). Plant growth and root system morphology of *Olea europea* L. rooted cuttings as influenced by arbuscular mycorrhizas. *Biotechnology* 73 (5): 647-654.

Clark R.B. and Zeto S.K. (1996). Growth and root colonization of mycorrhizal maize grown on acid and alkaline soil. *Soil Biology and Biochemistry* 28 (10-11): 10-11.

Cooke J.C. and Lefor M.W. (1998). The mycorrhizal status of selected plant species from Connecticut wetlands and transition zones. *Restoration Ecology* 6 (2): 214-222.

Cooper K.M. and Tinker P.B. (1978). Translocation and transfer of nutrients in vesicular-arbuscular mycorrhizas. II. Uptake and translocation of phosphorus, zinc and sulphur. *New Phytologist* 81 (1): 43-52.

Daft M.J. (1983). The influenced of mixed inocula on endomycorrhizal development. *Plant and Soil* 71 (1): 33-37.

Danneberg G., Latus C., Zimmer W., Hundeshagen B., Schneider-Poetsch H., and Bothe H. (1992). Influence of vesicular-arbuscular mycorrhiza on phytohormone balances in maize. *Journal of Plant Physiology* 141 (1): 33-39.

Davies F.T., Castro-Jimenez Y., and Duray S.A. (1987). Mycorrhizae, soil amendments, water relations and growth of *Rosa multiflora* under reduced irrigation regimes. *Scientia Horticulturae* 33 (3-4): 261-267.

Davies F.T., Grossi J.A.S., Carpio L., and Estrada-Luna, A.A. (2000). Colonization and growth effects of the mycorrhizal fungus *Glomus intraradices* in a commercial nursery container production system. *Journal of Environmental Horticulture* 18 (4): 247-251.

Debergh P.C. and Read P.E. (1991). Micropropagation. In P.C. Debergh and R.H. Zimmerman (eds.), *Micropropagation. technology and application* (pp. 1-14). Dordrecht, the Netherlands: Kluwer Academic Publishers.

Dexheimer J., Gerard J., Ayatti H., and Ghanbaja J. (1996). Study of origin and distribution of vacuolar granules in the hyphae of a vesicular-arbuscular endomycorrhiza. *Acta Botanica Gallica* 143(2-3): 167-180.

Diop T.A., Plenchette C., and Strullu D.G. (1994). Dual axenic culture of sheared root inocula of vesicular arbuscular mycorrhizal fungi asscociated with tomato roots. *Mycologia* 5 (1): 17-22.

Dixon R.K., Rao M.V., and Garg V.K. (1994). Water relations and gas exchange of mycorrhizal *Leucaena leucocephala* seedlings. *Journal of Tropical Forest Science* 6 (4): 542-552.

Druge U. and Schonbeck F. (1992). Effect of vesicular arbuscular mycorrhizal infection on transpiration photosynthesis and growth of flax (*Linum usitatissimum* L.) in relation to cytokinin levels. *Journal of Plant Physiology* 141 (1): 40-48.

Dutra P.V., Abad M., Almela V., and Agustí M. (1996). Auxin interaction with the vesicular-arbuscular mycorrhizal fungus *Glomus intraradices* Schenck & Smith improves vegetative growth of two citrus rootstocks. *Scientia Horticulturae* 66 (1-2): 77-83.

Edwards S.G., Young J.P.W., and Fitter A.H. (1998). Interactions between *Pseudomonas fluorescens* biocontrol agents and *Glomus mosseae*, an arbuscular mycorrhizal fungus, within the rhizosphere. *FEMS Microbiology Letters* 166 (2): 297-303.

Eissenstat D.M. and Newman E.I. (1990). Seedling establishment near large plants: Effects of vesicular-arbuscular mycorrhizas on the intensity of plant competition. *Functional Ecology* 4 (1): 95-99.

Elmes R.P., Hepper, C.M., Hayman D.S., and O'Shea J. (1984). The use of vesicular-arbuscular mycorrhizal roots by the nutrients film techniques as inoculum for field sites. *Annals of Applied Biology* 104: 437-441.

Estaun V., Calvet C., and Camprubi A. (1994). Arbuscular mycorrhizae and growth enhancement of micropropagated *Prunus* rootstock in different soilless potting mixes. *Agricultural Science in Finland* 3 (3): 263-267.

Ferguson J.J. and Menge J.A. (1986). Response of citrus seedlings to various field inoculation methods with *Glomus deserticola* in fumigated nursery soils. *Journal of the American Society of Horticultural Science* 111 (2): 288-292.

Fortuna P., Citernesi A.S., Morini S., Vitagliano C., and Giovannetti M. (1996). Influence of arbuscular mycorrhizae and phosphate fertilization on shoot apical growth of micropropagated apple and plum rootstocks. *Tree Physiology* 16 (9): 757-763.

Francis R. and Read D.J. (1984). Direct transfer of carbon between plants connected by vesicular-arbuscular mycorrhizal mycelium. *Nature* 307 (1): 53-56.

Frank A.B. (1885). Uber die Wurzelsymbiose beruhende Ernährung gewisser Bäume durch unterirdische Pilze [On the nutrition of certain trees based on root symbiosis of underground fungi]. *Ber. Deutsche Botanische Gesellschaft* 3: 128-145.

Frey B. and Schüepp H. (1993). Acquision of nitrogen by external hypahe of arbuscular mycorrhizal fungi associated with *Zea mays* L. *New Phytologist* 124 (2): 221-230.

Fries L.L.M., Pacovsky R.S., Safir G.R., and Kaminski (1998). Phosphorus effect on phosphatase activity in endomycorrhizal maize. *Physiological Plantarum* 103 (2): 162-171.

Gahoonia T.S. and Nielsen E.N. (1998). Direct evidence on participation of root hairs in phosphorus (^{32}P) uptake from soil. *Plant and Soil* 198 (2): 147-152.

George E., Häussler K., Kothari S.K., Li X.L., and Marschner H. (1992). Contribution of mycorrhizal hyphae to nutrient and water uptake of plants. In D.J. Read, D.H. Lewis, A.H. Fitter, and I.J. Alexander (eds.), *Mycorrhizae in ecosystems,* (pp. 42-47). Wallingford, UK: CAB International.

George E., Häussler K., Vetterlein D., George E., and Marschner H. (1992). Water and nutrient translocation by hyphae of *Glomus mosseae. Canadian Journal of Botany* 70 (11): 2130-2137.

George E., Romheld V., and Marschner H. (1994). Contribution of mycorrhizal fungi to micronutrient uptake by plants. In J.A. Manthey, D.E. Crowley, and D.G. Luster (eds.), *Biochemistry of metal micronutrients in the rhizosphere* (pp. 93-109). Boca Raton, FL: Lewis Publishers.

George E.F. and Sherrington, P.D. (1984). *Plant propagation by tissue culture: Handbook and directory of commercial laboratories.* Eversley, UK: Exegetic Ltd.

Gianinazzi S. (1991). Vesicular-arbuscular (endo-) mycorrhizas: Cellular, biochemical and genetic aspects. *Agriculture, Ecosystems and Environment* 35 (2-3): 105-119.
Gianinazzi S. and Gianinazzi-Pearson V. (1978). Enzymatic studies on the metabolism of vesicular-arbuscular mycorrhizal: III. Ultrastructural localization of acid and alkaline phosphatase in onion roots infected by *Glomus mosseae* (Nicol. & Gerd.). *New Phytologist* 82 (1): 127-1132.
Gianinazzi S., Gianinazzi-Pearson V., and Trouvelot A. (1990). Potentialities and procedures for the use of endomycorrhizas with special emphasis pH high value crops. In J.M. Whipps and B. Lumsden (eds.), *Biotechnology of fungi for improving plant growth* (pp. 41-54). Cambridge, UK: Cambridge University Press.
Gianinazzi S., Trouvelot A., and Gianinazzi-Pearson V. (1990). Role and use of mycorrhizas in horticultural crop protection. *Advances in Horticultural Science* 4 (1): 25-30.
Giovannetti M. and Mosse B. (1980). An evaluation of techniques for measuring vesicular arbuscular mycorrhizal infection in roots. *New Phytologist* 84 (3): 489-500.
Glenn M.G., Chew F.S., and Williams P.H. (1988). Influence of glucosinolate content of *Brassica* (Cruciferae) roots on growth of vesicular-arbuscular mycorrhizal fungi. *New Phytologist* 110 (2): 217-225.
Gnekow M.A. and Marschner H. (1989). Role of VA-mycorrhiza in growth and mineral nutrition of apple (*Malus pumila* var. domestica) rootstock cuttings. *Plant and Soil* 119 (2): 285-293.
Graham J.H., Linderman R.G., and Menge J.A. (1982). Development of external hyphae by different isolates of mycorrhizal *Glomus* sp. In relation to root colonization and growth of Troyer citrange. *New Phytologist* 91 (1): 183-189.
Grandmaison J., Benhamou N., Furlan V., and Visser S.A. (1988). Ultrastructural localization of *N*-acetylglucosamine residues in the cell of *Gigaspora margarita* through its life-cycle. *Biology of the Cell* 63 (1): 89-100.
Green H., Larsen J., Olsson P.A., Jensen D.F., and Jakobsen I. (1999). Suppression of the biocontrol agent *Trichoderma harzianum* by mycelium of the arbuscular mycorrhizal fungus *Glomus intraradices* in root-free soil. *Applied and Environmental Microbiology* 65 (4): 1428-1434.
Grinsted M.J., Hedley M.J., White R.E., and Nye P.H. (1982). Plant induced changes in the rhizosphere of rape (*Brassica napus* var. Emerald) seedlings. I. pH change and the increase in P concentration in the soil solution. *New Phytologist* 91 (1): 19-29.
Guillemin J.P. and Gianinazzi S. (1992). Fungicide interactions with VA fungi in *Ananas comosus* grown in a tropical enviroment. In D.P.S. Verma and T.H. Hohn (eds.), *Genes involved in plant-microbe interactions* (pp. 225-253). Vienna: Springer.
Guillemin J.P., Gianinazzi S., and Trouvelot A. (1992). Screening of VA endomycorrhizal fungi for establishment of micropropagated pineapple plants. *Agronomie* 12 (10): 831-836.

Gundlach H., Muller M.J., Kutchan T., and Zenk M.H. (1992). Jasmonic acid is a signal transducer in elicitor-induced plant cell cultures. *Proceedings of the National Academic of Sciences of USA* 89 (6): 2389-2393.

Guo Y., George E., and Marschner H. (1996). Contribution of an arbuscular mycorrhizal fungus to the uptake of cadmium and nickel in bean and maize plants. *Plant and Soil* 184 (2): 195-205.

Habte M. and Byappanahalli M.N. (1994). Dependency of cassava (*Manihot esculenta* Crantz) on vesicular-arbuscular mycorrhizal fungi. *Mycorrhiza* 4 (6): 241-245.

Hamel C., Barrantes-Cartin U., Furlan V., and Smith D. L. (1991). Endo-mycorrhizal fungi in nitrogen transfer from soybean to maize. *Plant and Soil* 138 (1): 33-40.

Harley J.L. (1989). The significance of mycorrhiza. *Mycological Research* 92 (2): 129-139.

Harley J.L. and Smith S.E. (1983). *Mycorrhizal simbiosis*. London: Academic Press Inc.

Harrison M.J. and Van Buuren M.L. (1995). A phosphate transporter from the mycorrhizal fungus *Glomus versiforme*. *Nature* 378 (6557): 626-629.

Hartmann H.T., Kester, D.E., and Davies F.T. Jr. (1990). *Plant propagation principles and practices*. Englewood Cliffs, NJ: Prentice-Hall.

Heap A.J. and Newman E.I. (1980). The influence of vesicular-arbuscular mycorrhiza on phosphorus transfer between plants. *New Phytologist* 85 (1): 173-179.

Hetrick B.A.D., Wilson G.W.T., and Leslie J.F. (1991). Root architecture of warm- and cool-season grasses: Relationship to mycorrhizal dependence. *Canadian Journal of Botany* 69 (1): 112-118.

Holford I.C.R. (1997). Soil phosphorus, its measurements and its uptake by plants. *Australian Journal of Soil Research* 35 (2): 227-239.

Hooker J.E., Gianinazzi S., Vestberg M., Barea J.M., and Atkinson D. (1994). The application of arbuscular mycorrhizal fungi to micropropagation systems: An opportunity to reduce chemical inputs. *Agricultural Science in Finland* 3 (3): 227-232.

Hooker J.E., Jaizme-Vega M., and Atkinson D. (1994). Biocontrol of plant pathogens using arbuscular mycorrhizal fungi. In S. Gianinazzi and H. Schuepp (eds.), *Impact of arbuscular mycorrhizas on sustainable agriculture and natural ecosystems* (pp. 191-200). Basel, Switzerland: ALS, Birkauser.

Hung L.L., O'Keefe D.M., and Sylvia D.M. (1991). Use of hydrogel as a sticking agent and carrier for vesicular-arbuscular mycorrhizal fungi. *Mycological Research* 95 (4): 427-429.

Jaizme-Vega M.C. and Azcon R. (1991). Effect of vesicular-arbuscular mycorrhizal fungi on pineapple [*Ananas comosus* (L.) Merr.] in the Canary Islands. *Fruits* 46 (1): 47-50.

Jaizme-Vega M.C. and Azcon R. (1995). Responses of some tropical and subtropical cultures to endomycorrhizal fungi. *Mycorrhiza* 5 (3): 213-217.

Jakobsen I. and Rosendahl L. (1990). Carbon flow into soil and external hyphae from roots of mycorrhizal cucumber plants. *New Phytologist* 115 (1): 77-83.

Jarstfer A.G. and Sylvia D.M. (1994). Aeroponic culture of VAM fungi. In A.K. Varma and B. Hock (eds.), *Mycorrhiza: Structure, function, molecular biology and biotechnology* (pp. 427-441). Berlin: Springer-Verlag.

Jasper D.A., Robson A.D., and Abbott L.K. (1979). Phosphorus and formation of vesicular-arbuscular mycorrhizas. *Soil Biology and Biochemistry* 11 (4): 501-505.

Johansen A., Jakobsen I., and Jensen E.S. (1992). Hyphal transport of ^{15}N-labelled nitrogen by a vesicular-arbuscular mycorrhizal fungus and its effect on depletion of inorganic soil. *New Phytologist* 122 (2): 281-288.

Johansen A., Jakobsen I., and Jensen E.S. (1993). External hyphae of vesicular-arbuscular mycorrhizal fungi associated with *Trifolium subterraneum* L. 3. Hyphal transport of ^{32}P and ^{15}N. *New Phytologist* 124 (1): 61-68.

Johansen A., Jakobsen I., and Jensen E.S. (1994). Hyphal N transport by a vesicular-arbuscular mycorrhizal fungus associated with cucumber grown at three nitrogen levels. *Plant and Soil* 160 (1): 1-9.

Johnson N.C. and Pfleger F.L. (1992). Vesicular-arbuscular mycorrhizae and cultural stress. In G.J. Bethlenfalvay and R.G. Linderman (eds.), *Mycorrhizae in sustainable agriculture* (pp. 71-99). ASA Special Publication No. 54. Madison, WI: Agronomy Society of America.

Joner E.J. and Jakobsen I. (1995). Growth and extracellular phosphatase-activity of arbuscular mycorrhizal hyphae as influenced by soil organic-matter. *Soil Biology and Biochemistry* 27 (9): 1153-1159.

Jones J.B., Wolf B., and Mills H.A. (1991). *Plant analysis handbook*. Athens, GA: Micro-Macro Publishing.

Karagiannidis N., Bletsos F., and Stavropoulos N. (2002). Effect of *Verticillium* wilt (*Verticillium dahliae* Kleb.) and mycorrhiza *(Glomus mosseae)* on root colonization, growth and nutrient uptake in tomato and eggplant seedlings. *Scientia Horticulturae* 94 (1-2): 145-156.

Killham K. and Firestone M.K. (1983). Vesicular arbuscular mycorrhizal mediation of grass response to acidic and heavy metal depositions. *Plant and Soil* 72 (1): 39-48.

Kloepper J.W., Zablotowicz R.M., Tipping E.M., and Lifshitz R. (1991). Plant growth promotion mediated by bacterial rhizosphere colonizers. *Beltsville Symposia in Agricultural Research* 14: 315-326.

Koide R.T. (1998). Ecological considerations of mycorrhizal symbiosis. In J.P. Lynch and J. Deikman (eds.), *Phosphorus in plant biology: Regulating roles in molecular, cellular, organismic and ecosystem processess* (pp. 17-25). Rockville, MD: American Society of Plant Physiologists.

Koide R.T., Landherr L.L., Besmer Y.L., Detweiler J.M., and Holcomb E.J. (1999). Strategies for mycorrhizal inoculation of six annual bedding plant species. *Hort Science* 34(7): 1217-1220.

Kormanik P.P., Schultz R.C., and Bryan W.C. (1982). The influence of vesicular-arbuscular mycorrhizae on the growth and development of eight hardwood tree species. *Forest Science* 28 (4): 531 -539.

Koske R.E. and Gemma J.N. (1995). *Scutellospora hawaiiensis:* A new species of arbuscular mycorrhizal fungus from Hawaii. *Mycologia* 87 (5): 678-683.

Lambert D.H., Baker D.E., and Cole H. Jr. (1979). The role of mycorrhizae in the interactions of phosphorus with zinc, copper, and other elements. *Soil Science Society of American Journal* 43 (6): 976-980.
Leyval C., Weissenhorn I., Glashoff A., and Berthelin J. (1994). Influence of heavy metals on germination of arbuscular-mycorrhizal fungal spores in soils. *Acta Botanica Gallica* 141 (4): 523-528.
Linderman R.G. (1992). Vesicular-arbuscular mycorrhizae and soil microbial interactions. In G.J. Bethlenfalvay and R.G. Linderman (eds.), *Mycorrhizae in sustainable agriculture* (pp. 45-70). ASA Special Publication No. 54. Madison, WI: Agronomy Society of America.
Linderman R.G. (1994). Role of VAM fungi in biocontrol. In F.L. Pfleger and R.G. Linderman (eds.), *Mycorrhizae and plant health* (pp. 1-26). St. Paul, MN: APS Press.
Linderman R.G. and Call C.A. (1977). Enhanced rooting of woody plant cuttings by mycorrhizal fungi. *Journal of the American Society for Horticultural Science* 102 (4): 629-632.
Loneragen J.P. (1997). Plant nutrition in 20th and perspectives for the 21st century. *Plant and Soil* 196 (2): 163-174.
Lovato P.E., Gianinazzi-Pearson V., Trouvelot A., and Gianinazzi S. (1996). The state of art of mycorrhizas and micropropagation. *Advances in Horticultural Science* 10 (1): 46-52.
Lovato P.E., Schuepp H., Trouvelot A., and Gianinazzi S. (1995). Application of arbuscular mycorrhizal fungi (AMF) in orchard and ornamental plants. In A. Varma and B. Hock (eds.), *Mycorrhiza structure, function, molecular biology and biotechnology* (pp. 521-559). Heidelberg: Springer-Verlag.
Maldonado-Mendoza I.E., Dewbre G.R., and Harrison M.J. (2001). A phosphate transporter gene from extra-radical mycelium of an arbuscular mycorrhizal fungus *Glomus intraradices* is regulated in response to phosphate in the environment. *Molecular Plant Microbe Interactions* 14 (10): 1140-1148.
Marin M., García-Férriz L., Ghorbel R., Ybarra M., and Marí, A. (2002). Effects of arbuscular mycorrhizae inoculation on micropropagated olive plants. *Acta Horticulturae* 586: 907-910.
Marin M., Ybarra M., Fe A., and Garcia-Ferriz L. (2002). Effect of arbuscular mycorrhizal fungi and pesticides on *Cynara cardunculus* growth. *Agricultural Science in Finland* 11 (3): 245-251.
Marschner, H. (1995). *Mineral nutrition in higher plants*. London: Academic Press Inc.
Marschner H. and Dell B. (1994). Nutrient uptake in mycorrhizal symbiosis. *Plant and Soil* 159 (1): 89-102.
Menge J.A., Labanauskas C.K., Johnson E.L.V., and Platt R.G. (1978). Partial substitution of mycorrhizal fungi for phosphorus fertilization in the greenhouse culture of citrus. *Soil Science Society of America Journal* 42 (5): 926-930.
Menge J.A., Lembright H., and Johnson E.L. (1977). Utilization of mycorrhizal fungi in citrus nurseries. *Proceedings International Society of Citriculture* 1: 129-132.

Menge J.A., Steirle D., Bagyaraj D.J., Johnson E.L.V., and Leonard R.T. (1978). Phosphorus concentrations in plants responsible for inhibition of mycorrhizal infection. *New Phytologist* 80 (4): 575-578.

Morton J.B. (1995). Taxonomic and phylogenetic divergence among five *Scutellospora* species based on comparative developmental sequences. *Mycologia* 87 (1): 127-137.

Mosse B. (1973). Plant growth responses to vesicular-arbuscular mycorrhizal: IV. In soil given additional phosphate. *New Phytologist* 72 (1): 127-136.

Mosse B. and Hepper C.M. (1975). Vesicular-arbuscular mycorrhizal infections in root organ cultures. *Physiological Plant Pathology* 5 (2): 215-223.

Mosse B. and Thompson J.P. (1984). Vesicular-arbuscular endomycorrhizal inoculum production. I. Exploratory experiments with beans *(Phaseolus vulgaris)* in nutrient flow culture. *Canadian Journal of Botany* 62 (10): 1523-1530.

Nemec S. (1986). VA mycorrhizae in horticultural systems. In G.R. Safir (ed.), *Ecophysiology of VA mycorrhizal plants* (pp. 193-211). Boca Raton, FL: CRC.

Nemec S. (1992). *Glomus intraradix* effects on citrus rootstock seedling growth in various potting media. *Journal of Agricultural Science* 118 (3): 315-323.

Newman E.I. (1966). A method of estimating the total lenght of root in a sample. *Journal of Applied Ecology* 3 (1): 139-145.

Niemi M. and Vestberg M. (1992). Inoculation of commercially grown strawberry with VA mycorrhizal fungi. *Plant and Soil* 144 (1): 133-142.

Olsen J.K., Schaefer J.T., Edwards D.G., Hunter M.N., Galea V.J., and Muller L.M. (1999). Effects of a network of mycorrhizae on capsicum (*Capsicum annuum* L.) grown in the field with five rates of applied phosphorus. *Australian Journal of Agricultural Research* 50 (2): 239-252.

Olsson P.A., Thingstrup L., Jakobsen I., and Baath E. (1999). Estimation of the biomass of arbuscular mycorrhizal fungi in a linseed field. *Soil Biology and Biochemistry* 31 (13): 1879-1887.

Ortas I., Harris P.J., and Rowell D.L. (1996). Enhanced uptake of phosphorus by mycorrhizal *Sorghum* plants as influenced by forms of nitrogen. *Plant and Soil* 184 (2): 255-264.

Osonubi O., Bakare O.N., and Mulongoy K. (1992). Interactions between drought stress and vesicular-arbuscular mycorrhiza on the growth of *Faidherbia albida* (syn. *Acacia albida*) and *Acacia nilotica* in sterile and non-sterile soils. *Biology and Fertility of Soils* 14 (3): 159-165.

Paradis R., Dalpe Y., and Charest C. (1995). The combined effect of arbuscular mycorrhizas and short-term cold exposure on wheat. *New Phytologist* 129 (4): 637-642.

Peterson R.L. and Farquhar M.L. (1994). Mycorrhizas—Integrated development between roots and fungi. *Mycologia* 86 (3): 311-326.

Pinochet J., Camprubi A., Calvet C., Fernandez C., and Rodríguez-Cabana R. (1998). Inducing tolerance to the root-lesion nematode *Pratylenchus vulnus* by early mycorrhizal inoculation of micropropagated Myrobalan 29 C plum rootstock. *Journal of the American Society of Horticultural Science* 123 (3): 342-347.

Pirozynski, K.A. and Dalpe Y. (1989). Geological history of the Glomaceae with particular reference to mycorrhizal symbiosis. *Symbiosis Rehovot* 7 (1): 1-35.
Pons F., Gianinazzi-Pearson V., Gianinazzi S., and Navatel J.C. (1983). Studies of VA mycorrhizae in vitro: Mycorrhizal synthesis of axenically propagated wild cherry (*Prunus avium* L.) plants. *Plant and Soil* 71 (3): 217-221.
Powell C.L and Santhanakrishnan P. (1986). Effect of mycorrhizal inoculation and phosphorus fertilizer on the growth of hardwood cuttings of kiwi (*Actinidia deliciosa* cv. Hayward) in containers. *New Zealand Journal of Agricultural Research* 29 (3): 263-268.
Prados-Ligero A.M., Bascon-Fernandez J., Calvet-Pinos C., Corpas-Hervias C., Lara-Ruiz A., Melero-Vara J.M., and Basallote-Ureba M.J. (2002). Effect of different soil and clove treatments in the control of white rot of garlic. *Annals of Applied Biology* 140 (3): 247-253.
Rambelli, A. (1973). The rhizosphere of mycorrhizae. In A.C. Marks and T.T. Kozlowski (eds.), *Ectomycorrhizae: Their ecology and physiology* (pp. 229-249). London: Academic Press.
Rapparini F., Baraldi R., Bertazza G., Branzanti B., and Predieri S. (1994). Vesicular arbuscular mycorrhizal inoculation of micropropagated fruit trees. *Journal of Horticultural Science* 69 (6): 1101-1109.
Ravolanirina F., Gianinazzi S., Trouvelot A., and Carre M (1989). Production of endomycorrhizal explants of micropropagated grapevine root-stocks. *Agriculture, Ecosystems and Environment* 29 (1-4): 323-327.
Ravolanirina F., Gianinazzi-Pearson V., and Gianinazzi S. (1987). Preliminary studies in in vitro endomycorrhizal inoculation of micropropagated tree species species of nutritional values. In F.S.P. Ng (ed.), *Trees and mycorrhiza* (pp. 91-101). Proceedings of the Asian Seminar, Forest Research Institute, Malaysia.
Regvar M., Gogala N., and Zalar P. (1996). Effects of jasmonic acid on mycorrhizal *Allium sativum. New Phytologist* 134 (4): 703-707.
Rillig M.C., Treseder K.K., and Allen M.F. (2002). Global change and mycorrhizal fungi. In M.G.A. van der Heijden and I.R. Sanders (eds.), *Ecological Studies Series* 157: *Mycorrhizal Ecology* (pp. 135-160). Heidelberg, Germany: Springer Verlag.
Ryan M.H. and Graham J.H. (2001). Is there a role for arbuscular mycorrhizal fungi in production agriculture? *Plant and Soil* 244 (1-2): 263-271.
Saito M. (1995). Enzyme-activities of the internal hyphae and germinated spores of an arbuscular mycorrhizal fungus, *Gigaspora margarita* Becker and Hall. *New Phytologist* 129 (3): 425-431.
Saito M. and Marumoto T. (2002). Inoculation with arbuscular mycorrhizal fungi: The status quo in Japan and the future prospects. *Plant and Soil* 244 (1-2): 273-279.
Sanders F.E. (1975). The effect of foliar-applied phosphate on the mycorrhizal infections of onion roots. In F.E. Sander, B. Mosse, P.B. Tinker, (eds.), *Endomycorrhizas* (pp. 261-276). London: Academic Press.
Sbrana C., Giovannetti M., and Vitagliano C. (1994). The effect of mycorrhizal infection on survival and growth renewal of micropropagated fruit rootstocks. *Mycorrhiza* 5 (2): 153-156.

Schellenbaum L., Berta G., Ravolanirina F., Tisserant B., Gianinazzi S., and Fitter A.H. (1991). Influence of endomycorrhizal infection on root morphology in a micropropagated woody plant species (*Vitis vinifera* L.). *Annals of Botany* 68 (2): 135-141.

Schreiner R.P. and Bethlenfalvay G.J. (1996). Mycorrhizae, biocides, and biocontrol. 4. Response of a mixed culture of arbuscular mycorrhizal fungi and host plant to three fungicides. *Biology and Fertility of Soils* 24 (1): 18-26.

Schreiner R.P., Mihara K.L., McDaniel H., and Bethlenfalvay G.J. (1997). Mycorrhizal fungi influence plant and soil functions and interactions. *Plant and Soil* 188 (2): 199-209.

Schubert A., Bodrino C., and Gribaudo I. (1992). Vesicular-arbuscular mycorrhizal inoculation of kiwifruit *(Actinidia deliciosa)* micropropagated plants. *Agronomie* 12 (10): 847-850.

Schubert A., Mazzitelli M., Ariusso O., and Eynard I. (1990). Effects of vesicular-arbuscular mycorrhizal fungi on micropropagated grapevines. Influence of endophyte strain, P fertilization and growth medium. *Vitis* 29 (1): 5-13.

Seymour N.P., Thompson J.P., and Fiske M.L. (1994). Phytotoxicity of fosetyl-Al and phosphonic acid to maize during production of vesicular-arbuscular mycorrhizal inoculum. *Plant Disease* 78 (5): 441-446.

Sharma A.K., Renu-Singh M., Singh U.S., and Singh R. (1988). Effect of vesicular-arbuscular mycorrhiza on uptake of phosphorus and zinc in rice (*Oryza sativa* L.). *Current Science* 57 (16): 901-902.

Singh S. (1999a). Role of mycorrhiza in tree nurseries —Part II. Inoculation of nursery soil/plants with mycorrhizal fungi. *Mycorrhiza News* 10 (4): 2-11.

Singh S. (1999b). Role of mycorrhiza in tree nurseries—Part III. Comparative performance of mycorrhizal fungi. *Mycorrhiza News* 11 (1): 2-9.

Singh S. (2002). Role of mycorrhiza in plants raised from cuttings or as micropropagated plants, Part II: Fruit trees; Part III: Ornamentals and other plants. *Mycorrhiza News* 14 (1): 2-9.

Smith S.E. and Gianinazzi-Pearson V. (1988). Physiological interactions between symbionts invesicular-arbuscular mycorrhizal plants. *Annual Review of Plant Physiology and Plant Molecular Biology* 39: 221-244.

Smith S.E. and Gianinazzi-Pearson V. (1990). Phosphate uptake and arbuscular activity in mycorrhizal *Allium cepa* L.: Effects of photon irradiance and phosphate nutrition. *Australian Journal of Plant Physiology* 17 (2): 177-188.

Smith S.E. and Read D.J. (1997). *Mycorrhizal symbiosis,* Second edition. London: Academic Press.

Sreenivasa M.N. and Bagyaraj D.J. (1989). Use of pesticides for mass production of vesicular-arbuscular mycorrhizal inoculum. *Plant and Soil* 119 (1): 127-132.

St-Arnaud M., Hamel C., Vimard B., Caron M., and Fortin J.A. (1997). Inhibition of *Fusarium oxysporum* f. sp. dianthi in the non-VAM species *Dianthus caryophyllus* by co-culture with *Tagetes patula* companion plants colonized by *Glomus intraradices*. *Canadian Journal of Botany* 75 (6): 998-1005.

Stahl P.D., Williams S.E., and Christensen M. (1988). Efficacy of native vesicular arbuscular mycorrhizal fungi after severe soil disturbance. *New Phytologist* 110 (3): 347-354.

Subhan S., Sharmila P., and Saradhi P.P. (1997). *Glomus fasciculatum* alleviates transplantation shock of micropropagated *Sesbania sesban*. *Plant Cell Reports* 17 (4): 268-272.

Sylvia D.M. (1998). Mycorrhizal symbioses. In D.M. Sylvia, J.J. Fuhrmann, P.G. Hartel, and D.A. Zuberer (eds.), *Principles and applications of soil microbiology* (pp. 408-426). Upper Saddle River, NJ: Prentice-Hall.

Sylvia D.M. and Jarstfer A.G. (1992). Sheared-root inocula of vesicular-arbuscular mycorrhizal fungi. *Applied and Environmental Microbiology* 58 (1): 229-232.

Sylvia D.M. and Jarstfer A.G. (1994). Production of inoculum and inoculation with arbuscular mycorrhizal fungi. In A.D. Robson, L.K. Abbott, and N. Malajczuk (eds.), *Management of mycorrhizas in agriculture, horticulture and forestry* (pp. 231-238). Dordrecht, the Netherlands: Kluwer Academic Press.

Thomas L., Mallesha B.C., and Bagyaraj D.J. (1994). Biological control of damping-off of cardamom by the VA mycorrhizal fungus, *Glomus fasciculatum*. *Microbiological Research* 149 (4): 413-417.

Thompson J.P. (1987). Decline of vesicular-arbuscular mycorrhizas in long fallow disorder of field crops and its expression in phosphorus deficiency in sunflower. *Australian Journal of Agricultural Research* 38): 847-867.

Thompson J.P. (1991). Improving the mycorrhizal condition of the soil through cultural practices and effects on growth and phosphorus uptake in plants. In C. Johansen, K.K. Lee, and K.L. Sahrawat (eds.), *Phosphorus nutrition of grain legumes in the semi-arid tropics* (pp. 117-137). Andhra Pradesh, India: ICRISAT.

Timmer L.W. and Leyden R.F. (1980). The relationship of mycorrhizal infection to phosphorus induced copper deficiency in sour orange seedlings. *New Phytologist* 85 (1): 15-23.

Tisserant B., Brenac V., Requena N., Jeffries P., and Dodd J.C. (1998). The detection of *Glomus* spp. (arbuscular mycorrhizal fungi) forming mycorrhizas in three plants, at different stages of seedling development, using mycorrhiza-specific isozymes. *New Phytologist* 138 (2): 225-239.

Tobar R., Azcon R., and Barea J.M. (1994). Improved nitrogen uptake and transport form ^{15}N-labelled nitrate by external hyphae of arbuscular mycorrhizal under water-stressed conditions. *New Phytologist* 126 (1): 119-122.

Trappe J.M. (1987). Phylogenetic and ecologic aspects of mycotrophy in the angiosperms from an evolutionary standpoint. In G.R. Safir (ed.), *Ecophysiology of VA mycorrhizal plants* (pp. 5-25). Boca Raton, FL: CRC Press.

Trappe J.M., Molina R., and Castellano M. (1984). Reactions of mycorrhizal fungi and mycorrhiza formation to pesticides. *Annual Review of Phytopathology* 22: 331-359.

Varma A. and Schuepp H. (1994). Infectivity and effectiveness of *Glomus intraradices* on micropropagated plants. *Mycorrhiza* 5 (1): 29-37.

Versaw W.K. (2001). Phosphate transporters of *Medicago truncatula* and arbuscular mycorrhizal fungi. *Plant and Soil* 244 (1-2): 239-245.

Vestberg M. (1992). The effect of vesicular-arbuscular mycorrhizal inoculation on the growth and root colonization of ten strawberry cultivars. *Agricultural Science in Finland* 1 (5): 527-535.

Vestberg, M. and Estaun V. (1994). Micropropagated plants, an opportunity to positively manage mycorrhizal activities. In S. Gianinazzi and H. Stuepp (eds.), *Impact of arbuscular mycorrhizas on sustainable agriculture and natural ecosystems* (pp. 217-226). Basel, Switzerland: Birkäuser.

Vestberg M., Palmujoki H., Parikka P., and Uosukainen M. (1994). Effect of arbuscular mycorrhizas on crown rot *(Phytophthora cactorum)* in micropropagated strawberry plants. *Agricultural Science in Finland* 3 (3): 289-295.

Vidal M.T., Azcon-Aguilar C., Barea J.M., and Pliego-Alfaro F. (1992). Mycorrhizal inoculation enhances growth and development of micropropagated plants of avocado. *Hort Science* 27 (7): 785-787.

Vierheilig H., Coughlan A.P., Wyss U., and Piche Y. (1998). Ink and vinegar, a simple staining method for arbuscular-mycorrhizal fungi. *Applied and Environmental Microbiology* 64 (12): 5004-5007.

Walker C. (1992). Sytematics and taxonomy of the arbuscular mycorrhizal fungi (Glomales)—a possible way forward. *Agronomy* 12 (10): 443-897.

Williams S.C.K., Vestberg M., Uosukainen M., Dodd J.C., and Jeffries P. (1992). Effects of fertilizers and arbuscular mycorrhizal fungi on the post-vitro growth of micropropagated strawberry. *Agronomie* 12 (10): 851-857.

Wright S.F. and Upadhyaya A. (1998). A survey of soils for aggregate stability and glomalin, a glycoprotein produced by hyphae of arbuscular mycorrhizal fungi. *Plant and Soil* 198 (1): 97-107.

Chapter 13

Interaction Between Arbuscular Mycorrhizal Fungi and Root Pathogens

Leonor Costa Maia
Norma Suely Sobral da Silveira
Uided Maaze Tiburcio Cavalcante

INTRODUCTION

Most plants live in symbiotic association with fungi, forming what was called by Frank (1885), more than a century ago, mycorrhizae. In this interaction, although the presence of the fungus is only at the root region, all parts of the plant are affected by the mycorrhizal colonization (Dugassa et al., 1996). Among the different types, the arbuscular mycorrhizae is the most common and widely distributed in the tropics.

The arbuscular mycorrhizae are formed by fungi recently classified in a new phylum: Glomeromycota, with only one class (Glomeromycetes) containing four orders, Archaeosporales, Diversisporales, Glomerales, and Paraglomerales, in which seven genera are included: *Acaulospora, Archaeospora, Entrophospora, Gigaspora, Glomus, Paraglomus,* and *Scutellospora* (Redecker et al., 2000; Morton and Redecker, 2001; Schüâler et al., 2001; Walker and Schüâler, 2004). More recently, a new family (Pacisporaceae) and genus *(Pacispora)* were proposed (Oehl and Sieverding, 2004).

The arbuscular mycorrhizal fungi (AMF) distribute their hyphae both inside and outside the roots, where they form what is known as the mycorrhizosphere, a term used to characterize the area that is affected by the mycorrhizae. To understand the concept of the mycorrhizosphere, one should consider the processes and components involved, which include the substrate, formation of the mycorrhizal association, and microbial dynamics in the rhizosphere over time (Linderman, 2000). In this environment, through competitive or mutualistic processes, the AMF interact in a positive or negative way with other soil organisms. The action of the mycorrhizal fungi is wide and, in most cases, brings advantage to their hosts, allowing survival in environments subject to biotic and/or abiotic stress, including stress produced by plant pathogenic organisms. Thus, the AMF have been suggested as candidates to be used in the biological control of plant diseases, considering that they are widely distributed and establish longtime relationships with the roots of most plant species (Traquair, 1995). As pointed out by Linderman (2000), the mycorrhizae "are perhaps the only stable microbial system in the rhizosphere," establishing a persistent interface between the roots and the soil. In this regard, an increased amount of research has been done to provide conditions for application of these organisms in plants of economic importance (Klironomos and Kendrick, 1993).

ROOT COLONIZATION

The life cycle of AMF is initiated with spore germination and, after formation of the germ tube, some hyphae start the process of host recognition and root penetration. There are indications that some phenolic compounds produced by the plants are involved in the communication between the host and the mycorrhizal fungus, but this hypothesis has yet to be confirmed (Lambais, 1996). Root penetration is initiated after differentiation of an infective hyphae from an apressorium at the root epidermis and occurs through mechanical pressure and/or partial enzymatic degradation of the cellular wall (Lambais, 1996). After that, hyphal proliferation occurs with penetration of the cortical cells of the host, with invagination of the plasma membrane and synthesis of material around the hyphae, forming an interface in which its structure is morphologically different from that of the cell wall (Bonfante and Perotto, 1995). In the host cell, the hyphae ramify, forming specialized structures known as arbuscules, which are the exchange sites between the plant and the fungus. The cell inhabited by the arbuscule is not affected by its presence and maintains its normal functioning. Some AMF also produce, at the host cortex, vesicles that apparently function as a reservoir, being rich in lipids. The structures formed by these fungi in the roots do not pass the endoderm and consequently do not invade the central cylinder.

CHANGES PRODUCED BY MYCORRHIZATION

In general, the mycorrhizal fungi are capable of producing morphological, physiological, and biochemical modifications in the root cells (see Table 13.1) that increase plant tolerance to root diseases, providing to the plant a better position in relation to the pathogen which, in that way, has its effect reduced or nullified.

TABLE 13.1. Changes induced by AMF in the host plants, independently of presence of root pathogens.

Mechanism	Reference
• Induction (transitory or not) of the activity or accumulation of enzymes, such as chitinase and β-1,3-endoglucanase; specific activation of the systems of enzymes (neutral phosphatase and Mg^{++} ATPase)	Serrano, 1990; Dumas-Gaudot, Grenier, et al., 1992; Gianinazzi-Pearson, 1996; Lambais and Mehdy, 1998
• Increase of lignification of the cell wall of the endoderm and stele and thickness of the cell wall by aposition of calose adjacent to intercellular hyphae	Dehne and Schönbeck, 1979; Cordier et al., 1998
• Increase of amino acid production (mainly arginine) and of reductor sugars	Dehne and Schönbeck, 1979
• Intensive synthesis of plasma membrane, increase of synthesis and volume cytoplasmatic, vacuole fragmentation, cytoskeletal reorganization, hypertrophy and movement of the nucleus, with general activation of the metabolism	Serrano, 1990; Bonfante and Perotto, 1995; Blee and Anderson, 2000
• Alteration of the phytohormone balance in shoots and roots; increase of levels of auxins, cytokinins, abscicic acid, and ethylene in the roots	Allen et al., 1980; Dugassa et al., 1996; Lambais, 2000
• Increase of the phenolic metabolism	Krishna and Bagyaraj, 1983; Grandmaison et al., 1993; Benhamou et al., 1994; Cordier et al., 1998
• Increase of phytoalexin-like isoflavonoid compounds	Sundaresan et al., 1993; Morandi, 1996
• Synthesis of proteins known as endomycorrhizines	Dumas et al., 1990
• Formation of an interface around the arbuscular hyphae	Serrano, 1990

It is clear that such modifications produce different responses which will depend on many factors, such as the species of plant, pathogen, and symbiotic fungus; inoculum level of the pathogen, period of inoculation of the mycorrhizal and pathogenic fungi; interactions with other organisms in the mycorhizosphere; and also the nutrient availability and soil humidity. As discussed by Blee and Anderson (2000), indirect mechanisms through the plant or direct mechanisms showing the interaction between the mycorrhizal fungi and the pathogen are possible (see Figure 13.1).

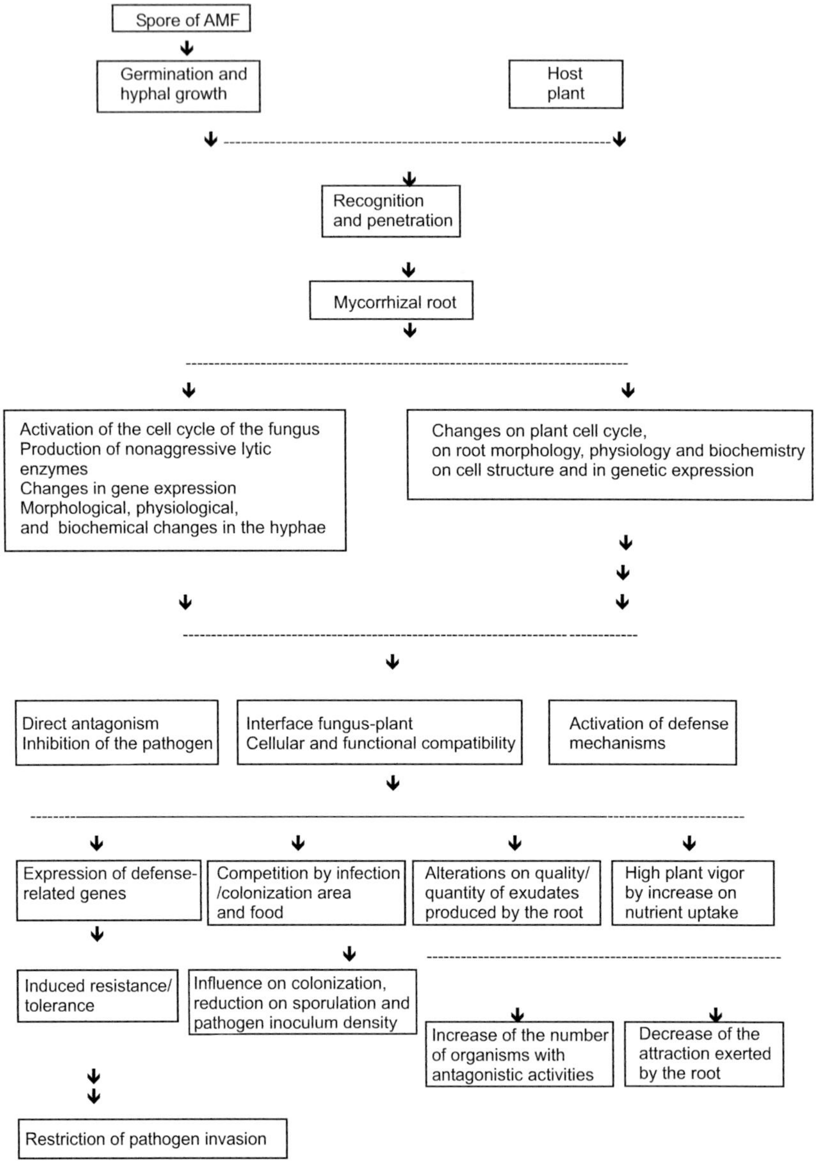

FIGURE 13.1. Key steps in the relationship of AMF, host plant, and plant pathogen. *Source:* Adapted from Bonfante and Perotto, 1995.

The protective effect of mycorrhizal symbiosis against pathogens of root systems has been widely demonstrated in many reviews, which emphasize mainly the pathogenic fungi (Bagyaraj, 1984; Zambolim, 1985; Jalali and Jalali, 1991; Hooker et al., 1994; Linderman, 1994; Schönbeck et al., 1994; Linderman, 2000; Azcón-Aguilar and Barea, 1997; Borowicz, 2001). It has been indicated that the AMF probably participate, with different strategies, for control of root diseases, mainly by the demonstrated capacity for inducing higher tolerance of the host against the pathogen. However, it should be considered that the effect of the mycorrhizal symbioses may decrease if the host is highly susceptible and/or the pathogen pressure of infection is high. In the same way, plant protection promoted by AMF can decrease as a result of the synergistic effect of pathogens, as observed by Greipsson and El-Mayas (2002).

INTERACTION: ARBUSCULAR MYCORRHIZAL FUNGI AND PLANT PATHOGENIC FUNGI

The studies regarding effect of AMF on reduction of root diseases produced by fungi have focused mainly on those rots produced by species of *Pythium, Phytophthora, Fusarium, Verticillium, Pyrenochaeta, Gaeumannomyces, Sclerotium,* and *Rhizoctonia* (Linderman, 1994). Most of the time, the plants associated with AMF have shown to be more tolerant/resistant to the attack of pathogenic fungi than those nonmycorrhizal (see Table 13.2). Among the AMF, species of *Glomus* have been the most studied with relation to their potential as defense plant organisms against root pathogens, although the benefits of species of other genera have also been recorded.

Different mechanisms have been proposed to explain the role of AMF on reduction of the effects of pathogenic fungi in many agriculturally important crops (see Table 13.3). However, such mechanisms are not well understood yet, mainly due to the large genetic diversity (Vandenkoornhuyse et al., 2001) and physiological response of the AMF (Smith and Read, 1996).

The increase in the absorption of nutrients, mainly phosphorus (P), promoted by the formation of the mycorrhizal association gives to the plant better conditions to stand the attack of pathogenic microorganisms. However, in some cases, the inhibition of root diseases produced by fungi in mycorrhizal plants is not related with the increase on P level, being produced by other factors, like competition among the microrganisms (St-Arnaud et al., 1994; Trotta et al., 1996). Controversy still exists as to whether competition is a key factor in plant disease control promoted by AMF. Linderman (2000) considers competition for the same nutrients or in-

TABLE 13.2. Reduction of the effect of plant pathogenic fungi in plants colonized by AMF.

Plant pathogen	Host	Mycorrhizal fungi	Reference
Aphanomyces euteiches	Pea *(Pisum sativum)*	*Glomus intraradices*	Rosendahl, 1985
Cladosporium sp., *C. herbarum, Fusarium* sp., *F. nivale, Phoma* sp.	Dunegrass *(Leymus arenarius)*	*Glomus caledonium, G. fasciculatum, G. mosseae*	Greipsson and El-Mayas, 2002
Cochliobolus sativus	Barley (*Hordeum* sp), corn *(Zea mays),* wheat *(Triticum sativum)*	*Glomus* spp.	Dehn and Dehne, 1986
Corticium rolfsii	Tomato *(Lycopersicon esculentum)*	*Gigaspora margarita, Glomus* spp.	Vargas, 1991
Fusarium moniliforme	Cardamom *(Elettaria cardomomum),* maize *(Z. mays)*	*Glomus fasciculatum*	Chhabra et al., 1992; Thomas et al., 1994
F. oxysporum	Asparagus *(Asparagus officinalis)*	*Glomus fasciculatum*	Wacker et al., 1990
F. oxysporum f.sp. *lini*	Flax *(Linum usitatissimum)*	*Glomus intraradices*	Dugassa et al., 1996
F. oxysporum f.sp. *lycopersici*	Tomato *(L. esculentum)*	*G. fasciculatum, G. mosseae, Scutellospora heterogama*	Dehne and Schönbeck, 1975
F. oxysporum f.sp. *radicis-lycopersici*	Tomato *(L. esculentum)*	*Glomus intraradices*	Caron et al., 1986
F. oxysporum f.sp. *medicaginis*	Alfalfa *(Medicago sativa)*	*Glomus* spp., *G. fasciculatum, G. mosseae*	Hwang, 1992
F. pallidoroseum	Amora (*Morus* sp.)	*G. fasciculatum, G. mosseae*	Sharma et al., 1995
F. solani f.sp. *piperis*	Black pepper *(Piper nigrum)*	*Entrophospora colombiana, Scutellospora* sp., *S. gilmorei, S. heterogama*	Chu et al., 1997
Gaeumannomyces graminis var. *tritici*	Wheat *(T. sativum)*	*Glomus fasciculatum*	Graham and Menge, 1982
Phytophthora nicotianae var. *parasitica*	Tomato *(L. esculentum)*	*Glomus mosseae*	Trotta et al., 1996

TABLE 13.2 *(continued)*

Plant pathogen	**Host**	**Mycorrhizal fungi**	**Reference**
P. parasitica	Citrus (*Citrus* sp.)	*Glomus fasciculatum, G. mosseae*	Davis and Menge, 1981
Pythium aphanidermatum	Tomato *(L. esculentum)*	*Glomus fasciculatum*	Hedge and Rai, 1984
P. ultimum	Marigold *(Tagetes erecta)*	*G. mosseae*	Calvet et al., 1993
Rhizoctonia solani	Bean *(Phaseolus vulgaris)*, tomato *(L. esculentum)*, potato *(Solanum tuberosum)*	*Acaulospora morrowiae, Gigaspora margarita, Glomus etunicatum, G. intraradices, G. leptothichum*	Cassiolato and Melo, 1991; Ahmed et al., 1994; Yao et al., 2002
Sclerotium rolfsii	Peanut *(Arachis hypogaea)*, chilli *(Capsicum annuum)*	*Glomus fasciculatum, G. macrocarpum, Sclerocystis dussii*	Krishna and Bagyaraj, 1983; Sreenivasa et al., 1992
Verticillium dahliae	Cotton *(Gossipium hirsutum)*, eggplant *(Solanum melongena)*, tomato *(L. esculentum)*	*G. margarita, Glomus etunicatum, G. mosseae, G. versiforme, Sclerocystis sinuosa*	Liu, 1995; Matsubara et al., 1995; Karagiannidis et al., 2002

fection site between the AMF and pathogens, unlikely. In fact, in some cases fungal root pathogens can infect tissues near to those colonized by AMF (Dehne, 1982), what might indicate that competition for food and/or space is not a rule.

Inoculation with *Glomus etunicatum* and *Glomus fasciculatum* protected strawberry plants against *Phytophthora fragariae,* and factors related with the increase of root ramification, promoted by improved nutrition, were indicated as involved in the process (Norman et al., 1996). However, increase in root ramification does not always occur, as observed in adventitious roots of tomato when colonized by *Glomus mosseae* and attacked by *Phytophthora nicotiana* var. *parasitica*. In this case, disease resistance was related to the associated effect of various factors, such as competition and biochemical changes in the host and/or in the enzymatic activity of the pathogen (Trotta et al., 1996).

TABLE 13.3. Mechanisms and effects directly related to the inhibition of plant pathogenic fungi in AMF colonized plants.

Mechanism	Effect	Reference
Higher uptake of nutrients, mainly phosphorus	Higher plant health to resist or tolerate the pathogen	Schenck, 1981; Linderman, 1994; Trotta et al., 1996
Changes in quality and/or amount of exudates produced by the root	Increase in the number of organisms with antagonistic properties against the pathogen	Linderman, 1991; Azcón-Aguilar and Barea, 1997
Microbial changes in the mycorrhizosphere	Alter the composition of functional groups of microbes, including the number and/or activity of pathogens; potential increased antagonistic activity against the pathogen	Azcón-Aguilar and Barea, 1997; Linderman, 2000
Competition by area of infection/colonization and food	Interference on colonization, reduction on sporulation and density of the pathogen inoculum	Trotta et al., 1996
Physical protection	Inhibition of pathogen penetration and infection	Benhamou et al., 1994; Fitter and Garbaye, 1994
Direct antagonism between the fungi	Pathogen inhibition	Liu, 1995
Increase in phenol concentration	Induced resistance or tolerance	Krishna and Bagyaraj, 1983; Chhabra et al., 1992
Gene expression with production of molecules related to plant defense	Restriction of pathogen invasion by induced resistance or tolerance	Dehne, 1982; Sundaresan et al., 1993; Benhamou et al., 1994; Gianinazzi-Pearson et al., 1996; Trotta et al., 1996; Azcón-Aguilar and Barea, 1997

The colonization of roots by symbiotic fungi affects the quality and quantity of exudates liberated by the roots, and exerts selective pressure on the microbial population of the rhizosphere, indirectly affecting the root pathogens. It was observed that the rhizosphere of a grass colonized by *G. fasciculatum* presented a higher number of actinomycetes antagonistic to *Ralstonia solanacearum* and *Fusarium solani* (Secilia and Bagyaraj, 1987). In the same way, a higher number of microorganisms with antagonistic properties against *Fusarium moniliforme* was observed in the rhizosphere of plants colonized by AMF (Thomas et al., 1994).

Mycorrhizal plants produce certain substances that have been indicated as copartners in the resistance to the attack of root pathogens. An example is the O-D phenol accumulated by the plant that may constitute one of the possible mechanisms of disease resistance, as observed on studies of the peanut-*S. rolfsii-G. fasciculatum* system, in which it was also verified that the compound was capable of inhibiting the growth of the pathogen in vitro (Krishna and Bagyaraj, 1983). Phenols, particularly flavonoids and isoflavonoids, are secondary metabolites involved in plant-microorganism interactions. Among such metabolites, the phytoalexins show antimicrobial activity, which is important in plant defense mechanisms (Mansfield, 1982). In roots colonized only by AMF, phytoalexins are accumulated in a lower amount and more slowly than in interactions that are exclusively pathogenic. However, other phenol compounds with antimicrobial properties, capable of stimulating growth of AMF and root colonization (example: phormononetina and biocanina A), are consistently found in higher amounts in roots of mycorrhizal plants (Lambais, 1996; Morandi, 1996). In some cases, the occurrence of mycorrhizal association promotes reduction of production of inoculum by the pathogen. The number of microsclerotia of *Verticillium dahliae* in the rhizosphere of cotton was reduced when plants were inoculated with AMF. On the other hand, mycorrhizal fungi and pathogens may mutually reduce their percentages of infection/colonization when simultaneously inoculated (Liu, 1995).

The hyphae of plant pathogenic and mycorrhizal fungi may be together in the same part of a root, which indicates that they are not mutually exclusive. However, the pathogen hyphae, will never invade a cell with arbuscles, and in this way these cells are not necrosed as occurs in those invaded by pathogenic fungi. This reaction may be related to expression of defense-related genes, which are induced by presence of the symbiont (Gianinazzi-Pearson et al., 1996), although in some cases this induction has not been proved (Tosi et al., 1993). Lambais (1996) mentioned that four groups of genes induced by the formation of mycorrhizal association are apparently related to plant defense. These are the genes

1. involved in the biosynthesis via of isoflavonoides phytoalexins;
2. that code hydrolases with antimicrobial activity;
3. that code proteins which contribute to increase rigidity of plant cell walls; and
4. that code pathogenesis-related protein.

Mycorrhizal colonization was shown to be responsible, at least in part, for the activation of defense mechanisms of carrot roots genetically transformed (Ri T-DNA) and infected by *Fusarium oxysporum* f.sp. *chrysanthemi*. The host reaction was shown by the production of compounds (probably phenols) in the intercellular spaces that, besides establishing a physical barrier, had a toxic activity against the pathogen (Benhamou et al., 1994).

It has been observed that transitory induction, followed or not by posterior supression of the activity of certain enzymes, such as chitinase, glucanase, and β-1,3-endoglucanase, is accumulated in higher amounts by the host as a defense response to the pathogenic microorganisms (Gianinazzi-Pearson and Gianinazzi, 1995). Although initially stimulated, the production of such enzymes decreases after the "recognition" of the symbiont. On the other hand, proteins related to pathogenesis of the PR-b1 group induced by pathogenic infection are produced in very low amounts when the plant forms a mycorrhizal association (Gianinazzi, 1984). The synthesis of other proteins, called endomycorrhizines, whose function is being studied, has also been mentioned as one of the responses to the beginning of colonization (Dumas et al., 1990). In the same way, new isoforms of chitinase, not induced during the infection by pathogens, were detected in mycorrhizal plants (Dumas-Gaudot, Furlan, et al., 1992).

The AMF are also capable of producing enzymes such as cellulases and pectinases that act in plant cell walls, although this activity is low when compared to the activity of enzymes produced by plant pathogenic fungi (Garcia-Garrido et al., 1992).

Other possibilities to be considered in the root pathogen-AMF relationship includes the neutral relationship. In this case, the pathogenic fungi are able to produce disease independently of the presence of the symbiont fungi (Table 13.4). Conversely, the severity of diseases may also increase in the presence of AMF, or these fungi may have their colonization rate reduced due to the presence of plant pathogenic fungi (see Table 13.4). The inoculation of *Glomus mosseae, G. hoi,* and *G. fistulosum* did not decrease the severity of root rot caused by *Phytophthora* in strawberry plants; indeed, the mycorrhizal colonization made the plants more susceptible to pathogen attack (Vestberg et al., 1994). *Glomus macrocarpum* was indicated as a pathogen to tobacco, with the colonization being related to occurrence of the disease and the populations of this AMF proportional to the incidence of disease (Guo et al., 1994).

ARBUSCULAR MYCORRHIZAL FUNGI × BACTERIA

The effect of AMF in relation to plant pathogenic bacteria has not been explored to a large extent, when compared to other pathogens that occur in the root system (Linderman, 1994). However, research mentioning the ben-

TABLE 13.4. No effect or negative response of the mycorrhizal colonization on the occurrence of root diseases caused by fungi in various crops.

Plant fungi	Pathogenichost	Arbuscular mycorrhizal fungi	Reference
No effect			
Fusarium oxysporum f. sp. *lycopersici*	Tomato *(L. esculentum)*	*Acaulospora scrobiculata, Glomus leptothichum, G. mosseae, Scutellospora heterogama*	Marchi and Costa, 1987; Melo, 1989
Fusarium solani	Soybean *(Glycine max)*	*Glomus mosseae*	Zambolim and Schenck, 1984
Macrophomina phaseolina			
Rhizoctonia solani			
Phytophthora fragariae	Strawberry (*Fragaria* sp.)	*Glomus caledonium, G. mosseae*	Baath and Hayman, 1984
Sclerotium rolfsii	Chilli *(C. annuum)*	*Acaulospora laevis, Gigaspora margarita*	Sreenivasa et al., 1992
Verticillium albo-atrum	Tomato *(L. esculentum)*	*G. caledonium, G. mosseae*	Baath and Hayman, 1983
Negative effect			
Phytophthora cactorum [a]	Strawberry cv. Jonsok (*Fragaria* sp.)	*Glomus fistulosum, G. hoi, G. mosseae*	Vestberg et al., 1994
Phytophthora megasperma [a]	Soybean *(G. max)*	Not indicated	Ross, 1972
Verticillium sp.[a]	Cotton *(Gossipium hirsutum)*	*G. fasciculatum*	Davis et al., 1979
Glomus macrocarpum [b]	Tobacco *(Nicotiana tabacum)*	*G. macrocarpum*	Guo et al., 1994

[a]Inoculation of AMF promoting increase of susceptibility or inducing disease.
[b]The AMF acting as pathogen and producing disease.

efit of the symbiosis in tomato has been published (Silveira and Maia, 1996). Looking at the effect of native mycorrhizal fungi in relation to *R. solanacearum* in tomato plants, Halos and Zorilla (1979) observed that mycorrhizal plants presented lower incidence of bacterial wilt than did nonmycorrhizal plants. The reduction on the wilt levels, in this case, was related to competition or to a barrier mechanism constituted by the structures formed by the AMF, which do not allow the penetration of the pathogenic bacteria in the host tissue. The incidence of wilt was reduced in this pathosystem (tomato-*R. solanacearum*) when plants were inoculated with *Scutellospora heterogama* or with a mixture of *Scutellospora heterogama* and *Acaulospora morrowiae* spores (Rego, 1991).

The studies to evaluate the interaction between AMF and nonpathogenic rhizospheric bacteria (mainly those nitrogen fixing, phosphate solubilizing and plant growth promoted bacteria) have shown that the occurrence of a synergic interaction between these organisms allows the increment of rhizosphere microbiota (Bagyaraj and Menge, 1978; Alten et al., 1991; Yuri et al., 1994; Barea et al., 1997; Dar et al., 1997). These results are important, considering that rhizosphere microorganisms have a significative effect on plant health.

ARBUSCULAR MYCORRHIZAL FUNGI × NEMATODES

AMF and plant parasitic nematodes may be simultaneously associated with roots; thus, the combined effect of these two groups of organisms on plant growth should be considered. In a review of studies from 1970 to early 1998, Borowicz (2001) observed that AMF tended to decrease the harmful effects of fungal pathogens but to increase those produced by nematodes; in this case, the outcome depended upon mode of feeding. Most of the studies related to the interaction of mycorrhiza-nematode were done with species of *Meloidogyne* (Reid, 1990), the gall root nematode. Little is known regarding the interactions involving endoparasitic migratory and ectoparasites nematodes (see Table 13.5).

In studies of this interaction, nematode inoculum density, species of AMF, period of mycorrhization in relation to inoculation of the nematode, natural resistance of the plant to the pathogen, and the cultivar of the tested plant are all important. The responses of the host might be evaluated by the influence of the interaction on plant growth and production, by the development of the AMF, or by suppression of the nematode parasitism. The results available regarding the role of AMF in the protection of plants against the nematodes have been diverse, but in general they indicate a benefit of the mycorrhization. Some authors have shown that the mycorrhizal association decreases the damage produced by plant pathogenic nematodes in some crops of economic importance, such as cotton (Smith et al., 1986), beans

TABLE 13.5. Effect of mycorrhization on reproduction of plant pathogenic nematodes.

Nematode	Host	AMF	Reference
Reduction of reproduction			
Heterodera cajani	Cowpea *(Vigna unguiculata)*	*Glomus fasciculatum*	Jain and Sethi, 1989
Heterodera solanacearum	Tobacco *(N. tabacum)*	*Gigaspora gigantea*	Hussey and Roncadori, 1982
Meloidogyne hapla	Carrot (*Daucus carota* ssp. *sativus*)	*Glomus mosseae*	Sikora, 1979
Meloidogyne incognita	Cotton *(G. hirsutum)*, oat *(Avena sativa)*, banana (*Musa* sp.), cowpea *(V. unguiculata)*, beans *(P. vulgaris)*, tobacco *(N. tabacum)*, peach *(Prunus persica)*, soybean *(G. max)*, tomato *(L. esculentum)*	*Glomus etunicatum, G. mosseae, G. fasciculatum, G. macrocarpum, Gigaspora margarita*	Sikora and Schönbeck, 1975; Sikora, 1979; Kellam and Schenck, 1980; Dehne, 1982; Hussey and Roncadori, 1982; Strobel et al., 1982; Jain and Sethi, 1989; Jaizme-Vega et al., 1997
Meloidogyne javanica	Soybean *(G. max)*, tomato *(L. esculentum)*	*Glomus etunicatum, G. macrocarpum*	Bagyaraj et al., 1979; Kellam and Schenck, 1980
Prathylenchus brachyurus	Ananas *(Ananas comosus)*, cotton *(G. hirsutum)*	*Glomus* sp., *Gigaspora margarita*	Hussey and Roncadori, 1982; Guillemin et al., 1994;
Prathylenchus vulnus	Apple *(Mallus domestica)*, pear *(Pyrus communis)*, peach *(Prunus persica)*	*Glomus intraradices, G. mosseae*	Lopez et al., 1997; Pinochet et al., 1993,1995a,b
Radopholus citrophylus	Citros (*Citrus* sp.)	*G. intraradices*	Smith and Kaplan, 1988
Radopholus similes	Banana (*Musa* sp.)	*G. intraradices*	Umesh et al., 1988
Rotylenchulus reniformis	Cowpea *(V. unguiculata)*	*G. fasciculatum*	Lingaraju and Goswami, 1994

TABLE 13.5 *(continued)*

Nematode	Host	AMF	Reference
Increase on reproduction			
Heterodera glycinis	Soybean (*G. max*)	*Glomus etunicatum*	Hussey and Roncadori, 1982
Meloidogyne arenaria	Peanuts *(A. hypogaea)*, grape *(Vitis vinifera)*	*Gigaspora margarita, Glomus etunicatum, G. fasciculatum*	Atilano et al., 1981; Hussey and Roncadori, 1982
Meloidogyne incógnita	Cotton *(G. hirsutum)*, soybean *(G. max)*	*Gigaspora margarita, Glomus macrocarpum, Scutellospora calospora, S. heterogama*	Schenck et al., 1975; Roncadori and Hussey, 1977
Pratylenchus coffeae	Coffee *(Coffea arabica)*	*Acaulospora mellea, Glomus clarum*	Vaast et al., 1998
Null effect			
Heterodera cajani	Cowpea *(V. unguiculata)*	*Glomus epigaeus*	Jain and Sethi, 1989
Meloidogyne incognita	Cotton *(G. hirsutum)*, cowpea *(V. unguiculata)*, peach *(P. persica)*, soybean *(G. max)*, tomato *(L. esculentum)*	*Gigaspora margarita, Glomus etunicatum, G. epigaeus, G. mosseae*	Hussey and Roncadori, 1982; Strobel et al., 1982; Jain and Sethi, 1989; Thompson and Hussey, 1993
Pratylenchus vulnus	Cherry (*Prunus* sp.), marmelade (*Cydonia* sp.)	*Glomus intraradices*	Calvet et al., 1995; Pinochet et al., 1995b
Radopholus similis	Citros (*Citrus* sp.)	*Glomus etunicatum*	O'Bannon and Nemec, 1979
Tylenchulus semipenetrans	Citros (*Citrus* sp.)	*Glomus fasciculatum, G. mosseae*	O'Bannon et al., 1979; Hussey and Roncadori, 1982

(Oliveira and Zambolim, 1986), citrus (Smith and Kaplan, 1988), cowpea (Santhi and Sundarababu, 1995), tomato (Sundarababu et al., 1995), and soybean (Price et al., 1995). However, in experiments with grapes, Atilano et al. (1981) registered an increase in damage caused by nematodes in mycorrhizal plants. On the other hand, Greipsson and El-Mayas (2002) referred to a synergistic effect of soil fungi and nematodes that break down the protection of AMF against pathogens.

The most common effect of AMF on plants susceptible to nematodes is promotion of tolerance, and the effects of the interaction can be grouped in the following categories: neutral, when effect on the fungus, host or nematode is not observed; positive, when the AMF compensate for the damage produced in the plant by the pathogen while development and reproduction of the nematodes are suppressed; negative, when the sporulation of the AMF and the development and/or plant production are suppressed and reproduction of the nematode is increased (Hussey and Roncadori, 1982).

When negatively affected by mycorrhization, the response of the nematode might be represented by reduction of number and size of galls; decrease on the number of eggs and larvae; inhibition of penetration and development of larvae; increase on mortality of infective larvae; reduction on size and number of giant cells; and loss of reproduction (see Table 13.5). Schenck (1981) observed that soybean roots colonized by *G. macrocarpum* presented a lower number of galls produced by *Meloidogyne incognita* than nonmycorrhizal roots.

In the same way as occurs with plant pathogenic bacteria and fungi, the mechanisms related to the reduction of nematodes in plants colonized by AMF are not clear, although the studies indicate diverse possibilities (Table 13.6). The interactions between plant, nematode, AMF, and other rhizosphere microorganisms, as well as the mechanisms involved in these processes, present high complexity, which demonstrates the need for integrated studies. Considering that use of AMF promises to decrease the damage produced by nematodes, studies should continue in a way to identify the best strategies for using the beneficial effect of mycorrhization. Considering that the control of nematodes is difficult and expensive, the inoculation of seedlings with AMF before transplant of plants to the field may be considered an important alternative as a component of integrated control, considering that in most of the studies there was reduction of damage. Besides, the AMF present an important characteristic: resistance to nematicides (Habte and Manjunath, 1988).

TABLE 13.6. Mechanisms that may act in the control of pathogenic nematodes on mycorrhizal plants.

Mechanism	Reference
Competition for space and food	Hussey and Roncadori, 1982; Pinochet et al., 1996
Chemical interactions and physical protection	Kellam and Schenck, 1980; Fitter and Garbaye, 1994
Production of phenolic compounds, hormones, and phytoalexins, affecting the nourishment of the nematode	Pinochet et al., 1996
Production of nematicide substances and/or increase in the concentration of phenylanine and serine	Suresh et al., 1985
Increase of lignification and suberization, causing damage to the migratory nematodes	Pinochet et al., 1996
Increase on uptake of nutrients, mainly phosphorus, contributing for high vigor of the host	Hussey and Roncadori, 1982; Cason et al., 1983
Alterations on root exudates, decreasing the attraction of the nematodes by the root and/or increasing the number of nematode-suppressing microbes that could interfere with their reproduction	Kloepper et al., 1991; Linderman, 1994

FUTURE PERSPECTIVES

Although general patterns of the effects of AMF on plant-pathogen relations have been determined (Borowicz, 2001), research including a broader diversity of species and a wider variety of environmental conditions is still needed for better understanding of mycorrhizal relationships. There is no doubt, however, that mycorrhization of plants should be considered in the management systems, as part of strategies for the integrated control of diseases. To have an effect on the plant, symbiotic fungi should be able to act under different environmental conditions, be competitive, and be capable to rapidly colonize the root to be there before the pathogen. In the same way, the synergistic effect of AMF and other biocontrol agents should be considered. A most pronounced effect of diseases produced by *Pythium ultimum* and *S. rolfsii* on roots of different hosts was observed with the combined use of AMF and *Trichoderma* (Calvet et al., 1993; Sreenivasa, 1994). The same was shown in studies with bacteria that live in the rhizosphere, some of which stimulates germination, favoring also the colonization of roots by mycorrhizal fungi (Alten et al., 1991).

Most of the work with AMF and plant pathogens was done in greenhouses, being necessary to develop of field experiments to try the general application of these symbiotic fungi in agricultural practices. As mentioned by Siqueira (1996), the advances in this field have been limited mostly because of the absence of inoculants in some local markets. However, although the condition of the obligate biotroph of the AMF is complicating their study and production in large scale, this should not be enough to stop studies and further utilization of these fungi. Besides the possibilities of use as inoculant of seedlings, before the transplant to the field or in protected cultures, production of inoculum of AMF is possible by using different methods, as already discussed (Jarstfer and Sylvia, 1992). One might also consider that the cultural practices could be manipulated to increase the incidence of native mycorrhizal fungi or maybe to change the predominant species, favoring the establishment and action of species with high potential to benefit the host plant that is being cultivated. Finally, we can remember that mycorrhizal fungi are useful not only to control plant pathogens. They can be also used as bioregulators, in cultures that depend on mycorrhization for complete development, and as biofertilizers, maximizing the growth of cultures in areas with nutritional limitations, mainly in phosphorus, and in other situations of plant stress. As early indicated by Schenck (1981), the AMF have high potential, especially if isolates that simultaneously promote higher growth and control of diseases on the host are selected. The use of microorganisms, such as mycorrhizal fungi, and of biological processes that occur in the soil constitute one of the strategies of sustainable agriculture, where there is a need to guarantee higher yield with reduction of the use of fertilizers that are risky to the environment and that increase costs (Siqueira and Franco, 1998). In this context, the biotechnological potential of the mycorrhizal association is evident, mainly by the impact that it can have in both levels: agricultural (increasing plant production) and environmental (promoting better quality of the ecosystems).

CONCLUSIONS

The behavior of mycorrhizal plants in regard to root pathogens is complex, being affected by diverse factors, including the density of inoculum of the pathogen, the species of AMF associated to the plant, and the levels of nutrients in the soil. There are indications that formation of mycorrhizae will induce plant resistance when the plant is challenged by a pathogen. The evidences show that mycorrhiza formation helps the host to suppress pathogens by several means, and it is probably a combination of mechanisms which respond. On the other hand, a better understanding of the relations of

AMF-host-pathogen-environment, as well as the most appropriate, fast, and efficient way to inhibit/suppress the plant pathogenic agent, are objectives that are still being pursued. Research should be directed to clarify the still unclear key points of the pathogen × host × mycorrhizal fungus relationship, under diverse environmental conditions, in order to avoid establishment of plant pathogens, and get the benefits from mycorrhization. It seems a general agreement that studies on the molecular aspects of symbiosis are essential. The results of many experiments have shown that each system should be considered separately, and that there is no model to be universally adopted. The simplest approach is the comparative study of defense-related genes on mycorrhizal and nonmycorrhizal roots, as has been done, but other techniques such as isolation of differentially expressed proteins, and even the analysis of proteomas of mycorrhizal and nonmycorrhizal plants, would provide new insights for understanding the mechanisms that regulate the symbiosis and the way it interferes in plant disease development. In natural situations, the host, AMF, and pathogen are together, living in equilibrium; to maintain the same, after introducing plant crops, is a challenge that should be pursued. Thus, researchers should be aware of the difficulties of working in this field and consider that an integrated approach, including use of mycorrhizal fungi, would be the most likely to guarantee better agricultural crops and maintenance of sustainability in agricultural systems.

REFERENCES

Ahmed M.A, Saleh E.A., and El-Fallal A.A. (1994). The role of biofertilizers in suppression of *Rhizoctonia* root-rot disease of broad bean. *Annals of Agricultural Science* (Cairo) 39: 379-395.

Allen M.F., Moore T.S., and Christensen M. (1980). Phytohormone changes in *Bouteloua gracilis* infected by vesicular-arbuscular mycorrhizae. I. Cytokinin increases in the host plant. *Canadian Journal of Botany* 58: 371-374.

Atilano R.A., Menge J.A., and van Gundy S.D. (1981). Interaction between *Meloidogyne arenaria* and *Glomus fasciculatus* in grape. *Journal of Nematology* 13: 52-59.

Azcón-Aguilar C. and Barea J.M. (1997). Arbuscular mycorrhizas and biological control of soilborne plant pathogens: an overview of the mechanisms involved. *Mycorrhiza* 6: 457-464.

Baath E. and Hayman D.S. (1983). Plant growth response to vesicular-arbuscular mycorrhiza: Interactions with *Verticillium* wilt on tomato plants. *New Phytologist* 95: 419-426.

Baath E. and Hayman D.S. (1984). No effect of VA mycorrhiza on red core disease of strawberry. *Transactions of the British Mycological Society* 82: 534-536.

Bagyaraj D.J. (1984). Biological interactions with VA mycorrhizal fungi. In C.L. Powell and D.J. Bagyaraj (eds.), *VA Mycorrhiza* (pp. 131-153). Boca Raton, FL: CRC Press.

Bagyaraj D.J., Manjunath A., and Reddy D.D.R. (1979). Interaction of vesicular-arbuscular mycorrhiza with root knot nematodes in tomato. *Plant and Soil* 51: 397-403.

Bagyaraj D.J. and Menge J.A. (1978). Interaction between a VA mycorrhiza and *Azotobacter* and their effects on rhizosphere microflora and plant growth. *New Phytologist* 80: 567-573.

Barea J.M., Azcón-Aguillar C., and Azcón R. (1997). Interactions between mycorrhizal fungi and rhizosphere micro-organisms within the context of sustainable soil-plant systems. In A.C. Gange and V.K. Brown (eds.), *Multitrophic interactions in terrestrial systems* (pp. 65-77). Cambridge, UK: Blackwell Science.

Benhamou N., Fortin J.Á., Hamel C., St-Arnaud M., and Shatilla S. (1994). Resistance responses of mycorrhizal Ri T-DNA- transformed carrot roots to infection by *Fusarium oxysporum* f.sp. *chrysanthemi. Phytopathology* 84: 958-968.

Blee K.A. and Anderson A.J. (2000). Defense responses in plants to arbuscular mycorrhizal fungi. In G.K. Podila and D.D. Douds Jr. (eds.), *Current advances in mycorrhizae research* (pp. 27-44). St. Paul, MN: APS Press.

Bonfante P. and Perotto S. (1995). Strategies of arbuscular mycorrhizal fungi when infecting host plants. *New Phytologist* 130: 3-21.

Borowicz V.A. (2001). Do arbuscular mycorrhizal fungi alter plant-pathogen relations? *Ecology* 82: 3057-3068.

Calvet C., Pera J., and Barea J.M. (1993). Growth response of marigold (*Tagetes erecta* L.) to inoculation with *Glomus mosseae, Trichoderma aureoviride* and *Pythium ultimum* in a peat-perlite mixture. *Plant and Soil* 148: 1-6.

Calvet C., Pinochet J., Camprubi A., and Fernandêz C. (1995). Increased tolerance to the root-lesion nematode *Pratylenchus vulnus* in mycorrhizal micropropagated BA-29 quince rootstock. *Mycorrhiza* 5: 253-258.

Caron M., Fortin J.A., and Richard C. (1986). Effect of *Glomus intraradices* on infection by *Fusarium oxysporum* f. sp. *radicis-lycopersici* in tomatoes over a 12-week period. *Canadian Journal of Botany* 64: 552-556.

Cason K.M.T., Hussey R.S., and Roncadori R.W. (1983). Interaction of vesicular - arbuscular mycorrhizal fungi and phosphorus with *Meloidogyne incognita* on tomato. *Journal of Nematology* 15: 410-417.

Cassiolato A.M.R. and Melo I.S. (1991). Interaction between *Rhizoctonia solani* and vesicular arbuscular mycorrhizal fungi in tomato. *Summa Phytopathologica* 17: 195-200.

Chhabra M.L., Bhatnagar M.K., and Sharma M.P. (1992). Influence of a vesicular-arbuscular (VA) mycorrhizal fungus on important diseases of maize. *Indian Phytopathology* 45: 235-236.

Chu E.Y., Endo T., Stein R.L.B., and Albuquerque F.C. (1997). Avaliação da inoculação de fungos micorrízicos arbusculares sobre a incidência da fusariose da pimenta-do-reino. *Fitopatolologia Brasileira* 22: 205-208.

Cordier C., Pozo M.J., Barea J.M., Gianinazzi S., and Gianinazzi-Pearson V. (1998). Cell defense responses associated with localized and systemic resistance to *Phytophthora parasitica* induced in tomato by an arbuscular mycorrhizal fungus. *Molecular Plant-Microbe Interactions* 11: 1017-1028.

Dar G.H., Zargar M.Y., and Beigh G.M. (1997). Biocontrol of *Fusarium* root rot in the common bean (*Phaseolus vulgaris* L.) by using symbiotic *Glomus mosseae* and *Rhizobium leguminosarum*. *Microbial Ecology* 34: 74-80.

Davis R.M. and Menge J.A. (1981). *Phytophthora parasitica* inoculation and intensity of vesicular-arbuscular mycorrhizae en citrus. *New Phytologist* 87: 705-715.

Davis R.N., Menge J.A., and Erwin D.C. (1979). Influence of *Glomus fasciculatus* and soil phosphorus on *Verticillium* wilt of cotton. *Phytopathology* 69: 453-456.

Dehn B. and Dehne H.W. (1986). Development of VA mycorrhizal fungi and interactions with *Cochliobolus sativus* in roots of gramineae. In V. Gianinazzi-Pearson and S. Gianinazzi (eds.), *Mycorrhizae: Physiology and genetics*. Ist. Europeum Symposium on Mycorrhizae, Dijon, 1985 (pp. 773-779). Paris: INRA.

Dehne H.W. (1982). Interaction between vesicular-arbuscular mycorrhizal fungi and plant pathogens. *Phytopathology* 72: 1115-1119.

Dehne H.W. and Schönbeck F. (1975). The influence of the endotrophic mycorrhiza on the fusarial wilt of tomato. *Zeitschrift für Pflanzenkrankheiten und Pflanzenschutz* 82: 630-632.

Dehne H.W. and Schönbeck F. (1979). The influence of the endotrophic mycorrhiza on plant disease. I. Colonization of tomato plants by *Fusarium oxysporum* f.sp. *lycopersici*. *Phytopathologische Zeitschrift* 95: 105-110.

Dugassa G.D., Von Alten H., and Schönbeck F. (1996). Effects of arbuscular mycorrhiza (AM) on health of *Linum usitatissimum* L. infected by fungal pathogens. *Plant and Soil* 185: 173-182.

Dumas E., Tahiri-Alaoui A., Gianinazzi S., and Gianinazzi-Pearson V. (1990). Observations on modifications in gene expression with VA mycorrhiza development in tobacco: qualitative and quantitative changes in protein profiles. In P. Nardon, V. Gianninazzi-Pearson, A.M. Grenier, L. Margulis, and D.C. Smith (eds.), *Endocytobiology* IV (pp. 153-157). Paris: INRA.

Dumas-Gaudot E., Furlan V., Grenier J., and Asselin A. (1992). New acidic chitinase isoforms induced in tobacco roots by vesicular-arbuscular mycorrhizal fungi. *Mycorrhiza* 1: 133-136.

Dumas-Gaudot E., Grenier J., Furlan V., and Asselin A. (1992). Chitinase, chitosanase and β-1,3-glucanase activities in *Allium* and *Pisum* roots colonized by *Glomus* species. *Plant Science* 84: 17-24.

Fitter A. and Garbaye J. (1994). Interactions between mycorrhizal fungi and other soil organisms. *Plant and Soil* 159: 123-132.

Frank A.B. (1885). Uber die ouf wurzelsymbiose behrulende Ernahrung gewissen Baume d urch Urterirdische Pilze. Der deutsch *Bot Gies.* 3: 128-145.

Garcia-Garrido J.M., Garcia-Romera I., and Ocampo J.A. (1992). Cellulase production by the vesicular-arbuscular mycorrhizal fungus *Glomus mosseae* (Nicol. and Gerd.) Gerd. and Trappe. *New Phytologist* 121: 221-226.

Gianinazzi S. (1984). Genetic and molecular aspects of resistance induced by infection or chemicals. In T. Nester and E.W. Kosuge (eds.), *Plant microbe interactions,* Volume 1 (pp. 321-342). New York: Macmillan.

Gianinazzi-Pearson V. (1996). Plant cell responses to arbuscular mycorrhizal fungi: Getting to the roots of the symbiosis. *Plant Cell* 8: 1871-1883.

Gianinazzi-Pearson V., Dumas-Gaudot E., Gollotte A., Tahiri-Alaoui A., and Gianinazzi S. (1996). Cellular and molecular defence-related root responses to invasion by arbuscular mycorrhizal fungi. *New Phytologist* 133: 45-57.

Gianinazzi-Pearson V. and Gianinazzi S. (1995). Proteins and protein activities in endomycorrhizal symbioses. In A. Varma and B. Hoch (eds.), *Mycorrhiza* (pp. 251-266). Berlin: Springer Verlag.

Graham J.H. and Menge J.A.. (1982). Influence of vesicular-arbuscular mycorrhizae and soil phosphorus on take-all disease of wheat. *Phytopathology* 72: 95-98.

Grandmaison I., Olah G.M., Van Calsteren M.-R., and Furlan V. (1993). Characterization and localization of plant phenolics likely involved in the pathogen resistance expressed by endomycorrhizal roots. *Mycorrhiza* 3: 155-164.

Greipsson S. and El-Mayas H. (2002). Synergistic effect of soil pathogenic fungi and nematodes reducing bioprotection of arbuscular mycorrhizal fungi on the grass *Leymus arenarius*. *Biocontrol* 47: 715-727.

Guillemin J.P., Gianinazzi S., Gianinazzi-Pearson V., and Marchal J. (1994). Control by arbuscular endomiccorrhizae of *Pratylenchus brachyurus* in pineapple microplants. *Agricultural Sciences of Finland* 3: 253-262.

Guo B.Z., Na Z.Q., and Hendrix J.W. (1994). A mycorrhizal pathogen (*Glomus macrocarpum* Tul. and Tul.) of tobacco: Effects of long- and short-term cropping on the mycorrhizal fungal community and stunt disease. *Applied Soil Ecology* 1: 269-276.

Habte M. and Manjunath A. (1988). Influence of phenamiphos on the vesicular-arbuscular mycorrhizal symbiosis in *Leucaena leucocephala*. *Biology and Fertility of Soils* 5: 313-316.

Halos P.M. and Zorilla R.A. (1979). Vesicular-arbuscular mycorrhizae increase growth and yield of tomato and reduce infection by *Pseudomonas solanacearum*. *Philippine Agriculturist* 62: 309-315.

Hedge S.V. and Rai P.V. (1984). Influence of *Glomus fasciculatum* on damping-off of tomato. *Current Science* 53: 588-589.

Hooker J.E., Jaime-Veja M., and Atkinson D. (1994). Biocontrol of plant pathogens using arbuscular mycorrhizal fungi. In S. Gianninazzi and H. Schuepp (eds.), *Impact of arbuscular mycorrhizas on sustainable agriculture and natural ecosystems* (pp. 191-200). Basel, Switzerland: Birkhauser Verlag.

Hussey R.S. and Roncadori R.W. (1982). Vesicular-arbuscular mycorhizae may limit nematode activity and improve plant growth. *Plant Disease* 56: 9-14.

Hwang S.F. (1992). Effects of vesicular-arbuscular mycorrhizal fungi on the development of *Verticillium* and *Fusarium* wilts of alfalfa. *Plant Disease* 76: 239-243.

Jain R.K. and Sethi C.L. (1989). Interaction between vesicular-arbuscular mycorrhiza and *Meloidogyne incognita* and *Heterodera cajani* on cowpea as influenced by time of inoculation. *Indian Journal of Nematology* 18: 263-268.

Jaizme-Vega M.C., Tenoury P., Pinochet J., and Jaumot M. (1997). Interactions between the root-knot nematode *Meloidogyne incognita* and *Glomus mosseae* in banana. *Plant and Soil* 196: 27-35.

Jalali B. and Jalali I. (1991). Mycorrhizae in plant disease control. In K. Arora, B. Rai, K.G. Mokerji, and G.R. Knudsen (eds.), *Handbook of applied mycology: Soil and plants* (pp. 131-154). New York: Marcel Dekker.

Jarstfer A.G. and Sylvia D. (1992). Inoculum production and inoculation strategies for vesicular-arbuscular mycorrhizal fungi. In F.B. Metting (ed.), *Soil microbial ecology* (pp. 349-377). New York: Marcel Dekker.

Karagiannidis N., Bletsos F., and Staavropoulos N. (2002). Effect of *Verticillium* wilt (*Verticillium dahliae* Kleb.) and mycorrhiza *(Glomus mosseae)* on root colonization, growth and nutrient uptake in tomato and eggplant seedlings. *Scientia Horticulturae* 94: 145-156.

Kellam M.K. and Schenck N.C. (1980). Interaction between a vesicular-arbuscular mycorrhizal fungus and root-knot nematode on soybean. *Phytopathology* 70: 293-296.

Klironomos J.N. and Kendrick W.B. (1993). Research on mycorrhizas: Trends in the past 40 years as expressed in the "Mycolit" database. *New Phytologist* 125: 595-600.

Kloepper J.W., Rodriguez-Kabana R., McInroy J.A., and Collin D.J. (1991). Analysis of populations and physiological characterization of microorganisms in rhizospheres of plants with antagonistic properties to phytopathogenic nematodes. *Plant and Soil* 136: 95-102.

Krishna K.R. and Bagyaraj D.J. (1983). Interaction between *Glomus fasciculatum* and *Sclerotium rolfsii* in peanut. *Canadian Journal of Botany* 61: 2349-2351.

Lambais M.R. (1996). Aspectos bioquímicos e moleculares da relação fungo-planta em micorrizas arbusculares. In J.O. Siqueira (ed.), *Avanços em Fundamentos e Aplicação de Micorrizas* (pp. 5-38). Lavras, Brazil: Universidade Federal de Lavras.

Lambais M.R. (2000). Regulation of plant defense-related genes in arbuscular mycorrhizae. In G.K. Podila and D.D. Douds Jr, (eds.), *Current advances in mycorrhizae research* (pp. 45-59). St. Paul, MN: APS Press.

Lambais M.R. and Mehdy M.C. (1998). Spatial distribution of chitinases and ß-1,3-glucanase transcripts in bean mycorrhizal roots under low aand high phosphate conditions. *New Phytologist* 140: 33-42.

Linderman R.G. (1991). Mycorrhizal interactions in the rhizosphere. In D.L. Keister and P.B. Cregan (eds.), *The Rhizosphere and plant growth* (pp. 343-348). Dordrecht, the Netherlands: Kluwer Academic Publishers.

Linderman R.G. (1994). Role of VAM fungi in biocontrol. In F.L. Pfleger and R.G. Linderman (eds.), *Mycorrhizae and Plant Health* (pp. 1-25). St. Paul: APS Press.

Linderman R.G. (2000). Effects of mycorrhizas on plant tolerance to diseases. Mycorrhiza-disease interactions. In Y. Kapulnik and D.D. Douds Jr. (eds.), *Arbuscular mycorrhizas: Physiology and function* (pp. 345-365). Dordrecht, the Netherlands: Kluwer Academic Publishers.

Lingaraju S. and Goswami B.K. (1994). Influence of a mycorrhizal fungus *Glomus fasciculatum* on the host-parasite relatiionship of *Rotylenchulus reniformis* in cowpea. *Indian Journal of Nematology* 23: 137-141.

Liu R.J. (1995). Effect of vesicular-arbuscular mycorrhizal fungi on *Verticillium* wilt of cotton. *Mycorrhiza* 5: 293-297.

Lopez A., Pinochet J., Fernández C., Calvet C., and Camprubi A. (1997). Growth response of OHF-333 pear rootstock to arbuscular mycorrhizal fungi, phosphorus nutrition and *Pratylenchus vulnus* infection. *Fundamental and Applied Nematology* 20: 87-93.

Mansfield J.W. (1982). The role of phytoalexins in disease resistance. In J.A. Bailey and J.W. Mansfield (eds.), *Phytoalexins* (pp. 253-288). Glasgow: Blackwell.

Marchi A.B. and Costa C.P. (1987). Interação entre micorrizas vesículo-arbuscular (MVA) e *Fusarium oxysporum* f.sp. *lycopersici* em tomateiro. Anais, *2ª Reunião Anual sobre Controle Biológico de Doenças de Plantas,* Piracicaba, Brazil.

Matsubara Y.I., Tamura H., and Harada T. (1995). Growth enhancement and *Verticillium* wilt control by vesicular-arbuscular mycorrhizal fungus inoculation in eggplant. *Journal of the Japanese Society for Horticultural Science* 64: 555-561.

Melo A.M.T. (1989). Reação de cebola e tomateiro à inoculação de fungos micorrízicos vesículo arbusculares e de *Pyrenochaeta terrestris* e *Fusarium oxysporum* f.sp. *lycopersici.* Master's thesis. Piracicaba, ESALQ-USP, Brazil.

Morandi D. (1996). Occurrence of phytoalexins and phenolic compounds in endomycorrhizal interactions, and their potential role in biological control. *Plant and Soil* 185: 241-251.

Morton J.B. and Redecker D. (2001). Two new families of Glomales, Archaeosporaceae and Paraglomaceae, with two new genera Archaeospora and Paraglomus, based on concordant molecular and morphological characteres. *Mycologia* 93(1): 181-195.

Norman J.R., Atkinson D., and Hooker J.E. (1996). Arbuscular mycorrhizal fungal-induced alteration to root architecture in strawberry and induced resistance to the root pathogen *Phytophthora fragariae. Plant and Soil* 185: 191-198.

O'Bannon J.H., Inserra R.N., Nemec S., and Vovlas N. (1979). The influence of *Glomus mosseae* on *Tylenchulus semipenetrans*-infected and uninfected *Citrus limon* seedlings. *Journal of Nematology* 11: 247-250.

O'Bannon J.H. and Nemec S. (1979). The response of *Citrus limon* seedlings to a symbiont, *Glomus etunicatus*, and a pathogen, *Radopholus similis. Journal of Nematology* 11: 270-275.

Oehl F. and Sieverding E. (2004). *Pacispora,* a new vesicular-abuscular mycorrhizal genus in the Glomeromycetes. *Journal of Applied Botany* 78: 72-82.

Oliveira A.A.R. and Zambolim L. (1986). Interação entre o fungo endomicorrízico *Glomus etunicatum* e o nematóide das galhas *Meloidogyne javanica* sob diferentes níveis de fósforo em feijão *(Phaseolus vulgaris). Fitopatolologia Brasileira* 11: 217-226.

Pinochet J., Calvet C., Camprubi A., and Fernandéz C. (1995a). Growth and nutritional response of Nemared peach rootstock infected with *Pratylenchus vulnus* and the mycorrhizal fungus *Glomus mosseae. Fundaments of Applied Nematology* 17: 205-210.

Pinochet J., Calvet C., Camprubi A., and Fernandéz C. (1995b). Interaction between the root-lesion nematode *Pratylenchus vulnus* and the mycorrhizal association of *Glomus intraradices* and Santa Lucia 64 cherry rootstock. *Plant and Soil* 170: 323-329.

Pinochet J., Calvet C., Camprubi A., and Fernandéz C. (1996). Interactions between migratory endoparasitic nematodes and arbuscular mycorrhizal fungi in perennial crops: A review. *Plant and Soil* 185: 183-190.

Pinochet J., Camprubi A., and Calvet C. (1993). Effects of the root-lesion nematode *Pratylenchus vulnus* and the mycorrhizal fungus *Glomus mosseae* on the growth of EMLA-26 apple rootstook. *Mycorrhiza* 4: 79-83.

Price N.S., Roncadori R.W., and Hussey R.S. (1995). The growth of nematode "tolerant" and "intolerant" soybeans as affected by phosphorus, *Glomus intraradices* and light. *Plant Pathology* 44: 597-603.

Redecker D., Morton J.B., and Bruns T.D. (2000). Molecular phylogeny of the arbuscular mycorrhizal fungi *Glomus sinuosum* and *Sclerocystis coremioides*. *Mycologia* 92(2): 282-285.

Rego I.C. (1991). Influência de fungos micorrízicos vesículo-arbusculares na murcha bacteriana [*Pseudomonas solanacearum* (Smith)] e na absorção de nutrientes em tomateiro. Master's thesis. Piracicaba, ESALQ-USP, Brazil.

Reid C.P.P. (1990). Mycorrhizas. In S.M. Lynch (ed.), *The rhizosphere* (pp. 282-315). London: Wiley-Liss.

Roncadori R.W. and Hussey R.S. (1977). Interaction of the endomycorrhizal fungus *Gigaspora margarita* and root-knot nematode on cotton. *Phytopathology* 67: 1507-1511.

Rosendahl S. (1985). Interactions between the vesicular-arbuscular mycorrhizal fungus *Glomus fasciculatum* and *Aphanomyces euteiches* root rot of peas. *Phytopathologische Zeitschrift* 114: 31-40.

Ross J.P. (1972). Influence of *Endogone* mycorrhiza on *Phytophtora* root of soybean. *Phytopathology* 62: 896-897.

Santhi A. and Sundarababu R. (1995). Effect of phosphorus on the interaction of vesicular arbuscular mycorrhizal fungi with *Meloidogyne incognita* on cowpea. *Nematologica Mediterranea* 23: 263-265.

Schenck, N.C. (1981). Can mycorrhizae control root disease? *Plant Disease* 65: 230-234.

Schenck N.C., Kinloch R.A., and Dickson D.W. (1975). Interaction of endomycorrhizal fungi and root-knot nematode on soybean. In F.E. Sanders, B. Mosse, and P.B. Tinker (eds.), *Endomycorrhizas* (pp. 607-617). London: Academic Press.

Schönbeck F., Grunewaldt-Stöcker G., and von Alten H. (1994). Distribution and ecological impact of the mycorrhizal symbiosis. In C.L. Campbell and D.M. Benson (eds.), *Epidemiology and management of root diseases* (pp. 65-81). Berlin: Springer-Verlag.

Schüâler A., Schwarzott D., and Walker C. (2001). A new fungal phylum, the Glomeromycota: Phylogeny and evolution. *Mycological Research* 105(12): 1413-1421.

Secilia J. and Bagyaraj D.J. (1987). Bacteria and actinomycetes associated with pot cultures of vesicular-arbuscular mycorrhizas. *Canadian Journal of Microbiology* 33: 1069-1073.

Serrano R. (1990). Structure and function of plasma membrane ATPase. Annual Review of Plant *Physiology and Molecular Biology* 40: 61-94.

Sharma D.D., Katiyar R.S., Das P.K., Bajpai A.K, Choudhury P.C., and Janardhan L. (1995). Effect of VA-mycorrhizal fungi on the incidence of major mulberry diseases. *Indian Journal of Sericulture* 34: 34-37.

Sikora R.A. (1979). Predisposition to *Meloidogyne* infection by the endotrophic mycorrhizal fungus *Glomus mosseae.* In F. Lamberti and C.E. Taylor (eds.), *Root-knot nematodes (Meloidogyne species): Systematic, biology and control* (pp. 399-404). New York: Academic Press.

Sikora R.A. and Schönbeck F. (1975). Effect of vesicular-arbuscular mycorrhiza (*Endogone mosseae*) on the population dynamics of the root-knot nematodes (*Meloidogyne incognita* and *Meloidogyne hapla*). Proceedings, *8 International Congress of Plant Protection,* Volume 5 (pp. 158-166). Bonn, Germany: Academic Press.

Silveira N.S.S. and Maia L.C. (1996). Fungos micorrízicos arbusculares em tomateiro (*Lycopersicon esculentum* Mill.). *Agrotrópica* 8: 53-60.

Siqueira J.O. (1996). (ed.). *Avanços em Fundamentos e Aplicação de Micorrizas.* Lavras: Universidade Federal de Lavras.

Siqueira J.O. and Franco A.A. (1998). *Biotecnologia do Solo—Fundamentos e Perspectivas.* Brasília: Ministério da Educação—ABEAS, ESAL/FAEPE.

Smith G.S. and Kaplan D.T. (1988). Influence of mycorrhizal fungi, phosphorus and burrowing nematode interactions on growth of rough lemon citrus seedlings. *Journal of Nematology* 20: 539-544.

Smith G.S., Roncadori R.W., and Hussey R.S. (1986). Interaction of endomycorrhizal fungi, superphosphate and *Meloidogyne incognita* on cotton in microplot and field studies. *Journal of Nematology* 18: 208-216.

Smith S.E. and Read D.J. (1996). *Mycorrhzial symbiosis,* Second edition. London: Academic Press.

Sreenivasa M.N. (1994). Biological deterrent activities of VA mycorrhiza and *Trichoderma harzianum* on *Sclerotium rolfsii* at different P-levels in chilli. *Environment and Ecology* 12: 319-321.

Sreenivasa M.N., Nirmalnath P.J., and Kulkarri S. (1992). Interaction between VA-mycorrhizal fungi and *Sclerotium rolfsii* in chilli (*Capsicum annuum* L.). *Zentralblatt fur Mikrobiologie* 147: 509-512.

St-Arnaud M., Hamel C., Caron M., and Fortin J.A. (1994). Inhibition of *Pythium ultimum* in roots and growth substrate of mycorrhizal *Tagetes patula* colonized with G*lomus intraradices. Canadian Journal of Plant Pathology* 16: 187-194.

Strobel N.E., Hussey R.S., and Roncadori R.W. (1982). Interactions of vesicular-arbuscular mycorrhizal fungi, *Meloidogyne incognita*,and soil fertility on peach. *Phytopathology* 72: 690-694.

Sundarababu R., Sankaranarayanan C., and Santhi A. (1995). Interaction between vesicular-arbuscular mycorrhizas and *Meloidogyne javanica* on tomato as influenced by time of inoculation. *Indian Journal of Nematology* 23: 125-126.

Sundaresan P., Raja N.U., and Gunasekaran P. (1993). Induction and accumulation of phytoalexins in cowpea roots infected with a mycorrhizal fungus *Glomus fasciculatum* and their resistance to fusarium wilt disease. *Journal of Biosciences* 18: 291-301.

Suresh C.K., Bagyaraj D.J., and Reddy D.D.R. (1985). Effect of vesicular arbuscular mycorrhiza on survival penetration and development of root-knot nematode in tomato. *Plant and Soil* 87: 305-308.

Thomas L., Mallesha B.C., and Bagyaraaj D.J. (1994). Biological control of damping-off of cardamom by the VA mycorrhizal fungus, *Glomus fasciculatum*. *Microbiological Research* 149: 413-417.

Thompson K.M. and Hussey R.S. (1993). Influence of vesicular-arbuscular mycorrhizae and phosphorus on root-knot on tomato. *Journal of Nematology* 13: 462.

Tosi L., Giovanetti M., Zazzerini A., and Sbrana C. (1993). Interactions between *Plasmopara helianthi* and arbuscular mycorrhizal fungi in sunflower seedlings susceptible and resistant to downy mildew. *Phytopathologia Mediterranea* 32: 106-114.

Traquair J.A. (1995). Fungal biocontrol of root diseases: Endomycorrhizal suppression of cylindrocarpon root rot. *Canadian Journal of Botany* 73: S89-S95.

Trotta A., Varese G.C., Gnavi E., Fusconi A., Sampo S., and Berta G. (1996). Interactions between the soilborne root pathogen *Phytophthora nicotianae* var. *parasitica* and the arbuscular mycorrhizal fungus *Glomus mosseae* in tomato plants. *Plant and Soil* 185: 199-209.

Umesh K.C., Krishhhnappa K., and Bagyaraj D.J. (1988). Interaction of burrowing nematode *Radopholus similis* (Cobb 18893) Thorne 1949 and VA mycorrhiza, *Glomus fasciculatum* (Taxter) Gerd. and Trappe in banana (*Musa accuminata* Colla.). *Indian Journal of Nematology* 18: 6-11.

Vaast P., Caswell-Chen E.P., and Zasoski R.J. (1998). Influence of a root lesion nematode, *Pratylenchus coffeae*, and two arbuscular mycorrhizal fungi, *Acaulospora mellea* and *Glomus clarum* on coffea (*Coffea arabica* L.). *Biology and Fertility of Soils* 26: 130-135.

Vandenkoornhuyse P., Leyval C., and Bonnin I. (2001). High genetic diversity in arbuscular mycorrhizal fungi: Evidence for recombination events. *Heredity* 87: 243-253.

Vargas R. (1991). Combate de *Corticium* en tomate y *Fusarium* en fresa mediante el uso de microorganisms antagonistas y/o hongos endomicorrizógenos (MVA). *Agronomia Costarricense* 15: 1-6.

Vestberg M., Palmujoki H., Parikka P., and Uosukainen M. (1994). Effect of arbuscular mycorrhizas on crown rot *(Phytophthora cactorum)* in micropropagated strawberry plants. *Agricultural Sciences of Finland* 3: 289-295.

von Alten H., Lindermann A., and Schönbeck F. (1991). Increasing VA-mycorrhization with application of rhizosphere bacteria. In D.L. Keister and P.B. Cregan (eds.), *The rhizosphere and plant growth* (p. 381). Dordrecht, the Netherlands: Kluwer Academic Publishers.

Wacker T.L., Safir G.R., and Stephens C.T. (1990). Effect of *Glomus fasciculatum* on the growth of asparagus and the incidence of *Fusarium* root rot. *Journal of American Sciences and Horticultural Sciences* 115: 550-554.

Walker C. and Schüâler A. (2004). Nomenclatural clarifications and new text in the Gomeromycyota. *Meleoroglopica* 108(9): 979-982.

Yao M.K., Tweddell R.J., and Desilets H. (2002). Effect of two vesicular-arbuscular mycorrhizal fungi on the growth of micropropagated potato plantlets and on the extent of disease caused by *Rhizoctonia solani*. *Mycorrhiza* 12: 235-242.

Yuri M.M., Machado J.O., Churata-Masca M.G.C., and Andrioli J.L. (1994). Respostas do tomateiro a inóculo de solo rizosférico de *Paspalum notatum*, fungus MVA e *Azotobacter paspali*. *Científica* 22: 53-56.

Zambolim L. (1985). Como plantas micorrizadas comportam-se em relação aos fitopatógenos. Anais, *1a Reunião Brasileira sobre Micorrizas* (pp. 76-99). Brazil: Universidade Federale de Lavras.

Zambolim L. and Schenck N.C. (1984). Effect of *Macrophomina, Rhizoctonia, Fusarium* and the mycorrhizal fungus *Glomus mosseae* on nodulated and nonnodulated soybeans. *Fitopatologia Brasileira* 9: 129-138.

Chapter 14

Production of Seedlings Inoculated with Arbuscular Mycorrhizal Fungi and Their Performance After Outplanting

Orivaldo José Saggin-Júnior
Eliane Maria Ribeiro da Silva

INTRODUCTION

Presently, one agricultural area of great commercial interest involves the production of quality seedlings. These seedlings must have a guarantee that they do not carry pathogens. Therefore, many seedling producers disinfect the subsoil in order to diminish the risk of spreading diseases. Producers now utilize the production of in vitro vegetatively cloned seedlings under axenic conditions, maintaining the characteristics obtained through genetic improvement and eliminating diseases. These practices improve the commercial quality of the seedling but usually eliminate or reduce drastically the population of the majority of organisms present in the soil, including such beneficial ones as arbuscular mycorrhizal fungi (AMF).

AMF are involved in the most durable, intimate, and important symbiosis on the earth (Allen, 1996), and they are considered an ancestral characteristic that evolved together with terrestrial vegetation (Trappe, 1987; Simon et al., 1993).

To this symbiotic relationship is attributed the capacity to improve the nutritional state of the plants, to widen their ability to adapt to different ecosystems, and to enhance the tolerance to stressful biological and nonbiological factors (Siqueira and Saggin Júnior, 1995). This is reflected in a higher productive capacity and survival of the seedlings (Saggin Júnior and Lovato, 1999). The most important benefit of mycorrhizae is the increase in nutrient absorption, mostly those transported by diffusion, such as $H_2PO_4^-$, Zn^{++}, Cu^{++}, K^+, and NH_4^+. For these, movement from soil solution toward the root surface is much slower than is absorption, which creates a zone of exhaustion of these nutrients around the roots. The extraradical mycelia of the AMF have an important role; they go beyond the zone of exhaustion and thereby reduce the time required for transport and increase the absorption of nutrients. For example, the transport of phosphate to the roots via mycorrhizal hyphae can be 1,000 times faster than through soil diffusion (Bieleski, 1973), and it can contribute with 75 percent of the total phosphorus absorbed (Kothari et al., 1990). Besides this, when there is a water deficit, the direct transportation of water to the roots can be interrupted and, in this case, the external hyphae also acquire an important role in the absorption of water and of nutrients transported by mass flux (Allen and Allen, 1986). Whenever the transportation of a nutrient is faster by the fungal hyphae than by the soil, the AMF will be benefiting from this absorption (Pedersen and Sylvia, 1996).

Besides the nutritional benefits, numerous nonnutritional benefits have been attributed to the AMF, including

1. improvement of plant-water relationship (Nelsen, 1987);
2. higher tolerance to draught (Augé et al., 1992; Duan et al., 1996);
3. higher tolerance to pathogens (Azcón-Aguilar and Barea, 1996);
4. higher tolerance to soil abiotic stresses such as erosion, compaction, acidity, salinity, and heavy metals (Sylvia and Williams, 1992);
5. modification of the composition of the rhizosphere community (Linderman, 1992);
6. protection against herbicides (Siqueira et al., 1991); and
7. aggregation of the soil particles (Miller and Jastrow, 1992; Tisdall, 1994).

These nonnutritional benefits can be related to alterations in the metabolism and physiology of mycorrhizal plants (Cooper, 1984; Smith and Gianinazzi-Pearson, 1988; Schwab et al., 1991). Mycorrhizal plants can enhance the accumulation of secondary metabolites (Peipp et al., 1997), increase enzymatic activity (Ruiz-Lozano and Azcón, 1996), as well as cause alterations in the content of hormones (Duan et al., 1996; Goicoechea et al., 1997), amino acids (Johansen et al., 1996), and carbohydrates (Thomson et al., 1990). Mycorrhizal fungi can cause physiological alterations, increasing the stomatal conductance, respiration, transpiration, absorption of CO_2, photosynthesis (Mathus and Vyas, 1995; Ruiz-Lozano et al.,1995; Lovelock et al., 1997; Goicoechea et al.,1997), radicular exudation (Tawaraya et al., 1996), and efficiency in the use of water (Ruiz-Lozano and Azcón, 1995).

As they promote these nutritional and nonnutritional benefits to the plants, the AMF have become tools with an enormous potential to be explored in the production of seedlings. They can allow a greater development of seedlings before transplanting and help them improve their establishment and adaptation field condition after they are transplanted. In the case of reforestation of degraded areas, the inoculation of AMF in plants is essential, as the success of the enterprise will depend on it, making the return of the cycling of nutrients and the plant succession possible.

The importance of the reintroduction of the AMF in the disinfected subsoil to produce seedlings was first demonstrated in citrus (Kleinschmidt and Gerdemann, 1972). Until then, blight and the leaf symptoms of nutritional deficiency in citrus seedlings that grew in disinfected subsoil had been attributed to a toxic effect that possibly inhibited the absorption of phosphorus (Martin, 1948; Martin et al., 1953). The importance of AMF in the production of seedlings has been demonstrated in a great number of plant species (Gianinazzi et al., 1990; Azcón-Aguilar and Barea, 1997).

The objective of the current chapter is to present important information about the effect of AMF in the production of seedlings and the performance of the seedlings after transplanting in the field.

PREREQUISITES FOR THE INOCULATION OF SEEDLINGS WITH AMF

Several factors must be considered before the inoculation of seedlings with AMF, but the most important ones are the mycorrhizal dependence of the plant to be inoculated and the symbiotic effectiveness of the AMF used to inoculate the growth medium. Therefore, it is necessary to evaluate the mycorrhizal dependence of the seedlings and the selection of efficient AMF fungi.

Evaluation of the Mycorrhizal Dependence of the Host Plant

The first step for the inoculation of seedlings is to know how much the plant depends on the mycorrhizal association in order to grow normally. Experiments have already shown host dependency to mycorrhizae, for example, seedlings of citrus (Menge et al., 1978) and coffee (Saggin Júnior and Siqueira, 1996). However, mycorrhizal dependence of the majority of plants, particularly native woody species of interest to reforestation, is not known. We know that mycorrhizal dependence (MD) is an intrinsic characteristic of the plant (Hetrick, Wilson, and Cox, 1992; Hetrick et al., 1993, 1995, 1996), which depends on nutritional requirements; and efficiency of nutrient absorption by roots, growth index, nutritional reserves of the seed, and abundance, distribution, and morphology of the roots (Graham and Syvertsen, 1985; Janos, 1987, 1988).

But MD cannot be adequately described by one of these characteristics alone. At first, researchers tried to relate MD to the morphology of roots (Baylis, 1975); however, it has been verified that the differences of MD due to effect of morphological characteristics of the roots on MD is subtler than was predicted by Baylis (Hetrick et al., 1991; Hetrick, Wilson, and Todd, 1992). The literature suggests that the relationship between MD and radicular morphology is strong only when the plant is adapted to soils of high fertility, such as those in temperate ecosystems (Reinhardt and Miller, 1990; Hetrick et al., 1991). This indicates that the set of genetic characteristics which influence the MD is important, particularly in plants adapted to tropical soils of low fertility (Saggin Júnior, 1997; Siqueira and Saggin Júnior, 2001). On the other hand, the productivity that results from the inoculation, the choice of the most appropriate fungus, and the correct dosage of fertilizers are related to its response to the mycorrhizae.

The first definition of MD was presented by Gerdemann (1975) as being the "degree to which the plant depends on the mycorrhizal condition to obtain its maximum growth and production at a certain level of soil fertility." This definition was based on already existing knowledge—that the majority of plant species can grow without mycorrhizae if they are thoroughly fertilized, mainly with phosphorus (P) (Gerdemann, 1968; Mosse, 1973). The higher the level, the more dependent on mycorrhizae the plant is. The MD can also be categorized in different degrees as proposed by Habte and Manjunath (1991), based on response to the mycorrhizae in different conditions of availability of P (see Figure 14.1). The categorization elaborated by these authors, even though imperfect, particularly for some native species (Siqueira and Saggin Júnior, 2001), can be improved and used as a good and practical indicator of MD. The response to the mycorrhizae can be quantified by the expressions of Menge et al. (1978) or Plenchette et al. (1983) even though these were originally proposed to quantify MD.

Therefore, the inoculation of seedlings require basic studies on mycorrhizal dependency of the plant to be inoculated. The knowledge of its MD allows us to elaborate the inoculation strategy of the seedlings. The plants that do not depend on the mycorrhizae present little or no root colonization by AMF, and they do not present response to the mycorrhizae (Figure 14.2, part a). When seedlings of the wood species native to Brazil are not dependent, they do not respond to higher levels of P in the soil (Figure 14.2, part b), as they probably have other nutritional strategies for their growth (Chapin, 1980). The seedlings that have low mycorrhizal dependency usually present high responses to the mycorrhizae when they grow in soil with low availability of phosphorus, but they do not respond to symbiosis with high avail-

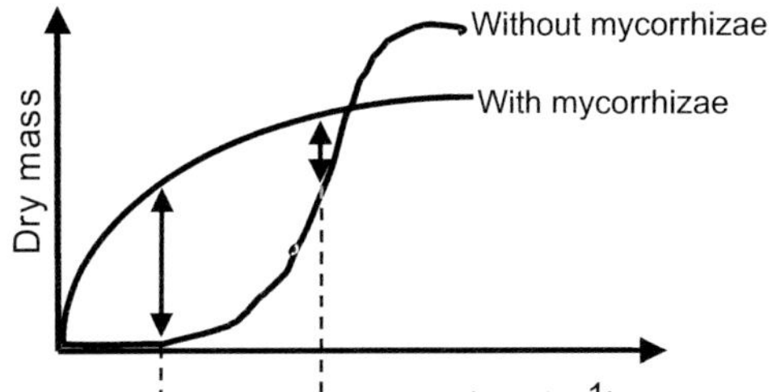

MD categories	0.02	0.2
Very highly dependent	>75%	> 0%*
Highly dependent	50-75%	= 0%
Moderately dependent	25-50%	= 0%
Marginally dependent	<25%	= 0%
Independent	0%	= 0%

FIGURE 14.1. Mycorrhizal dependence (MD) categories proposed by Habte and Manjunath, 1991, based on response to mycorrhizal inoculation in two levels of P in the soil solution. *Response to mycorrhiza inoculation quantified as Plenchette et al., 1983.

ability of this nutrient (Figure 14.2, part c). So they are very responsive to phosphorus when they are not mycorrhizal (Figure 14.2, part d). Seedlings of species with high mycorrhizal dependency respond to the mycorrhizae even when the availability of P in the soil is high. However, these responses can be small (Figure 14.2, part e). They might present low responses to P (Figure 14.2, part f), but even so the introduction of AMF is essential for their survival. Seedlings without MD need not be inoculated, although some species may become dependent on mycorrhizae in a later phase, after the formation of the seedlings. This can occur in species that have large seeds with sufficient reserves of nutrients which can maintain the plant independent from AMF in its initial phase, becoming dependent after these reserves have been exhausted. In this case, the inoculation can be done during the transplant to the field, without harming the formation of the seedling. In the case of seedlings with low MD we can opt between an inoculation and a sufficient fertilization with phosphorus. Seedlings formed only with P, if they are transplanted to an area of low fertility and with few propagules of AMF, might show reduced growth and might not be able to survive. Seedlings with high MD inoculation are essential, and we must be careful with the factors that can harm the mycorrhization of the seedlings, such as making use of certain pesticides and excessive fertilization.

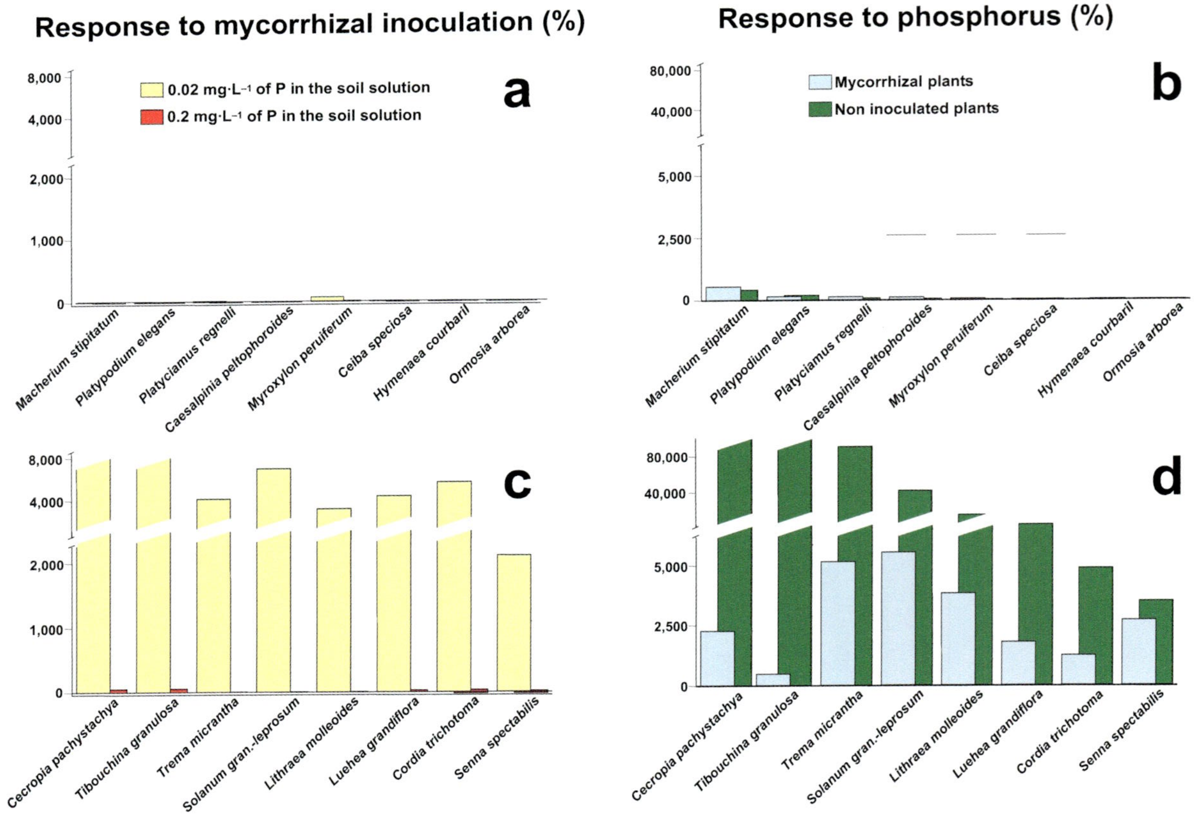
Response to mycorrhizal inoculation (%)
Response to phosphorus (%)
a
0.02 mg·L⁻¹ of P in the soil solution
0.2 mg·L⁻¹ of P in the soil solution
8,000
4,000
2,000
1,000
0
Macherium stipitatum
Platypodium elegans
Platyciamus regnelli
Caesalpinia peltophoroides
Myroxylon peruiferum
Ceiba speciosa
Hymenaea courbaril
Ormosia arborea
b
Mycorrhizal plants
Non inoculated plants
80,000
40,000
5,000
2,500
0
c
Cecropia pachystachya
Tibouchina granulosa
Trema micrantha
Solanum gran.-leprosum
Lithraea molleoides
Luehea grandiflora
Cordia trichotoma
Senna spectabilis
d

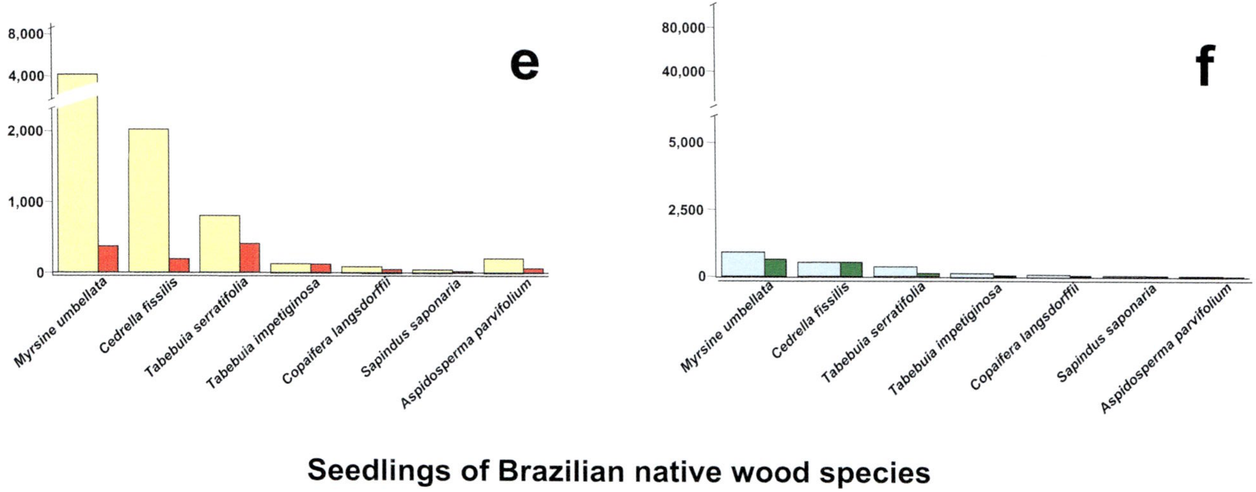

FIGURE 14.2. Response to mycorrhizae (shoot dry mass of mycorrhizal plant- shoot dry mass of noninoculated ones/shoot dry mass of noninoculated × 100) in two P levels in soil solution and response to P (shoot dry mass in 0.2 $mg{\cdot}L^{-1}$ of P in soil solution-shoot dry mass in natural level of P in soil solution (0.002 $mg{\cdot}L^{-1}$) /shoot dry mass in natural level of P in soil solution ×100) of mycorrhizal and non-inoculated plants. The seedlings on parts a and b are considered not dependent of mycorrhizae because they do not respond to mycorrhizal inoculation. The seedlings on parts c and d are considered facultative mycorrhizal because they respond to it only with low P in soil solution. The seedlings on parts e and f are considered obligate mycorrhizal because they still respond to it in high P in soil solution and only respond to P with mycorrhizae. *Source:* Modified from Saggin Júnior, 1997.

Selection of Efficient Fungi for a Specific Plant

The AMF are not host specific. The same fungus can colonize different plants, but the capacity of different fungi to promote growth of the same species of plant is variable (Gianinazzi-Pearson, 1985; Smith and Gianinazzi-Pearson, 1988). A fungus is considered efficient when in a wide range of environmental conditions it colonizes the roots rapidly and extensively; it competes with other microorganisms for sites of infection and absorption of nutrients; it rapidly forms an extensive extraradical mycelium; it absorbs and transfers nutrients to the plant; and it promotes nonnutritional benefits to the plant, such as aggregation and stabilization of the soil, among others. Rarely will only one species of AMF be efficient in all of these characteristics, which could, this way, inoculate multiple species to show better results than would inoculation of only one species, since this would show evidence that two or more fungi can coexist harmoniously in the root (Tiwari and Adholeya, 2002). For a direct evaluation of the symbiotic effectiveness (SE) of an AMF, it would be necessary to evaluate the physiological and biological mechanisms involved in the absorption and functioning of symbiosis. Evaluations as kinetic parameters of absorption of the external mycelium, of the consumption of photosynthates by the fungus, or of quantification of the nutrients absorbed via mycelium would be forms of measuring the efficiency of different AMF. However, they are variables not yet thoroughly studied and of difficult practical evaluation (Smith et al., 1994), which implies that in the majority of studies the SE is evaluated indirectly through the responses of the plant to the inoculation of different fungi.

With the clear intention of inoculating seedlings with AMF, one should prioritize the selection of fungi with high SE. AMF adapted to edaphoclimatic and cultural conditions of the plant in question are normally selected among fungi indigenous to the ecosystem in which we intend to apply them (Menge, 1983; Saggin Júnior and Siqueira, 1996). The isolated indigenous fungi must be evaluated in relation to the response their inoculation promotes in plant growth under different soil conditions. Sieverding (1991) suggested that the tests of SE be done at first under competition with other soil organisms, including other AMF. Later, the isolated ones that were selected for competing efficiently will be evaluated under different edaphic conditions (pH, P, N, K, salinity, temperature, and water stress) in order to obtain specific fungi that have characteristics compatible with the objective of the selection. The wider the adaptation of the AMF to different conditions of soil, the larger its SE, and those with high SE can be indicated for inoculation of seedlings for whom condition of soil after transplant is difficult to predict. Normally AMF colonize the seedling under the condition of a fertile substrate and adapt to transplant to a low fertility condition.

The spore production capacity is considered to be an important criterion for the selection of AMF (Abbott and Robson, 1982, 1985) because sporulation is linked to the AMF capacity of persistence after inoculation. Persistence is considered an advantage because, most of the time, the inoculation of seedlings is the only opportunity for the introduction of AMF in a culture. Moreover, the selected fungus must produce enough propagule for inoculant production, and the spores are resistant to the dryness and to the storing and application of the inoculant techniques. Therefore, besides the evaluation of the SE in different conditions, persistence has been used also as a criterion of selection of AMF. Persistence is the result of competition, of tolerance to factors of the soil and of the environment, and of the capacity of the fungus to spread itself out in the soil, sporulating and producing propagules. However, the selection of a very persistent fungus can bring future problems such as the ones presently observed in relation to rhizobia in Brazil. In the selection of rhizobia, they searched for competitive and aggressive strains, so the majority of rhizobia commercial inoculants for soybean in Brazil contains strains of *Bradyrhizobium japonicum* and *B. elkanii*. But when we are looking for the introduction of other strains which are efficient in nitrogen fixation, such as the strains of fast growth of *Sinorhizobium fredii*, these cannot compete with the strain of *B. elkanii* SEMIA 5019 in acid soils, making impossible its recommendation for use as a commercial inoculant (Hungria et al., 2001).

As the availability of phosphorus in the soil is one of the most important factors in the functioning of symbiosis, it is ideal that the SE of different isolated fungi be evaluated under different levels of this soil nutrient (Saggin Júnior and Siqueira, 1995). These authors suggest that the wider the range of P in the soil in which the mycorrhizal fungus maintains the symbiosis with the mutualist nature, the greater is its potential to be used in agriculture. This range was designated by the authors as the "mutualism range" (see Figure 14.3).

Saggin Júnior and Siqueira (1995) proposed a methodology to evaluate the SE of isolates of AMF, evaluating it in the presence of AMF indigenous to the soil. The SE of each isolate is estimated by the relative percentage between the benefit it promotes to the plant and the benefit that the plant without the fungus obtains from the addition of phosphorus, called the mycorrhizal benefit (MB) and the P benefit (PB). These values are estimated mathematically through the integration of the curves of response to P from the mycorrhizal plants and those without, respectively. These values are represented by the shaded areas in Figure 14.3. This estimation of SE allows for a differentiation of the isolates of AM fungi that belong to the same species (Saggin Júnior and Siqueira, 1996). Moreover, the SE values make it possible to categorize the AMF as to their degree of effectiveness, which

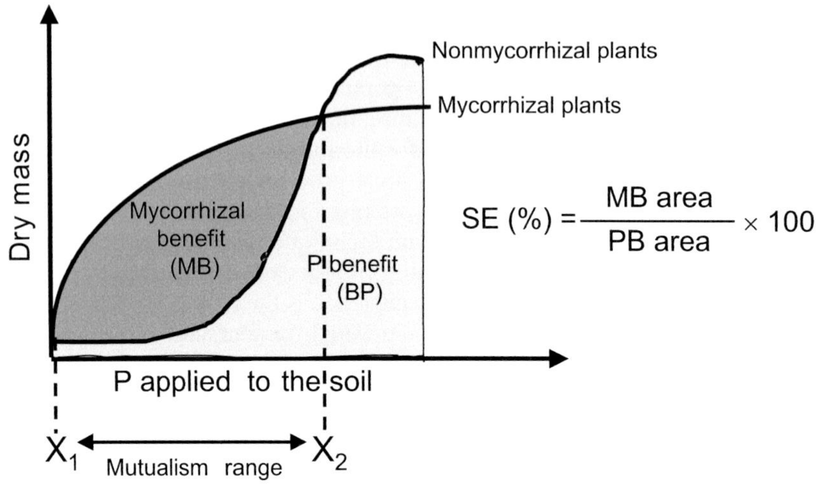

FIGURE 14.3. Representation of areas of mycorrhizal benefit and P benefit that can be estimated by integral calculus from the curves of the mycorrhizal plants and the nonmycorrhizal ones in response to P applied to the soil. The range of mutualism represents the range of P applied to the soil where the fungus maintains the mutualistic nature of the symbiosis. SE (%) is an estimated percentage of the symbiotic effectiveness of the mycorrhizal fungus based on the areas of MB and PB.

varies from low to high. Fungal isolates with high SE will adapt to edaphic conditions with a great variability of available P, as it usually occurs in agricultural ecosystems, and they will promote the growth of the plant with high as well as low availability of phosphorus in the soil. This is important in order to have more assurance in recommending the inoculation of the isolates in different areas. The adaptation of the fungi in different conditions of P is also important when we inoculate seedlings where high dosages of P is frequently applied to the substrate but then the seedlings are transplanted to soil of lesser fertility.

PRODUCTION OF INOCULANTS FOR SEEDLINGS

The mycorrhizal inoculant is quite difficult to find in the open market in most countries, because production is difficult as the AMF do not multiply in the absence of metabolically active roots. These fungi are obligate biotrophs, that is, they do not complete their reproductive cycle without being associated to a living host root. Therefore, their multiplication in a totally sterile environment (which is necessary to produce an inoculant of microorganisms with quality) is expensive and difficult. For this reason, in most inoculants, the AMF are multiplied in a live host plant cultivated in a substrate that has been fumigated or sterilized or treated with a nutritional solution under controlled conditions.

This difficulty, added to a lack of technology to produce inoculants on a large scale, has immensely delayed the practical application of these fungi to agriculture. It has created an impasse difficult to be overcome: "The inoculant was not being used in large scale because we did not have it available and it was not available because it was not used." However, some commercial inoculants of AMF were recently launched in several countries, which might break this cycle. A short list of commercial inoculant producers can be found online at <http://dmsylvia.ifas.ufl.edu/commercial.htm>. This list was maintained by Dr. Sylvia for a program of quality control of commercial inoculants.

Techniques to produce AMF inoculants have been researched and even patented using plants or cultures of roots as hosts and means of culture, nutritional solution, or subsoil for the cultivation of hosts. The main objective of the research has been to obtain a purer and lighter inoculant than those produced by cultivating the host plant directly in the soil.

Multiplication of AMF in Monoxenic Cultures

The ideal inoculant should be produced in axenic conditions, free of any unknown microorganisms and pathogens. In searching for this ideal model, we first tried to establish a mixed cultivation in vitro of AMF and complete plants (Mosse, 1962; Allen et al., 1979; Hepper, 1981) and later in root cultures (Mosse and Hepper, 1975).

A new perspective on axenic multiplication of AMF appeared with the utilization of root organ cultures genetically modified (transgenic) by *Agrobacterium rhizogenes* (Mugnier and Mosse, 1987). These roots proliferated abundantly in a minimal cultural medium that allowed the fungus to grow and sporulate (Bécard and Fortin, 1988). Even though this has opened up an excellent possibility for multiplication of AMF, this methodology is still not being applied for production of inoculants on a large scale due to some problems such as the long time for multiplication, low production of spores of some species, difficulty maintaining some isolates under continuous cultivation in vitro, and the possibility of loss of physiological and genetic characteristics of the isolates due to continuous cultivation. Recently, the possibility of obtaining 20,000 spores of *Glomus intraradices* every two months through repeated harvests was reported (Douds, 2002).

The technique of multiplication of AMF in monoaxenic culture has been employed to obtain material for research in the field of molecular genetics of these fungi and also for physiological and ontogenetic studies. Recently, a germplasm bank of AMF has been created in which isolates are maintained through this system of multiplication (http://www.mbla.ucl.ac.be/ginco-bel/).

Multiplication of AMF in Nutritional Solution

Methods of producing inoculum of AMF, through previously colonized plants and cultivated in hydroponic systems, have been investigated, and several models were evaluated in the 1980s. Among the hydroponics techniques, the one that presents the greater possibility of use in the mass production of inoculum is a system called aeroponics (Sylvia and Hubbell, 1986). In this system the nutrient solution (http://dmsylvia.ifas.ufl.edu) is intermittently sprinkled over the roots colonized inside a growth chamber (water tank) promoting an aerial environment, efficient for the multiplication of AMF without a physical substrate. This system is quite promising for the multiplication of some species of *Glomus,* but it has not been efficient for the general multiplication of the species of AMF. The causes of this limitation have been attributed to the lack of adaptation of the species of AMF in the absence of a solid substrate and/or associated microorganisms and to the lack of knowledge of an ideal nutrient solution and/or specific light conditions, temperature, and oxygenation.

In an aeroponic system, host plants are germinated in vermiculite containing disinfected spores of the AMF to be multiplied. After this phase, the presence of colonization of AMF in the roots is confirmed by microscopic techniques and the plants are then transferred to the chambers where the nutrient solution will be sprinkled over the roots. The host plants suggested by Jarstfer and Sylvia (1995) to be utilized for aeroponic system are sweet potato *(Ipomea batatas),* bahia grass *(Paspalum notatum),* corn, and the hybrid sorghum/Sudan grass.

Three methods were tested for sprinkling the roots with a nutrient solution: an atomizing disc, a pressurized spray through nozzles, and an ultrasonic generated fog. The atomizing disc is located at the bottom of the chamber. It rotates, throwing the nutrient solution against the sidewalls, pulverizing and generating a flux of air that distributes the mist inside the chamber. This method can present a series of problems, for example, the roots can break and get tangled in the equipment and it takes room inside the chamber, reducing the space allowed for the roots. For the pressurized spray through nozzles, they are distributed around the chamber to obtain maximum moistening of the roots with the nutrient solution that leaks from them and drips down to the roots. This system presents drops that are larger and can harm the extraradial growth of the AMF. In the first two systems, the nutrient solution is stored at the bottom of the chamber and can be reutilized, as long as periodical corrections to the pH with 20 percent H_2SO_4 or 10 percent KOH are made, but it should be changed every two weeks to avoid the depletion of or unbalanced nutrients and the accumulation of toxins. In the third system, the ultrasonic device forms a fine fog with droplets between 1 and 10 μm in diameter but produces controversial results, and it has been reported that it does not promote enough humidity for the plants, harming growth and the sporulation (Jarstfer and Sylvia, 1995).

Multiplication of AMF in Substrate

The methods mostly utilized for the production of inoculants of AMF, and which are most technically viable, consist of the multiplication of these fungi in host plants cultivated in substrate. The multiplication of AMF through these methods generally is done in containers kept inside the greenhouse, called "culture pots," although it can also be done in seedbeds, in nurseries, or on farms.

The environment of the greenhouse provides us with more control over the contamination of the culture pots with other AMF that we do not wish to multiply. These can be brought in by the wind, insects, or larger animals (e.g., rodents or frogs). Even inside a greenhouse the control for contamination must be very careful, as species of AMF such as *Paraglomus occultum* and *Glomus etunicatum* present a very high dissemination capacity, passing easily from one pot to another, dominating and inhibiting the culture of other species of AMF. Therefore, the counters must always be cleaned thoroughly and kept without any traces of soil. Irrigation must be done carefully, avoiding spatters. Ventilation must be moderate. The spacing between the pots must be fairly wide. Access to the plants by insects must be controlled, particularly ants that, being miniscule, are often ignored and pass freely from one pot to another. To avoid ants, we use grease on the legs of the counters or recepticles containing oil are placed at the feet of the counters.

The most common plants generally used for the multiplication of AMF in tropical environments are brachiaria grasses (*Brachiaria decumbens* Stapf and other species), corn (*Zea mays* L.), millet (*Panicum miliaceum* L.), sorghum [*Sorghum bicolor* (L.) Moench], Sudan grass [*Sorghum sudanense* (Piper) Stapf], and the hybrid sorghum/Sudan grass, Bahia grass (*Paspalum notatum* Flüggé), andropogon grass (*Andropogon gayanus*

Kunth), guinea colonial grass (*Panicum maximum* Jacq.), tropical kudzu [*Pueraria phaseoloides* (Roxb.) Benth.], centrosema (*Centrosema pubescens* Benth.), and stylosanthes [*Stylosanthes guianensis* (Aubl.) Sw. var. *vulgaris*]. Gramineous plants produce a larger volume of colonized roots, which bear important propagules, but the leguminous plants generally produce more spores (Sieverding, 1991). We should favor plants that are well adapted to the location of the multiplication and, as the majority of the species that spread through seedlings are dicotyledonous, we should choose to multiply the inoculant of seedlings in gramineous plants in order to avoid multiplication of typical pathogens of dicotyledons in the leguminous plants. We should also avoid host plants that can be potential weeds or keep them from producing seeds through periodic pruning.

AMF have been multiplied in plants cultivated in the most different substrates, among them soil, sand (St. John et al., 1981), peat (Ponton et al., 1990), expanded clay (Feldmann and Idczak, 1992), perlite, vermiculite (Sreenivasa and Bagyaraj, 1988), asbestos wool (Heinzemann and Weritz, 1990), microspheres of glass (Redecker et al., 1998), and diverse organic compounds and their mixtures (Zambolim et al., 1992). The choice of the substrate will depend on the regional availability, the cost, the weight, the fungus and the plant adaptation, the perspective of its direct use as part of the substrate for seedlings, and, essentially, whether it has low levels of nutrients, particularly of P which inhibits the establishment of symbiosis and multiplication of AMF.

Soil, sand, and their mixture, as they are very conducive to the AMF, have been traditionally and largely utilized, as long as they have thick texture, which facilitates drainage and airing. Gaur and Adholeya (2000) obtained a better multiplication of *G. intraradices* utilizing grains of sand with diameters between 0.25 and 0.50 mm. Sand can be utilized as substrate for plants maintained with a nutritional solution in a semihydroponics system of multiplication of AMF (Ojala and Jarrell, 1980). The weight of these materials is the most limiting factor for the production and commercialization of this type of inoculum.

Soil, sand, or their mixture are widely employed for the multiplication of AMF to be used in research, but it is also viable to produce inoculants on a commercial scale if they are produced by the nursery owner themselves. In Colombia, Sieverding (1991) successfully developed a practical experiment to produce inoculants in seedbeds. The seedbeds were cultivated with *Brachiaria decumbens* with a commercial inoculant of *Glomus manihotis* (Manihotina) that cannot be found in the market today. After six months of cultivation, 41 percent (nonfumigated soil) and 95 percent (soil fumigated with methyl bromide) of *Glomus manihotis* spores and the rest of other native species of AMF were recovered. Production of 40 to 183 spores of *Glomus manihotis* per gram of soil in the seedbeds was obtained.

To obtain a better inoculum, substrates of mineral and organic origin have been evaluated. AMF can be multiplied on several organic substrates such as tree bark, shale (Maronek et al., 1980), charcoal, and peat (Ponton et al., 1990), among others. However, some organic substrates can inhibit the formation of mycorrhizae (Guttay, 1982, 1983). Many peats inhibit the AMF due to low pH, high level of organic matter, and available phosphorus (Graham and Timmer, 1984). This effect depends on the nature of the peat, as a more fibrous substrate of peat and moss will result in larger root colonization (Wang et al., 1993). Restrictions on the use of organic substrates for the production of inoculants also include a greater possibility of multiplying and carrying pathogens, the regional offer of materials, and the great variation of chemical composition that the same material might present.

The substrates of mineral origin have been utilized successfully in the multiplication of AMF, even though they might also, sometimes, inhibit these fungi. Diverse substrates have been evaluated and recommended for the production of inoculant and vehicle for the propagules, among them calcinated montmorilonite (Plenchette et al., 1982) and light aggregated expanded clay (Dehne and Backhaus, 1986). In the latter, fungus sporulates inside the aggregates increase the survival of propagules and maintain its vitality at 25 to 30°C for more than 5 years. An adaptation of this substrate for the mass production of inoculant, which can be used in nurseries with tropical species from the Amazon, is described by Feldmann and Idczak (1992).

The advantages of utilizing substrates of mineral origin for the production of inoculant are

1. it is an inorganic material, industrialized, with no pathogens;
2. the survival of the AMF propagules inside the porous spaces is wide and extends for long periods of time;
3. the inoculant produced is light;
4. the substrate has several defined granulation measurements available in the market, which can favor the inoculation and the mechanization;
5. they are relatively cheap;
6. they can be stored for long periods of time under several conditions;
7. they can be washed and disinfected; and
8. they can be utilized as a component of the substrate for seedlings.

Mixtures of soil, sand, mineral, and organic substrates can be dosed to promote the multiplication of a particular species of AMF. Zambolim et al.

(1992) verified that the mixtures peat:vermicule:soil:chicken manure (4:4:1:1); peat:vermicule (1:1) and peat:vermicule:chicken manure (4.5:4.5:1) promoted excellent sporulation in *Glomus etunicatum.* Inert materials added to the substrate can promote the multiplication of certain AMF species by making it more appropriate to the fungus. For example, by placing a fiberglass screen close to the substrate, Huang and Tang (1988) were able to get a better multiplication of *Glomus aggregatum*. The spores of this species are frequently found in organic fragments of the soil, such as dead seed peels and exoskeletons of insects.

The locally available materials are preferred for multiplication of AMF inoculum. For the production of commercial inoculum, one should look for light composition of the substrate that will facilitate transportation and commerce. If the objective is multiplication on a small scale, soil and sand should be included as a substrate since they facilitate the multiplication of the AMF and decrease the cost. A small amount of organic matter aids in the multiplication of AMF. In Embrapa Agrobiology, Seropédica, RJ, Brazil, we have successfully used a substrate prepared with a mixture of three types of soil in equal parts: clay soil collected from a ravine, sandy soil (quartz sand), and soil under forest vegetation with a high proportion of organic matter. The soil was sterilized in autoclave for the multiplication of several species of AM fungi.

PRODUCTION OF SEEDLINGS INOCULATED WITH AMF

Plants multiplied by seedlings such as vegetables, fruits, forest, and ornamental plants can be inoculated with AMF due to the possibility of using a small amount of inoculum and to the easy handling, disinfecting and adjustment of the fertility of the substrate. The growth of the seedling in semicontrolled conditions promotes the establishment of inoculated AMF. Furthermore, the use of subsoil and organic material in the composition of the substrate, which have few or no AMF propagules, decrease the competition of the native AMF with the inoculated fungus. In many cases, there is also a disinfecting of the substrate to avoid pathogens and weeds killing the native AMF. This makes inoculation of the seedlings viable and necessary when it is related to plants that present high mycorrhizal dependency or to seedlings that will be planted in areas that have few AMF propagules, such as those that are degraded or where there was a longtime culture of a nonmycorrhizal arbuscular species, such as *Pinus.*

There are many reports on nutritional improvement after inoculation of AMF and on the increase in the growth of seedlings, accelerating their formation and reducing the time and cost of nursery. There are, however, other less observed benefits promoted by the inoculation of seedlings. In the production of seedlings, both by conventional and cloning techniques, a difficult factor to control is the nonuniformity of plant growth. Even plants cloned in vitro show variation among the individuals in the formation of seedlings. This variability can be diminished with the inoculation of AMF (Saggin Júnior and Lovato, 1999). However, few works evaluate the uniformity of the seedlings, which should be considered to evaluate mycorrhizal seedlings, since it is an important characteristic that affects the production costs and the commercial value of the seedlings. Another benefit that seedling inoculation can bring is the increase in tolerance and even in protection against attacks by soil pathogens (Saggin Júnior and Lovato, 1999). This happens due to the promotion of a better nutrition of plants with roots damaged by fungi or nematodes, and by direct or indirect antagonistic effects on the pathogens (Sylvia and Chellemi, 2001).

Several technical aspects should be considered in the inoculation of seedlings with AMF, from the choice of the inoculum to the care that should be taken in the handling of the inoculated seedlings. Technology should be adapted to the local conditions of seedling production areas.

AMF Inoculum Types

AMF inoculum carries propagules of these fungi, such as pieces of mycelium, fragments of colonized roots and spores. This inoculum is usually solid, which creates a practical limitation for its application. This problem exists because the AMF propagules are relatively large when compared to those of other microorganisms, rapidly decanting in aqueous suspension, so that the use of inoculants in aqueous suspension is limited to research work.

Several types of inoculants have been tested. Initially, colonized roots and mixed inocula, consisting of the soil where the fungus was multiplied, containing spores, hyphae, and mycorrhizal roots, were tested. After that, techniques of spore extraction from the soil were developed (Gerdemann and Nicolson, 1963), facilitating the application of only these as inoculants. From these three basic types other different types of inoculants were developed, such as the use of an inert substrate in place of soil (Dehne and Backhaus, 1986; Sreenivasa and Bagyaraj, 1988); the lyophilization of spores and colonized roots (Tommerup and Kidby, 1979); the pelletization of mixed inoculum (Hall, 1979; Hall and Kelson, 1981); the pelletization of seeds with mixed inoculum (Hattingh and Gerdemann, 1975; Powell, 1979); the suspension of spores and seeds in gel for fluidic sowing (Hayman et al., 1981); the use of fiberglass screens as substrate of multiplication and excipient of the inoculant produced (Huang and Tang, 1988); and the use of these screens as a support for the spores adhered to alginate films (Calvet et al., 1996). According to the authors, this last type of inoculant offers remarkable advantages, since the fiberglass screen has low mass; thus the number of spores inoculated is easily measured and the alginate film promotes protection of the spores against drying.

The Choice of Fungus for Each Seedling

The inocula of AMF should be adapted to the plants and to the environment in which they will be transferred in the field. The best way to guarantee the adaptation to field environment would be to isolate a mixture of native AMF species from the same soil to be planted, multiply them, and use them as inoculants for the seedlings. Unfortunately, many native fungi are difficult to multiply and the procedure does not guarantee that they would be adapted to the new agricultural conditions of the soil.

Therefore, very often, the producers of commercial inocula or research institutions have a very short list of AMF species available for inoculation. In fact, there is a great need to study each AMF for its efficacy. A minimum selection of species of AMF should be created based on the pH of the soil in which the seedling will be sown. There are AMF species clearly adapted to acid pH (*Gigaspora margarita* and *Entrophospora colombiana*), slightly acid pH (*Glomus etunicatum* and *Glomus clarum*), and neutral to slightly alkaline pH (*Glomus mosseae* and *Glomus intraradices*).

There has been a search for efficient fungi to promote the growth and nourishment of the seedlings. Several species of efficient AMF have been isolated from the coffee tree (Saggin Júnior and Siqueira, 1996). However, AMF can play different beneficial roles other than nourishing, including antagonism to pathogens and favoring beneficial microorganisms. Thus, the efficiency in these several roles will vary with the species and/or isolates of the fungus. A single fungus will hardly be efficient for every plant species. On the contrary, some AM fungi may have different roles according to the environment. This idea has attracted many scientists, and it may lead to a future trend of the constitution of AMF having several species, even with other beneficial rhizospheric microorganisms associated. It is necessary, however, that the aspects linked to the competition among different fungi be evaluated.

Inoculation of Seedlings with AMF

The inoculation of seedlings can be done in four different stages:

1. at the time of sowing;
2. during the transplantation of pregerminated seedlings;
3. during transplantation to the field; and
4. during the in vitro propagation or acclimatization, the latter being specific to micropropagated plantlets/seedlings.

The inoculation of seedlings during the sowing or transplantation of plantlets is a common practice. In these stages the form of application of the inocula apparently has little effect on the final colonization of the roots. In coffee seedlings, the application of 50 AMF spores on the surface of the substrate, in orifices next to the plantlet or directly over the roots, resulted in a colonization similar to that of the radial system at the end of 5 months (Dias et al., 1987). However, we prefer to place the inoculum under a thin layer of soil to avoid its drying, oxidation, or even solarization. The inoculant should be in contact with the roots of the transplanted plantlets or below the seeds at the point where the radial system will develop. In Table 14.1, comparison of different types of inocula and applications in citrus, using direct sowing or the transplantation of pregerminated plantlets, has been made. It was observed that in transplanted plantlets, the application of inocula in orifices next to the roots resulted in a greater growth of the aerial part. This type of inocula has a greater effect over the results. The use of lyophilized roots and spores suspended in methyl cellulose 2 percent did not result in efficient radial colonization and benefits for the plant growth.

Both in recipients and in seedbed, inoculation can be practiced during sowing in species that present fast germination, because if germination is slow, the viability of the inoculum may be reduced until it takes place. The germination technique in seedbed is mainly used in plants that have slow, not uniform, germination or that have very small seeds. When the germination is slow, inoculation in sowing should be avoided; it can be practiced after the emergence or during the transplantation. In case of a nonuniform germination, the plantlets are transplanted in groups of different ages, and preferably inoculated during this operation. If the inoculation takes place in the seedbed, the plants should remain at least 30 days in contact with the inoculum before being transplanted to guarantee the root colonization. If plants have very small seeds, they usually germinate fast, but they have a very long seedbed stage until they reach suitable size for transplantation. In these plants inoculation should be done at the moment of sowing, since they

TABLE 14.1. Types of inoculant and application forms used during the sowing or transplanting of citrus plantules to fumigated soil of a nursery and their effects in the shoot dry mass (SDM) and root colonization (Col.).

Inoculation		Sowing[a]		Transplanting[b]	
Types of inoculant	**Application form**	**SDM (g)**	**Col.(%)**	**SDM (g)**	**Col.(%)**
Autoclaved soil inoculum (noninoculated)	Broadcast over plots and rototilled (30 cm depth)	0a	0a	0.6a	0a
Lyophilized roots	3 g/plant deposited in hole bored into root zone[c]	5.8b	37b	0.7a	37b
Spores separated from soil inoculum	Seed and 20,000 spores suspended in 2 percent methyl cellulose per m^2	6.9b	55c	1.7b	49c
Spores separated from soil inoculum	400 spores/plant deposited in hole bored into root zone[c]	13.0c	87d	2.1c	70de
Soil inoculum (16 spores/g)	400 spores/plant deposited in hole bored into root zone[c]	12.1c	83d	2.1c	66de
Soil inoculum (16 spores/g)	5,000 spores/m^2 banded 15 cm deep after planting[c]	10.9c	72cd	1.7b	57cd
Soil inoculum (16 spores/g)	5,000 spores/m^2 banded 15 cm deep 14 days before planting	11.5c	70c	1.6b	60cde
Soil inoculum (16 spores/g)	4,800 spores/m^2 broadcast over plots and rototilled (30 cm depth)	11.5c	76d	1.6b	60cde
Soil inoculum (16 spores/g)	14,400 spores/m^2 broadcast over plots and rototilled (30 cm depth)	13.7c	88d	1.7b	57cde

Source: Fergunson and Menge, 1986. Copyright American Society for Horticultural Science. Reprinted with permission.

[a]Mean of 120 plants; [b]Mean of 60 plants; [c]Applied 4 months after sowing or 14 days after transplanting the plantules.

Mean separation within columns by 5 percent Duncan's multiple range test.

are born with a small cotyledonous reserve and need mycorrhizal symbiosis to reach transplantation size. Therefore, to maximize the viability of the propagules in the inoculant, the ideal stage on formation of the seedlings for inoculation should be observed (Table 14.2).

The inoculation of seedlings in the transplantation to the field can be efficient depending on the species of the plant. Uninoculated seedlings of coffee, and inoculated with *Gigaspora margarita* and *Glomus clarum* during the transplantation to cement pots (0.6 × 1m), showed a growth similar to seedlings inoculated during formation and were superior to the noninoculated seedlings (Siqueira et al., 1993). Similar results were obtained when seedlings from nurseries were inoculated during transplantation to the field (Siqueira, Aylas, et al., 1994). However, the response to the inoculation in the transplantation will depend on the number of viable propagules in the soil to which the seedlings will be transplanted. When coffee plant seedlings were transplanted to pots with soil containing AMF, the effects of the inoculation in the transplantation were detected only until 60 days after the transplantation (Saggin Júnior et al., 1992). The inoculation in the transplantation is more laborious and certainly more expensive than the one carried out during the formation of the seedling, and the benefits of the AMF for the growth and the valuation of the seedling are lost. It should be favored as the ideal stage for inoculation only in the case of plants with very large

TABLE 14.2. Ideal phases for seedling inoculation to avoid viability loss of mycorrhizal inoculant.

Seed size[a]	Uniformity of germination	Rate of germination	Sowing location	Ideal phase for inoculation[b]
Small	Any	Fast	Seedbed	On sowing
Small	Any	Slow	Seedbed	On transplant
Big	Nonuniform	Any	Seedbed	On plantlet transplant or seedling transplant to the field
Big	Uniform	Fast	Container	On sowing or seedling transplant to the field
Big	Uniform	slow	Seedbed	On plantlet transplant or seedling transplant to the field

[a]Small is difficult to handle on sowing; big is easy to handle on sowing.
[b]The inoculation during seedling transplant is ideal when big seeded plants manifest mycorrhizal dependence only after the exhaustion of the cotyledonary reserve.

seeds that may not show dependence on mycorrhizae during the seedling stage, due to food reserves in the seeds.

The inoculation of seedlings during the micropropagation in vitro or at the stage of readaptation to the outside environment, called the acclimatization stage, has been important to help most plants adapt again to the ex vitro environment; only some plants can overcome this process of associating themselves to these fungi. The micropropagation of plants has had an increasing use in the production of seedlings of selected and pathogen-free plants. However, as pathogens are eliminated, micropropagation also eliminates the AMF, which should necessarily be inoculated in the production of micropropagated plantlets as early as possible.

In the production of micropropagated seedlings, after multiplication of the selected material in vitro, the rooting is hormonally promoted. Thus, the inoculation of AMF is possible only after this stage. Generally, the inoculation of AMF is done at the beginning of the acclimatization process. This process is difficult for the plant since it involves the activation of photosynthesis to obtain energy and the activation of absorption to obtain nutrients. Furthermore, the plant comes in contact with innumerable microorganisms and variations of temperature, light, and humidity. Thus, there are considerable benefits from the inoculation of AMF at this stage, for it helps the plant to adapt to and resist the stress, in an association of nutritional and non-nutritional effects.

Amount of Inocula for Inoculation of Seedlings

The amount of mycorrhizal inocula needed to inoculate seedlings in nurseries or seedbeds is not considered a limiting factor. The dose of inoculum depends on the number of viable propagules that it contains. The number of spores may have little significance, since they may be alive, dormant, parasitical, or dead. Pieces of roots and mycelia present may also be alive or dead. The number of viable propagules can be estimated by performing serial dilution tests of the inoculant followed by the culture of a host plant, with an estimate of the most probable number (MPN) of propagules (Feldmann and Idczak, 1992). They verified that the inoculants in expanded clay produced by them contained 30 to 50 viable propagules per cm^3, and that when applied in the 2 to 5 percent range of the volume of substrate, they promoted a good mycorrhization of the seedlings. David Sylvia (http://www.agronomy.psu.edu/People/Faculty/SylviaD.html), working at University of Florida, had an evaluation program of commercial inoculants made through a test with corn cultivated with 5 g of inoculant during 28 days. On his Web site <http://dmsylvia.ifas.ufl.edu/> he points out the commercial inoculants that promoted a root colonization rate of corn above 25 percent during the conduction of the test.

Generally, excessive amount of propagules have been used in research to guarantee the success of inoculation. However, the inoculum dose in commercial production of seedlings should be ideal as the success of inoculation and its cost will depend on it. It is known that a few spores can be sufficient to guarantee adequate levels of colonization. Daft and Nicolson (1969) verified that only three spores of *Glomus macrocarpum* located over the roots of tomato plantlets promoted in 12 weeks a radial colonization superior to 40 percent and similar to the one promoted by 226 spores. Siqueira, Colozzi-Filho, and Saggin Júnior (1994) verified 50, 100, 200, 400, and 800 spores of *Gigaspora margarita* over roots of coffee plantlets which resulted in similar colonization 90 days after the inoculation. The application of mixed inoculum containing 200 spores along with other propagules had an effect similar to the application of 400 spores on the increase of colonization as time went by. Spores are considered the most resistant propagule, but inoculants containing other types of propagules generally establish and multiply the AMF more rapidly than would using only spores.

Thus it is estimated that for most cases 50 to 100 spores, placed in contact with the roots, are sufficient to guarantee a good colonization of seedlings. Mixed inoculum that contain 100 spores/g can be easily obtained for some species of AMF. It would be necessary to apply 1 g of this inoculant near the roots to guarantee good colonization of seedlings. Then, for 1,000 seedlings, only 1 kg of this inoculant would be necessary. The inoculant can also be incorporated directly to the substrate, thus reducing expenses with labor. Saggin Júnior et al. (1992) verified that the soil infestation with only 0.5 spore/g guaranteed good development and colonization of coffee seedlings. Based on this fact, for the inoculant with 100 spores/g, it would be necessary to incorporate only 5 kg per ton of substrate.

For inoculation of seedlings during transplantation to the field, 100 g of mixed inoculum was applied (Siqueira, Aylas, et al., 1994) with success in the inoculation. There are no reports about inoculant doses applied during the transplantation of seedling to field. The size of the seedling and the pit, as well as the quality of the inoculant, are the factors that will determine the amount to be applied during transplantation.

Cost of Inoculation of Seedlings with AMF

The cost should be low to make the inoculation attractive to producers and nursery owners since, most times, the aim should not be the benefit of inoculation in seedling production, and one should search for the valuation of the produced seedling and the posttransplantation benefit.

Sieverding (1991) estimates the cost of inoculation be lower than 1 cent per seedling. This cost will depend on the technology of production of the inoculant, on the inoculation practice used, and on the amount of inoculant used.

In commercial research carried out in a firm in the United States (see Table 14.3), it is seen that the cost of seedling inoculation may be between less than 1 cent when it is done during sowing up to more than 16 cents per seedling when it is done during transplantation to the field.

TABLE 14.3. Cost of the inoculation with arbuscular mycorrhizal fungi inoculant commercialized in the United States for a producer company.

Type of inoculant	Indications	Application recommendation	Use rate	Cost per 1,000 seedlings
Spores of AMF[a] 50 spores per cm^3	Sowing (for sowing cultures)	Seeds are coated with the inoculant.	1,360 g of seeds for planting 1 ha	—
	Transplanting (for small seedlings, such as strawberries)	The roots are slightly dirty with inoculant.	5,000 seedlings	US$ 15
	Bareroot transplanting (for grapes and others seedlings)	The inoculant is placed on the roots or dispersed in the planting hole.	500 seedlings	US$ 150
Spores of AMF[a] + hydrogel 50 spores per cm^3	Transplanting (for small seedlings, such as strawberries)	The roots are placed into inoculant suspension.	5,000 seedlings	US$ 16
	Bareroot transplanting (for grapes and others seedlings)	The roots are placed into inoculant suspension.	500 seedlings	US$ 160

[a]Species mixture (*Paraglomus brasilianum, Glomus clarum, Glomus deserticola, Glomus intraradices, Glomus monosporum, Glomus mosseae,* and *Gigaspora margarita*).

Substrates for the Formation of Mycorrhizal Seedlings

The substrates used for the production of seedlings contain different materials in their composition, usually using materials with regional availability. Furthermore, the characteristics of the same material vary a lot due to its origin, resulting in substrates with different physical and chemical properties that may affect differently the mycorrhization of the seedlings and their benefits.

The substrate for sowing should generally contain much sand or be made up of only sand. This material provides a good mycorrhization of plants due to good drainage and aeration. Besides that, they have advantages of a low cost, absence of invading plants, and facilitating the transplantation of the plantlets. However, when the sand is washed, the inoculation of AMF should be performed because of the absence of the propagules of these fungi.

The substrate for containers or mobile seedbeds must be made up of light material and good drainage. This also favors the colonization of the seedlings. Usually, ready-made commercial substrates are used, of mixtures containing vermiculite, perlite, or expanded clay; these materials do not have AMF propagules, so inoculation is necessary. When an AMF inoculant has already been produced in these types of materials, it is more easily incorporated to the seedling substrate.

The substrates for plastic bags or other large containers are usually based in a soil to which, according to the texture, sand, organic matter, and fertilizers are mixed. The substrate containing soil, if fumigated or subsoil having been used, will certainly need inoculation with AMF. In this kind of substrate the various forms and procedures of inoculation discussed before can be adapted and used.

The main characteristics of the substrate of the seedling that will affect mycorrhization are the aeration, the proportion of organic matter, and the availability of nutrients, especially phosphorus. Aeration has great influence on AMF because they are obligate aerobics. An ideal aeration to promote the colonization of seedlings should provide an adequate supply of water, a large supply of oxygen, and proper porosity for the growth of the roots and hyphae (Saif, 1981).

The effects of organic matter on the mycorrhizal association vary. They are generally synergistic, although inhibiting effects are also reported. Increase in the density of the spores and in the radial colonization are observed when organic matter is incorporated to the substrate (Nappi et al., 1981; Zambolim et al., 1992), probably because of aeration and to the greater root development. However, the effects of organic matter may be depressive, especially if the organic compound is not mature, containing a high proportion of ammonia (Sieverding, 1991; Roldan and Albaladejo, 1993) or high availability of nutrients, mainly phosphorus (Menge et al., 1982).

Different sources and doses of organic matter for the substrate of seedlings diversely influence the response to the AMF and the establishment of symbiosis. Mycorrhized seedlings of yellow passion fruit showed a linear growth increase with doses of cow manure or waste compound up to a proportion of 40 percent of the substrate, but the mycorrhizal colonization reached the maximum between doses 10 to 20 percent of the substrate, becoming inhibited in larger doses (Bento, 1997). When the cow manure dose is equal to or greater than 20 percent of the substrate, noninoculated passion fruit seedlings grow more than do inoculated ones. However, when a waste compound is used, only mycorrhizal plants respond in growth to the increase of organic material in the substrate, which seems to show that this compound is quite inhibitive to the seedlings, and that the AMF helped the seedlings to overcome this stress. Generally, in the yellow passion fruit, incorporation of 10 percent of the substrate with several sources of organic matter allows good growth for the seedlings and a good response to the inoculation with AMF, except for organic compounds very rich in nutrients, such as earthworm humus, which tend to reduce the mycorrhizal colonization of the seedlings (see Figure 14.4).

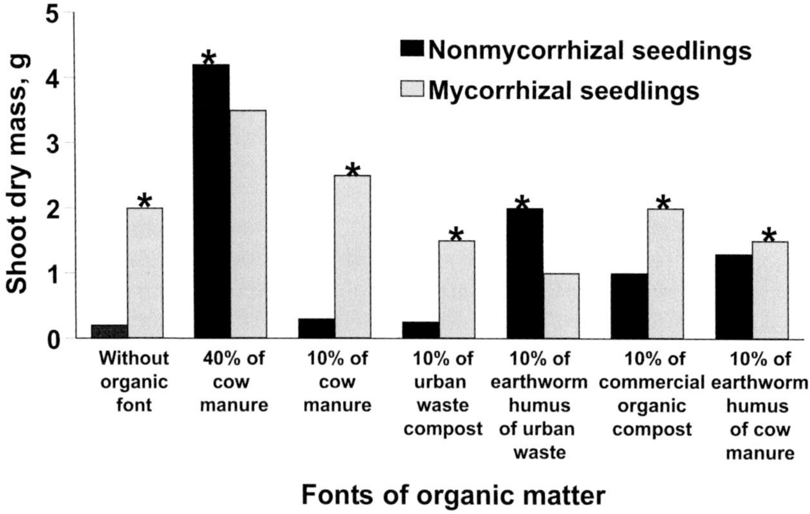

FIGURE 14.4. Shoot dry mass of mycorrhizal and nonmycorrhizal yellow passion fruit seedlings grown on basic substrate amended with different fonts of organic matter. Asterisks indicate superiority among mycorrhizal and nonmycorrhizal plants for Tukey 5 percent. *Source:* Modified from Bento, 1997.

A substrate made up of 10 percent of organic matter promotes a good growth for the seedlings and a good colonization of the root system in most of the cases studied. However, besides the source of organic matter, the ideal dosage of organic material depends on the plant and its dependence to mycorrhizae. For example, 25 percent of cow manure in the substrate of coffee seedlings still provokes a response in growth to inoculation of various AMF species (Siqueira et al., 1995). In yellow passion fruit seedlings, this effect depends on the species of fungus inoculated (Silveira et al., 1993). In papaya trees, it has been verified that no species of AMF presents a beneficial effect of inoculation in the presence of 25 percent of cow manure in the substrate (Silva et al., 1993).

In the same manner as the organic material, the use of mineral fertilizers in the substrate of seedlings may inhibit the mycorrhizal colonization and diminish the response to inoculation. Many studies relate fertilization to mycorrhization. Among the nutrients, those that most influence these symbioses are N and P. It is known that the availability of P in the soil controls, through the plant, the colonization of the radial system (Saggin Júnior et al., 1994). The effects of N are not as well-known as those of P, and increases or decreases of colonization have been reported. The source of N influences colonization; preference should be given to a mixture of sources with a balance of NH_4^+ and NO_3^-, since the use of ammoniacal N alone tends to reduce mycorrhizal colonization.

Thus the fertilization of the seedling substrate with P and N should be well balanced to promote a fast and vigorous development and not reduce the benefits gained through symbioses. Ideal levels of N for the development of mycorrhizal seedlings of *Leucaena leucocephala* are around 50 $mg{\cdot}kg^{-1}$ of N applied to nonfumigated soil, or of 100 $mg{\cdot}kg^{-1}$ when the soil was fumigated not forming N_2-fixing nodules (Onguene and Habte, 1995). The inoculation of nitrogen-fixing bacteria in seedlings of leguminous plants can avoid the need for nitrogen fertilization of the seedlings. In relation to the availability of P in the substrate of seedlings, Siqueira and Colozzi-Filho (1986) suggested that it should be between 10 and 100 $mg{\cdot}kg^{-1}$ for coffee seedlings, since between these values mycorrhizal symbiosis presents a mutualistic nature and reaches the maximum benefit for the host plant. This range of value varies according to the efficiency of the fungus and to the mycorrhizal dependence of the plant. In several studies with different plant species, particularly application of P in the substrate of mycorrhizal seedlings, it is verified that maximum benefit of inoculation in growth and maximum mycorrhizal colonization are obtained with the P available in the substrate, in the range between 9 and 27 $mg{\cdot}kg^{-1}$ (Saggin-Júnior and Lovato, 1999). This suggests that for most seedlings, in this P range that is available in the substrate, a greater benefit from inoculation can be obtained.

Handling of Mycorrhizal Seedlings

When seedlings are mycorrhizal, besides the substrate and the fertilization, other techniques should be considered in the handling of seedlings, such as the use of pesticides, irrigation, and illumination. One of the most important factors in the handling of seedlings is the use of pesticides that can affect AM fungus, the formation, and the functioning of the symbiosis. The use of fumigating biocides, water vapor, or sunlight in the disinfestations of the substrate or of the seedbed, aiming at the elimination of phytopathogenic organisms, nematodes, and weeds, is common in the production of seedlings. These treatments greatly reduce the native AMF population, possibly resulting in blight and in symptoms of nutrient deficiency in plants with high mycorrhizal dependence, making it necessary to inoculate the seedlings.

The use of fungicides in seedlings and seeds is frequent, mainly to control soilborne diseases and sometimes foliar diseases. Fungicides can severely affect AMF and the formation of mycorrhizae, or sometimes have a fungistatic effect and, surprisingly, can even foster the development of AMF (Trappe et al., 1984). Among the fungicides that are very toxic to the AMF are Benomyl, Pentachloronitrobenzene, also called PCNB or Quintozene (Nesheim and Linn, 1969; Menge et al., 1979) and Triadimefon (Jalali and Domsch, 1975; Bärtschi, 1982). Benomyl is the fungicide whose action on most AMF has been studied, probably due to its great toxicity which affects all aspects of the development of the mycorrhizae. There are reports on reductions of sporulation, radicular colonization, intensity of infection, length of active external hyphae, number of intercellular hyphae, and number of arbuscules (Jalali and Domsch, 1975; Nemec, 1980; Ocampo and Hayman, 1980; Hale and Sanders, 1982; Kough et al., 1987; Carey et al., 1992; Sukarno et al., 1993). The reductions of the external hyphae and of the fungus-plant interfaces diminish the absorption of P and its transference to the plant, reducing the beneficial effect of the AMF (Hale and Sanders, 1982). Since the use of Benomyl by nursery owners is common to control damping off of the plantlets, one should highlight the importance of not using this fungicide in seedlings that depend on mycorrhizae or in those inoculated with AMF. Fungicides such as Captan, Metalaxyl, Fosetyl-Al, and Thiram, have shown little or no depressive action on the AMF; however, there are many conflicting results attributing a depressive effect or no effect on the AMF of a same fungicide active principle (see for example Manjunath and Bagyaraj, 1984, and Kough et al., 1987). Thus, along with the active principle of the product and the way it acts, the effect of a specific fungicide on AMF depends on plant tolerance and on the species of fungus. Fungicides have different effects on AMF species, altering the original compositions of the population (Schreiner and Bethlenfalvay, 1996). Thus, the specific interactions among active principles, AMF species, and plant species need to be better understood so that viable control practices of pathogens in seedlings can be established without affecting mycorrhizal association.

The use of herbicides in seedlings is less than other pesticides, because the control of weeds in the substrate is basically done through disinfestations and, when necessary, by manual pulling. However, the use of herbicides, such as Simazine and Trifluraline, in the formation of seedlings of certain species is common, such as in citrus. Grafting is done directly in the soil and only at the time of commercialization are the seedlings transferred to the containers. It is also common to use nonselective herbicides, such as Paraquat, to clean the nursery area. Trifluraline does not seem to affect AMF, while Simazine and Paraquat show a toxic effect on AMF (Smith et al., 1981; Nemec and Tucker, 1983), although this has not been proved in studies with growth of these fungi in vitro (Hamel et al., 1994). Generally, herbicides have little effect on AMF (Trappe et al., 1984), with more indirect effects on these fungi due to the herbicide effects on the metabolism of the host plant (Ocampo and Barea, 1982). Plants with metabolisms altered by herbicides lose control over symbiosis; for example, Simazine may promote the colonization of *Chenopodium quinona,* a nonhost species of AMF (Schwab et al., 1982). However, mycorrhization can increase the plant tolerance to herbicide residues (Siqueira et al., 1991), although some reports support that the toxic effects of herbicides are increasing in the plant (Hamel et al., 1994). Therefore, due to the conflicting information available, one should avoid the use of herbicides in a nursery.

The use of insecticides, acaricides, and nematicides in seedlings is wide and their effects on AMF are less studied than other pesticides (Trappe et al., 1984). Most tested insecticides and nematicides have little or no effect on the formation of spores and on root colonization.

Other factors that can affect AMF include irrigation and illumination, which are less often studied. The irrigation of mycorrhizal seedlings should be controlled to avoid anaerobic conditions, even if only for short periods of time, for the AMF demand high concentrations of O_2 for their development, physiological activity, and colonization (Saif, 1981, 1984). This care should be taken particularly in the acclimatization of micropropagated seedlings due to the plantlets' great need for humidity at this stage. This problem in acclimatization can be solved with the use of very porous substrates that promote fast drainage (Wang et al., 1993).

Illumination in the nursery indirectly affects the AMF through the effects on the photosynthetic rate of the host plant. The amount of photosynthates that move to the roots controls the infection of the roots by the AMF, through a mechanism not yet made clearly evident (Siqueira et al., 1984; Schwab et al., 1991; Koide and Schreiner, 1992). Thus during the formation of seedlings its acclimatization to sunlight is necessary not only because of the adaptation to the conditions of the fields but also to benefit the establishment of the AMF, providing them with a greater amount of photosynthates. In this aspect, the control of the seedling density by area and the selection of seedlings as to their size are also important practices in the handling of mycorrhized seedlings, for they influence illumination. The density of the seedlings should be particularly controlled if they are produced in containers, since they are determinant factors for the establishment of the density, the architecture of the seedling, the size of the leaves, and the rate of growth. The selection of seedlings in the nursery as to the uniformity of size, along with promoting perfect illumination, allows the elimination of plants with genetic problems and diseases.

FUTURE PERSPECTIVES OF MYCORRHIZAL SEEDLING PRODUCTION

The application of AMF in the production of seedlings is promising as the beneficial effects have become evident in the initial stages of seedling production and transplantation to the field, being able to increase productivity. Inoculated seedlings have superior growth and survival in the field over seedlings not inoculated with AMF. Saggin Júnior and Lovato (1999) reviewed that these results can be verified up to the beginning of the production stage in coffee plants and in walnut trees. In coffee plants, field studies show that the mycorrhization of seedlings fosters an increase in productivity in the first years of production (Saggin Júnior and Siqueira, 1996). However, the field benefits of the mycorrhization of the coffee seedlings tend to diminish with time due to a complex interaction of the biological and chemical factors involved (Siqueira et al., 1998). This shows the need to carry out research and technological development programs for the use of mycorrhizal seedlings in the field that have been done for many years.

The most promising field for use of seedlings inoculated with AMF is in those seedlings that are meant for reforestation of degraded areas (Souza and Silva, 1996). In this case, the AMF inoculation is essential for the seedling to survive in the stressful environment and for reestablishment of the nutrient recycling process. The AMF enable leguminous plants in degraded soil to be better nourished so they can be able to fix atmospheric N_2, which is the main source of entrance of N in the system, increasing the accumulation of biomass and organic waste (Bethlenfalvay, 1992). Besides that, the AMF are important so that the process of plant succession in the area to be replanted be activated, for they allow the entrance of plant species that are very dependent on this symbiosis and modify the competitive relations among plant species (Flores-Aylas et al., 2003).

In the seedlings, there is still the possibility of inoculating other beneficial organisms to the plants along with the AMF. In the case of leguminous plants seedlings, joint inoculation of rhizobia and AMF is largely used for reforestation of degraded areas (Souza and Silva, 1996), and it can be used in fruit plants of fabaceae, such as the tamarind (*Tamarindus indica* L.) and the inga (*Inga edulis* Mart.). Introduction of other beneficial organisms, such as the rhizobacteria that promote plant growth (Germida and Walley, 1996) and are antagonistic to pathogens (Camprubí et al., 1995), with the effect of fostering greater growth of plants and biological control, must be tried. In addition, phosphate-solubilizing microbes can be inoculated to increase the nutritional effect of AMF (Azcón-Aguilar and Barea, 1996).

Therefore, in the future the production of seedlings should be directed to a new focus that will search for the commercial valuation of seedlings through the use of biotechnologies of inoculation of microorganisms. Each seedling producer will try to offer the buyer a seedling with unique characteristics of the attendant microorganisms, competing for biodiversity and a better combination of beneficial microorganisms. Ideal combinations of selected AMF with other beneficial microorganisms should be investigated and even patented. The seedling should have a rhizospheric community of associated beneficial microorganisms that will give specific characteristics to different environmental conditions and objectives of planting. In this way, the applied research of AMF may open up new ways for using combinations of inoculants of rhizospheric microorganisms.

CONCLUSIONS

The inoculation of seedlings with AMF is considered a safe and eco-friendly biotechnology, since these fungi occur naturally in the soils of most land ecosystems. Moreover, it is low-cost technology, which makes its application viable even if the economic return is little. For its use, it is necessary to have information on the mycorrhizal dependence of the plants whose seedlings will be produced and on the species of AMF adapted to this plant and to the environment in which the seedling will be planted. It is also necessary to adapt the seedling substrate to make it noninhibitive to the establishment of symbiosis and to the adjustment of handling practices to the incorporation of the inoculant and preservation of the inoculated fungus. Furthermore, studies that search ideal combinations of AMF with other rhizospheric microorganisms will be able to foster the production of seedlings with high performance and of greater commercial value.

The great benefits promoted by the inoculation of AMF in seedlings, such as the ease and low cost of its inoculation, show that this biotechnology can be easily adopted by seedling producers in countries where it is still little used, such as Brazil, giving inertia to this segment of agribusiness.

REFERENCES

Abbott L.K. and Robson A.D. (1982). The role of vesicular arbuscular mycorrhizal fungi in agriculture and the selection of fungi for inoculation. *Australian Journal of Agriculture Research* 33:389-408.

Abbott L.K. and Robson A.D. (1985). Selection of "efficient" VA mycorrhizal fungi. In R. Molina (ed.), *Proceeding of the 6th North American Conference on Mycorrhizae* (pp. 89-90). Forest Research Laboratory, Bend.

Allen E.B. and Allen M.F. (1986). Water relations of xeric grasses in the field: Interactions of mycorrhizas and competition. *New Phytologist* 104:559-571.

Allen M.F. (1996). The ecology of arbuscular mycorrhizas: A look back into the 20th century and a peek into the 21st. *Mycological Research* 100:769-782.

Allen M.F., Moore T.S., Christensen M., and Stanton N. (1979). Growth of vesicular-arbuscular mycorrhizal and non-mycorrhizal *Bouteloua gracilis* in a defined medium. *Mycologia* 71:666-669.

Augé R.M., Stodola A.J.W., Brown M.S., and Bethlenfalvay G.J. (1992). Stomatal response of mycorrhizal cowpea and soybean to short-term osmotic stress. *New Phytologist* 120:117-125.

Azcón-Aguilar C. and Barea J.M. (1996). Arbuscular mycorrhizas and biological control of soil-borne plant pathogens—An overview of the mechanisms involved. *Mycorrhiza* 6:457-464.

Azcón-Aguilar C. and Barea J.M. (1997). Applying mycorrhiza biotechnology to horticulture: Significance and potentials. *Scientia Horticulturae* 68:1-24.

Bärtschi H. (1982). Influence de quelques pesticides sur l'infection endomycorhizienne. In S. Gianinazzi, V. Gianinazzi-Pearson, and A. Trouvelot (eds.), *Mycorrhizae, an integral part of plants: Biology and perspectives for their use* (pp. 259-266). Paris: INRA.

Baylis G.T.S. (1975). The magnolioid mycorrhiza and mycotrophy in root systems derived from it. In F.E. Sanders, B. Mosse, and P.B. Tinker, (eds.), *Endomycorrhizas* (pp. 373-389). London: Academic Press.

Bécard G. and Fortin J.A. (1988). Early events of vesicular-arbuscular mycorrhiza formation on Ri T-DNA transformed roots. *New Phytologist* 108:211-218.

Bento M.M. (1997). Fontes de matéria orgânica na composição do substrato para produção de mudas micorrizadas de maracujazeiro, ESALQ, Piracicaba. (Dissertação de mestrado), p. 59.

Bethlenfalvay G.J. (1992). Mycorrhizae and crop productivity. In G.J. Bethlenfalvay and R.G. Linderman (eds.), *Mycorrhizae in sustainaible agriculture* (pp. 1-27). Madison, WI: ASA/CSSA/SSSA.

Bieleski R.L. (1973). Phosphate pools, phosphate transport, and phosphate availability. *Annual Review of Plant Physiology* 24:225-252.

Calvet C., Camprubí A., and Rodríguez-Kábana R. (1996). Inclusion of arbuscular mycorrhizal fungi in alginate films for experimental studies and plant inoculation. *HortScience* 31:285.

Camprubí A., Calvet C., and Estaun V. (1995). Growth enhancement of *Citrus reshni* after inoculation with *Glomus intraradices* and *Trichoderma aureoviride* and associated effects on microbial populations and enzyme activity in potting mixes. *Plant and Soil* 173(2):233-238.

Carey P.D., Fitter A.H., and Watkinson A.R. (1992). A field study using the fungicide benlate to investigate the effect of mycorrhizal fungi on plant fitness. *Oecologia* 90:550-555.

Chapin, F.S. III. (1980). The mineral nutrition of wild plants. *Annual Review of Ecology and Systematics* 11:233-260.

Cooper K.M. (1984). Physiology of VA mycorrhizal associations. In C.L. Powell and D.J. Bagyaraj (eds.), *VA Mycorrhiza* (pp. 155-186). Boca Raton, FL: CRC Press.

Daft M.J. and Nicolson T.H. (1969). Effect of *Endogone* mycorrhiza on plant growth. III. Influence of inoculum concentration on growth and infection in tomato. *New Phytologist* 68:953-963.

Dehne H.W. and Backhaus G.F. (1986). The use of vesicular-arbuscular mycorrhizal fungi in plant production. I. Inoculum production. *J. Plant Dis. Prot., Stuttgart,* 92:415-424.

Dias G.C., Oliveira E., and Souza P. (1987). Efetividade de métodos de inoculação de fungos micorrízicos vesicular-arbusculares (MVA) em mudas de cafeeiro. In *Programas e resumos da Reunião Brasileira Sobre Micorrizas,* Volume 2 (pp. 4-5). São Paulo: Secretaria do Meio Ambiente e da Agricultura/USP.

Douds Jr. D.D. (2002). Increased spore production by *Glomus intraradices* in the split-plate monoxenic culture system by repeated harvest, gel replacement, and resupply of glucose to the mycorrhiza. *Mycorrhiza* 12:163-167

Duan X., Neuman D.S., Reiber J.M., Green C.D., Saxton A.M., and Augé R.M. (1996). Mycorrhizal influence on hydraulic and hormonal factors implicated in the control of stomatal conductance during drought. *Journal of Experimental Botany* 47:1541-1550.

Feldmann F. and Idczak E. (1992). Inoculum production of vesicular-arbuscular mycorrhizal fungi for use in tropical nurseries. In J.R. Norris, D.J. Read, and A.K. Varma (eds.), *Methods in microbiology* (pp. 339-357). San Diego: Academic Press.

Fergunson J.J. and Menge J.A. (1986). Response of citrus seedlings to various field inoculation methods with *Glomus deserticola* in fumigated nursery soils. *Journal of The American Society of Horticultural Science* 111:288-292.

Flores-Aylas W.W., Saggin-Júnior O.J., Siqueira J.O., and Davide, A.C. (2003). Efeito de *Glomus etunicatum* e fósforo no crescimento de espécies arbóreas em semeadura direta. *Pesquisa Agropecuária Brasileira* 38(2):257-266.

Gaur A. and Adholeya A. (2000). Effects of the particle size of soil-less substrates upon AM fungus inoculum production. *Mycorrhiza* 10:43-48.

Gerdemann J.W. (1968). Vesicular-arbuscular mycorrhiza and plant growth. *Annual Review of Phytopathology* 6:397-418.

Gerdemann J.W. (1975). Vesicular-arbuscular mycorrhizal. In J.G. Torrey and D.T. Clarkson (eds.), *The development and function of roots* (pp. 575-591). London: Academic Press.

Gerdemann J.W. and Nicolson T.H. (1963). Spores of mycorrhizal *Endogone* species extracted from soil by wet sieving and decanting. *Transactions of British Mycological Society* 46:235-244.

Germida J.J. and Walley F.L. (1996). Plant growth-promoting rhizobacteria alter rooting patterns and arbuscular mycorrhizal fungi colonization of field-grown spring wheat. *Biology and Fertility of Soils* 23:113-120.

Gianinazzi S., Gianinazzi-Pearson V., and Trouvelot A. (1990). Potentialities and procedures for the use of endomycorrhizas with emphasis on high value crops. In J.M. Whips and B. Lumsden (eds.), *Biotechnology of fungi for improving plant growth* (pp. 41-54). Cambridge, UK: Cambridge University Press.

Gianinazzi-Pearson, V. (1985). Mycorrhizal effectiveness in phosphate nutrition: How, when and where? In R. Molina (ed.), *Proceeding of the 6th North American Conference on Mycorrhizae* (pp. 150-154). Forest Research Laboratory, Bend.

Goicoechea N., Antolín M.C., and Sánchez-Díaz M. (1997). Gas exchange is related to the hormone balance in mycorrhizal or nitrogen-fixing alfalfa subjected to drought. *Physiolgia Plantarum* 100:989-997.

Graham J.H. and Syvertsen J.P. (1985). Host determinants of mycorrhizal dependency of citrus rootstock seedlings. *New Phytologist* 101:667-676.

Graham J.H. and Timmer L.W. (1984). Vesicular-arbuscular mycorrhizal development and growth response of rough lemon in soil and soilless media: Effect of phosphorus source. *Journal of The American Society of Horticultural Science* 109:118-121.

Guttay A.J.R. (1982). The growth of three woody plant species and the development of their mycorrhizae in three different plant composts. *Journal of the American Society of Horticultural Science* 107:324-327.

Guttay A.J.R. (1983). The interaction of fertilizers and vesicular-arbuscular mycorrhizae in composted plant residues. *Journal of the American Society of Horticultural Science* 108:222-224.

Habte M. and Manjunath A. (1991). Categories of vesicular-arbuscular mycorrhizal dependency of host species. *Mycorrhiza* 1(1):3-12.

Hale K.A. and Sanders F.E. (1982). Effects of benomyl on vesicular-arbuscular mycorrhizal infection of red clover (*Trifolium pratense* L.) and consequences for phosphorus inflow. *Journal of Plant Nutrition* 5:1355-1367.
Hall I.R. (1979). Soil pellets to introduce vesicular-arbuscular mycorrhizae into soil. *Soil Biology and Biochemistry* 11:85-86.
Hall I.R. and Kelson A. (1981). An improved technique for the production of endomycorrhizal infested soil pellets. *New Zealand Journal of Agricultural Research* 24:221-222.
Hamel C., Morin F., Fortin A., Granger R.L., and Smith D.L. (1994). Mycorrhizal colonization increases herbicide toxicity in apple. *Journal of the American Society of Horticultural Science* 119:1255-1260.
Hattingh M.J. and Gerdemann J.W. (1975). Inoculation of Brazilian sour orange seed with endomycorrhizal fungus. *Phytopathologia* 65:1013-1016.
Hayman D.S., Morris E.J., and Page R.J. (1981). Methods for inoculating field crops with mycorrhizal fungi. *Annals of Applied Biology* 99:247-253.
Heinzemann J. and Weritz J. (1990). Steinwoole: Ein neuer tragerstoff zur vermehrung von vesikular-arbuskularen mykorrhizapilzen. *Angewandte Botanik* 64(3-4):271-274.
Hepper C.M. (1981). Techniques for studying the infection of plants by vesicular-arbuscular mycorrhizal fungi under axenic conditions. *New Phytologist* 88:641-647.
Hetrick B.A.D., Wilson G.W.T., and Cox T.S. (1992). Mycorrhizal dependence of modern wheat varieties, landraces, and ancestors. *Canadian Journal of Botany* 70:2032-2040.
Hetrick B.A.D., Wilson G.W.T., and Cox T.S. (1993). Mycorrhizal dependence of modern wheat cultivars and ancestors: A synthesis. *Canadian Journal of Botany* 71:512-518.
Hetrick B.A.D., Wilson G.W.T., Gill B.S., and Cox T.S. (1995). Chromosome location of mycorrhizal responsive genes in wheat. *Canadian Journal of Botany* 73:891-897.
Hetrick B.A.D., Wilson G.W.T., and Leslie J.F. (1991). Root architecture of warm- and cool-season grasses: Relationship to mycorrhizal dependence. *Canadian Journal of Botany* 69:112-118.
Hetrick B.A.D., Wilson G.W.T., and Todd T.C. (1992). Relationships of mycorrhizal symbiosis, rooting strategy, and phenology among tallgrass prairie forbs. *Canadian Journal of Botany* 70:1521-1528.
Hetrick B.A.D, Wilson G.W.T., and Todd T.C. (1996). Mycorrhizal response in wheat cultivars: relationship to phosphorus. *Canadian Journal of Botany* 74:19-25.
Huang R.S. and Tang C.S. (1988). Production of vesicular-arbuscular mycorrhizal sporocarps, *Glomus aggregatum,* on fiberglass screens. *Plant Soil* 108:233-235.
Hungria M., Campo R.J., Chueire L.M.O., Grange L., and Megias M. (2001). Symbiotic effectiveness of fast-growing rhizobial strains isolated from soybean nodules in Brazil. *Biology and Fertility of Soils* 33(5):387-394.
Jalali B.L. and Domsch K.H. (1975). Effect of systemic fungitoxicants on the development of endotrophic mycorrhiza. In F.E. Sanders, B. Mosse, and P.B. Tinker (eds.), *Endomycorrhizas* (pp. 619-626). London: Academic Press.

Janos D.P. (1987). VA mycorrhizas in humid tropical ecosystems. In G.R. Safir (ed.), *Ecophysiology of VA mycorrhizal plants* (pp. 107-134). Boca Raton, FL: CRC Press.

Janos D.P. (1988). Mycorrhiza applications in tropical forestry: Are temperate-zone approaches appropriate? In F.S.P. Ng (ed.), *Trees and mycorrhiza* (pp. 133-188). Kuala Lumpur, Malaysia: Forest Research Institute.

Jarstfer A.G. and Sylvia D.M. (1995). Aeroponic culture of VAM fungi. In A. Varma and B. Hock (eds.), *Mycorrhiza* (pp. 427-441). Berlin-Heidelberg: Springer-Verlag.

Johansen A., Finlay R.D., and Olsson P.A. (1996). Nitrogen metabolism of external hyphae of the arbuscular mycorrhizal fungus *Glomus intraradices*. *New Phytologist* 133:705-712.

Kleinschmidt G.D. and Gerdemann J.W. (1972). Stunting of citrus seedlings in fumigated nursery soils related to the absence of endomycorrhizae. *Phytopathology* 62:1447-1453.

Koide R.T. and Schreiner R.P. (1992). Regulation of vesicular-arbuscular mycorrhizal symbiosis. *Annual Review of Plant Physiology and Plant Molecular Biology* 43:557-581.

Kothari S.K., Marschner H., and Römheld V. (1990). Direct and indirect effects of VA mycorrhizal fungi and rhizosphere microorganisms on acquisition of mineral nutrients by maize (*Zea mays* L.) in a calcareous soil. *New Phytologist* 116: 637-645.

Kough J.L., Gianinazzi-Pearson V., and Gianinazzi S. (1987). Depressed metabolic activity of vesicular-arbuscular mycorrhizal fungi after fungicide applications. *New Phytologist* 106:707-715.

Linderman R.G. (1992). Vesicular-arbuscular mycorrhizae and soil microbial interactions. In G.J. Bethlenfalvay and R.G. Linderman (eds.), *Mycorrhizae in sustainable agriculture* (pp. 45-70). Madison, WI: ASA/CSSA/SSA.

Lovelock C.E., Kyllo D., Popp M., Isopp H., Virgo A., and Winter K. (1997). Symbiotic vesicular-arbuscular mycorrhizal influence maximum rates of photosynthesis in tropical tree seedlings grown under elevated CO_2. *Australian Journal of Plant Physiology* 24:185-199.

Manjunath A. and Bagyaraj D.J. (1984). Effect of fungicides on mycorrhizal colonization and growth of onion. *Plant and Soil* 80:147-150.

Maronek D.M., Hendrix J.W., and Kiernan J. (1980). Differential growth response to the mycorrhizal fungus *Glomus fasciculatum* of southern magnolia and bar harbor juniper grown in containers in composted hardwood bark-shale. *Journal of The American Society of Horticultural Science* 105:206-208.

Martin J.P. (1948). Effect of fumigation, fertilization, and various other soil treatments on growth of orange seedlings in old citrus soils. *Soil Science* 66:273-288.

Martin J.P., Aldrich D.G., Murphy W.S., and Bradford G.R. (1953). Effect of soil fumigation on growth and chemical composition of citrus plants. *Soil Science* 75:137-151.

Mathus N. and Vyas A. (1995). Influence of VA mycorrhizae on net photosynthesis and transpiration of *Ziziphus mauritiana*. *Journal of Plant Physiology* 147:328-330.

Menge J.A. (1983). Utilization of vesicular-arbuscular mycorrhizal fungi in agriculture. *Canadian Journal of Botany* 61:1015-1024.

Menge J.A., Jarrell W.M., Labanauskas C.K., Ojala J.C., Huszar C., Johnson E.L.V., and Sibert D. (1982). Predicting mycorrhizal dependency of troyer citrange on *Glomus fasciculatus* in California citrus soils and nursery mixes. *Soil Science Society of America Journal* 46:762-768.

Menge J.A., Johnson E.L.V., and Minassian V. (1979). Effect of heat treatment and three pesticides upon the growth and reproduction of the mycorrhizal fungus *Glomus fasciculatus. New Phytologist* 82:473-480.

Menge J.A., Johnson E.L.V., and Platt R.G. (1978). Mycorrhizal dependency of several citrus cultivars under three nutrient regimes. *New Phytologist* 81:553-559.

Miller R.M. and Jastrow J.D. (1992). The role of mycorrhizal fungi in soil conservation. In G.J. Bethlenfalvay and R.G. Linderman (eds.), *Mycorrhizae in sustainable agriculture* (pp. 29-44). Madison, WI: ASA/CSSA/SSSA.

Mosse B. (1962). The establishment of vesicular-arbuscular mycorrhiza under aseptic conditions. *Journal of General Microbiology* 27:509-520.

Mosse B. (1973). Advances in the study of vesicular-arbuscular mycorrhiza. *Annual Review of Phytopathology* 11:171-196.

Mosse B. and Hepper C.M. (1975). Vesicular-arbuscular mycorrhizal infections in root organ cultures. *Physiological Plant Pathology* 5:215-223.

Mugnier J. and Mosse B. (1987). Vesicular-arbuscular mycorrhizal infection in transformed root-inducing T-DNA roots grown axenically. *Phytopathology* 77: 1045-1050.

Nappi P., Jodice R., and Kofler A. (1981). Micorrize vesicolo-arbuscolare in vigneti dell'alto Adige sottoposti a differenti tecniche di lavorazione del suolo. *Allionia* 24:27-42.

Nelsen C.E. (1987). The water relations of vesicular-arbuscular mycorrhizal systems. In G.R. Safir (ed.), *Ecophysiology of VA mycorrhizal plants* (pp. 71-91). Boca Raton, FL: CRC Press.

Nemec S. (1980). Effects of 11 fungicides on endomycorrhizal development in sour orange. *Canadian Journal of Botany* 58:522-526.

Nemec S. and Tucker D. (1983). Effects of herbicides on endomycorrhizal fungi in Florida citrus (*citrus* ssp.) soils. *Weed Science* 31:427-431.

Nesheim O.N. and Linn M.B. (1969). Deleterious effect of certain fungitoxicants on the formation of mycorrhizae on corn by *Endogone fasciculata* and on corn root development. *Phytopathology* 59:297-300.

Ocampo J.A. and Barea J.M. (1982). Depressed metabolic activity of VA mycorrhizal fungi by photosynthesis inhibitor herbicides. In S. Gianinazzi, V. Gianinazzi-Pearson, and A. Trouvelot (eds.), *Mycorrhizae, an integral part of plants: Biology and perspectives for their use* (pp. 267-270). Paris: INRA.

Ocampo J.A. and Hayman D.S. (1980). Effects of pesticides on mycorrhiza in field-grown barley, maize and potatoes. *Transactions of the British Mycologial Society* 74:413-416.

Ojala J.C. and Jarrell W.M. (1980). Hydroponic sand culture systems for mycorrhizal research. *Plant and Soil* 57:297-303.

Onguene N.A. and Habte M. (1995). Nitrogen and phosphorus requirements for raising mycorrhizal seedlings of *Leucaena leucocephala* in containers. *Mycorrhiza* 5:347-356.

Pedersen C.T. and Sylvia D.M. (1996). Mycorrhiza: Ecological implications for plant interactions. In K.G. Mukerji (ed.) *Concepts in mycorrhizal research* (pp. 195-222). Dordrecht, the Netherlands: Kluwer Academic Publishers.

Peipp H., Maier W., Schmidt J., Wray V., and Strack D. (1997). Arbuscular mycorrhizal fungus induced changes in the accumulation of secondary compounds in barley roots. *Phytochemistry* 44:581-587.

Plenchette C., Fortin J.A., and Furlan V. (1983). Growth responses of several plants species to mycorrhizae in a soil of moderate P-fertility I. Mycorrhizal dependency under field conditions. *Plant and Soil* 70:199-209.

Plenchette C., Furlan V., and Fortin J.A. (1982). Effects of different endomycorrhizal fungi on five host plants grown on calcined montmorillonite. *Journal of the American Society of Horticultural Science* 107:535-538.

Ponton F., Piché Y., Parent S., and Caron M. (1990). The use of vesicular-arbuscular mycorrhizae in Boston fern production: I. Effect of peat-based mixes. *HortScience* 25:183-189.

Powell C. L. (1979). Inoculation of white clover and ryegrass seed with mycorrhizal fungi. *New Phytologist* 83:81-85.

Redecker D., Thierfelder H., and Werner D. (1998). Production of biomass of arbuscular mycorrhizal fungi in the glass bead compartment system. In A. Varma (ed.), *Mycorrhiza manual* (pp. 495-498). Berlin: Springer Verlag.

Reinhardt D.R. and Miller R.M. (1990). Size class of root diameter and mycorrhizal fungal colonization in two temperate grassland communities. *New Phytologist* 116(1):129-136.

Roldan A. and Albaladejo J. (1993). Vesicular-arbuscular mycorrhiza (VAM) fungal population in a xeric torriorthent receiving urban refuse. *Soil Biology and Biochemistry* 25:451-456.

Ruiz-Lozano J.M. and Azcón R. (1995). Hyphal contribution to water uptake in mycorrhizal plants as affected by the fungal species and water status. *Physiologia Plantarum* 95:472-478.

Ruiz-Lozano J.M. and Azcón R. (1996). Mycorrhizal colonization and drought stress as factors affecting nitrate reductase activity in lettuce plants. *Agriculture, Ecosystem and Environment* 60:175-181.

Ruiz-Lozano J.M., Azcón R., and Gomez M. (1995). Effects of arbuscular mycorrhizal *Glomus* species on drought tolerance: Physiological and nutritional plant responses. *Applied Environmental Microbiology* 61:456-460.

Saggin Júnior O.J. (1997). Micorrizas arbusculares em mudas de espécie arbóreas nativas do Sudeste brasileiro, UFLA, Lavras. (Tese de Doutorado), p. 120.

Saggin Júnior O.J. and Lovato P.E. (1999). Aplicação de micorrizas arbusculares na produção de mudas e plantas micropropagadas. In J.O. Siqueira, F.M.S. Moreira, A.S. Lopes, L.R.G. Guilherme, V. Faquin, A.E. Furtini Neto, and J.G. Carvallo (eds.), *Inter-relação fertilidade, biologia do solo e nutrição de plantas* (pp. 725-733). Lavras: SBCS, Viçosa/ UFLA/DCS.

Saggin Júnior O.J. and Siqueira J.O. (1995). Avaliação da eficiência simbiótica de fungos endomicorrízicos para o cafeeiro. *Revista Brasileira de Ciência do Solo* 19:221-228.

Saggin Júnior O.J. and Siqueira J.O. (1996). Micorrizas arbusculares em cafeeiro. In J.O. Siqueira (ed.), *Avanços em fundamentos e aplicação de micorrizas* (pp. 203-254). Lavras: UFLA.

Saggin Júnior O.J., Siqueira J.O., Colozzi-Filho A., and Oliveira E. (1992). A infestação do solo com fungos micorrízicos no crescimento pós-transplante de mudas de cafeeiro não micorrizadas. *Revista Brasileira de Ciência do Solo* 16:39-46.

Saggin Júnior O.J., Siqueira J.O., Guimarães P.T.G., and Oliveira E. (1994). Interação fungos micorrízicos *versus* superfosfato e seus efeitos no crescimento e teores de nutrientes do cafeeiro em solo não fumigado. *Revista Brasileira de Ciência do Solo* 18:27-36.

Saif S.R. (1981). The influence of soil aeration on the efficiency of vesicular-arbuscular mycorrhizas. I. Effect of soil oxygen on the growth and mineral uptake of *Eupatorium odoratum* L. inoculated with *Glomus macrocarpus*. *New Phytologist* 88:649-659.

Saif S.R. (1984). The influence of soil aeration on the efficiency of vesicular-arbuscular mycorrhizas. III. soil carbon dioxide and growth and mineral uptake in mycorrhizal and non-mycorrhizal plants of *Eupatorium odoratum* L., *Guizotia abyssinica* (L.f) cass. and *Sorghum bicolor* (L.) moench. *New Phytologist* 96:429-435.

Schreiner R.P. and Bethlenfalvay G.J. (1996). Mycorrhizae, biocides, and biocontrol. 4. Response of a mixed culture of arbuscular mycorrhizal fungi and host plant to three fungicides. *Biology and Fertility of Soils* 23:189-195.

Schwab S.M., Johnson E.L.V., and Menge J.A. (1982). Influence of simazine on formation of vesicular-arbuscular mycorrhizae in *Chenopodium quinona* Willd. *Plant and Soil* 64:283-287.

Schwab S.M., Menge J.A., and Tinker P.B. (1991). Regulation of nutrient transfer between host and fungus in vesicular-arbuscular mycorrhizas. *New Phytologist* 117:387-398.

Sieverding E. (1991). Vesicular-arbuscular mycorrhiza management in tropical agrosystems, Deutsche Gesellschaft für Technische Zusammenarbeit (GTZ), Eschborn, pp 371.

Silva L.R.C., Silveira A.P.D., Azevedo I.C., Oliveira E., and Foltran D.E. (1993). Influência de fungos micorrízicos arbusculares no desenvolvimento de mudas de mamoeiro, em substrato com e sem matéria orgânica. In *Resumos, Congresso Brasileiro De Ciência Do Solo,* 24, Goiânia, Vol. I, SBCS, Goiânia, 365-366.

Silveira A.P.D., Silva L.R.C., Azevedo I.C., Oliveira E., and Meletti L.M.M. (1993). Seleção de fungos micorrízicos arbusculares para produção de muda de maracujazeiro, substrato com e sem matéria orgânica. In *Resumos, Congresso Brasileiro De Ciência Do Solo,* 24, Goiânia, Vol. I, SBCS, Goiânia, 363-364.

Simon L., Bousquet J., Lévesque R.C., and Lalonde M. (1993). Origin and diversification of endomycorrhizal fungi and coincidence with vascular land plants. *Nature* 363:67-69.

Siqueira J.O., Aylas W.W.F., Saggin Júnior O.J., and Colozzi-Filho A. (1994). Métodos de inoculação com fungos micorrízicos e seus efeitos na produção do cafeeiro em Minas Gerais. In *Resumos, Reunião Brasileira Sobre Micorrizas,* Universidade Federal de Santa Catarina, Florianópolis, 66.

Siqueira J.O. and Colozzi-Filho A. (1986). Micorrizas vesículo-arbusculares em mudas de cafeeiro II. Efeito do fósforo no estabelecimento e funcionamento da simbiose. *Revista Brasileira de Ciência do Solo* 10:207-211.

Siqueira J.O., Colozzi-Filho A., and Saggin Júnior O.J. (1994). Efeitos da infecção de plântulas de cafeeiro com quantidades crescentes de esporos do fungo endomicorrízico *Gigaspora margarita. Pesquisa Agropecuária Brasileira* 29:875-883.

Siqueira J.O., Colozzi-Filho A., Saggin Júnior O.J., Guimarães P.T.G., and Oliveira E. (1993). Crescimento de mudas e produção do cafeeiro sob influência de fungos micorrízicos e superfosfato. *Revista Brasileira de Ciência do Solo* 17:53-60.

Siqueira J.O., Hubbell D.H., and Valle R.R. (1984). Effects of phosphorus on formation of the vesicular-arbuscular mycorrhizal symbiosis. *Pesquisa Agropecuária Brasileira* 19:1465-1474.

Siqueira J.O., Safir G.R., and Nair M.G. (1991). VA-mycorrhizae and mycorrhiza stimulating isoflavonoid compounds reduce plant herbicide injury. *Plant and Soil* 134:233-242.

Siqueira J.O. and Saggin Júnior O.J. (1995). The importance of mycorrhizae association in natural low-fertility soils. In A.T. Machado, R. Magnavaca, S. Pandey, and A.F. Silva, (eds.), *Proceedings of International Symposium on Environmental Stress: Maize in perspective* (pp. 239-280). México: EMBRAPA/CNPMS, Sete Lagoas/CIMMYT/UNDP.

Siqueira J.O. and Saggin Júnior O.J. (2001). Dependency on arbuscular mycorrhizal fungi and responsiveness of some Brazilian native woody species. *Mycorrhiza* 11:245-255.

Siqueira J.O., Saggin Júnior O.J., Colozzi-Filho A., and Oliveira E. (1995). Influência do substrato de formação e da micorriza no crescimento de mudas de cafeeiro transplantadas. *Pesquisa Agropecuária Brasileira* 30:1417-1425.

Siqueira J.O., Saggin Júnior O.J., Flores-Aylas W.W., and Guimarães P.T.G. (1998). Arbuscular mycorrhizal inoculation and superphosphate application influence plant development and yield of coffee in Brazil. *Mycorrhiza* 7:293-300.

Smith, S.E., Dickson, S., Morris, C., and Smith, F.A. (1994). Transfer of phosphate from fungus to plant in VA mycorrhizas: Calculation of the area of symbiotic interface and of fluxes of P from two different fungi to *Allium porrum* L. *New Phytologist* 127(1):93-99.

Smith S.E. and Gianinazzi-Pearson V. (1988). Physiological interactions between symbionts in vesicular-arbuscular mycorrhizal plants. *Annual Review of Plant Physiology and Plant Molecular Biology* 39:221-244.

Smith T.F., Noack A.J., and Cosh S.M. (1981). The effect of some herbicides on vesicular-arbuscular endophyte abundance in the soil and on infection of host roots. *Pesticide Science* 12:91-97.

Souza F.A. and Silva E.M.R. (1996). Micorrizas arbusculares na revegetação de áreas degradadas. In J.O. Siqueira (ed.), *Avanços em fundamentos e aplicação de micorrizas* (pp. 255-290). Lavras: Universidade Federal de Lavras/DCS e DCF.

Sreenivasa M.N. and Bagyaraj D.J. (1988). Selection of a suitable substrate for mass multiplication of *Glomus fasciculatum*. *Plant and Soil* 109:125-127.

St. John T.V., Hays R.I., and Reid C.P.P. (1981). A new method for producing pure vesicular-arbuscular mycorrhiza-host cultures without specialized media. *New Phytologist* 89:81-86.

Sukarno N., Smith S.E., and Scott E.S. (1993). The effect of fungicides on vesicular-arbuscular mycorrhizal symbiosis. I. The effects on vesicular-arbuscular mycorrhizal fungi and plant growth. *New Phytologist* 25:139-147.

Sylvia D.M. and Chellemi D.O. (2001). Interactions among root-inhabiting fungi and their implications for biological control of root pathogens. *Advances in Agronomy* 73:1-33.

Sylvia D.M. and Hubbell D.H. (1986). Growth and sporulation of vesicular-arbuscular mycorrhizal fungi in aeroponic and membrane systems. *Symbiosis* 1:259-267.

Sylvia D.M. and Williams S.E. (1992). Vesicular-arbuscular mycorrhizae and environmental stress. In G.J. Bethlenfalvay and R.G. Linderman (eds.), *Mycorrhizae in sustainable agriculture* (pp. 101-124). Madison, WI: ASA/CSSA/SSSA.

Tawaraya K., Watanabe S., Yoshida E., and Wagatsuma T. (1996). Effect of onion *(Allium cepa)* root exudates on hyphal growth of *Gigaspora margarita*. *Mycorrhiza* 6:57-59.

Thomson B.D., Robson A.D., and Abbott L.K. (1990). Mycorrhizas formed by *Gigaspora calospora* and *Glomus fasciculatum* on subterranean clover in relation to soluble carbohydrate concentrations in roots. *New Phytologist* 114:217-225.

Tisdall J.M. (1994). Possible role of soil microorganisms in aggregation in soils. *Plant and Soil* 159:115-121.

Tiwari P. and Adholeya A. (2002). In vitro co-culture of two AMF isolates *Gigaspora margarita* and *Glomus intraradices* on Ri T-DNA transformed roots. *FEMS Microbiology Letters* 206:39-43.

Tommerup I.C. and Kidby D.K. (1979). Preservation of spores of vesicular-arbuscular endophytes by L-drying. *Applied Environmental Microbiology* 37: 831-835.

Trappe J.M. (1987). Phylogenetic and ecologic aspects of mycotrophy in the angiosperms from an evolutionary standpoint. In G.R. Safir (ed.), *Ecophysiology of VA mycorrhizal plants* (pp. 5-25). Boca Raton, FL: CRC Press.

Trappe J.M., Molina R., and Castellano M. (1984). Reactions of mycorrhizal fungi and mycorrhiza formation to pesticides. *Annual Review of Phytopathology* 22:331-359.

Wang H., Parent S., Gosselin A., and Desjardins Y. (1993). Vesicular-arbuscular mycorrhizal peat-based substrates enhance symbiosis establishment and growth of three micropropagated species. *Journal of the American Society of Horticultural Science Alexandria* 118:896-901.

Zambolim L., Reis M.A. Dos, and Costa L.M. (1992). Substratos para multiplicação de inóculo do fungo micorrízico vesículo-arbuscular *Glomus etunicatum. Fitopatologia Brasileira* 17:28-31.

Chapter 15

Status of Endomycorrhizal (AMF) Biofertilizer in the Global Market

Polachi Raja

INTRODUCTION

The term *mycorrhiza* (fungus-root) was first coined by a German scientist to describe the symbiotic association of specialized soil fungi with plant roots (Frank, 1885). He also realized that the association was natural and probably beneficial to the host plant. Mycorrhizae were not considered by the scientific community to be more than a curiosity until the 1950s. Harley's book, *The Biology of Mycorrhiza,* published in 1959, became a major foundation for the science of these amazing fungi. Since then several articles have been published in this subject (Mosse, 1973). The field has generally been divided in two main types of mycorrhizae: Ectomycorrhizae, which occur on trees such as pines, firs, oaks, eucalyptus, alder, and birches, and endomycorrhizae, which occur on the rest of the plants on earth except members of a few families that do not form any mycorrhizae. The agriculturally produced crop plants that form endomycorrhizae of the vesicular-arbuscular mycorrhiza type (VAM) are now called arbuscular mycorrhizal fungi (AMF).

The importance of mycorrhizal fungi in sustainable agriculture is based on their role as a link between plant and soil (Bethlenfalvay, 1992). As agents of nutrient transport between plant and soil, they have an impact on soil conservation (Elliott and Coleman, 1988) and soil nutrition (Bethlenfalvay and Newton, 1991), as well as on plant nutrition (Reid, 1990). The integration of the soil biota into mycorrhizal symbiosis calls for a new definition of mycorrhizal fungi as colonizers of both host plant and host soil. Recognition of the critical importance of soil in sustainable agriculture (Reganold et al., 1990) provides a new outlook on the study and utilization

The author would like to thank group of MYAGRI company generous support for the prepartion of this manuscript.

of mycorrhizal fungi (Sieverding, 1991). Such a concept would focus on the root and the collective soil biota as the symbionts, with mycorrhizal fungi as a major link between them. More than 100 years of mycorrhizal research has demonstrated that knowledge of the full range of the functions of mycorrhizae is essential for an understanding of ecosystem dynamics (Read, 1991), and yet the selection of suitable fungi for use in particular agricultural applications remains a matter of guesswork (Mosse, 1986).

The practice and philosophy of agriculture is changing from conventional to sustainable approaches (Reganold et al., 1987; Poincelot, 1990). Hopefully, all methods holding promise to maintain and restore the resource base will be explored. An examination of the potential of mycorrhizal biotechnology to balance the requirements of plant yield and those of soil conservation is therefore in order.

The aim of this chapter is to highlight recent world fertilizer consumption rates, alternative farming systems, sustainable agriculture practices, effective microorganisms, and lists of mycorrhizal producers and formulaters in the global market.

WORLD FERTILIZER CONSUMPTION

World fertilizer consumption has continued to increase in recent years. In fact, consumption has increased continuously since records were first kept, except during a hiatus from 1988-1989 to 1993-1994 because of the political and economic turmoil in Eastern Europe that limited its ability to import more. The recovery, which started in 1994-1995, has continued to date. All the three main plant nutrients—nitrogen (N), phosphate (P_2O_5), and potash (K_2O)—moved roughly at random, except that potash fell marginally in 1998-1999 while the others rose. However, from a little further back, in 1980-1981, N is still rising strongly (+36.3 percent in 1998-1999), whereas both phosphate (+4.4 percent) and potash (–9.8 percent) have stagnated. This is not surprising because N is the nutrient most needed by almost all plants and the one they are most responsive to. Fertilizer consumption was once the prerogative of the developed world for the simple reason that only rich countries could afford its use. But with the West curtailing its agricultural production, sometimes even paying farmers not to produce, its need for fertilizer has fallen. On the other hand, the developing world, perennially short of food but slowly growing richer, has been trying very hard to increase its agricultural production and is using more and more fertilizer. As a result, the developing countries in 1999 accounted for 61 percent of world NPK consumption, with developing Asia (excluding Japan) alone consuming 48 percent. Nevertheless, because of the low Western population and densely populated developing countries, per capita consumption is still heavily skewed toward the rich countries.

ALTERNATIVE FARMING SYSTEMS

There are four established techniques to alternative farming in the United States. A general trend in all four schools is an emphasis on the use of biological systems to supply fertility and pest control rather than chemical inputs. Alternative farming includes organic farming, biodynamic farming, biological farming, and nature farming.

Organic farming is the most widely recognized alternative farming system. Modern organic farming evolved as an alternative to chemical agriculture in the 1940s, largely in response to the publications of J. I. Rodale in the United States, Lady Eve Balfour in England, and Sir Albert Howard in India.

In 1980, the U.S. Department of Agriculture released a landmark *Report Titled Report and Recommendations on Organic Farming* in which organic farming was defined as such: Organic farming is a production system that avoids or largely excludes the use of synthetically compounded fertilizers, pesticides, growth regulators, and livestock feed additives. To the maximum extent feasible, organic farming systems rely upon crop rotations, crop residues, animal manures, legumes, green manures, off-farm organic wastes, mechanical cultivation, mineral-bearing rocks, and aspects of biological pest control to maintain soil productivity and tilt, to supply plant nutrients, and to control insects, weeds, and other pests.

Biodynamic farming evolved in Europe in the 1920s following lectures on agriculture by the Austrian anthroposophist Rudolf Steiner. Biodynamic farming parallels organic farming in many ways but places greater emphasis on the integration of animals to relate a closed nutrient cycle, the effect of crop planting dates in relation to the calendar, and the awareness of spiritual forces in nature. A unique feature of this system is the use of eight specific preparations derived from cow manure, silica, and herbal extracts to treat compost piles, soils, and crops.

Nature farming was developed in Japan in the 1930s by Mokichi Okada, who later formed the Mokichi Okada Association (MOA). Nature farming parallels organic farming in many ways but includes a special emphasis on soil health through composts rather than organic fertilizers, when possible. Kyusei Nature Farming, a branch group, emphasizes use of microbial preparations in addition to traditional nature farming. Nature farming is most used in the Pacific Rim, including California and Hawaii.

Biological farming has become synonymous with farmers using the Reams fertility system as the basis for crop production. The Reams system is based on the LaMotte-Morgan soil test and the use of rock phosphate, calcium carbonate, and compost to achieve nutrient ratios of 7:1 calcium to magnesium, 2:1 phosphorus to potassium, and so on. Biological farming allows the use of selected chemical fertilizers (avoiding disruptive materials such as anhydrous ammonia and potassium chloride) and adopts low-input approaches to the use of herbicides and insecticides. The Pandol Brothers, a large commercial fruit and vegetable operation in California, reduced their annual pesticide bill from $500,000 to $50,000 per year after adopting a biological fertility program (Higa, 1991).

SUSTAINABLE AGRICULTURE SYSTEMS

The increase in crop production and land productivity in agriculture is one of the necessary steps to meet the food demand of the growing human population in these areas. This is limited by a series of socioeconomic and edaphoclimatical constraints of the tropics, which are explained elsewhere (Rehm, 1986). Among the soil-related constraints on crop production, low chemical soil fertility is the most severe on more than half of the arable land in the tropics (Sanchez and Salinas, 1981). Soil crop production can only be attained when limited amounts of fertilizers in either organic or inorganic form are applied and soil amendments are combined with improved crop production technologies. With this in mind, a number of soil management technologies have been proposed. On more fertile soils the main problem is to maintain soil productivity through suitable crop production systems.

Agricultural research in the past decade has shown that crop production constraints in the tropics are most likely to be circumvented by technologies based on biological processes. Therefore, the efforts of international agricultural research centers are concentrated on the development of improved crops by using biologically based crop production systems. Emphasis is laid on enhanced production through increased resource productivity on farms with limited resources and on underutilized land areas. It is recognized that, on a long-term basis, soil productivity can be maintained or increased only via biological technologies and in this way food can be produced at lower costs per unit.

Sustainable agriculture is an essential element of the overall effort to make human activities compatible with the demands of the earth's ecosystem (Sanchez and Salinas, 1981). Thus, an understanding of the different approaches to ecological agriculture is necessary if we want to use the planet's resources wisely. Although sustainable agriculture is based on long-term goals and not a specific set of farming practices, it is usually accompanied by a reduction of purchased inputs in favor of managing on-farm resources. A good example is reliance on biologically fixed nitrogen from legumes versus manufactured nitrogen fertilizers. Low-input agriculture is one of several alternative-farming systems whose methods are adaptable to sustainable agriculture.

Low-input farming is based on a reduction—but not necessarily elimination—of chemical fertilizers, insecticides, and herbicides. Farmers are adopting these practices not only to reduce the costs but also to minimize impact on the environment or because they perceive future pesticide regulations.

In a search for information on how to farm with fewer chemicals, it is helpful to examine alternative farming systems in existence that largely exclude chemicals in favor of biological farming practices. Experiences of producers who have successfully practiced these methods are valuable to farmers considering a transition to low-input sustainable agriculture.

Evolution of Sustainable Agriculture

In the 1960s and 1970s, a growing environmental agriculture movement evolved in response to increasing soil erosion, pesticide use, and groundwater contamination (Dudal, 1976). Simultaneously, economic conditions for farmers were becoming more stressful and the number of family farms declined.

In 1989, the National Research Council of the National Academy of Sciences issued a highly significant report on alternative agriculture, which was defined as a system of food and fiber production that applies management skills and information to reduce costs, improve efficiency, and maintain production levels through such practices as crop rotations, proper integration of crops and livestock, nitrogen-fixing legumes, integrated pest management, conservation tillage, and recycling of on-farm wastes as soil conditioner and biofertilzers.

Again, in 1993 the National Academy of Sciences left no doubt as to these earlier concerns when the National Research Council released a report, *Pesticides in the Diets of Infants and Children,* which concluded that people in this age group could be at considerable health risk from consumption of foods containing pesticide residues.

Both of these reports have raised considerable speculation about the future of our chemical-based agricultural production systems. A growing consensus of consumers, environmentalists, legislators, and many farmers is that our current farming practices must change considerably to achieve a significant reduction in pesticide usage in global agriculture. The ultimate goal of sustainable agriculture, according to the National Research Council and other sources as well, is to develop farming systems that are productive, profitable, energy conserving, environmentally sound, conserving of natural resources, and ensure food safety and quality. Consequently, the leading question that U.S. farmers are asking is, "How can I make these changes, reduce my chemical inputs, and achieve an acceptable level of economic and environmental sustainability?" (Higa, 1995).

A successful transition from chemical-based farming systems to more sustainable agriculture will largely depend on what farmers do to improve and maintain the quality of their agricultural soils. Indeed, soil quality is the key to sustainable agriculture. Not surprisingly, the alternative agricultural practices advocated by the National Research Council are mainly those that can improve and maintain soil quality. Experience has shown that the transition from conventional agriculture to nature farming or organic farming can involve certain risks, such as lower yields initially and increased pest problems. Once through the transition period, which might take several years, most farmers will find their new farming systems to be stable, productive, and manageable, using low-input chemical fertilizers, and profitable without pesticides.

EFFECTIVE MICROORGANISMS

The remarkable microorganisms and their often unpredictable nature and biosynthetic capabilities, given a specific set of environmental and cultural conditions, has made them likely candidates for solving particularly difficult problems in the life sciences and other fields as well. The various ways in which microorganisms have been used over the past 50 years to advance medical technology, human and animal health, food processing, food safety and quality, genetic engineering, environmental protection, agricultural biotechnology, and more effective treatment of agricultural and municipal wastes provide a most impressive record of improvement. Many of these technological advances would not have been possible using straightforward chemical and physical engineering methods or, if they were, they would not have been practically or economically feasible.

Microbial technologies have been applied to various agricultural and environmental problems with considerable success in recent years. They have not been widely accepted by the scientific community because it is often difficult to consistently reproduce their beneficial effects (Nishio, 1996). Microorganisms are effective only when they are presented with suitable and optimum conditions for metabolizing their substrates, including available water, oxygen, pH, and temperature of their environment. Meanwhile, the various types of microbial cultures and inoculants available on the market today have rapidly increased because of these new technologies. Significant achievements are being made in systems where technical guidance is coordinated with the marketing of microbial products. Since microorganisms are useful in eliminating problems associated with the use of chemical fertilizers and pesticides, they are now widely applied in nature farming and organic agriculture (Higa, 1991; Parr et al., 1994).

For many years, soil microbiologists and microbial ecologists have tended to differentiate soil microorganisms as beneficial or harmful according to their functions and how they affect soil quality, plant growth, yield, and plant health. Beneficial microorganisms are those that can fix atmospheric nitrogen, decompose organic wastes and residues, detoxify pesticides, suppress plant diseases and soilborne pathogens, enhance nutrient cycling, and produce bioactive compounds such as vitamins, hormones, and enzymes that stimulate plant growth. Harmful microorganisms are those that can induce plant diseases, stimulate soilborne pathogens, immobilize nutrients, and produce toxic and putrescent substances that adversely affect plant growth and health (Marumoto et al., 1996).

A more specific classification of beneficial microorganisms has been suggested by Higa (1991, 1994, 1995), which he refers to as "effective microorganisms." His reports present some new perspectives on the role and application of beneficial microorganisms, including EM, as microbial inoculants for shifting the soil microbiological equilibrium in ways that can improve soil quality, enhance crop production and protection, conserve natural resources, and ultimately create a more sustainable agriculture and environment. The reports also discuss strategies on how beneficial microorganisms, including EM, can be more effective after inoculation into soils.

EVOLUTION OF AMF

In the early 1950s, there was a controversy on AMF that could be observed in roots or their spores in soil but could not be cultured. Their study was greatly hindered until spores were used to establish "pure" pot cultures that resulted in a few spores multiplying into thousands under controlled conditions of living plant roots. This observation that AMF fungi were obligate symbionts, i.e., they cannot be cultured artificially, has limited studies to those not requiring culturing on aseptic, artificial media (Hepper, 1984; Raja and Mahadevan, 1991). However, the spores were large enough to be physically extracted from soil by sieving and starting new cultures. These studies became the cornerstone for nearly all subsequent studies on AMF. A number of earlier studies also demonstrated that AMF made plants grow better, largely due to increased phosphorus uptake from soil (Bolan, 1991). During the 1960s and 1970s, many research papers reported the increased growth of plants with AMF compared to control, but most of these studies were done in greenhouse pot experiments where the soil or potting mix had been sterilized or fumigated to eliminate indigenous microbes that might have confounded the results (Fitter, 1986; Abbot and Robson, 1991). Though it is difficult to compare the plants with and without AMF in the field due to colonization of all treatments by indigenous AMF, when indigenous AMF have been eliminated by soil fumigation, growth effects from inoculation have been astounding, especially in soils with low P or high P fixing capacity, or where some other soil factor limits plant growth (Sieverding, 1991; Barea and Jeffries, 1994). Plants with coarse, nonfibrous roots, such as citrus and grapes, were shown to be the most AMF dependent. Also, plants that were inoculated with AMF before transplant exhibited better survival and faster regrowth in the field.

From 1960s to the present the number of AMF species increased from just a few to more than 150, and still several more are being discovered. All the research to date indicates that AM fungi exhibit very little host specificity. It was shown that some isolates or strains of species function more effectively on some plants in some soils than do other isolates or strains. It was shown in the early 1980s that the effectiveness of AMF was highly correlated to the level of fungal hyphae that extend out from the roots into the bulk soil and scavenge water and nutrients from places that roots could never reach (Visser, 1985; Schroth and Weinhold, 1986). Thus, AMF can increase the capacity for absorption, especially for immobile elements such as P, Cu, and Zn, remarkably (Fitter, 1991).

The benefits of AM fungi are not only limited to improved uptake of nutrients and water from soil. In recent years, it has been shown that these fungi interact with other soil microorganisms with the same special positive attitude. For example, it has been shown that the conversions of atmospheric nitrogen from a gas into a form of fertilizer nitrogen (biological nitrogen fixation) by soil bacteria can be greatly enhanced by the presence of AMF (Jakobsen and Rosendahl, 1990; Camel et al., 1991; Finlay and Sonderstrom, 1992). Myriad interactions can occur in the soil around roots with AMF, but recent results indicate that microbial associations with AMF play significant roles in the benefits observed by inoculation with AM fungi. Establishing the mycorrhiza "team" on plant roots in early stages of the growth cycle is clearly the right approach for exploiting these remarkable fungi in agriculture. Inoculation with mycorrhizal fungi and their microbial associates can maximize crop yields and quality while minimizing application of chemical fertilizers and pesticides that can be harmful to people and the environment.

AMF Morphology and Nutrition

Arbuscular mycorrhizal fungi establish a network of microscopic hyphal filaments in the soil. When these filaments come in contact with a young root, they thread their way between the cortical cells and quickly propagate, forming intracellular arbuscules and in some cases intercellular vesicles. These arbuscular fungi are so called because of the treelike structures that are differentiated in the roots. Spores are also differentiated in the soil and roots. They act as reserve propagation organs and as a reference structure for species identification. Thus far we have no serious clues as to the sexuality of AMF. Their association with the Zygomycetes is based on the similarity of their spores to the spores of other known representatives of this class.

The close association created between plant and fungus through the hyphal network allows the exchange of nutrients for the survival and growth of the two partners. First, the wide dispersal of the fungus in the soil through its large filament network gives the fungus access to a much larger volume of soil than the root system itself. The fungal hyphae act more or less as a pump, supplying the root with a supplement of water and mineral salts to which it normally would not have access. In return, the fungus receives metabolized nutrients from the plant which it is unable to synthesize by itself, such as sugars, amino acids, and secondary metabolites.

AMF Taxonomy

AMF belong to a very old category of fungus, the Zygomycetes, and have been recently regrouped in a single order, the Glomales (Morton and Benny, 1990), which include all species capable of living in symbiosis with plants. The bulk of known species belong to the family Glomaceae (Pirozynski and Dalpe, 1989), which includes the genera *Glomus* and *Sclerocystis*. The AMF consist of approximately 155 species belonging to 3 families and 6 genera and have a worldwide distribution (Newsham et al., 1995). The bulk of known species have been described over the past two decades, which indicates the increased interest in these organisms and also the difficulty inherent in their taxonomic treatment. In fact, the greatest difficulty is that the entire taxonomy of these organisms is currently based on the morphological characteristics of the spores. They are single-celled structures, of generally globoid shape, with thick walls made up of several layers of different textures, connected to the filamentous network by a suspensor hypha of varied morphology. Since the morphological characters are reduced and often variable depending on the maturity of the spores studied, ultrastructural studies serve to support observations made previously in optic microscopy. In addition, although still exploratory, molecular approaches can be used to detect, with the help of specific probes and only for some species, the qualitative and quantitative presence of a fungus directly in its substrate.

TECHNIQUES OF AMF INOCULUM PROPAGATION

A major obstacle that slows work on biodiversity and the utilization of these soil microorganisms is their obligate symbiont status. These fungi can grow only in the presence of a living plant, which means that their propagation requires the use of pot cultures, either in greenhouses or in growth chambers. To overcome this barrier, an in vitro culture method on excised roots was developed, but only half a dozen strains are currently available in monaxenic culture. This fairly recent methodology, though, represents a very promising avenue for the development of studies on biodiversity, taxonomy, ontogeny of spores, and verification of the mycorrhizal potential of the isolated AMF strains. One favorable and most positive point is that the AMF are not specific in terms of the partner plant they choose, which means that the same fungus can be grown on a large number of plant species.

Soil Pot Culture

Sterilized soil has been regularly used for AMF pot culture technique to increase spore counts with the help of a highly susceptible host plant (Bagyaraj and Manjunath 1980; Menge, 1984). The host plant sprouts the massive roots and tolerates the high light conditions required for rapid AMF production. Several host plants such as *Allium cepa, Zea mays, Sorghum bicolor, Paspalum notatum, Cenchrus ciliaris, Trifolium subterraneum, Panicum maximum, Coleus* sp., and *Coprosma robusta* have been used for their suitability to propagate AMF inoculum (Bagyaraj, 1992). *Chloris gayana* is the best host for mass multiplication of *Glomus fasciculatum* was reported (Sreenivasa and Bagyaraj, 1988a). The purpose of soil sterilization is to kill the existing AM fungi, pathogenic organisms, and weed seeds. Ferguson and Menge (1982) found that using large containers may give a higher spore count. Light quality and irradiant soil water contend and temperature influence root colonization and spore production (Menge, 1984). Sporulation is positively correlated with temperature from 15 to 30ºC for many AM fungi and hence a warm growth environment should be maintained (Schenck and Smith, 1982).

Soilless Pot Culture

Soilless media avoids weed seeds and pathogenic organisms, and provides uniform physical and chemical characteristics of this propagation system. The soilless mixture should hold sufficient water for plant growth but should also have good aeration. Vermiculite, bark, calcined clay, expanded clay, and perlite provide good aeration. Peat and vermiculite hold more water but allow air to penetrate better than sand. Some of the strategies for management in soilless media are the addition of suitable soluble nutrients solutions (Douds and Schenck, 1990) and incorporation of slow-release fertilizers (Coltman et al., 1988). The commercial preparation with perlite, vermiculite, and peat moss (1:1:1) is the best substrate for mass propagation of *Glomus fasciculatum* (Sreenivasa and Bagyaraj, 1988b). Jarstfer and Sylvia (1992) cultured 15 species of AM fungi of *Acaulospora, Gigaspora, Glomus,* and *Scutellospora* in different soilless media.

Aeroponic and Hydroponic Culture

A fine mist of suitable nutrient solution that sprays the roots of the host plant is termed *aeroponic culture* (Zobel et al., 1976). Aeroponic propagation systems were adapted for growing AM fungi (Hung and Sylvia, 1988). In general plants are inoculated in sand or vermiculite before they are transferred to these systems. Plants have also been inoculated directly in the aeroponic system (Hung et al., 1991). Higher spore count has been produced in an aeroponic system than in that of soil-based pot cultures. Since no substrate is present with the inoculum, with aeroponic culture of roots it is possible to produce inoculum with hundreds of thousands of propagules per dry gram of roots (Sylvia and Jarstfer, 1992). The AMF inoculum can be produced using the nutrient-film technique (NFT) by growing precolonized plants in a suitable nutrient solution, which flows over the host plant roots (Elmes and Mosse, 1984; Mosse and Thompson, 1984).

Root Organ Culture

AM fungi have not been successfully cultured on artificial media. AM fungal spores are easily germinated, but hyphae cannot be cultured. The various tissue-culture techniques have been established and proposed the contaminant-free AMF inoculum. The Ri T-DNA transformed root cultures offer an efficient method for growing AMF fungal colonized roots (Mosse and Hepper, 1975; Mugnier and Mosse, 1987; Becard and Fortin, 1988; Chabot et al., 1992). However, there are some fundamental obstacles, for example, conditions which stimulated root growth suppressed the germination of spores in agar to be overcome before this technique would culminate an economical technology for mass production of AMF inocula. Raja and Mahadevan (1991) have reviewed the prospects of this technique in the mass multiplication of AMF.

Biodiversity and Agricultural Potential

The fact that colonized plants are better able to obtain their nourishment in the soil and resist environmental stresses gives fungal symbionts a biofertilizing and crop protection role. In agriculture, the increased uptake of soil minerals by colonized plants means that it is possible to consider substantially reducing applications of fertilizers and pesticides and at the same time obtain equivalent or even higher crop yields (Abbott and Robson, 1991). Through appropriate management of mycorrhizae in agriculture, it is also possible to maintain soil quality and sustainability while protecting the environment over the long term and reducing costs of production.

AMF BENEFITS OF SYMBIOSIS

The colonized plant is better nourished and better adapted to its environment. It obtains increased protection against environmental stresses (Sylvia and Williams, 1992), including drought (Subramanian et al., 1995), cold (Charest et al., 1993; Paradis et al., 1995), salinity (Davis and Young, 1985), and pollution (Leyval et al., 1994; Shetty et al., 1995). In addition, symbiosis tends to reduce the incidence of root diseases and minimizes the harmful effect of certain pathogenic agents (Dehne, 1982). By and large, the growth and health of colonized plants is improved. At the same time they obtain increased protection against environmental conditions detrimental to their survival.

IMPACT OF AMF IN AGRICULTURE

AMF biotechnology in agriculture can gradually be integrated with conventional agricultural practices. Over the medium term, it is quite possible to restore soil quality and improve soil sustainability. It has been shown that appropriate management of mycorrhizae in agriculture allows a substantial reduction in the use of chemicals, thus lessening the level of pollution of surface waters and reducing farm labor and costs of production, while maintaining yields at their highest levels. The objective in maintaining and developing a high-quality germplasm bank is to support and accelerate the use of mycorrhizae as biofertilizing agents in agriculture and in turn contribute to the development of a healthy and sustainable environment. The advantages and benefits of adopting mycorrhizae in agriculture allows us to better visualize the scope of this phenomenon at the crop level and, in turn, the impact of its long-term adoption on the quality of life (see Table 15.1).

AVAILABILITY AND COST OF MYCORRHIZAL INOCULUM

The sources for mycorrhizal inoculants are increasing day by day. Several years ago Native Plants Inc., Salt Lake City, Utah, and Jira-Agroindustrias, Tulua, Colombia, were the two sources for mycorrizal inoculum. Recently, several companies in North America began selling it in one form or another. An incomplete list of different types of inoculum available and their suppliers are listed below. Costs can range from US$2 to US$40 per kg of AMF inoculum. The costs and applications vary in trademark of products. AMF inoculation is profitable when the monetary equivalent of the yield response to inoculation is greater than the costs of the inoculation process (Sieverding, 1991). The overall costs of the AMF inoculation technology consists of the price of the inoculum, especially the period of propagation (four to six months), the transport expenses, and the crop application method.

PRODUCERS AND FORMULATORS OF AMF INOCULUM

This information is provided for the convenience of the reader and does not constitute an endorsement of any product by the author of this chapter. Any claim should be independently validated. The producers, formulators, and suppliers of commercial AMF inoculum in the global market are listed in Table 15.2.

FUTURE PERSPECTIVES

Future needs are of two major kinds. The mechanism by which AMF influence root systems needs to be understood at a basic level so that root system design and modification can be used as a management tool, especially within intensive horticultural systems. In addition, the use of AMF within

TABLE 15.1. Advantages and benefits of adopting AMF in an agriculture system.

Field	Advantages	Benefits
Agriculture	Agrosystem stability	Improved plant roots, soil microflora and the abiotic geochemical soil matrix
	Plant production	Increase shoot and root biomass
	Stress resistance	Drought, cold, and pollution
	Resistance to pathogens	Crop protection, reduced use of pesticides, fungicides, improvement in animal, plant, human health and product quality
	Value added to product	Increased synthesis of primary and secondary metabolites
Plant physiology	Improvement in nutrition	Reduced chemical fertilizers (10 to 30 percent)
	Carbohydrate level	Improved photosynthetic activity
	Tolerance to water stress	Cultivation of arid soils or soils unfit for agriculture and altered hormonal balance
	Resistance to low temperatures	Diversity of crops in inhospitable areas
Plant morphology	Transformation of root system and root architecture	Adaptation to stress, increased resistance to erosion, fixation of soils, increased numerous and branched roots, root apices and improved production with certain root crops
Plant community	Microorganism diversity in the subsoil	Reestablishment of soil microflora and improvement of soil quality
	Habitat restoration	Stabilize the biogeochemical cycling
	Survival of partners	Improvement in yields, better acclimatization at transplanting and diversity of plant cover
Micropropagation plants	High-value plants	Uniform size, better growth, shorten nursery period

sustainable systems needs to be better understood in relation to nutrient cycling.

It is important that AMF should not be seen as the single and final solution for the control of plant pathogens, chemical fertilizers, pesticides, and insecticides but as part of an integrated approach. In open systems manage-

TABLE 15.2. Current global mycorrhizal product.

Name of company and address	Product name	Benefits
Accelerator Horticultural Products	—	—
AgBio Inc 9915 Raleigh St., Westminster, CO 80030	AgBio-Endos	• Increase yields • Improve transplant success • Reduce effects from drought, high salts, disease
Becker-Underwood Inc 801 Dayton Avenue Ames, IA 50010	Rhizanova	• Reduced tree loss • Improved growth rates • Enhanced adaptability • Drought resistance • Increased flowering • Better water and nutrient uptake
Bio-Organics 53606 Bridge Drive LaPine, Oregan 97739	Endomycorrhizal inoculant	• Better growth • Increase yield • Reduced the transplant shock
Biological Crop Protection Ltd Wye, Ashford, Kent, UK	—	—
Biorize Dijon, France	Endorize	• Better nutrition • Better rooting • Pyto-protection
BioScientific Inc 4405 S. Litchfield Road Avondale, Arizona 85323	BuRize	• Nutrient and water uptake • Root health and disease resistance • Tolerance to environmental stresses
BioTerra Technologies Inc 9491 W. Pioneer Avenue Las Vegas, NV 89117	Bioterra Plus	• 100 percent natural and organic • Accelerates and increases plant growth • Increases plant yield • Generates more vigorous soil • Reduces expensive chemical need and increases profits • Significantly improves transplant success • Increases absorption of water and nutrients • Promotes plant tolerance to disease, drought and salinity
BioTrack Technology (P) Ltd., 325/36 Kathiravan Colony Anna Nagar West Chennai, Tamil Nadu India	RHIZAgold	• Increases nutrient and water uptake • Improves soil and plant ecosystem • Increases plant growth and establishment • Reduces transplant stress and plant loss • Improves soil structure and porosity • Reduces chemical fertilizer • Improves drought tolerance • Improvd ressistance to soil borne pathogens
Central Glass Co., Ltd. Kowa-Hitotsubashi Building 7-1 Kanda-Nishikicho3-Chome Chiyoda Ku Tokyo, 101-0054 Japan	Cerakinkong	• The absorption of phosphoric acid is improved • Promote faster rooting to raise healthy seedlings • Improve rate of successful rooting at the time of planting • A plant is helped with the growth and it improves the yield.

TABLE 15.2. *(continued)*

Name of company and address	Product name	Benefits
EcoLife Corporation Moorpark, CA	—	—
First Fruits Triadelphia WV	—	—
Global Horticare Lelystad, the Netherlands	—	—
Horticultural Alliance Inc		
Idemitsu Kosan Co. Japan	—	—
J.H. Biotech Inc Ventura, CA	—	—
Malaysian Agri Hi-Tech Sdn., Bhd., Block: Stroma UKM-MTDC Smart Technology Centre, 43600 Bangi, Selangor Malaysia	Mycogold Myco V	• Increase efficiency in fertilizer application • Rehabilitation and prevention from soil borne pathogen • Tolerance to water stress • Increase quality of yield • Promote beneficial microorganisms • Promote maximization of plant potential
MicroBio Ltd Royston, Herts, United Kingdom	—	—
Mikro-Tek Inc Timmins, Ontario, Canada	—	—
Mycorrhizal applications Inc PO Box 1181, Grants Pass Oregon, 97528	BioGrow	• Improved nutrient and water uptake • Improved root growth • Improved plant growth and yield • Improved disease resistance • Reduced transplant shock • Reduced drought stress
Osaka Gas Co. Ltd.,	—	—
Plant Health Care Inc 440 William Pitt Way Pittsburgh, PA 15238	Mycor	• Better survival and growth of new plantings • Reduces risk of stress damage due to heat and drought • Reduces labor and replacement costs • Less dead-heading • Increase feeder root growth capacity • Improves soil borne disease and insect resistance
PlantWorks Ltd 1/19 Innovation Buildings Sittingbourne Research Center Sittingbourne, Kent, ME9 8HL United Kingdom	TerraVital	• Reduce plant mortality • Reduce plant maintenance costs • Increase tolerance to water stress • Increase plant biodiversity • Increase plant growth • Increase tolerance to root pathogens • Increase soil stability

TABLE 15.2. *(continued)*

Name of company and address	Product name	Benefits
Premier Horticulture	Pro-Mix 'BX'	• Minimizes fertilizer costs upto 30 percent • Improves crop uniformity • Reduces fungicide usage and plant root diseases • Reduces plants culls and losses • Helps overcome difficult growing conditions • Provides superior quality plants
Premier Tech Quebec, Canada	Pro-Mix 'BX'	• Minimizes fertilizer costs upto 30 percent • Improves crop uniformity • Reduces fungicide usage and plant root diseases • Reduces plants culls and losses • Helps overcome difficult growing conditions • Provides superior quality plants
Reforestation Technologies International 1341 Dayton Street, Unit G Salinas, CA 93901	AM120	• Enhanced soil nutrient uptake • Increased drought and pest tolerance • Increased tolerance to toxic heavy metals • Improved soil aggregation • Enhance biodiversity of soil microorganisms
Roots Inc Independence, MO		
T&J Enterprises 2328 W. Providence Ave. Spokane, Washington 99205	BioVam	• Increased yields • Faster growing plants • Reduction in transplant shock • Higher resistance to heat, cold and drought • Prolific, vibrant growth with enhanced Coors • Reduced spoilage in fruits and vegetables and longer lasting flowers • Plants that have higher natural resistant to diseases and destructive insects
TIPCO Inc, USA	Diehard	
Tree of Life Nursery, USA	Grow life	
Tree Pro	Mycorrhizal plant inoculant and root dip	• Improves soil and plant ecosystem • Increases plant establishment and growth • Reduces transplanting stress and plant loss • Increases nutrient and water uptake • Improves oil structure and porosity • Reduces fertilizer use
Triton Umweltschutz GmbH	Tri-Ton	• Highly improved provision of nutrients • An improved tolerance of the plant against drought stress, unfavorable pH values and soils with high salinity • Modified attitude of the plants in resisting the attacks of the excitants of diseases

ment is likely to be the way forward, with practices shaped by research aimed at a greater understanding of the role of AMF in plant-soil systems. However, in closed systems inoculation is likely to be the way forward for many cases. For many years difficulties associated with inoculum production have prevented large-scale use. However, inoculum is now available commercially throughout Western Europe, Japan, the United States, Malaysia, India, and South America. It is certain that only by adopting these approaches it is likely that the benefits of AMF to agricultural production will be optimized, leading to a reduction in chemical inputs and increased sustainability.

CONCLUSIONS

Beneficial effects of mycorrhizae, including plant growth, are well established in controlled conditions and field trials. Nevertheless, although frequently reported as mycorrhizal effects, it should not be excluded that other microorganisms may also be involved in such positive interactions. Mycorrhizae are commonly defined as plant-fungus associations; in functional terms, however, they may better be regarded as a value regulating plant-soil interaction, possibly increasing carbon sink to roots and carbon efflux to soil (Garbaye, 1991). Their management in consideration of other soil microorganisms is a challenge and an opportunity for sustainable plant-soil systems (Staley et al., 1992).

Further research is necessary, both at a basic and applied level, i.e., regarding the maintenance of inoculated organisms in the field. It is worth noting, however, that the number of field experiments is increasing every year. It is therefore to be hoped that, not withstanding the difficulties, and even some failures and contradictions, this body of experience will soon allow a wider utilization of mycorrhizas and associated organisms in plant generations.

REFERENCES

Abbott, L.K. and A.D. Robson (1991). The effect of root density, inoculum placement and infectivity of inoculum on the development of vesicular-arbuscular mycorrhizas. *Phytologist* 97: 285-299.

Bagyaraj, D.J. (1992). Vesicular arbuscular mycorrhiza: Application in agriculture. In J.R. Norris, D.J. Read, and A.K. Varma. (eds.), *Methods in microbiology,* Volume 24. *Techniques for the study of mycorrhiza* (pp. 359-373). London: Academic Press.

Bagyaraj, D.J. and A. Manjunath (1980). Selection of a suitable host for mass production of vesicular arbuscular mycorrhizal inoculum. *Plant and Soil* 55: 495-498.

Barea, J.M. and P. Jeffries (1994). Arbuscular mycorrhizas in sustainable plant-soil systems. In A. Varma and B. Hock (eds.), *Mycorrhiza: Function, molecular biology and biotechnology* (pp. 35-48). Heidelberg: Pringer-Verlag.

Becard, G. and J.A. Fortin (1988). Early events of vesicular arbuscular mycorrhiza formation on RiT-DNA transformed roots. *New phytologist* 108: 211-218.

Bethlenfalvay, G.W. (1992). Mycorrhizae and crop productivity. In G.J. Bethenfalvay and R.G. Linderman (eds.), *Mycorrhizae in sustainable agriculture* (pp. 1-27). Madison, WI: American Society of Agronomy.

Bethlenfalvay, G.J. and W.E. Newton (1991). Agro-ecological aspects of the mycorrhizal nitrogen fixing legume symbiosis. In D.L. Keister and P.B. Cregan (eds.), *The rhizosphere and plant growth: Proceedings of a symposium* (pp. 349-354). Dordrecht, the Netherlands: Kluwer Academic Publishers.

Bolan, N.S. (1991). A critical review on the role of mycorrhizal fungi in the uptake of phosphorus by plants. *Plant and Soil* 134: 189-207.

Camel, S.B., M.G. Reyes-Solis., R. Ferrera-Cerrat, R.L. Franson, M.S. Brown, and G.J. Bethlenfalvay (1991). Growth of vesicular-arbuscular mycorrhizal mycelium through bulk soil. *Soil Science Society American Journal* 55: 389-393.

Chabot, S., G. Becard, and Y. Piche (1992). Life cycle of *Glomus intraradix* in root organ culture. *Mycologia* 83: 315-321.

Charest, C., Y. Dalpe, and A. Brown (1993). The effect of vesicular-arbuscular mycorrhizae and chilling on two hybrids of *Zea mays* L. *Mycorrhizae* 4: 89-92.

Coltman, R.R., D.R. Waterer, and R.S. Huang (1988). A simple method for production of *Glomus aggregatum* inoculum using controlled-release fertilizer. *Horticulture Science* 23: 213-215.

Davis, E.A. and J.L. Young (1985). Endomycorrhizal colonization of glass-house grown wheat as influenced by fertilizer salts when banded or soil-mixed. *Canadian Journal of Botany* 63: 1199-1203.

Dehne, H.W. (1982). Interaction between vesicular arbuscular mycorrhizal fungi and plant pathogens. *Phytopathology* 72: 1115-1119.

Douds, D.D. and N.C. Schenck (1990). Increased sporulation of vesicular arbuscular mycorrhizal fungi by manipulation of nutrient regimes. *Applied Environmental Microbiology* 56: 413-418.

Dudal, R. (1976). Inventory of the major soils of the world with special refeences to mineral stress hazards. In M.J. Wright (ed.), *Plant adaptation to mineral stress in problem soils* (pp. 3-13). Ithaca, NY: Cornell University.

Elliott, E.T. and D.C. Coleman (1988). Let the soil work for us. *Ecology Bulletin* 39: 23-32.

Elmes, R.P. and B. Mosse (1984). Vesicular–arbuscular endomycorrhizal inoculum production. II. Experiments with main *(Zea mays)* and another hosts in nutrient flow culture. *Canadian Journal of Botany* 62: 1531-1536.

Ferguson, J.J. and J.A. Menge (1982). Factors that affect production of endomycorrhizal inoculum. *Proceedings Florida State Horticultural Society* 95: 35-39.

Finlay, R.D. and B. Sonderstrom (1992). Mycorrhiza and carbon flow to the soil. In M.F. Allen (ed.), *Mycorrhizal functioning* (pp. 134-160). New York: Chapman and Hall.

Fitter, A.H. (1986). Effect of benomyl on leaf phosphorus concentration in alpine grasslands: A test of mycorrhizal benefit. *New Phytologist* 103: 767-776.

Fitter, A.H. (1991). Costs and benefits of mycorrhizas: Implications and functioning under natural conditions. *Experientia* 47: 350-355.

Frank, A.B. (1885). Ueber die aufWurzelsymbiose beruhende Ernahrung gewisser Baume durch unterirdische Pilze, Ber. *Dtsch. Bot. Ges.* 3: 128-145.

Garbaye J. (1991). Biological interaction in the mycorrhizosphere. *Experentia* 47: 370-375.

Hepper, C.M. (1984). Isolation and culture of VA mycorrhizal (VAM) fungi. In C.L. Powell and D.J. Bagyaraj (eds.), *VA mycorrhiza* (pp. 95-112). Boca Raton, FL: CRC Press.

Higa, T. (1991). Effective microorganisms: A biotechnology for mankind. In J.F. Parr, S.B. Hornick, and C.E. Whitman (eds.), *Proceedings of the First International Conference on Kyusei Nature Farming* (pp. 8-14). Washington, DC: U.S. Department of Agriculture.

Higa, T. (1994). Effective microorganisms: A new dimension for nature farming. In J.F. Parr, S.B. Hornick, and C.E. Whitman (eds.), *Proceedings of the First International Conference on Kyusei Nature Farming* (pp. 20-22). Washington, DC: U.S. Department of Agriculture.

Higa, T. (1995). Effective microorganisms: Their role in Kyusei nature farming and sustainable agriculture. In J.F. Parr, S.B. Hornick, and C.E. Whitman (eds.), *Proceedings of the First International Conference on Kyusei Nature Farming* (pp. 23-31). Washington, DC: U.S. Department of Agriculture.

Hung, L.L., D.M. O'Keefe, and D.M. Sylvia (1991). Use of a hydrogel as a sticking agent and carrier of vesicular arbuscular mycorrhizal fungi. *Mycological Research* 95: 427-429.

Hung, L.L. and D.M. Sylvia (1988). Production of vesicular arbuscular mycorrhizal fungus inoculum in aeroponic culture. *Applied Environmental Microbiology* 54: 353-357.

Jakobsen, I. and L. Rosendahl (1990). Carbon flow into soil and external hyphae from roots of mycorrhizal cucumber plants. *New Phytologists* 115: 77-83.

Jarstfer, A.G. and D.M. Sylvia (1992). Inoculum production and inoculation strategies for vesicular arbuscular mycorrhizal fungi. In B. Metting (ed.), *Soil microbial technologies: Application in agriculture, forestry and environmental management* (pp. 349-377). New York: Marcel Dekker.

Leyval, C., I.I. Weissenhorn, A. Glashoff, and J. Berthelin (1994). Influence of heavy metal on germination of arbuscular-mycorrhizal fungal spores in soils. *Acta Botanica Gallica* 141: 523-528.

Marumoto, T., H. Okabe, T. Ezaki, M. Nishiyama, and K. Yamamoto (1996). Application of symbiotic microorganisms to soil conservation and reforestation. *Proceedings of International Symposium, Bio-Japan Tokyo* 242-250.

Menge, J.A. (1984). Inoculum production. In C.L. Powell and D.J. Bagyaraj (eds.), *VA mycorrhiza* (pp. 187-203). Boca Raton, FL: CRC Press.
Morton, J.B. and G.L. Benny (1990). Revised classification of arbuscular mycorrhizal fungi. (Zygomycetes): A new order, Glomales, two new suborders, Glomineae and Gigasporineae, with an emendation of Glomaceae. *Mycotaxon* 37: 471-491.
Mosse, B. (1973). Advances in the study of vesicular-arbuscular mycorrhiza. *Annual Review Phytopathology* 11: 171-196.
Mosse, B. (1986). Mycorrhiza in a sustainable agriculture. *Biological Agricultural Horticulture* 3: 191-209.
Mosse, B. and C.M. Hepper (1975). Vesicular arbuscular mycorrhizal infections in root organ cultures. *Physiological Plant Pathology* 5: 215-223.
Mosse, B. and J.P. Thompson (1984). Vesicular arbuscular endomycorrhizal inoculum production. I. Exploratory experiments with beans *(Phaseolus vulgaris)* in nutrient flow culture. *Canadian Journal of Botany* 62: 1523-1530.
Mugnier, J. and B. Mosse (1987). Vesicular arbuscular mycorrhizal infection in transformed root inducing T-DNA roots grown axenically. *Phytopathology* 77: 1045-1050.
Newsham, K.K., A.H. Fitter, and A.R. Watkinson (1995). Multi-functionality and biodiversity in arbuscular mycorrhizas. *Tree* 10: 407-411.
Nishio, M. (1996). Microbial fertilizers in Japan. *Food and Fertilizer Technology Centre* 430: 1-13.
Paradis, R., Y. Dalpe, and C. Charest (1995). The combined effect of arbuscular mycorrhiza and short-term cold exposure on wheat. *New Phytologists* 129: 637-642.
Parr, J.F., S.B. Hornick, and D.D. Kaufman (1994). Use of microbial inoculants and organic fertilizers in agricultural production. In proceedings of the International seminar on the use of microbial and organic fertilizers in agricultural production. *Food and Fertilizer Technology Center, Taipei, Taiwan:* 12-21.
Pirozynski, K.A. and Y. Dalpe (1989). The geological history of the Glomaceae with particular reference to mycorrhizal symbiosis. *Symbiosis* 7: 1-36.
Poincelot, R.P. (1990). Agriculture in transition. *Journal of Sustainable Agriculture* 1: 9-40.
Raja, P. and A. Mahadevan (1991). Axenic cultivation of VAM fungi—A review. *Journal of Plant Science Research* 7: 1-6.
Read, D.J. (1991). Mycorrhizas in ecosystems. *Experientia* 47: 376-391.
Reganold, J.P., L.F. Elliott, and Y.L. Unger (1987). Long-term effects of organic and conventional farming on soil erosion. *Nature(London)* 330: 370-372.
Reganold, J.P., R.I. Papendick, and J.F. Parr (1990). Sustainable agriculture. *Science America* 262: 112-120.
Rehm, S. (1986). Handbuch der landwirtschaft und ernahrung in den entwicklungslandern. Band 3: Grundlagen des pflanzenbaus in den tropen und subtropen (pp. 53-61). Stuttgart: Ulmer.
Reid, C.P.P. (1990). Mycorrhizas. In J.M. Lynch (ed.), *The rhizosphere* (pp. 281-315). Chichester, UK: John Wiley and Sons.

Sanchez, P.A. and J.G. Salinas (1981). Low-input technology for managing oxisols and ultisols in tropical America. *Advances in Agronomy* 34: 279-406.

Schenck, N.C. and G.S. Smith (1982). Response of six species of vesicular-arbuscular mycorrhizal fungi and their effects on soybean at four temperatures. *New Phytologist* 92: 193-201.

Schroth, M.N. and A.R. Weinhold. (1986). Root-colonizing bactgeria and plant health. *Horticulture Science* 21: 1295-1298.

Shetty, K.G., B.A.D. Herick, and A.P. Schwab (1995). Effects of mycorrhizae fertilizer amendments on zinc tolerance of plants. *Environmental Pollution* 88: 307-314.

Sieverding, E. (1991). *Vesicular-arbuscular mycorrhiza management in tropical agrosystems* (pp. 223-246). Eschborn: Deutsche Gesellschaft fur Technische Zusammenarbeit, Technical Cooperation, Federal Republic of Germany.

Sreenivasa, M.N. and D.J. Bagyaraj (1988a). *Chloris gayana* (Rhodes grass), a better host for the production of *Glomus fasciculatum* inoculum. *Plant and Soil* 106: 289-290.

Sreenivasa, M.N. and D.J. Bagyaraj (1988b). Selection of a suitable substrate for mass multiplication of *Glomus fasciculatum. Plant and Soil* 109: 127-132.

Staley, T.E., E.G. Lawrence, and E.L. Nance (1992). Influence of a plant growth promoting pseudomonad and vesicular arbuscular mycorrhizal fungus on alfalfa and birdsfoot trefoil growth and nodulation. *Biology Fertility Soils* 14: 175-180.

Subramanian, K.S., C. Charest, L.M. Dwyer, and R.I. Hamilton (1995). Arbuscular mycorrhizae and water relations in maize under drought stress at tasseling. *New Phytologist* 129: 643-650.

Sylvia, D.M. and A.G. Jarstfer (1992). Sheared-root inocula of vesicular arbuscular mycorrhizal fungi. *Applied Environmental Microbiology* 58: 229-232.

Sylvia, D.M. and S.E. Williams (1992). Vesicular arbuscular mycorrhizae and environmental stress. In G.J. Bethlenfalvay and R.G. Linderman (eds.), *Mycorrhizae in sustainable agriculture,* Special Publication 54 (pp. 101-124). Madison, WI: ASA.

USDA (1980). *Report and recommendations on organic farming.* Washington, DC: U.S. Department of Agriculture.

Visser, S. (1985). Role of the soil invertebrates in determining the composition of soil microbial communities. In A.H. Fitter, D. Atkinson, D.J. Read, and M.B. Usher (eds.), *Ecological interactions in soil, plants, microbes and animals* (pp. 297-317). Boston: Blackwel Scientific Publications.

Zobel, R.W., P. Del Tredici, and J.G. Torrey (1976). Method of growing plants aeroponically. *Plant Physiology* 57: 344-346.

Chapter 16

Role of Cyanobacteria As Biofertilizers: Potentials and Limitations

Pilar Irisarri

INTRODUCTION

Biofertilizers are preparations containing viable forms of beneficial microorganisms intended for seed or soil application. Cyanobacteria are oxygenic photosynthetic bacteria, and some of them are able to fix dinitrogen. They have inhabited much of the surface of the earth for billions of years and today they are responsible for a significant amount of biological nitrogen fixation (BNF). Cyanobacteria's trophic independence made them suitable for use as biofertilizers. Cyanobacteria are widely distributed organisms all over the world and can be found in extreme habitats, from hot springs to artic regions. Among the ecosystems in which they can be found, wet soils provide an ideal environment for cyanobacteria to grow.

The existence of an agronomic potential for cyanobacteria in rice cultivation was recognized by De (1939), who attributed the natural fertility of tropical rice fields to BNF by those organisms. Since then, long-term fertility experiments and nitrogen-fixation measurements have confirmed the importance of cyanobacteria and other nitrogen-fixing organisms in maintaining moderate but consistent rice production in fields receiving no nitrogen fertilizer (Roger, 1991).

Nitrogen (N) is one of the most important crop nutrients, and in countries of the developing world nitrogen fertilizers are not always available or are so only to a limited extent. Increased use of chemical fertilizer is also contributing to the degradation of the biological properties of the soil and, indirectly, environmental conditions. Thus, some ways and means must be developed to minimize the use of chemical fertilizers. Energy crises com-

I am grateful to Dr. P. A. Roger who sent me his bibliographical references. The previously unpublished results from our laboratory mentioned in this chapter were made possible by grants from IFS (C/2747) and INIA-FPTA N° 108.

pelled us to think about using renewable natural resources and BNF. Cyanobacteria contribution to soil fertility in many agroecosystems deserves more attention due to their ecological and economical advantages. Use of certain nitrogen-fixing cyanobacteria as partial substitutes for chemical fertilizer seems to be one of the most important approaches to BNF technology. Hence, cyanobacteria study has been pursued from both theoretical and applied perspectives.

It is well established that the best results of inoculation with cyanobacteria are obtained when indigenous species or strains are applied; exotic species are hardly able to survive when introduced to a habitat of tough competition and zooplankton predation, so they are generally eliminated.

The main limiting factor for a technology allowing the enhancement of the presence of indigenous cyanobacteria is certainly the lack of data about the ecology and environmental elements promoting or modulating the growth increase in biomass. Some limitations for its use as a biofertilizer has been discussed in the present work.

CURRENT STATUS OF CYANOBACTERIAL USE AS BIOFERTILIZERS

Earlier reports indicated that cyanobacteria were not abundant in rice fields and so it was justified to inoculate. The effects of soil inoculation with cyanobacteria on rice yield were first reported by Watanabe et al. (1951), with a 25 percent increase in yield after inoculation of a poorly drained paddy with *Tolypothrix tenuis*.

Floating gelatinous colonies of *Nostoc* were added to some fields in China, with populations developed in ponds and released into paddy fields when the rice was planted (Whitton and Roger, 1989).

Studies on inoculation with *Tolypothrix tenuis* have been discontinued in Japan, but subsequently there have been many reports from India and a limited number from other countries such as the Philippines, Burma, Thailand, Vietnam, Russia (Hashem et al., 1996), and Egypt (Yanni, 1996).

In India, where the technique of “algalization” (inoculation with cyanobacteria) was developed, cyanobacteria were provided to rice farmers by the All India Coordinated Project on Algae (Venkatamaran, 1981).

In Thailand, since 1992, the pilot plant at the Technopolis has supplied more than 20,000 tons of biofertilizer to the Department of Agricultural Extension. The technology is now being transferred from the ministry to the private sector for commercial production.

Most published data of inoculation with cyanobacteria refer to tropical rice fields, which are different in characteristics and agricultural management from temperate ones. BNF is far more diverse and complex in the tropics than under temperate conditions (Balandreau and Roger, 1996). Assays of cyanobacterial inoculation in temperate climates were performed in the United States (Reynaud and Metting, 1988), Spain (Fernández-Valiente et al., 1996), Italy, Argentina (Zaccaro, 2000), and Uruguay (Irisarri et al., 1999).

Rice grows in flooded conditions during part of all the cropping period (lowland rice or deepwater rice) in about 88 percent of rice land (Roger, 1995). Information is scarce about the use of cyanobacteria as a biofertilizer for nonflooded cultures. However, Alfassi et al. (1997) found that *Nostoc commune* and *Anabaena cylindrica* could supply wheat plant (in pots) with atmospheric nitrogen to maintain its vegetative growth.

AVAILABILITY OF FIXED NITROGEN TO RICE PLANTS

The extent to which cyanobacteria may contribute to the N requirements of the rice crop is determined by a number of factors, the more obvious including the standing crop, the rate of nitrogen fixation per unit area, the turnover of the fixed nitrogen, and the extent to which any N released becomes available to the rice plant (Whitton and Roger, 1989).

Nitrogen fixed by cyanobacteria may become available to rice plants only after its release into their surroundings, by mineralization of their intracellular content, and/or as extracellular products. So it was supposed that a successful setup of desired cyanobacteria in the rice fields would be a source of slow release of nitrogen according to plant demands.

Some studies using ^{15}N have shown that the nitrogen fixed by the cyanobacteria is actually taken up by the crop plants. When fresh *Nostoc* inoculum, composed mainly of vegetative cells, was incorporated to the soil, about 38 percent of the ^{15}N was recovered in the first crop instead of 28 percent when dried material was used (Tirol et al., 1982). This pot experiment demonstrated that cyanobacterial N was less available than ammonium sulfate for the first crop. But for the other two crops in the same assay, its availability was very similar, indicating the slow release of cyanobacterial nitrogen, a finding that agrees with the cumulative effect of inoculation. However, its very low C:N ratio (from 5:1 to 6:1) gave a better N availability than that of organic fertilizers such as farmyard manure.

Ghosh and Saha (1997) reported that inoculation significantly increased the release of inorganic N either in soil or in soil solution but such increase rarely exceeded 4 percent of total N gained in the soil system during the vegetative growth stages of the rice plant, when the nutritional requirements of the plants are at their maximum. However, the same authors reported that inoculation caused an increase in nitrogen uptake by the grain (30 percent) and in total uptake when urea was not applied. Inoculation improved the N status of soil by nearly ten times more than that utilized by the crop. These results indicated a transformation of a very small fraction of cyanobacterial fixed N_2 to the plant-available form in the soil system during the vegetative growth period of the crop.

Saha and Mandal (1980) observed that most of the residual N from cyanobacteria remaining in wet soil after the crop harvest did not persist with air drying of the soil. In other assays with cyanobacterial N labeled with ^{15}N, it was mineralized quickly enough to be available during the growing season of a rice crop (Thind and Rowell, 1999). Depending on field conditions, cyanobacteria may have a role in assimilating nitrogen and protecting it from being lost as well as being a major driving force for NH_3 volatilization through diurnal increases in pH. Cyanobacteria may also cause N loss through stimulation of nitrification-denitrification processes as they affect the depth of the aerobic soil layer through their O_2 input, but this needs confirmation (Mandal et al., 1999).

CULTIVATION OF CYANOBACTERIA FOR BIOFERTILIZER

The basic principle is to grow cyanobacteria using natural sunlight under conditions simulating the rice field. The open-air soil culture of inocula is simple and of low cost but has climatic limitations. It was estimated that a 2 m^2 tray might produce enough cyanobacteria to inoculate 1 ha in 2 to 3 months under sunny summer conditions (Roger, 1995).

In India, farmers multiplied the inoculum in shallow tanks with a thin layer of soil covered by water and superphosphate. The "starter culture" recommended was a soil-based mixture of *Aulosira, Tolypothrix, Scytonema, Nostoc, Anabaena,* and *Plectonema*. A copious growth of the inoculated species was expected in a week or two. The capacity of these organisms to stand desiccation made it possible to preserve the product in a sundried form. Under such culture conditions, Roger et al. (1987) observed that the final amount of individual strains in the algal flakes was unpredictable. Soil-based inocula were most frequently unbalanced and generally dominated by one or two strains, mostly a *Nostoc*. They suggested that a suitable method for producing a multistrain balanced inoculum of known quality would be to produce monostrain inocula of various strains, evaluate their concentrations after drying, and mix them according to the colony forming units (CFU) contained to obtain a balanced mixture.

In a two-year assay in Uruguay, fresh laboratory-produced inocula could not proliferate in the field, and inocula produced in tanks with soil and water presented less than 1 percent of heterocystous cyanobacteria. In Japan, nurturing inoculum growth was often difficult (Grant et al., 1986). Roger and Kulasooriya (1980) explained how to produce inoculum under artificially controlled conditions, but large-scale biofertilizer production under these conditions seems difficult to accomplish.

The efficiency of inoculum can be increased if it contains a heavy load of the perennating bodies such as akinetes and hormocysts. Many species of *Anabaena* and *Nostoc* form a long chains of spores, and sometimes the entire trichome gets transformed into spores. A culture containing such organisms will multiply faster when inoculated into the field (Hori et al., 2002).

Fresh inoculum applied at a high density might be more efficient than a dry soil-based inoculum to provide nitrogen to the current crop (Roger, 1991).

Various inert supports have been tested for growth and conservation of inocula—pumice stone, sand, porous gravel, synthetic sponge blocks (2 to 4 cm)—and the material efficiency was verified even after two years of storage (Roger et al., 1987).

A multivariate analysis of data from West African sites (Reynaud and Roger, 1978) showed that the best time for inoculation was at the beginning of tillering.

INOCULATION AND CROP YIELD

Inoculation with cyanobacteria has failed to increase rice yield consistently (Roger et al., 1993; Watanabe, 1986), which may explain why farmers have not adopted this technology. In an analysis of 634 experiments, Balandreau and Roger (1996) reported yield increases after inoculation. Yet in most experiments, the only variable measured was grain yield (Reddy and Roger, 1988); therefore, reasons for the presence or absence of a yield increase after cyanobacteria inoculation are poorly understood.

Yanni's study (1996) of inoculation with cyanobacteria in Egypt resulted in a yield increase of around 10 percent. In a two-year field assay in Uruguay, where rice seeds are sown in dry soil and followed by three years of forage species culture, grain yield was unaffected by inoculation but the number of heterocystous cyanobacteria was not different from the inoculated plots. The amount of total nitrogen in the soil did not vary at the end of the crop season. These results were obtained using inocula produced in the laboratory with the three more abundant isolates, which had not a chance to compete with algae and grazers prior to inoculation.

Roger (1995) reported small and variable responses of yield to inoculation and an experimental error frequently larger than the response. He also affirmed that many unsuccessful results of inoculation are not published.

In a survey of published results without inoculation, Roger (1991) shows that the contribution of cyanobacteria to the N balance of rice fields is significant, with average values around 13 kg $N \cdot ha^{-1}$. In temperate rice fields from Uruguay a gross input of 1 kg $N \cdot ha^{-1}$ per crop was estimated by the acetylene reduction activity assay (Irisarri et al., 2001). This value of a photo-dependent BNF contribution could be underestimated as cyanobacterial blooms were not considered.

REASONS FOR THE FAILURE OF INOCULATION

In many cases no benefits in yield were found after inoculation. The low content in cyanobacteria propagules of the inoculum might be one of the reasons for the failure of inoculation. This is not surprising, as the recommended rate of inoculum application will provide on average less than one propagule for every 130 indigenous cyanobacteria already present (Roger and Kulasoriya, 1980). Inoculated strains could not even be detected immediately after inoculation, presumably because they constituted such a small portion of the total population. In other studies, inoculated strains did multiply but rarely dominated the native population of cyanobacteria (Reddy and Roger, 1988).

Many rice fields visually show obvious growth of cyanobacteria but typically only about half of the cyanobacteria genera present are heterocystous and thus nitrogen fixers (Whitton and Roger, 1989).

When beneficial effects on plant growth due to inoculation have been found it is likely that establishment of large populations of cyanobacteria has been possible due to the large phosphorous content of propagules in the inocula, which will give the introduced cells a substantial growth advantage over indigenous strains (Roger et al., 1986). In any case, whether the inoculum strains succeed in becoming established or not, there is little evidence indicating that they can fix N_2 more effectively than indigenous cyanobacteria.

Reddy and Roger (1988) observed a longer persistence of some foreign inoculated strains and the late establishment of blooms of inoculated *Aulosira fertilissima* with predators control, showing some potential for inoculation of rice soils with foreign strains.

The fate of cyanobacteria in the rice field ecosystem depends on their ability to grow, colonize, and survive in the soil (Tomaselli and Giovannetti, 1993). Survival capability did not appear to be closely correlated to the colonization process. This process depends mainly on the growth rate and spread of the organisms during favorable conditions. Survival depends on the capability to overcome adverse growth conditions.

Different strains have different capabilities to survive in soils under drought conditions. Survival seems to be related to the existence of resistant structures that are unaffected by desiccation, such as akinetes or sheaths, well represented in *Nostoc* and *Scytonema* (Tomaselli and Giovannetti, 1993). When appropriate conditions for growth are restored, the quiescent cells of the drought-resistant strains fully recover their functions.

Grazing of inoculum by invertebrates was detrimental to inoculum establishment (Grant et al., 1986). Primary consumers that proliferate at the expense of cyanobacteria are a major limiting factor in paddy fields. A large positive N balance was obtained over three rice crops by using neem (*Azadirachta indica* A. Juss) seeds to control grazing of cyanobacteria. The dominance of mucilaginous cyanobacteria strains, which happens to be a common observation for rice fields (Begum et al., 1996), should indicate that grazing and drought may be limiting factors for cyanobacterial growth.

Surface broadcast application of nitrogen fertilizer not only inhibits photodependent BNF but also favors the growth of green algae, which increases the pH of the floodwater and causes fertilizer losses by ammonia volatilization (Fillery et al., 1986). Combined sources of nitrogen, which are commonly used in rice production, constitute an important stress on microbial systems. Cyanobacteria could be able to revert to nitrogen-fixing mode, when exogenous nitrogen is depleted. Our results indicated lack of significance of nitrogen treatment at sowing over nitrogen fixation rate measured as acetylene reduction activity (ARA) (Irisarri et al., 2001). High ammonium concentrations can be expected in rice fields when fertilizer is added, but such fertilizer is diluted or consumed in a short time, particularly if it is applied before flooding, as in our conditions.

Delaying N fertilizer application could possibly allow the growth of N_2-fixing cyanobacterial blooms at the early stages of the crop (Yanni, 1991), but the resulting effects on N losses from fertilizer applied in a cyanobacterial-rich water are unknown. Deep placement of N fertilizer, instead of broadcasting, favored cyanobacterial growth (Roger, 1995).

Acidic soils represent another problem in obtaining better results from inoculation with cyanobacteria (Roger et al., 1980). The majority of cyanobacteria grow through a wide pH range (from 6.5 to 8.5), but some species are sensitive to low pH. Uruguayan rice fields presented an average heterocystous cyanobacterial density lower than tropical ones (Irisarri et al., 2001). The slightly acidic pH of the soil could explain the abundant green algae and aquatic weed development which competed for light with cyanobacteria.

Solar ultraviolet-B (UVB) radiation at the earth's surface is increasing due to a continued depletion of the stratospheric ozone layer, and the survival of some cyanobacteria is severely affected by UV light (Sinha and Häder, 2000). N_2-fixing cyanobacteria strains having photoprotective compounds may be potent candidates as biofertilizers.

Pesticides may have adverse effects on the growth of cyanobacteria. Sensitivity of cyanobacteria to herbicides varies according to the species (Leganés and Fernández-Valiente, 1992). The addition of field-recommended doses of quinclorac and propanil, two herbicides commonly used in Uruguay as cyanobacterial isolates, affected oxygen photoevolution but not nitrogenase activity (Irisarri et al., 2001). On the contrary, some pesticides have been found to accelerate the growth of cyanobacteria (Wright et al., 1977).

More attention should be given to the impact of physical, chemical, and biological limiting factors occurring in the field, to improve the biofertilizer benefit:cost ratio (Mulongoy et al., 1991).

BENEFITS OF CYANOBACTERIA BEYOND NITROGEN INPUT TO THE AGROECOSYSTEM

Biofertilizers can affect growth and yield of plants either directly or indirectly through different mechanisms. For example, a buildup of organic matter takes place due to cyanobacteria inoculation (Das et al., 1991). The amount of addition of organic C to soils by the growth of cyanobacteria varies from 0.45 $t{\cdot}ha^{-1}$ to 41.6 $t{\cdot}ha^{-1}$ (Rao and Burns, 1990), contributing to humus formation despite the absence of typical lignin in them. A good cyanobacterial bloom and persistence as soil organic matter depends on decomposability, which in turn depends on the biodegradability of cell wall compounds and their physiological growth stages. Data of long-term experiments are lacking (Mandal et al., 1999), but inoculation may bring about, directly or indirectly, certain changes in the physical, chemical, and biological properties of the soil and soil-water interface in rice fields, which are of agronomic importance. The extracellular organic compounds liberated by the cyanobacteria; the O_2 released due to photosynthesis during their active growth period; and the subsequent addition of biomass after their death are likely to cause some important changes in the physicobiochemical properties of the soil, including retardation of NH_3 volatilization, mobilization of fixed phosphates, and regulation of micronutrients (particularly Fe, Mn, and Zn), affecting their availability. Cyanobacteria are known to have the ability to mobilize bound phosphates in soil sediments or in pure culture (Roger, 1995), possibly because of organic acid excretion. There is increasing evidence that crops grown in organic-rich and biologically active soils are less susceptible to pest attack (Altieri, 2002).

However, these benefits can be achieved only if there is a good growth of cyanobacteria in rice fields.

IS INOCULATION WITH CYANOBACTERIA WORTHWHILE?

It is important to verify that N is actually limiting before improving BNF. Significant yield responses to applied N are obtained on nearly all lowland rice soils when irrigation is adequate and weeds or pest problems do not limit growth (Cassman et al., 1998). Because knowledge of factors that allow cyanobacteria to establish and bloom in rice fields is limited, inoculation is conducted on a trial-and-error basis. No method for estimating the chance of success of inoculation in a given agroecosystem is available.

The ubiquity of heterocystous cyanobacteria in rice soils does not imply that inoculation is not needed. The extent to which inoculation will eventually prove worthwhile will depend on local circumstances. The response is likely to be most evident in nonacidic soils with moderate to high phosphorous availability where nitrogen fertilizer is not broadcast. Since spore germination is photodependent, inoculated propagules applied on the soil surface might germinate better than indigenous propagules mixed with the soil (Roger, 1991). Agrosystems with a long dry fallow or an upland crop before rice may be especially responsive to inoculation (Roger, 1991). A low incidence of grazers and a lower indigenous density of viable propagules of cyanobacteria after the dry period would explain these systems' inoculation dependence.

During unfavorable seasons most heterocystous cyanobacteria can form perennating bodies which germinate with the onset of favorable conditions. Under such circumstances, inoculation should reduce the lag time necessary to reach an appropiate cyanobacteria population density (Roger, 1991).

Data appear to be lacking to support the claim by Agarwal (1979) that the introduced cyanobacteria can establish themselves if inoculation is done repeatedly for three to four cropping seasons.

In general, cyanobacteria have less potential in terms of nitrogen fixation than legumes or *Azolla,* but inoculation, when successful, is a low-cost technology with a cost:benefit ratio far more favorable than that of green manure (Roger, 1991). This may ensure a much better survival rate of the microorganisms and easy transportation and application of the biofertilizer.

Roger (1995) proposed to enhance the growth of indigenous cyanobacteria in rice fields as an alternative strategy to inoculation for improving the inputs of N_2 fixation. According to Balandreau and Roger (1996), the goal of developing BNF is not to achieve the maximum nitrogen input; it is to achieve the maximum income (money and/or food) for farmers. Increasing BNF is desirable not only because the cost of synthetic fertilizer is unpredictable but also because it contributes to the sustainability of soil fertility. An indispensable requirement of sustainable agriculture is the continuous renewal of soil structure and fertility with renewable resources, lessening the need for chemical fertilizers, thus reducing their environmental burden.

Inoculation is advisable with local strains to deal with bacteria adapted to the plant and to the local edaphic and climatic conditions. Introduced cyanobacteria are difficult to produce in large quantities and are rarely well established.

Defining environmental conditions in which natural or inoculated cyanobacteria can add onto current mineral N fertilizer applications is the strategy for the future.

INTEGRATED SOIL FERTILITY MANAGEMENT

In the 1960s and 1970s, the green revolution increased agricultural output with the assistance of chemical fertilizers and pesticides, high yielding seed varieties, and intensive land use. Although contributing to production, these new technologies also had potential risks: unbalanced, excessive, or untimely application of agrochemicals could deplete soil, cause water pollution, render commonly used pesticides ineffective, and harm human health and environment. The environmental impact and sustainability of development have become an international priority issue. More prudent agricultural cultivation practices, for example, integrated soil fertility management and green and organic agriculture have generated worldwide attention.

To improve agricultural production, we must not disregard the development of a more self-sufficient and sustaining agriculture with a deeper understanding of the nature of agroecosystems and the principles by which they function (Altieri, 2002).

A good example of this approach is the results obtained by Yanni and Osman (1990). Natural infection with leaf and neck blast caused by *Pyricularia oryzae* in rice became elevated with increasing N fertilization. Inoculation with fresh cyanobacteria decreased leaf blast. Yanni and Sehly (1991) suggested that inoculation gave rise to a slow release of biologically fixed N_2. Such a mechanism could regulate the N status of the rice ecosystem without exceeding the critical level at which susceptibility to blast may dramatically increase.

On the other hand, demand is growing for organically grown food products. In this situation the ecological and economic merits of biofertilizers have been highlighted and the production of organic rice should depend on BNF.

FUTURE PERSPECTIVES

The efficiency of nitrogen inputs must be significantly improved for two reasons: first, the manufacture of inorganic fertilizer N requires fossil fuels, which are a nonrenewable resource; second, inefficient use of both inorganic fertilizer N and organic nitrogen leads to gaseous emission of N to the atmosphere which may contribute to global change.

The challenges for the future are to employ more suitable cyanobacteria to produce biofertilizers and to better understand rice agroecosystems so as to make the use of cyanobacteria profitable.

Efforts have been made to obtain cyanobacteria strains with especially high nitrogen-fixing ability to be incorporated into the inocula, but there is no evidence that such mutant strains have a particular advantage in the field.

Another approach to achieve this can be through the development of ammonia-leaking strains. *Anabaena azollae* actually makes available the nitrogen fixed by it to *Azolla* as ammonia. *Anabaena azollae* and *A. variabilis,* when immobilized in polyurethane and polyvinyl foams or calcium alginate beads, have been shown to release ammonia extracellularly (Kannaiyan et al., 1994). Algal mutants with defects in the regulation of the enzyme glutamine synthetase can be made to liberate ammonia. As genetically improved cyanobacteria face competition with the predominant native species, it may be advisable to use somaclones of *Anabaena-Azollae* (Sinha et al., 1998).

It may prove useful to introduce fast-growing strains, although Roger (1996) stated that most fast-growing strains with doubling times of 5 to 12 h belong to *Anabaena* genus and do not form mucilaginous colonies, so they would certainly be susceptible to grazing.

Whitton and Roger (1989) pointed out that introduced strains with high nitrogen-fixing ability did not survive unless there was a simultaneous change in the environment, which gave them a competitive advantage.

Modern agricultural fields are generally treated with high doses of N fertilizers and pesticides despite the adverse effects these may exert on the soil microflora of naturally ocurring N-fixing cyanobacteria. Vaishampayan et al. (1998) indicated the need to develop ammonia-derepressible pesticide-resistant cyanobacterial mutants as a viable and efficient photobiological N fertilizer phycotechnology. Although some of these strains perform well under in situ conditions, they need further improvement to adapt to the various biotic and abiotic stress factors operative in the fields, particularly with respect to strain competition during changes in edaphic and climatic conditions. A knowledge of the various biotic and abiotic stress factors that may be operative in the field is a prerequisite for mass cultivation of mutant cyanobacteria for field experiments. Some factors may modify cyanobacterial physiology in the field while others may either cause leaching, photodecomposition, volatilization, or possible microbial degradation of the pesticides or conversion of the same to more powerful multiple inhibitors in the field (Vaishampayan et al., 1998). The genetically manipulated cyanobacterial species may act more efficiently as the self-renewable constitutive N supply in wet agriculture, producing ammonia at the sole expense of the plentiful solar energy, atmospheric CO_2, N_2, and water, irrespective of the pre- or posttreatment of the fields with pesticides and/or synthetic N fertilizers. These may not completely replace the use of the latter but may contribute in considerably reducing their use; these may help in making up the significant loss of the available fixed N from the soil.

The survival of the transgenic microbial inoculant population, its monitoring, and positive and negative interactions with indigenous microbial populations are the challenges that face this emerging technology (Tengerdy and Szakács, 1998).

CONCLUSIONS

In Asia, 24 percent of the increase in rice production from 1965 to 1980 was attributed to better crop nutrition due primarily to the increased use of N fertilizers. Today, more than 20 percent of all N fertilizers produced worldwide are applied to the rice fields of Asia. The densely populated and land-scarce nations of Asia, where more than half of the earth's people live and rice is a staple food, are dependent on petroleum-based fertilizers. Any reduction in the consumption of this fertilizer without affecting productivity as well as the introduction of a cyclic nutrient supply system through biological sources would be ecologically and economically advantageous.

In temperate rice cultivation, biofertilizers appeared also as ecologically friendly and economically important alternatives for organic rice production.

The challenge is to increase our knowledge about limiting factors for cyanobacteria contribution to the N status of rice ecosystem. It is of no use to transfer technologies from one site to another, if the set of ecological interactions associated with such techniques cannot be replicated.

In the context of biological management of soil fertility, we must consider that the advantages of slow N release might not be apparent in the first crop after inoculation and that the feasibility of the technique is not assured.

There is still a lot to investigate about the multiple effects of natural resources management, some of which may even be undesirable. Perhaps biofertilizers may not be an alternative to synthetic fertilizers, but they can enhance the availability of plant nutrients from the soil.

REFERENCES

Agarwal A. (1979). Blue-green algae to fertilize Indian rice paddies. *Nature* 279: 181.

Alfassi F.A., Algarni S.M., and Kabli S.M. (1997). Efficiency of cyanobacteria from soil of western region in the kingdom of Saudi-Arabia as biofertilizer for wheat. *Arab Gulf Journal of Scientific Research* 15: 481-503.

Altieri M.A. (2002). Agroecology: The science of natural resource management for poor farmers in marginal environments. *Agriculture, Ecosystems and Environment* 93: 1-24.

Balandreau J. and Roger P. (1996). Some comments about a better use of biological nitrogen fixation in rice cultivation. In M. Rahnan, A. Kumar, C. van Hove, Z.N.T. Begum, T. Heulin, and Hartmann A. (eds.), *Biological nitrogen fixation associated with rice production* (pp. 1-12). Dordrecht, the Netherlands: Kluwer Academic Publishers.

Begum Z.N.T., Mandal R., Khan U.M.Z., and Hossain M.Z. (1996). Prospects and potentiallity of cyanobacteria as an alternative source of nitrogen fertilizer in Bangladesh rice cultivation. In M. Rahnan, A. Kumar, C. van Hove, Z.N.T. Begum, T. Heulin, and Hartmann A. (eds.), *Biological nitrogen fixation associated with rice production* (pp. 119-131). Dordrecht, the Netherlands: Kluwer Academic Publishers.

Cassman K.G., Peng S., Olk D.C., Ladha J.K., Reichardt W., Dobermann A., and Singh U. (1998). Opportunities for increased nitrogen-use efficiency from improved resource management in irrigated rice systems. *Field Crop Research* 56: 7-39.

Das S.C., Mandal B., and Mandal L.N. (1991). Effect of growth and subsequent decomposition of blue-green algae on the transformation of iron and manganese in submerged soils. *Plant and Soil* 138: 75-84.

De P.K. (1939). The role of blue green algae in nitrogen fixation in rice fields. *Proceedings of Research Society of London* 127: 121-139.

Fernández-Valiente E., Quesada A., Prosperi C., Trillo M., Carreres R., and Gonzáles T. (1996). Contribution of biological nitrogen fixation to rice production in Mediterranean rice fields. In M. Rahnan, A. Kumar, C. van Hove, Z.N.T. Begum, T. Heulin, and A. Hartmann (eds.), *Biological nitrogen fixation associated with rice production* (pp. 147-157). Dordrecht, the Netherlands: Kluwer Academic Publishers.

Fillery I.R.P., Roger P.A., and De Datta S.K. (1986). Ammonia volatilization from nitrogen sources applied to rice fields: II. Floodwater properties and submerged photosynthetic biomass. *Soil Sci. Soc. Am. J.* 50: 86-91.

Ghosh T.K. and Saha K.C. (1997). Effects of inoculation of cyanobacteria on nitrogen status and nutrition of rice (*Oryza sativa* L.) in an Entisol amended with chemical and organic sources of nitrogen. *Biol Fertil Soils* 24: 123-128.

Grant I.F., Roger P.A., and Watanabe I. (1986). Ecosystem manipulation for increasing biological N2 fixation by blue-green algae (cyanobacteria) in lowland rice fields. *Biological Agriculture and Horticulture* 3: 299-315.

Hashem M.A., Islam M.R., Jahiruddin M., and Asaduzzaman M. (1996). Enrichment of indigenous blue-green algal population in rice fields and evaluation of its effect in rice culture. In M. Rahnan, A. Kumar, C. van Hove, Z.N.T. Begum, T. Heulin, and A. Hartmann (eds.), *Biological nitrogen fixation associated with rice production* (pp. 159-167). Dordrecht, the Netherlands: Kluwer Academic Publishers.

Hori K., Ishii S., Ikeda G., Okamoto J., Tanji Y., Weeraphasphong C., and Unno H. (2002). Behavior of filamentous cyanobacterium *Anabaena* spp. in water column and its cellular characteristics. *Biochemical Engineeering Journal* 10: 217-225.

Irisarri P., Gonnet S., Deambrosi E., and Monza J. (1999). Diversity of heterocystous cyanobacteria in rice soils. *Agrociencia* 3: 31-37.

Irisarri P., Gonnet S., and Monza J. (2001). Cyanobacteria in Uruguayan rice fields: Diversity, nitrogen fixing ability and tolerance to herbicides and combined nitrogen. *Journal of Biotechnology* 91: 95-103.

Kannaiyan S., Aruna S.J., Kumari S.M.P., and Hall D.O. (1994). Immobilized cyanobacteria as a biofertilizer for rice crops. *Journal of Applied Phycology* 9: 167-174.

Leganés F. and Fernández-Valiente E. (1992). Effects of phenoxy-acetic herbicides on growth, photosynthesis and nitrogenase activity in the cyanobacteria from rice fields. *Arch. Environ. Contam. Toxicol.* 22: 130-134.

Mandal B., Vlek P.L.G., and Mandal L.N. (1999). Beneficial effects of blue-green algae and *Azolla*, excluding supplying nitrogen, on wetland rice fields: A review. *Biol. Fertil. Soils* 28: 329-342.

Mulongoy K., Gianinazzi S., Roger P.A., and Dommergues Y. (1991). Biofertilizers: Agronomical and environmental impacts and economics. In E.J. DaSilva, C. Ratledge, and A. Sasson (eds.), *Microbial technology: Economical and social aspects* (pp. 1-13). Cambridge: Cambridge University Press.

Rao D.L.N. and Burns R.G. (1990). The effect of surface growth of blue-green algae and bryophytes on some microbiological, biochemical and physical soil properties. *Biol. Fertil. Soils* 9: 239-244.

Reddy P.M. and Roger P.A. (1988). Dynamics of algal populations and acetylene-reducing activity in five rice soils inoculated with blue-green algae. *Biol. Fert. Soils* 6: 14-21.

Reynaud P.A. and Metting B. (1988). Colonization potential of cyanobacteria on temperate irrigated soils in Washington State, USA. *Biological Agriculture and Horticulture* 5: 197-208.
Reynaud P.A. and Roger P.A. (1978). N_2-fixing algal biomass in Senegal rice fields. *Ecological Bulletin* 26: 148-157.
Roger P.A. (1991).Reconsidering the utilization of blue-green algae in wetland rice cultivation. In S.K. Dutta and C. Sloger (eds.), *Biological N_2 fixation associated with rice production* (pp. 119-141). New Dehli, India: Oxford & IBH Pub.
Roger, P.A. (1995). Biological N_2-fixation and its management in wetland rice cultivation. *Fertilizer Research* 42: 261-276.
Roger P.A. (1996). Fauna. In W.H. Smith (ed.), *Biology and management of the floodwater ecosystem in rice fields* (pp. 57-78). IRRI, ORSTOM.
Roger P.A. and Kulasooriya S. (1980). *Blue green algae and rice.* Manila, Philippines: IRRI.
Roger P.A., Kulasooriya S., Tirol A.C., and Craswell E.T. (1980). Deep placement: A method of nitrogen fertilizer application compatible with algal nitrogen fixation in wetland rice soils. *Plant and Soil* 57: 137-142.
Roger P.A., Santiago-Ardales S., Reddy P.M., and Watanabe I. (1987). The abundance of heterocystous blue-green algae in rice soils and inocula used for application in rice fields. *Biol. Fertil. Soils* 5: 98-105.
Roger P.A., Santiago-Ardales S., and Watanabe I. (1986). Chemical composition of cultures and natural samples of N_2-fixing blue-green algae from rice fields. *Biol. Fertil. Soils* 2: 131-146.
Roger P.A., Zimmerman W.J., and Lumpkin T. (1993). Microbiological management of wetland rice fields. In B. Metting (ed.), *Soil microbial technologies* (pp. 417-455). New York: Marcel Dekker.
Saha K.C. and Mandal L.N. (1980). Agrochemical studies on the effect of inoculation of N-fixing blue-green algae oion an alluvial soil treated with P and Mo on the yield of rice and changes in the N content of the sol. *Plant and Soil* 57: 23-30.
Sinha R.P. and Häder D.-P. (2000). Effects of UV-B radiation on cyanobacteria. *Recent Res. Devel. Photochem. Photobiol.* 4: 239-246.
Sinha R.P., Vaishampayan A., and Häder D.-P. (1998). Plant cyanobacterial symbiotic somaclones as a potential bio-nitrogen fertilizer for paddy agriculture biotechnological approaches. *Microbiol. Res.* 153: 297-307.
Tengerdy R.P. and Szakács G. (1998). Perspectives in agrobiotechnology. *Journal of Biotechnology* 66: 91-99.
Thind H.S. and Rowell D. (1999). The fate of algal nitrogen in a flooded soil system. *Nutrient Cycling in Agroecosystems* 55: 89.94.
Tirol A.C., Roger P.A., and Watanabe I. (1982). Fate of nitrogen from blue-green algae in a flooded rice soil. *Soil Sci. Plant Nutr.* 28: 559-569.
Tomaselli L. and Giovannetti L. (1993). Survival of diazotrophic cyanobacteria in soil. *World J. Microbiol. Biotechnol.* 9: 113-116.
Vaishampayan A., Sinha R.P., and Häder D.-P. (1998). Use of genetically improved nitrogen-fixing cyanobacteria in rice paddy fields: Prospects as a source material for engineering herbicide sensitive and resistance in plants. *Bot. Acta* 111: 176-190.

Venkatamaran G.S. (1981). Blue-green algae: A possible remedy to nitrogen scarcity. *Curr Sci* 50: 253-256.

Watanabe I. (1986). Nitrogen fixation by non-legumes in tropical agriculture with special reference to wetland rice. *Plant and Soil* 90: 343-357.

Watanabe I., Nishigaki S., and Konishi C. (1951). Effect of nitrogen-fixing blue-green algae on the growth of rice plants. *Nature* 168: 748-749.

Whitton B.A. and Roger P.A. (1989). Use of blue-green algae and *Azolla* in rice culture. In R. Campbell and R.M. Macdonald (eds.), *Microbial inoculation of crop plants,* Volume 25 (pp. 89-100). Oxford: IRL Press at Oxford University Press.

Wright S.J.L., Stainhorpe A.F., and Downs J.D. (1977). Interactions of the herbicide propanil and a metabolite, 3,4 dichloroaniline, with blue-green algae. *Acta Phytophatol. Acad. Sci. Hung.* 12: 51-60.

Yanni Y.G. (1991). Potential of indigenous cyanobacteria to contribute to rice performance under different schedules of nitrogen application. *World J. Microbiol. Biotechnol.* 7: 48-52.

Yanni Y.G. (1996). Contribution of cyanobacterization to rice growth and performance under different field stand densities and levels of combined nitrogen. In M. Rahnan, A. Kumar, C. van Hove, Z.N.T. Begum, T. Heulin, and A. Hartmann (eds.), *Biological nitrogen fixation associated with rice production* (pp. 113-139). Dordrecht, the Netherlands: Kluwer Academic Publishers..

Yanni Y.G. and Osman Z.H. (1990). Contributions of algalization to rice growth, yield, N attributes and incidence of infestation with the blast fungus *Pyricularia oryzae* under different fungicidal treatments. *World J. Microbiol. Biotechnol.* 6: 371-376.

Yanni Y.G. and Sehly M.R. (1991). Rice performance and natural infection with blast (*Pyricularia oryzae* Cav.) under different algalization techniques and rates of fertilizer nitrogen. *World J. Microbiol. Biotechnol.* 7: 43-47.

Zaccaro M.C. (2000). Plant growth-promoting cyanobacteria. In *Proceedings of the 5th International PGPR Workshop, Córdoba, Argentina.* Available online at <http://www.ag.auburn.edu/argentina/pdfmanuscripts/zaccaro.pdf>.

Chapter 17

Cyanobacterial Biofertilizers for Rice: Present Status and Future Prospects

Siba Prasad Adhikary
Bagmi Pattanaik

INTRODUCTION

In man's continued effort to enhance food production, new high yielding crop varieties and new techniques for crop and field management are constantly evolving, resulting in the parallel increase of chemical fertilizer used to boost the yield. But continuous use of chemical fertilizers lead to environmental pollution and loss of soil fertility. These chemical fertilizers destroy natural soil microflora and result in a reduced humus layer of soil, affecting natural fertility and sustainability of crop yield. To sustain agriculture production of crops, maintenance of soil fertility and soil health is essential. In this regard, microorganisms play an important role in the cycling of nutrients, decomposition of waste residues, and other related activities to maintain productivity. These organisms are generally called biofertilizers or, more precisely, bioinoculants (Singh, 1998). Biofertilizers are preparations containing viable forms of beneficial microorganisms intended for seed, plant, or soil applications. They can affect growth and yield of plants either directly or indirectly through different mechanisms. These are eco-friendly, self-replicating, and easily manageable inputs that augment and increase the availability of crop nutrients.

Rice is grown on about 43 million hectares (ha) in India alone, with production of about 81 million tons (mt) (43 percent of the total food grain production of the country). However, the demand for food grains is continuously increasing due to increases in world population. Among the various fertilizers used for higher rice yield, nitrogenous fertilizers are required the most. Rice, however, is not exclusively dependent on fertilizer or FYM (farmyard manure) for external N supplies because the crop receives sizable nitrogen input from green manures and certain biofertilizers, such as BGA (blue-green algae/cyanophytes/cyanobacteria), which grow along with rice.

The waterlogged conditions of rice fields are conducive to the growth of cyanobacteria and the fertility of rice field soils is attributed to the presence of these photosynthetic organisms.

During the past two decades tremendous interest has risen to study the cyanobacterial biofertilizers for rice cultivation for obvious basic and applied reasons (Venkataraman, 1981; Motsara et al., 1995). These organisms provide a simple and model system for the study of fundamental cellular processes, such as macromolecular synthesis, enzyme regulation, cell differentiation and developmental pattern formation, regulation of gene expression, and the mechanism of stress and adaptive responses under various environmental regimes (Adhikary, 2000). Thus, their input and ecological role in the natural environment as initial colonizers and primary producers of organic matter, and also their part in biological nitrogen fixation, are of increasing interest for agriculture microbiologists.

CYANOBACTERIA (BLUE-GREEN ALGAE)

Cyanobacteria are a fascinating group of photosynthetic gram-negative bacteria that can fix atmospheric dinitrogen. These were formerly included in algae, as they have chlorophyll-a and perform oxygenic photosynthesis. They originated in the early precambrian period of the earth, over 2.5 billion years ago (Schoof, 1983). They can exist in almost all conceivable habitats. They need simple inorganic nutrients, a little moisture, and diffuse light for growth. In addition, they have the ability to grow in extreme habitats of the environment where light and temperature is high, for example, hot springs (Castenholz, 1969), desert soils (Mazor et al., 1996), on exposed rocks (Hoffmann, 1989; Krumbein, 1987; Tripathy et al., 1999; Adhikary, 2000; Pattanaik and Adhikary, 2002), and on rice field soils in dry conditions after harvesting (Adhikary and Sahu, 2000b). The cosmopolitan distribution of cyanobacteria is attributed to their double trophic nature, in other words, most cyanobacterial species have the ability to fix atmospheric nitrogen in addition to engaging in photosynthesis (Fay, 1983; Whitton, 1992).

Morphology of Cyanobacteria

Nitrogen-fixing cyanobacteria were broadly divided into three major groups: unicelluar, filamentous nonheterocystous, and filamentous heterocystous. Rippka et al. (1979) subdivided cyanobacteria into five groups. Some cyanobacterial species are single and spherical or oval in structure, for example, *Synechocystis* and *Synechococcus*. In other species, vegetative cells are enclosed by an additional fibrous layer, resulting in an aggregate of vegetative cells in definite numbers: *Gloeocapsa, Gloeothece, Glocobacter,* etc. In some others, the aggregation is irregular due to difference in their division, for example, *Dermocarpa, Myxosarcina, Chroococcidiopsis,* and the *Pleurocapsa* group. In filamentous forms, cells are placed one above the other to form a trichome. Trichomes with an outer mucilagenous sheath layer are called filaments. Cells in a filament are characteristic of *Lyngbya, Phormidium, Plectonema,* and *Oscillatoria*. Heterocystous filamentous organisms are *Nostoc* (see Figure 17.1a), *Anabaena* (see Figure 17.1b), *Cylindrospermum* (see Figure 17.1c), and *Calothrix* (see Figure 17.1d), which are unbranched. False branching occurs in *Aulosira* (see Figure 17.1e), *Scytonema* (see Figure 17.1f), and *Tolypothrix* (see Figure 17.1g). Branched heterocystous cyanobacteria are *Westiellopsis* (see Figure 17.1h) and *Stigonema* (see Figure 17.1i). Usually nitrogen fixation occurs in the heterocystous cyanobacteria under phototrophic condition. Recent findings showed that even nonheterocystous cyanobacteria can also fix nitrogen in anaerobic conditions, and some of the species even under aerobic conditions (Gallon et al., 1988).

Heterocysts and Nitrogen Fixation

In the filaments of cyanobacteria, thick-walled cells found in the intercalary or terminal in position are known as heterocysts. Under combined nitrogen-depleted conditions in the growth medium, the gradual transformation of certain vegetative cells in the filaments leads to development of morphologically, physiologically, biochemically, and functionally specialized heterocysts (Stewart, 1980; Wolk, 1982). Fay et al. (1968) reported that the heterocyst is the sole site of aerobic N_2 fixation in heterocystous cyanobacteria.The heterocystous forms can fix N_2 in the air because they differentiate heterocysts from vegetative cells designed to exclude oxygen and therefore conducive to the activities of the oxygen-sensitive nitrogenase enzyme (Wolk, 1982). In filamentous heterocystous cyanobacteria, the vegetative cells supply the photosynthate to heterocysts and the latter returns the favor in the form of fixed nitrogen (Haselkorn, 1978; Stewart, 1980; Apte, 1992). Oxygen physically degrades the oxygen labile nitrogenase proteins, destroys the anoxic conditions essential for the enzyme activity, and represses the transcription of the *nif* operon. Photosynthesis, which requires N_2 fixations, is also a cause of its inhibition. The biochemical modification of heterocysts to create an anoxic niche is done by preventing O_2 evolution of PSII through degradation of biliproteins (Fay, 1969; Wolk and Simon, 1969; Thomas, 1970). Heterocystous cyanobacteria are capable of reducing

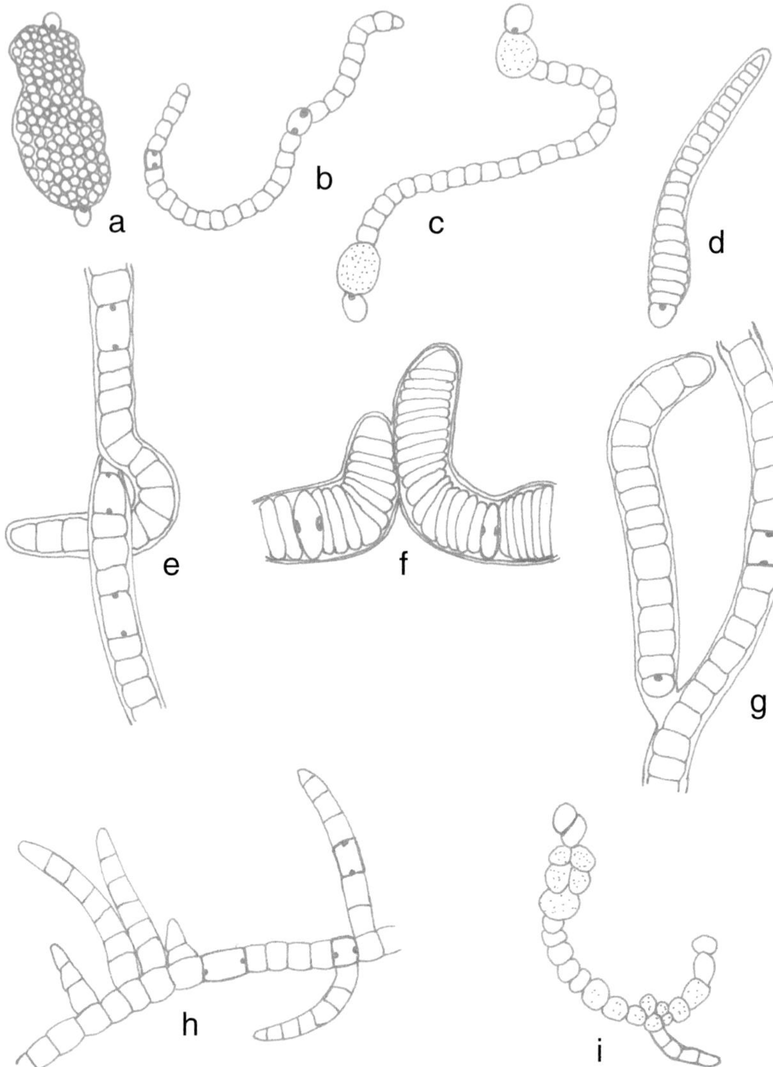

FIGURE 17.1. (a) *Nostoc,* a heterocystous filamentous cyanobacteria; (b) *Anabaena,* unbranched; (c) *Cylindropspermum,* unbranched; (d) *Calothrix,* unbranched; (e) *Aulosira,* with false branching; (f) *Scytonema,* with false branching; (g) *Tolypothrix,* with false branching; (h) *Westiellopsis,* a branched heterocystous cyanobacteria; and (i) *Stigonema,* a branched heterocystous cyanobacteria.

molecular N_2 to NH_3. This reduction is catalyzed by the enzyme nitrogenase found in the heterocyst. Nitrogenase consists of two dissociating protein components, component I (containing Mo, Fe, S^{-2}) and component II (containing Fe and S^{-2}). Glutamate is converted to glutamine in the heterocyst in the presence of nitrogenase, glutamine synthetase, ATP, and molecular nitrogen. Glutamine with 2-oxoglutarate and GOGAT (glutamine oxoglutarate amino transferase) forms glutamate again.

Stewart and colleagues (Stewart and Rowell, 1975; Stewart et al., 1975) reported that NH_3 is an important regulator of nitrogen fixation, which in high concentration inhibits nitrogenase synthesis. To overcome the repression of nitrogenase synthesis by NH_3, at least three processes have been proposed:

1. Subcellular compartmentation of the site of nitrogenase synthesis and the site of its activity
2. The NH_3 transported rapidly from the nitrogen-fixing sites and at sites removed from the loci of nitrogenase synthesis assimilated into general cell metabolism
3. Presence of a very efficient ammonia-scavenging mechanism which at the site of nitrogenase activity assimilates the NH_3 into a noninhibitory form

Usually there are at least four ways by which cyanobacteria may overcome inhibition of nitrogenase.

1. Production of physical barriers to inward diffusion of O_2 in the heterocysts (Lambein and Wolk, 1973; Wolk et al., 1988; Murry and Wolk, 1989)
2. Rapid and continuous biochemical consumption of O_2 before it can inhibit nitrogenase (Winkenbach and Wolk, 1973; Neuer et al., 1983)
3. Spatial or temporal separation of N_2 fixation and the O_2 producing component of photosynthesis (Mullineaux et al., 1981; Paerl and Bland, 1982; Stal and Krumbein, 1987; Gallon et al., 1988)
4. Creation of an environment of low partial pressure of O_2 frequently brought about by cellular aggregation (Hallenbeck, 1987; Haselkorn et al., 1987)

In the nitrogen-fixing cyanobacteria the arrangement of *nif* gene varies within genomes of different species. The major *nif* cluster comprises several open-reading frames (ORFs), 2 interrupting DNA elements, and 4 operons dispersed over nearly 80 kb of DNA (Golden et al., 1985). At least

14 genes associated with nitrogen fixation have been identified in *Anabaena* sp. PCC 7120. *nif* K, D,H operon, the *nif* B, *fdx*N, *nif*S, *nif*U operon, a *fdx*H gene which encodes a heterocyst-specific ferrodoxin, *xis*A and *xis*F, which reside on the interrupting DNA elements and encode for two site-specific recombinases absolutely essential for nitrogen fixation (Apte and Prabhavathi, 1994).

Cyanobacteria play an important role in nutrient cycling in waterlogged rice field environments largely to fix atmospheric nitrogen with the help of enzyme nitrogenase. The first report on the occurrence of cyanobacteria in rice fields was detected by Fritsch (1907). However, cyanobacterial contributions to the N content of soil by fixing nitrogen were demonstrated by De (1939). Cyanobacteria constitute the major component of rice field ecosystem. These organisms multiply rapidly in the floodwater in neutral to alkaline soils and release nitrogen slowly for meeting nutrient requirements of rice (Tirol et al., 1982; Whitton, Aziz, Kawecka, and Rother, 1988; Whitton, Aziz, Irancis, et al., 1988). The luxuriant growth of cyanobacteria in rice fields is due to high humidity, temperature, light, water, and nutrients (Roger and Reynaud, 1979). Thus, biological nitrogen fixation by cyanobacteria has been the most effective system for sustaining production in low-input traditional rice cultivation.

Secondary Metabolites for Soil Quality Improvement

The role of cyanobacteria in rice field ecosystems was manifold. Nitrogen is fixed by these organisms, and ammonia is released either through exudation or microbial decomposition after the death and autolysis of cells (Malliga and Subramanian, 2001). Besides increasing nitrogen fertility, they produce growth-promoting substances (Roger and Kulasooriya, 1980). Cyanobacterial exudates contain such substances as vitamin B_{12}, gibberellic acid, indole acetic acid, and amino acids (Okuda and Yamaguchi, 1952, 1956), niacin, panthothenic acid, folic acid, cyanocobalamin, and pyridoxine (Mishra and Kaushik, 1989). Besides, they liberate a variety of amino acids such as aspartic acid, glutamic acid, and alanine (Whitton, 1965; Fogg, 1971; Adhikary and Pattnaik, 1981). These substances are easily soluble in water. By extracellular carbohydrate production from the cyanobacteria, they form cyanobacterial crusts on the soil surface in the field which checks soil erosion, increases the storage of rainwater and reduces the loss of water (Van den Ancker et al., 1985). They also enrich the soil with organic matter and act as a storehouse of such nutrients as nitrogen, phosphorus, and other micronutrients (Marathe, 1972; Roychoudhury et al., 1979). Under field conditions, a 5 to 32 percent increase in soil organic carbon has been observed as a result of algal inoculation (Singh and

Bisoyi, 1989). Qualitatively, the polysaccharide is composed of galactose/fructose, arabinose, maltose, N-acetyl-N glucosamin. The synthesis and excretion of polysaccharide is also regarded as a contrivance to avoid salt injury (Saxena and Kaushik, 1992).

Use as Biofertilizer

Cyanobacteria (BGA) can generate photosynthate from CO_2 and water. This trophic independence makes BGA especially attractive as a biofertilizer. The natural fertility of tropical rice fields due to N_2-fixing BGA with agronomic potential was recognized in by De (1939).

By now, the role of N_2-fixing cyanobacteria in the maintenance of the fertility of rice fields has been well documented (Watanabe and Yamamato, 1971; Singh, 1985; Venkataraman, 1981; Tripathi et al., 2001). Nitrogen fixed by BGA may become available to rice plants only after its release extracellularly into the surroundings, either as an extracellular product or by mineralization of their intracellular content through microbial decomposition after their death (Tripathi et al., 2001).

All India Coordinated Project on Algae (AICPA), Department of Biotechnology, government of India, has been assessing the effect of algalization on rice fields in different parts of the country. The results of these investigations suggest that

1. BGA inoculation enhances the yield from 4 percent to a maximum of 32.8 percent in absence of chemical nitrogen fertilizer;
2. BGA increases yield by 8.85 percent with a recommended level of chemical N fertilizer;
3. BGA inoculation gives 25 to 30 kg $N \cdot ha^{-1}$ and an additional 10 to 15 percent increase in rice yield with full recommended does of chemical N fertilizer. In areas where no fertilizers are used for various reasons, use of algal biofertilizers can result in a benefit of 30 kg $N \cdot ha^{-1}$.

Biologically fixed nitrogen by the cyanobacteria increases the crop yield and, in addition, produces growth-promoting substances that have been shown to (1) accelerate root growth (Venkataraman and Neelakanthan, 1967; Marsalek et al., 1992); (2) excrete polysaccharides thereby improving soil aggregation, stimulating some beneficial soil microorganisms, improving soil water holding capacity, and increasing soil organic matter (Roger and Kulasooriya, 1980; Singh and Bisoyi, 1989; Das et al., 1991); and (3) improve phosphate solubilizing ability (Bisoyi and Singh, 1988) and physical properties of soil (Bertocchi et al., 1990; Roger and Burns, 1994).

Roger and Kulasooriya (1980) reported that presoaking of seeds or seedlings in BGA culture or extracts results in enhanced germination, faster growth, early seedling recovery, increased root growth, prolonged tillering, and increased protein content of the grain. Cyanobacteria bring significant changes in soil aggregation properties and also increase the dehydrogenase, urease, and phosphatase enzyme activities in soil. BGA, while growing in flooded rice soil, affect such trace elements as Mn, Fe, and Zn through influencing the forms in which they occur in soil (Rai, 2001). There are also reports of a 5 to 32 percent increase in soil organic carbon as a result of algal inoculation (Singh and Bisoyi, 1989). BGA also plays an important role in solubilizing phosphorus and making it available to the plants. Thus, application of BGA biofertilizer to rice fields is beneficial to soil quality, plant growth, N_2 availability, yield percentage, and the environment.

FACTORS GOVERNING DISTRIBUTION OF CYANOBACTERIA DURING THE CROP GROWTH CYCLE

Growing rice has become one of the most important uses of land around the world, covering nearly 140 million ha (Watanabe et al., 1987). Rice production is favored by high temperature and therefore restricted to warm regions of the world—especially Asia, for which rice is the staple food (Grist, 1975). Paddy soils are influenced by the seasonal cycles of flooding and dry conditions. During flooding, the microbial properties of soil change, pH of the soil becomes neutral, and an increase in the availability of nutrients from both soil and irrigation water occurs and enhances rice production (Watanabe et al., 1987). The cyanobacteria grow together with rice plants and form patches of thick yellowish or blue-green mats on the water surface. Species composition varies according to different physiochemical factors in different regions (Whitton, Kawecka, and Rother, 1988). Depending on the season, latitude, cloud cover, plant canopy, and water turbidity, the light intensity supports enough photosynthesis for maintaining algal growth, which is inhibitory for photodependant biological nitrogen fixation (Roger, 1991). Usually nitrogen-fixing cyanobacteria *Cylindrospermum, Nostoc, Anabaena, Calothrix, Rivularia, Aulosira, Scytonema, Tolypothrix,* and *Westiellopsis* species are dominant in rice field soils. From these in general *Nostoc, Anabaena,* and *Cylindrospermum* species possess higher growth rates and less nitrogen-fixing capacity, contrary to *Calothrix, Rivularia, Tolypothrix, Aulosira,* and *Westiellopsis,* which have a relatively slower growth rate and possess a higher nitrogen-fixing capacity (Nayak et al., 1996).

SEASONAL VARIATION OF CYANOBACTERIAL GROWTH IN PADDY FIELDS

Seasonal variation affects the growth of cyanobacteria in various ways in different regions. In countries such as India, the growth of nitrogen-fixing cyanobacteria occurs particularly during the mid-September to mid-December. In equatorial Senegal, the seasonal pattern is different from the rest of the world (Roger and Reynaud, 1976, 1977). In high temperatures and high light intensities, chlorophytes were dominant. In low light intensity, heterocystous and nonheterocystous cyanobacteria were dominant. Singh (1961) and Roger and Reynaud (1979) reported substantial variation in cyanobacterial species' biomass and abundance in soils during the paddy cultivation cycle. Singh (1961) detected high growth of cyanobacteria in the early part of the rainy season; 95 percent of flora recorded in dry periods were spores of heterocystous cyanobacteria (Roger and Reynaud, 1976). This shows that growth of cyanobacteria in the rice field depends upon several seasonal factors. Seasonal variation in climatic regime influences the properties of the rice field soils, hence showing their overall influence on the occurrence of cyanobacteria in rice field soils.

Light

The growth of cyanobacteria in rice fields usually depends on light. Being phototrophic, they are restricted to photic zones. Besides, some cyanobacterial species are able to live in dim light and are also facultative heterotrophically using organic substances in darkness. Usually they are found on the upper 0.5 cm horizon of the soil and at the deeper horizons in dormant conditions as spores or filaments (Chapman and Chapman, 1973; Roger and Reynaud, 1976). In temperate zones, cyanobacteria can grow easily at a high rate in warm conditions (Gupta, 1966) but can limit their growth in wet conditions where light intensity is low (Ichimura, 1954; Watanabe, 1961). Photoinhibition of photosynthetic organisms is dependent upon light intensity, duration of light exposure, and light quality. Blue light is particularly involved in photoinhibition of photosynthesis (Samuelsson et al., 1985). Under transplanted rice, light is reduced by 50 percent after 15 days, 85 percent after 30 days, and 95 percent after 60 days (Kurosawa, 1956; Roger and Reynaud, 1982), which influences the distribution of cyanobacteria during paddy cultivation cycles.

Temperature

Temperature is rarely a limiting factor for cyanobacteria in paddy fields, because the range of temperatures permitting growth of cyanobacteria is larger than that required by rice. However, it influences both cyanobacterial biomass composition and productivity (Roger and Reynaud, 1982). Cyanobacteria in rice fields showed better growth at a temperature range of 30 to 40°C (Singh, 1975). Temperatures higher than 40°C have a deleterious effect on cyanobacterial growth in rice fields (Venkataraman, 1964; Roger and Kulasooriya, 1980). Low temperatures decrease the growth of the organisms and favor the growth of eukaryotic algae. As high temperature is associated with high light intensities, it is easy to relate the effect of high irradiance on occurrence of cyanobacteria. A survey of 102 rice soils showed a higher incidence of spore-forming *Nostoc* in dry soils (80 percent) than in wet soils (47 percent) (Roger et al., 1987).

pH

Cyanobacteria can grow in a wide range of pH variations. In rice fields pH variation occurs due to photosynthetic removal of HCO_3^- (Whitton, Aziz, Irancis, et al., 1988). pH depends on the time of day and on crop phase. Variation of pH affects occurrence and abundance of nitrogen fixation of cyanobacteria (Okuda and Yamaguchi, 1956; Roger and Reynaud, 1977; Roger et al., 1987). In natural conditions, they grow in a neutral to alkaline range. A clear correlation of pH of soil and water with cyanobacteria abundance has been reported. Correlation was observed in

1. water pH and cyanobacteria number (Garcia et al., 1973),
2. soil pH and N_2-fixing cyanobacteria biomass (Roger et al., 1987),
3. soil pH and number of spores of N_2-fixing cyanobacteria during the dry season (Garcia et al., 1973), and
4. soil pH and growth of cyanobacteria (Okuda and Yamaguchi, 1952).

Roger et al. (1987) showed that pH lower than 6.5 has a significant influence on the growth of cyanobacteria. In culture media the optimal pH for cyanobacterial growth ranged from 7.5 to 10, and the lower limit was about 6.5 to 7 (Chapman et al., 1972; Prasad et al., 1978; Roger and Reynaud, 1979). In the rice fields of Spain cyanobacterial growth was favored at a pH of 8 (Quesada et al., 1995). However, cyanobacteria can also grow at low pH values, between 5 and 6 (Prasad et al., 1978) and at pH lower than 5 (Aiyer, 1965; Mastuguchi and Yoo, 1979; Roger et al., 1987).

Chemical Characteristics of Soil

Sixteen elements are required for growth of living organisms. Of these carbon, hydrogen, and oxygen not considered as mineral nutrients are obtained from atmosphere. The remaining 13, grouped as macro- or micronutrients depending on the quantitative requirement, are available in the soil. For growth of cyanobacteria in the rice fields, all the nutrients are available in the soil. The essential nutrients are available in ionic form, such as P as $H_2PO_4^-$, N as NO_3^- or NH_4^+, K as K^+, and Mg as Mg^{+2}. Of these, phosphate nitrogen and iron are the major nutrients which influence cyanobacterial growth under field conditions.

PHOSPHORUS

Phosphorus is commonly available in the form of phosphate. It helps in the biosynthesis of nucleotides, nucleic acids, lipids, and covalently modified proteins in cyanobacterial metabolism. The abundance of cyanobacteria in the field depends upon the P availability (Roger and Kulasooriya, 1980; Roger et al., 1987). Application of P in rice fields stimulates cyanobacterial growth and photodependent biological nitrogen fixation (De and Mandal, 1956; Wilson and Alexander, 1979; Roger and Kulasooriya, 1980; Roger et al., 1987). In rice fields 40 to 80 kg $P_2O_5 \cdot ha^{-1}$ is required for cyanobacteria. P availability in soil was greatly limited by the soil pH. In soil it tends to become bound as calcium phosphate at high pH (Bisoyi and Singh, 1988).

The rate of phosphate uptake depends on many factors. Rate of phosphate uptake was optimal between pH 7.5 and 9 and declined in the acidic range (Healy, 1973; Ullrich-Eberius and Yingchol, 1974; Lawry and Jensen, 1979). Optimal activity was observed at pH 8 and 10 (Healy and Hendzel, 1979). P uptake by cyanobacteria is also influenced by temperature, with higher uptake rate at 35°C and declining rate below 30°C (Healy, 1973).

Nitrogen

Nitrogen plays a crucial role in sustaining crop productivity. Cyanobacteria mainly use inorganic nitrogen (nitrate, ammonium, and dinitrogen), but urea and other organic sources of N, such as amino acids, are also assimilated (Kartz and Myers, 1955; Neilson and Larsson, 1980; Rawson, 1985). Cyanobacteria use fixed nitrogen (e.g., nitrate or ammonium) and support the system for N_2 fixation only in the absence of a source of combined N (Stewart and Rowell, 1975).

Nitrate is the most abundant source of combined N for cyanobacterial nutrition. Nitrate assimilation by cyanobacteria involves nitrate uptake and reduction of intracellular nitrate to ammonium. Nitrate uptake takes place through a transport system that exhibits a high affinity for nitrate (Flores et al., 1983). Nitrate-promoted inhibition of nitrogenase synthesis operates at the level of *nif* HDK (Fujita et al., 1991; Huang and Chou, 1991; Martin-Nieto et al., 1991).

Ammonium has negative effects on nitrogenase synthesis and thus affects heterocyst development. Ammonium present in the growth medium did not allow the development of heterocysts (Singh, 1975; Anand and Karuppusamy, 1987) and also inhibited nitrogenase in fully developed heterocysts (Rowell et al., 1977; Ramos et al., 1985). At a high pH, in other words, at pH 10, ammonia concentration at the range 0.5 to 1 mM provokes an inactivation of nitrogenase through a modification of Fe protein (dinitrogenase reductase) (Böhm et al., 1992).

Iron

Iron normally exists in two ionic forms, Ferrous (Fe II) and Ferric (Fe III). Under anaerobic conditions, iron is present as the ferrous (Fe II) ion. In aerobic conditions, iron is present as ferric (Fe III), which is biologically available iron utilized by microorganisms, though some photosynthetic microorganisms can meet their iron requirement from Fe II (Anderson and Morel, 1980; Finden et al., 1984). Iron is involved in a wide variety of biochemical processes, both as a cofactor and a catalyst. These include photosynthesis, chlorophyll synthesis, nitrate reduction, N_2 fixation, and other biochemical redox reactions. Nitrogen metabolism and electron transport are extremely sensitive to iron limitation. Iron deficiency in several cyanobacteria resulted in a decrease in thylakoid membrane, phycobilisome, carboxysomes (Sherman and Sherman, 1983), alteration of the thylakoid protein (Guikema and Sherman, 1984), and peak shift in chlorophyll pigment (Pakrasi et al., 1985).

Salinity

Sodium (Na) exists in soils as solution in the form of NaCl, Na_2SO_4, and Na_2CO_3. It is absorbed by plants and photosynthetic microorganisms as the Na^+ ion. Salinity is one of the most important factors in nature leading to severe crop loss. Many cyanobacteria exhibit considerable tolerance to salt

(NaCl) (Padhi et al., 1998). Cyanobacteria usually require Na^+ for growth (Allen and Arnon, 1955), N_2 fixation (Apte and Thomas, 1980), nitrate reduction, and intracellular pH regulation (Kaplan et al., 1984). High concentration of NaCl has a much more inhibitory effect on the cyanobacterial cells, inhibition of growth (Batterton and Van Baalen, 1971; Anantani and Vaidya, 1983), declining rate of photosynthesis (Vonshak and Richmond, 1981) and respiration (Vonshak et al., 1988), decrease in protein synthesis (Kumar and Kaushik, 1994), and has an adverse effect on N_2 fixation (Apte and Thomas, 1997). Due to an increase of salinity in heavy irrigated soils, there is a potential threat on the decrease of salt-sensitive cyanobacterial forms which might influence the overall nitrogen economy of rice fields.

CYANOBACTERIA BIOFERTILIZER FOR INCREASED CROP YIELD AND MAINTENANCE OF SOIL QUALITY

It is well established that waterlogged conditions are conducive for the growth of blue-green algae (cyanobacteria), and fertility of rice field soils is attributed to the presence of these phototrophic organisms. More specifically the heterocystous forms of cyanobacteria fix atmospheric nitrogen even in the presence of sunlight that is made available to the plants during its life cycle period or after its death by the decomposition of the cells. Thus cyanobacteria play an important role in cycling nutrients, decomposition of the waste residues, and many other related activities to maintain soil health and productivity. Hence, when such organisms are inoculated into the crop field for increasing productivity they are referred to as biofertilizers or microbial inoculants. Biofertilizers are preparations of live or latent cells that are used for application of seed, soil, or composting in the field with the objective of increasing the number of such organisms in that particular system to accelerate their metabolic activity to augment the extent of availability of nutrients in a form which can be easily assimilated by plants.

Cyanobacterial biofertilizer is an easily manageable and self-generating system which not only provides nutrients to plants in terms of nitrogen, amino acids, and growth-promoting substances but also improves soil health and texture (Bertocchi, et al., 1990; Roger and Burns, 1994). In India, De (1939) and Singh (1961) suggested that cyanobacteria play an important role in maintaining fertility of tropical rice fields. Since then a number of investigators carried out work at different agricultural universities and the ICAR Institute of India to study the role of BGA inoculation to the rice crop (Goyal and Venkataraman, 1970, 1971; Aiyer et al., 1972; Manhdane et al., 1990; Adhikary and Sahu, 2000b). The inoculation of rice fields with cyanobacteria (BGA), also termed as algalization, was carried out by

Venkataraman and his associates (Venkataraman, 1961, 1972; Goyal and Venkataraman, 1971) and convincingly showed beneficial effects on grain yield and nitrogen savings in general. Studies by various authors have shown that BGA, besides contributing nitrogen to the tune of 25 to 30 $kg \cdot ha^{-1}$ per season, increase the productivity of rice up to 10 to 15 percent (Singh, 1985; Singh and Bisoyi, 1989; Adhikary and Sahu, 2000a; Dwivedi et al., 2000). Therefore, in areas where chemical fertilizers were not used or marginal farmers cannot afford to use these commercial inputs, cyanobacterial biofertilizer can give a benefit of 25 to 30 kg $N \cdot ha^{-1}$. Even in the presence of chemical fertilizers this biofertilizer can provide supplementary effects leading to a reduction in the need for doses of chemical nitrogen by 25 to 30 kg per crop. There are several other benefits which have been found out due to cyanobacterial inoculation in the field. These include addition of organic matter to the soil (Singh et al., 1981), mobilizing bound phosphate (Bisoyi and Singh, 1988), improving aeration and waterlogging capacity of soil, and increasing soil aggregation (Goyal et al., 1997).

ALGALIZATION TECHNOLOGY

Based on the natural ecology of cyanobacteria in rice field ecosystems, a cost-effective and easily adoptable biofertilizer technology was developed by Venkataraman (1978) at IARI, New Delhi, India. This basic method involves mass production of a mixture of cyanobacteria constituting nitrogen-fixing species of *Anabaena, Nostoc, Plectonema, Tolypothrix,* and *Aulosira* in shallow trays or cemented tanks in open air using soil as the carrier material. These organisms have been screened as better in terms of growth rate and nitrogen-fixing capacity from among the several species isolated out of more than 2,000 soil samples collected from different part of India through the All India Coordinated Project on Algae at IARI. For mass culture in the trays and tanks, for each square meter 2 kg of soil and 100 g of single superphosphate are added. To this, 5 mL of Malathion or 25 mg of Furadon granule (Carbofuran, 3 percent) is used to prevent breeding of mosquitoes and other insects. The contents are thoroughly mixed and allow to settle. Cyanobacteria culture containing all the strains is sprinkled on the surface. These organisms multiply in the tanks when exposed to direct sunlight and form a thick mat over the surface in 10 to 12 days. The contents are allowed to dry and the dried flakes are collected, packed, and used to inoculate at the rate of 10 $kg \cdot ha^{-1}$ in the rice fields after transplantation. The advantage of the technology is that farmers, after getting the soil-based culture from IARI or any of the recognized cyanobacteria multiplication centers, can produce these biofertilizers on their own as per the requirement with minimum and easily manageable inputs. Later on this technology was further elaborated and, depending on the requirement of the biofertilizers, four different production methods have been developed, including the tank method, pit method (making pits of variable size with a polythene lining at the base), and field method or nursery cum field production method (cultivated in leveled fields covering large areas (Venkataraman, 1981). All these methods have been practiced extensively since 1978; however, they have many limitations. The main limitation is essentially climatic, as production is affected by low temperatures in winters and by washing out of the organisms in rainy season, resulting in the product being highly unpredictable in terms of amount of cyanobacterial biomass. Further, the open-air tank culture method is quickly and easily contaminated by other organisms that may outgrow the desired ones. Infestation by insects and other predators often results in a product during such cultivation practices that does not contain desired organisms in an appropriate quantity. Thus, it was felt that though algalization technology certainly holds promise and the economics also favorable, large-scale production and harnessing maximum benefits from cyanobacterial biofertilizers is possible only through resolution of these major constraints during production.

Upgrading the Technology

During the past decade, in a number of ways the problems concerning quality biofertilizer production have been worked out. It was felt that large-scale production of cyanobacterial biofertilizers with equal proportion, definite population size, and good quality is possible if multiplication is carried out under controlled conditions within defined parameters. Many such approaches were taken by different scientists to achieve the goal. These are discussed in the following sections.

Production of Cyanobacteria in Polybags

Mixing fresh cyanobacteria biomass with carrier material yielded good quality inoculum with longer shelf life. Cyanobacterial strains were cultivated in polybags separately, mixed together after harvest, filtered, and mixed with carrier before packing. Such biofertilizers are free from contaminants and therefore perform better after inoculation.

Development of Alternate Carrier

Carrier is important for biofertilizer formulation. The shelf life of the biofertilizers depends on the quality of the carrier. A few characteristics of the carrier were considered before selecting. It should not be toxic to cyanobacteria and should have good surface area, high absorption capacity, and good buffering capacity; and it should be able to carry high microbial load, be easily available, and be cost-effective. Wheat and rice straw, and also coconut coir cut into small pieces, have been successfully developed to be used as a carrier. Individual cyanobacteria is grown separately in alternate cheap media (Prasanna et al., 1998) in shallow tanks and the fresh harvested mass is properly mixed with a definite quantity of carrier. The cyanobacterial load in the carrier is maintained at 10^6 cells per gram carrier so that only 400 g of biofertilizer is necessary to inoculate 1 acre of rice field. Straw carrier has also been used in a different way in which desired organisms were grown with pulverized and sterilized straw under controlled conditions in poly alkene bags (Shanmugasundaram, 1996). These methods resulted in a product of high quality and, to a large extent, free from other contaminating microorganisms. In these methods, since different cyanobacteria are grown individually and then mixed with carriers in the desired ratio, it ensures an appropriate share of all the organisms used in the inoculum. Unlike soil-based inoculum, this biofertilizer is supplied as ready-to-use quality material; therefore farmers can get it directly from their local markets.

Multanimetti As Carrier

Recently, Goyal et al. (1997) has reported that *multanimetti,* a montmorillonite clay, can be used as a suitable carrier for cyanobacteria biofertilizers. It fulfills most of the qualities of a good carrier. It is a powder form of carrier with small particle size, good buffering capacity, and highly absorptive nature. It is also microbially inert. Under this method cultures were grown separately under controlled conditions in polyhouses or culture rooms. The production medium in the outdoor cultivation contained tapewater, superphosphate, and a small amount of insecticide. The cyanobacteria in the growing state are taken out from the tanks and mixed with the clay in the ratio of 1:1 and dried in the sun. The dried material is finely powdered and packed. This biofertilizer in multanimetti carrier possesses a longer shelf life and can be used even after two years of storage at ambient conditions.

Besides these methods, different pond designs, simple raceways fitted with agitators, production under controlled conditions in polyhouses, and growing in alternate cheap media have also been worked out to attain many generations of quality biomass of cyanobacterial biofertilizers.

FUTURE PERSPECTIVES

The ability of cyanobacterial biofertilizer to contribute nitrogen for rice crop productivity and maintenance of soil quality has been proven by several scientists in the past through extensive field trials. But the practice of mass production of cyanobacterial biofertilizer for field use in polythene lined pits/tanks in the open air, as developed earlier, is not scientifically viable. Despite the simplicity of this easily manageable technology, a number of technical, environmental, and socioeconomic constraints limit the wider adaptation of this technology. The major problems, which need to be addressed for wider acceptance of the technology, are as follows:

- Due to wide variation in agroclimatic conditions, a single biofertilizer product developed at a particular place cannot be effective throughout a region/country. Hence, biofertilizer products containing region-specific strains should be developed for use at the field level.
- The cyanobacteria in biofertilizer packs usually shows variable properties to survive and compete with native flora for establishment, growth, and nitrogen fixation. Hence, analysis of ecological factors that influence the fate and performance of inoculated cyanobacteria should be understood, and accordingly ecologically compatible species/strains should be used for production of region-specific biofertilizers.
- To meet current market demand, there are lack of adequate mass production units, nonavailability of quality materials, and lack of organized network for maintaining and distributing the biofertilizers. Hence cyanobacterial biofertilizer production units using region-specific and ecologically compatible species with the provision of quality control by a microbiologist/agricultural graduate should be set up at the block level, and even at grampanchayat level in intensive rice-grown areas. Further, the old method of production of cyanobacterial biofertilizer in tanks/pits should be discouraged and instead open-air cemented tanks-cum-polyhouses used for production of quality biofertilizers under controlled conditions. Biofertilizer packs should be prepared using appropriate carrier materials without compromising quality.
- Presently general awareness about the use of cyanobacterial biofertilizer and its benefits in rice cultivation is lacking. Information in the form of literature, posters, and other aids should be provided at the user level. The gap between the scientists, agriculture extension workers, and farmers must be bridged. Appropriate measures are to be

taken to train agriculture extension officials from time to time to make them aware of the latest developments in the subject and to upgrade the usage of the technique at the field level. Extensive field-level training should form an integral component for popularizing biofertilizer technology. An organizational setup at the regional level for production, quality control, distribution, and marketing for easy availability of biofertilizers at the farmer's doorstep is essential.

There is immense market potential of cyanobacterial biofertilizer in view of the 43 million ha of land under rice cultivation in India alone. This can be achieved by producing quality biofertilizers using region-specific and ecologically compatible species/strains following appropriate production techniques and adapting an effective popularization campaign, making it a people's movement.

CONCLUSIONS

The refined BGA biofertilizer technology developed in the recent past has immense market potential considering that 43 million ha are under rice cultivation in India alone. The demand of cyanobacterial biofertilizer is estimated to be 2,600 tons (Motsara et al., 1995), which might be much higher in the twenty-first century. However, its production irrespective of the quality and method of production is only a fraction of the demand. To date cyanobacterial biofertilizer is produced only by government organizations following pit, tank, or field-cum-nursery-production methods developed 20 years ago which have been proved ineffective in subsequent years. Even the commercial houses have not come forward to adopt the recently developed refined production technology for generating quality biofertilizer to overcome the difficulties in the technology. Only when the recently upgraded technology for cyanobacterials biofertilizer production in polyhouses under controlled conditions using quality inoculum from authentic sources, together with the quality assurance of biofertilizers products, will be adopted will the technology become common for farmers and commercial houses as well. For an extension of the technology, farmers in the rural areas need to be educated through extensive demonstration trials on the use of biofertilizers. Cost of the biofertilzers should not be emphasized, as was done earlier in case of soil-based BGA inoculum that it is a cheap product. Once the farmers come forward to adopt the refined cyanobacteria biofertilizer technology, being convinced on its performance in the field, it would lead to large-scale commercialization of this technology.

REFERENCES

Adhikary S.P. (2000). Ecophysiological studies of cyanobacteria under diverse environmental conditions. Doctoral thesis, Utkal University, Bhubaneswar, India.

Adhikary S.P. and Pattnaik H. (1981). Effect of organic carbon sources on the liberation of extracellular amino acids by a nitrogen fixing blue-green alga *Westiellopsis prolifica* Janet in light and dark. *Indian Journal of Botany* 4: 60-69.

Adhikary S.P. and Sahu J.K. (2000a). Studies on the establishment and nitrogenase activity of inoculated cyanobacteria in the field and their effect of yield of rice. *Oryza* 37(1): 39-43.

Adhikary S.P. and Sahu J.K. (2000b). Survival strategies of cyanobacteria occurring as crust in the rice fields under drought conditions. *Indian Journal of Microbiology* 40: 53-56.

Aiyer R.S. (1965). Comparative algological studies in rice fields in Kerala state. *Agricultural Research Journal Kerala* 3: 100-104.

Aiyer R.S., Salabudeen S.A., and Venkataraman G.S. (1972). Longterm algalization field trial with high yielding varieties of rice (*Oryza sativa* L.). *Indian Journal of Agricultural Science* 42: 380-383.

Allen M.B. and Arnon D.I. (1955). Studies on nitrogen fixing blue green algae I. Growth and nitrogen fixation by *Anabaena cylindrica. Plant Physiology* 30: 366-372.

Anand N. and Karuppusamy A. (1987). Growth of nitrogen fixing and non-nitrogen fixing blue green algae in presence of some common fertilizers. *Phykos* 26: 22-26.

Anantani Y.S. and Vaidya B.S. (1983). Halotolerance in a blue-green alga: Shifts in electrophoretic patterns of soluble proteins and enzyme peroxidase under salt stress. *Phykos* 22: 108-112.

Anderson M.A. and Morel F.M.M. (1980). Uptake of Fe (III) by a diatom in oxic culture medium. *Moecular Biological Letter* 1: 263-268.

Apte S.K. (1992). Molecular biology of cyanobacterial nitrogen fixation: Recent advances. *Indian Journal of Microbiology* 32: 103-126.

Apte S.K. and Prabhavathi N. (1994). Rearrangements of nitrogen fixation *(nif)* genes in the heterocystous cyanobacteria. *Journal of Biosciences* 19: 579-602.

Apte S.K. and Thomas J. (1980). Sodium is required for nitrogenase activity in cyanobacteria. *Current Microbiology* 3: 291-293.

Apte S.K. and Thomas J. (1997). Possible amelioration of coastal soil salinity using halotolerant nitrogen fixing cyanobacteria. *Plant and Soil* 189: 205-211.

Batterton J.C. Jr. and Van Baalen C. (1971). Growth responses of blue-green algae to sodium chloride concentrations. *Archives Microbiology* 76: 151-165.

Bertocchi C., Navarini L., Cesaro A., and Anastasio M. (1990). Polysaccharides from cyanobacteria. *Carbohydrate Polymers* 12: 127-153.

Bisoyi R.N. and Singh P.K. (1988). Effect of seasonal changes on cyanobacterial production and nitrogen yield. *Microbial Ecology* 16: 149-154.

Böhm T., Halbherr A., Smaglinski S., Ernst A., and Böger P. (1992). In vitro activation of dinitrogenase reductase from the cyanobacterium *Anabaena variabilis* (ATCC 29413). *Journal of Bacteriology* 174: 6179-6183.

Castenholz R.W. (1969). The thermophilic cyanophytes of iceland and the upper temperature limit. *Journal of Phycology* 5: 360-368.

Chapman R.L., Bayer D.E., and Lang N.J. (1972). Observations on the dominant algae in experimental California rice fields. *Journal of Phycology* Suppl. 8: 17.

Chapman V.J. and Chapman D.J. (1973). Soil algae and symbiosis. In D.J. Chapman (ed.), *The algae* (pp. 381-387). London: Mac Millan Publication.

Das S.C., Mandal B., and Mandal L.N. (1991). Effect of growth and subsequent decomposition of blue-green algae on the transformation of iron and manganese in submerged soils. *Plant and Soil* 138: 75-84.

De P.K. (1939). The role of blue-green algae in nitrogen fixation in rice fields. *Proceedings of Royal Society London* 127: 129-139.

De P.K. and Mandal L.N. (1956). Fixation of nitrogen by algae in rice fields. *Soil Science* 81: 453-458.

Dwivedi C.P., Sinha S.K., Ashraf S.S., Verma D.C., and Dubey A.K. (2000) Application of blue green algae (BGA) biofertilizer in paddy cultivation under farm conditions on Uttar Pradesh. In P.K. Singh, D.W. Dhar, S. Pubbi, and R. Parasanna (eds.), *Biofertilizer—Blue-green algae and Azolla* (pp. 133-141). New Delhi, India: Venus Printers and Publishers.

Fay P. (1969). Cell differentiation and pigment composition in *Anabaena cylindrica. Archives Microbiology* 67: 62-70.

Fay P. (1983). *The blue greens.* Arnold-Heinemann Press.

Fay P., Stewart W.D.P., Walsby A.E., and Fogg G.E. (1968). Is the heterocyst the site of nitrogen fixation in blue-green algae? *Nature* (London) 220: 810-812.

Finden D.A.S., Tippings E., Jaworeki G.H.M., and Reynod C.S. (1984). Light-induced reduction of natural iron (III) oxide and its relevance to phytoplankton. *Nature* (London) 309: 783-784.

Flores E., Guerrero M.G., and Losada M. (1983). Photosynthetic nature of nitrate uptake and reduction in the cyanobacterium *Anacystis nidulans. Biochemica Biophysica Acta* 722: 408-426.

Fogg G.E. (1971). Extracellular products of algae in fresh water. *Archives Hydrobiology* 5: 1-25.

Fritsch F.E. (1907). The substantial and freshwater algal flora of the tropics. *Annals Botany* 30: 235-275.

Fujita Y., Takahasi Y., Shonai F., Ogura Y., and Matsubara M. (1991). Cloning, nucleotide sequences and differential expression of the *nif* H and H-like (frx c) genes from the filamentous nitrogen-fixing cyanobacterium *Plectonema boryanum. Plant Cell Physiology* 32: 1093-1106.

Gallon J.R., Perry S.M., Rajab T.M.A., Flayeh K.A.M., Junes J.S., and Chaplin A.E. (1988). Metabolic changes associated with the diurnal pattern of N_2-fixation in *Gloeothece. Journal of General Microbiology* 134: 3079-3087.

Garcia J.L., Raimbault M., Jacq V., Rinaudo G., and Roger P. (1973). Microbial activities in paddy fields in Senegal: Influence of physico-chemical properties of soils and of the rhizosphere effect. *Review of Ecology and Biology* 11: 169-185.

Golden J.W., Robinson S.J., and Haselkorn R. (1985). Rearrangement of nitrogen fixation genes during differentiation in the cyanobacterium *Anabaena. Nature* (London) 314: 419-423.

Goyal S.K., Singh B.V., Nagpal V., and Marwaha T.S. (1997). An improved method for production of algal biofertilizer. *Indian Journal of Agricultural Science* 67: 314-315.
Goyal S.K. and Venkataraman G.S. (1970). Effect of algalization on high yielding rice varieties. *Phykos* 9: 137-138.
Goyal S.K. and Venkataraman G.S. (1971). Effect of algalization on high yielding rice varieties II. *Phykos* 10: 32-33.
Grist D.H. (1975). *Rice*. London: Longman Press.
Guikema J.A. and Sherman L.A. (1984). Influence of iron deprivation on the membrane composition of *Anacystis nidulans. Plant Physiology* 74: 90-95.
Gupta A.B. (1966). Algal flora and its importance in the economy of rice fields. *Hydrobiologia* 28: 518-523.
Hallenbeck P.C. (1987). Molecular aspects of nitrogen fixation by photosynthetic prokaryotes. *CRC Critical Review of Microbiology* 14: 1-48.
Haselkorn R. (1978). Heterocysts. *Annual Review of Plant Physiology* 29: 319-344.
Haselkorn R., Golden J.W., Lammens P.J., and Mulligan M.E. (1987). Rearrangement of *nif* genes during cyanobacterial heterocysts differentiation. *Philippines Trans Royal Society* (London) 317: 173-181.
Healy F.P. (1973). Characteristics of phosphorus deficiency in *Anabaena. Journal of Phycology* 9: 383-394.
Healy F.P. and Hendzel L.L. (1979). Fluorometric measurement of alkaline phosphatase activity in algae. *Freshwater Biology* 9: 429-439.
Hoffmann L. (1989). Algae of terrestrial habitats. *Botanical Review* 55: 77-105.
Huang T.C. and Chou W.M. (1991). Setting of the circadian N_2-fixing rhythm of the prokarytic *Synechococcus* sp. RF-1 white its *nif* gene is repressed. *Plant Physiology* 96: 324-326.
Ichimura S. (1954). Ecological studies on the plankton in paddy fields. I. Seasonal fluctuations in the standing crop and productivity of plankton. *Japanese Journal of Botany* 14: 269-279.
Kaplan A., Volokita M., Zenvirth O., and Reinhold L. (1984). An essential role for sodium in the HCO_3-transporting system in the cyanobacterium *Anabaena variabilis. FEBS Lettesr* 176: 166-168.
Kartz W.A. and Myers J. (1955). Nutrition and growth of several blue-green algae. *American Journal of Botany* 42: 282-287.
Krumbein W.E. (1987). Microbial interactions with mineral materials. In D.R. Houghton and H.O.W. Eggins (eds.), *Biodeterioration* (pp. 78-100). New York: Elsevier Applied Science.
Kumar H. and Kaushik B.D. (1994). Response of *Calothrix brevissima* to salt. *Indian Journal of Microbiology* 34: 37-41.
Kurosawa H. (1956). The weekly succession in the standing crop of plankton and zoobenthos in the paddy field Part I and Part II. *Bulletin Research Science Japan* 41: 86-98; 45: 73-84.
Lambein F. and Wolk C.P. (1973). Structural studies on the glycolipids from the envelope of the heterocyst of *Anabaena cylindrica. Biochemistry* 12: 791-798.

Lawry N.H. and Jensen T.E. (1979). Deposition of condensed phosphate as an effect of varying sulfur deficiency in the cyanobacterium *Synechococcus* sp. *(Anacystis nidulans). Archives Microbiology* 120: 1-7.

Malliga P. and Subramanian G. (2001). Usefulness of coir waste as a carrier for BGA biofertilizer. In P.K. Singh, D.W. Dhar, S. Pabbi, R. Prasana, and A. Arora (eds.), *Recent advances in the exploitation of blue-green algae and azolla* (pp. 59-66). New Delhi, India: Venus Printers and Publishers.

Manhdane V.K., Patil P.L., and Bhanvase D.B. (1990) Effect of different levels of blue green algal and nitrogen on yield of paddy. In B.D. Kaushisk (ed.), *Cyanobacterial nitrogen fixation* (pp. 315-320). New Delhi, India: Associated Publishers Company.

Marathe K.V. (1972). Role of some BGA in soil aggregation. In T.V. Desikachary (ed.), *Taxonomy and biology of blue green algae* (pp. 328-331). Madras, India: Madras University Press.

Marsalek B., Zahradnickova M., and Heonkova M. (1992). Extracellular abscisic acid produced by cyanobacteria under salt stress. *Journal of Plant Physiology* 139: 506-508.

Martin-Nieto J., Herrero A., and Flores E. (1991). Control of nitrogenase mRNA levels by products of nitrate assimilation in the cyanobacterium *Anabaena* sp. strain PCC 7120. *Plant Physiology* 97: 825-828.

Matsuguchi T. and Yoo I.D. (1979). Stimulation of phototrophic N_2 fixation in paddy fields through rice straw applications. In R. Wetselaar, J.R. Simpson, and T. Rosswall (eds.), *Proceedings of nitrogen cycling in South-east Asian wet monsoonal ecoystems* (pp. 18-25). Canberra: Australian Academy of Science.

Mazor G., Kidron G.J., Vonshak A., and Abeliovich A. (1996). The role of cyanobacterial exopolysaccharides in structuring desert microbial crusts. *FEMS Microbiol Ecology* 21: 121-130.

Mishra S. and Kaushik B.D. (1989). Growth promoting substances of cyanobacteria I. Vitamins and their influence on rice plant. *Proceedings of Indian National Science Academy B* 55: 295-300.

Motsara M.R., Bhattacharyya P., and Srivastava B. (1995). *Biofertilizer technology, marketing and usage—A source book-cum-glossary.* New Delhi, India: Fertilizer Development and Consultation Organisation.

Mullineaux P.M., Gallon J.R., and Chaplin A.E. (1981). Acetylene reduction (nitrogen fixation) by cyanobacteria grown under alternating light dark cycles. *FEMS Microbiology Letter* 10: 245-247.

Murry M.A. and Wolk C.P. (1989). Evidence that the barrier to the penetration of oxygen into heterocysts depends on two layers of the cell envelope. *Archives Microbiology* 151: 469-474.

Nayak H., Sahu J.K., and Adhikary S.P. (1996). Blue-green algae of rice fields of Orissa State II. Growth and nitrogen fixing potential. *Phykos* 35: 111-118.

Neilson A.H. and Larsson T. (1980). The utilisation of organic nitrogen for growth of algae: Physiological aspects. *Physiologia Plantarum* 48: 542-553.

Neuer G., Papen H., and Bothen H. (1983). Heterocyst biochemistry and differentiation. In G.C. Papageorgiou and L. Packer (eds.), *Photosynthetic prokaryotes:*

Cell differentiation and function (pp. 219-242). New York: Elsevier Science Publishing Inc.

Okuda A. and Yamaguchi M. (1952). Algae and atmospheric nitrogen fixation in paddy soils. II- Relation between the growth of blue-green algae and physical or chemical properties of soil and effect of soil treatments and inoculation on the nitrogen fixation. *Memoir of Research Institute of Food Science* 4: 1-11.

Okuda A. and Yamaguchi M. (1956). Nitrogen fixing microorganisms in paddy soils. II. Distribution of blue-green algae in paddy soils and the relationship between the growth of them and soil properties. *Soil and Plant Food* 2: 4-7.

Padhi H., Rath B., and Adhikary S.P. (1998). Tolerance of nitrogen-fixing cyanobacteria to NaCl. *Biologia Plantarum* 40: 261-268.

Paerl H.W. and Bland P.T. (1982). Localized tetrazolium reduction in relation to N_2-fixation, CO_2 fixation and H_2 uptake in aquatic filamentous cyanobacteria. *Applied Environmental Microbiology* 43: 218-226.

Pakrasi H.B., Riethman H.C., and Sherman L.A. (1985). Organization of pigment proteins in the photosystem II complex of the cyanobacterium *Anacystis nidulans* R_2. *Proceedings of National Academy of Science*, USA 82: 6903-6907.

Pattanaik B. and Adhikary S.P. (2002). Blue-green algal flora at some archaeological sites and monuments of India. *Feddes Repertorium* 113: 289-300.

Prasad B.N., Mehrotra R.K., and Singh Y. (1978). On pH tolerance of some soil BGA. *Acta Botanica Indica* 6: 130-138.

Prasanna R., Sharma B.K., Sharma R.K., and Kaushik B.D. (1998). Standardization of growth parameters and formulation of medium for cyanobacterial biofertilizer strains. *Indian Journal of Microbiology* 38: 211-215.

Quesada A., Sanchez-Maeso, E., and Fernadez-Valiente E. (1995). Seasonal variations in the physical and chemical characteristics of a shallow water ecosystem, the rice fields of Valencia, Spain. *Archives Hydrobiology* 132: 494-511.

Rai L.C. (2001). Cyanobacterial ecology and environmental management. In P.K. Singh, D.W. Dhar, S. Pabbi, R. Prasanna, and A. Arora (eds.), *Recent advances in the exploitation of blue-green algae and* azolla (pp. 31-44). New Delhi, India: Venus Printers and Publishers.

Ramos J.L., Maduero F., and Guerrero M.G. (1985). Regulation of nitrogenase levels in *Anabaena* sp. ATCC 33047 and other filamentous cyanobacteria. *Archives Microbiology* 141: 105-111.

Rawson D.M. (1985). The effects of exogenous amino acids on growth and nitrogenase activity in the cyanobacterium *Anabaena cylindrica* PCC 7122. *Journal of General Microbiology* 131: 2549-2554.

Rippka R., Deruelles J., Waterbury J.B., Herdman M., and Stanier R.Y (1979). Generic assignments, strain histories and properties of pure cultures of cyanobacteria. *Journal of General Microbiology* 111: 1-61.

Roger P.A. (1991). Reconsidering the utilization of cyanobacteria in wetland rice cultivation. In S.K. Dutta and C. Sloger (eds.), *Biological N_2 fixation associated with rice production* (pp. 119-141). New Delhi, India: Oxford and IBH.

Roger P. and Reynaud P. (1976). Dynamics of the algal populations during a culture cycle in a Sahel rice field. *Review Ecology and Biology of Soil* 13: 545-560.

Roger P. and Reynaud P. (1977). Algal biomass in rice field of Senegal: relative importance of cyanobacteria that fix nitrogen. *Review Ecology and Biology of Soil* 14: 519-530.

Roger P. and Reynaud P. (1979). Ecology of blue-green algae in paddy fields. In P.A. Roger (ed.), *Nitrogen and rice* (pp. 289-309). Los Banos, Philippines: IRRI.

Roger P. and Reynaud P. (1982). Free-living blue-green algae in tropical soils. In Y.R. Dommergues and H. Diem (eds.), *Microbiology of tropical soils and plant productivity* (pp. 147-168). The Hague, the Netherlands: Martinus Nijhoff/Dr. W. Junk.

Roger P.A. and Kulasooriya S.A. (1980). *Blue-green algae and rice.* Los Banos, Laguna, Philippines: IRRI.

Roger P.A., Santiago-Ardales S., Reddy P.M., and Watanabe I. (1987). The abundance of heterocystous blue-green algae in rice soils and inocula used for application in rice fields. *Biology and Fertility of soils* 5: 98-105.

Roger S.L. and Burns R.G. (1994). Changes in aggregate stability, nutrient status, indigenous microbial populations and seedling emegence following inoculation of soil with *Nostoc muscorum. Biology and Fertility of Soils* 18: 209-215.

Rowell P., Enticott S., and Stewart W.D.P. (1977). Glutamine synthase and nitrogenase activity in the blue-green alga *Anabaena cylindrica. New Phytology* 79: 41-54.

Roychoudhury P., Kaushik B.D., Krishnamurthy G.S.R., and Venkataraman G.S. (1979). Effect of BGA and *Azolla* application on the aggregation status of the soil. *Current Science* 48: 454-455.

Samuelsson G., Lonneborg A., Rosenquist E., Gustafsson P., and Oquist G. (1985). Photoinhibition and reactivation of photosynthesis in the cyanobacterium *Anacystis nidulans. Plant Physiology* 79: 992-995.

Saxena S. and Kaushik B.D. (1992). Polysaccharides (biopolymers) from halotolerant cyanobacteria. *Indian Journal of Experimental Biology* 30: 433-434.

Schoof J.W. (1983). *Earth earliest biosphere—Its origin and evolution.* Princeton, NJ: Princeton University Press.

Shanmugasundaram S.S. (1996). *Consolidated Report—Mission mode project on Technology Development and Demonstration of Algal Biofertilizer.* Madurai, India: Madurai Kamraj University.

Sherman D.M. and Sherman L.A. (1983). Effect of iron deficiency and iron restoration on ultrastructure of *Anacystis nidulans. Journal of Bacteriology* 156: 393-401.

Singh P.K. (1975). Fertilizer tolerance of blue-green algae and their effect on heterocyst differentiation. *Phykos* 14: 81-88.

Singh P.K. (1985). Nitrogen fixation by the blue green algae in paddy fields. In P.N. Jaiswal (ed.), *Rice research in India* (pp. 344-362). New Delhi, India: Indian Council of Agriculture Research.

Singh P.K. (1998). Algal biofertilizer production: Current status. In P.K. Singh and S. Pabbi (eds.), *Mass production of blue green algal biofertilizer* (pp. 1-6). New Delhi, India: Indian Agricultural Research Institute.

Singh P.K. and Bisoyi R.N. (1989). Blue-green algae in rice fields. *Phykos* 28: 181-195.

Singh P.K., Panigrahi B.C., and Satapathy K.B. (1981). Comparative efficiency of *Azolla,* blue-green algae and other organic manures in relation to N and P availability in flooded rice soil. *Plant and Soil* 62: 35-44.

Singh R.N. (1961). *The role of blue green algae in nitrogen economy of Indian Agriculture.* New Delhi: Indian Council of Agriculture Research.

Stal L.J. and Krumbein W.E. (1987). Temporal separation of nitrogen fixation and photosynthesis in the filamentous non-heterocystous cyanobacterium *Oscillatoria* sp. *Archives Microbiology* 149: 76-80.

Stewart W.D.P. (1980). Some aspects of structure and function in N_2 fixing cyanobacteria. *Annual Review of Microbiology* 34: 497-536.

Stewart W.D.P., Haystead A., and Dharmawardene W.N. (1975). Nitrogen assimilation and metabolism in blue green algae. In W.D.P. Stewart (ed.), *Nitrogen fixation by free-living microorganisms* (pp. 129-158). London: Cambridge University Press.

Stewart W.D.P. and Rowell P. (1975). Effects of L-methionine-DL-Sulphoximine on the assimilation of newly fixed NH_3, acetylene reduction and heterocyst production in *Anabaena cylindrica. Biochemistry Biophysics Research Communication* 65: 846-856.

Thomas J. (1970). Absence of pigments of photosystem II of photosynthesis in heterocysts of a blue-green algae. *Nature* (London) 228: 181-183.

Tirol A.C., Roger P.A., and Watanabe I. (1982). Fate of nitrogen from a blue-green alga in flooded rice soil. *Soil Science and Plant Nutrition* 28: 559-569.

Tripathi U., Pabbi S., and Singh P.K. (2001). Status of blue-green algal biofertilizer in India. In P.K. Singh, D.W. Dhar, S. Pabbi, R. Prasanna, and A. Arora (eds.), *Recent advances in the exploitation of blue-green algae and* Azolla (pp. 45-57). New Delhi, India: Venus Printers and Publishers.

Tripathy P., Roy A., Anand N., and Adhikary S.P. (1999). Blue-green algal flora on the rock surface of temples and monuments of India. *Feddes Repertorium* 110: 133-144.

Ullrich-Eberius, C.I. and Yingchol Y. (1974). Phosphate uptake and its pH dependence in halophytic and glycophytic algae and higher plants. *Oecologia* 17: 17-26.

Van der Ancker J.A.M., Jungerius P.D., and Mur L.R. (1985). The role of algae in the stabilization of coastal dune blow outs. *Earth Surface Process Landforms* 10: 189-192.

Venkataraman G.S. (1961). The role of blue-green algae in agriculture. *Science and Culture* 27: 9-13.

Venkataraman G.S. (1964). Thermal resistance and viabililty of microalgae. *Oikos* 64: 26-30.

Venkataraman G.S. (1972). *Algal biofertilizer and rice cultivation.* New Delhi, India: Today and Tomorrow's Printers and Publishers.

Venkataraman G.S. (1978). *Algal biofertilizer for rice.* New Delhi, India: Information Mannual, Shiv Narayan Printers.

Venkataraman G.S. (1981). *Blue-green algae for rice production a manual for its promotion.* FAO field document No.2 (RAS/75/004) FAO, Rome, p. 102.

Venkataraman G.S. and Neelakanthan S. (1967). Effect of cellular constituents of nitrogen fixing blue-green alga *Cylindrospermum* on root growth of rice plants. *Journal of General and Applied Microbiology* 13: 53-62.

Vonshak A., Guy R., and Guy M. (1988). The response of filamentous cyanobacterium *Spirulina platensis* to salt stress. *Archives Microbiology* 150: 417-420.

Vonshak A. and Richmond A. (1981). Photosynthetic and respiratory activity in *Anacystis nidulans* adapted to osmotic stress. *Plant Physiology* 68: 504-511.

Watanabe A. (1961). Collection and cultivation of nitrogen-fixing blue green algae and their effect on the growth and crop yield of rice plants. *Stud Rokugawa Institute,* Tokyo 9: 162-166.

Watanabe A. and Yamamoto Y. (1971). Algal nitrogen fixation in the tropics. *Plant and Soil* (special volume): 403-413.

Watanabe I., De Datta S.K., and Roger P.A. (1987). Nitrogen cycling in wetland rice soils. In J.R. Wilson (ed.), *Advances in nitrogen cycling in agricultural ecosystems* (pp. 239-256). London: CAB International.

Whitton B.A. (1965). Extracellular products of blue-green algae. *Journal of General Microbiology* 40: 1-11.

Whitton B.A. (1992). Diversity, ecology and taxonomy of the cyanobacteria. In N.H. Mann and N.G. Carr (eds.), *Photosynthetic prokaryotes* (pp. 1-51). New York: Plenum Press.

Whitton B.A., Aziz A., Irancis P., Rother J.A., Simon, J.W., and Tahmida Z.N. (1988). Ecology of deepwater rice fields in Bangladesh 1. Physical and Chemical environment. *Hydrobiologia* 169: 3-22.

Whitton B.A., Aziz A., Kawecka B., and Rother J.A. (1988). Ecology of deep water rice fields in Bangladesh 3. Associated algae and macrophytes. *Hydrobiologia* 169: 31-42.

Wilson J.T. and Alexander M. (1979). Effect of soil nutrient status and pH on nitrogen-fixing algae in flooded soils. *Soil Science American Journal* 43: 936-939.

Winkenbach F. and Wolk C.P. (1973). Activities of enzymes of the oxidative and reductive pentose phosphate pathways in heterocysts of a blue-green algae. *Plant Physiology* 52: 480-483.

Wolk C.P. (1982). Heterocysts. In N.G. Carr and B.A. Whitton (eds.), *The biology of cyanobacteria* (pp. 359-386). Oxford: Blackwell.

Wolk C.P., Cai Y., Cardemil L., Flores E., Hohn B., Murry M., Schmetterer G., Schrautemeier B., and Wilson R. (1988). Isolation and complementation of mutants of *Anabaena* PCC 7120 unable to grow aerobically on dinitrogen. *Journal of Bacteriology* 170: 1229-1244.

Wolk C.P. and Simon R.D. (1969). Pigments and lipids of heterocysts. *Planta* 97: 126-134.

Chapter 18

A Comparative Study on Nitrogen-Fixing Cyanobacteria in South American and European Rice Fields

Carlos Hugo Prosperi

INTRODUCTION

Cyanobacteria, also called blue-green algae (BGA), have been well-known since ancient times as important contributors to rice field fertilization. They are distributed worldwide, and their capability as fertilizers derives from their potential to fix atmospheric nitrogen and elaborate nitrogenous compounds, either as free-living organisms or in symbiotic association with the water fern *Azolla* (Agarwal, 1979; Fay, 1983; Singh and Singh, 1983; Shi and Hall, 1988). The existence of an agronomic potential for cyanobacteria in rice cultures was first recognized with scientific bases by De (1939), who noticed that the natural fertility of several rice fields was due to biological nitrogen fixation performed by these microorganisms (Roger, 1989b).

Some rice fields frequently show visually obvious growths of cyanobacteria in the form of a greenish scum over the water surface, although eukaryotic green algae (Chlorophycae) may be more abundant if high quantities of nitrogenous fertilizers are added (Whitton and Roger, 1989). Research on the practical utilization of cyanobacteria as biofertilizers in rice fields has been focused in general on the inoculation with these microorganisms produced in soil (Venkataraman, 1981).

Cyanobacteria usually grow naturally in the same fields that are used for rice culture. In spite of this fact, inoculation of cyanobacteria in order to increase yields is a very common practice in most countries, especially in those cases where they are not abundant (Gupta, 1966). Ancient farmers in

I wish to express my gratitude to Drs. Eduardo Fernandez Valiente and Antonio Quesada del Corral (Autonomous University of Madrid, Spain) for their cooperation.

Southeast Asia empirically knew that fields forming a blue-green scum after irrigation were better than others for rice. The abundance of cyanobacteria increases with the rise in pH and phosphorous soil content, and they are also very dependent upon other environmental factors such as light intensity, water temperature, and grazing by aquatic zooplankton or small invertebrates that feed on them (Prosperi and Bellone, 1994; Prosperi, 2000).

Inoculation techniques are mainly based on the use of foreign strains with well-known nitrogen-fixing capacity tested in laboratory cultures, but after inoculation in the field they have to compete with indigenous strains, which are presumably better adapted to local environmental conditions (Sharma and Gupta, 1983).

The first quantitative study to establish the behavior of foreign strains after being inoculated in a rice field is that of Reddy and Roger (1988). In this study, the fate of five laboratory-grown heterocystous strains was studied during one month in plots of several soils. During that month, the inoculated strains grew satisfactorily in all kinds of soils but did not become dominant over the indigenous strains. Grazers also played an important role in controlling the development of the foreign strains, which formed blooms very rarely and contributed poorly to soil fertilization (Roger, 1988).

In natural field conditions, with a rich indigenous flora, it is very likely that local strains will outgrow all other cyanobacterial populations derived from laboratory isolates (Roger, 1988).

This has an important implication: that the number and contribution of inoculated strains to total nitrogen fixation is normally lower than that of indigenous strains (Reddy and Roger, 1988). As a consequence, agricultural practices have been proposed that enhance the growth of local strains for producing inocula, as the better way to improve this capability of cyanobacteria for biofertilization (Prosperi et al., 1992, 1993).

The selection of local or indigenous strains for this purpose should take into account the response of their nitrogen-fixing capacity to environmental conditions that can limit or modify it (Roger and Kulasooriya, 1980). Thus, the proper taxonomy of the organisms involved is a condition sine qua non for accomplishing of this purpose, considering that the positive or negative reactions to environment may vary significantly in relation to the species or strain selected as a potential biofertilizer (Fernandez Valiente et al., 1996, 1997).

This is the main reason why it is not possible to establish comparisons between the physiological or ecological behavior of cyanobacterial strains without perfect identification of such strains or species and sharp knowledge of its taxonomic position.

The agronomic potential of nitrogen-fixing cyanobacteria for a cultural cycle of rice has been calculated by Roger (1989b) in about 30 Kg $N \cdot ha^{-1}$. Even being moderate as compared with industrial fertilizers, this potential is much better than nothing for farmers that cannot afford the high costs of agrochemicals. Such could be the case in many countries, such as India, Bangladesh, Vietnam, or China, where cyanobacteria naturally growing in wetlands have been a blessing for those populations with high rice consumption in their everyday diet (Watanabe and Roger, 1984; Prosperi, 2000).

In many Western countries, attempts have been made to use cyanobacteria as biofertilizers, but the reasons are economical or ecological. It has been demonstrated that the intense use and even abuse of different nitrogen compounds, such as urea or nitrates in order to get higher yields, has an undesirable consequence: the excess of nitrate or ammonium is a strong contaminant for continental waters, either shallow or underground. The visible results are the eutrophication of freshwater bodies that become unfit for human use, and the death of fishes.

Modern agronomic practices are founded on the sustainable use of agroecosystems, which means a rational exploitation of the fields without serious damage or alteration of the natural environment (Prosperi, 2000). The use of biofertilizers instead of agrochemicals is an important development in that way.

In Europe and in America, nitrogen-fixing cyanobacteria are under study as a replacement for industrial fertilizers because of their nonpolluting properties and high potential for sustainable cropping systems. Spain is the most important rice producer in the European Community, with intensive practices of rice cultivation in the area of Valencia, and a high population density. In South America, Argentina is a country with a low population density, and rice is cultivated extensively mainly in the areas of Corrientes and Entre Ríos, also known as Mesopotamia, because it is surrounded by two large rivers, Parana and Uruguay (Prosperi et al., 1996; Maggi and Prosperi, unpublished).

In Valencia, intensive rice culture is the consequence of the addition of huge amounts of chemical fertilizers to the soils. As a result, it is quite common to see dead fish floating on the surface of the Albufera, a lake located close to the Mediterranean coast that collects the waste waters from the rice fields, which are loaded with trouble-causing nitrogen compounds (Prosperi et al., 1992).

In Argentina, both of the mentioned rice-cultivating areas use the Parana River as their main water source, the most important affluent of La Plata River. The amount of organic pollution in this river is not considerable at the moment, because fertilizers are seldom used and the river is very large. But this situation may change in future years if rice production increases. The use of cyanobacteria seems to be a good alternative.

The main objectives of this work were the taxonomic study of indigenous cyanobacterial strains growing naturally in rice fields located both in Spain and in Argentina, the evaluation of their capabilities as nitrogen fixers and, thus, as potential biofertilizers, and finally a comparative study of the results obtained from both countries.

MEASURING NITROGEN FIXATION IN THE FIELD

A new incubation device was used for in situ measurement of nitrogen fixation in rice fields. This device consists of a rigid transparent plastic bottomless bottle provided with an agitation system, and was developed by members of the Autonomous University of Madrid (Spain) and is used both in Valencia and in Corrientes (Quesada et al., 1989).

As it is well-known, the most widely used method to assess nitrogen fixation in field and laboratory studies is the measurement of acetylene-reducing activity, which is directly proportional to nitrogenase activity (Dennis and Turpin, 1990). In rice fields, one of the methods used is the collection of soil or water samples and their later incubation under laboratory conditions, and the other possibility is the direct evaluation by means of in situ assays. This second option has the advantage that it does not disturb the natural conditions of soil and/or water, but it has some problems that arise mainly from the slow diffusion of acetylene and ethylene in the floodwater (Lee and Watanabe, 1977; Roger and Kulasooriya, 1980).

In order to avoid these inconveniences, the previously mentioned device has an assay chamber. This chamber comprises a rigid, transparent, bottomless 5 L plastic bottle with an internal system of agitation consisting of a metal rod and a clamp holder specially designed to support a small engine with a screw propeller connected to a battery. The system permitted the adjustment of the propeller in such a way that the blades were slightly submerged, thus permitting agitation of both air and water during incubation time but without increasing turbidity. The device is easily portable and very inexpensive, thus allowing to perform several simultaneous measurements. The use of a rigid bottle eliminates the problems derived from the use of nonrigid bags, such as changes in internal volumes and leakage of gases, and the use of an agitation system overcomes problems caused by the slow diffusion of gases in the water (Prosperi et al., 1993).

A rubber stopper at the neck of the bottle allowed the injection of acetylene and removal of air and ethylene by means of a syringe. This gas mixture was then reinjected in a vacuum tube with a rubber stopper and transported to the laboratory to be tested via gas chromatograph (Quesada et al., 1989).

DETERMINATION OF NITROGENASE ACTIVITY IN VIVO

Nitrogenase activity under laboratory conditions was measured by the acetylene reduction technique, in 15 mL aliquots of cell suspensions placed in stoppered 35 mL vials. To start the assay, 10 percent of the air was withdrawn and replaced with the same volume of acetylene. Incubation was carried out in a shaking water bath, illuminated with a quantum flux density of 100 μmol photon·m^{-2}·s^{-1}. After acetylene addition, 0.5 mL samples were taken at time intervals with gas-tight syringes, and the ethylene concentration was determined by peak area integration in a gas chromatograph (Minchin, 1986; Prosperi, 1994a).

SAMPLES COLLECTION AND IDENTIFICATION

The most used methods for evaluating algal abundance in rice fields or soils in general are (1) plating on soil suspension, (2) direct counting, or (3) pigment analysis. The three methods have advantages and disadvantages.

The last one, pigment analysis, gives no identification of species. It is useful only for quantitative purposes and, because of contamination by humic acid and chlorophyll degradation products, can be used only to compare the effects of different treatments on the total flora in the same soil. It is a relatively easy test to perform, but the results are not as useful when applied to soils than when applied, for example, in limnology (Roger and Watanabe, 1982).

The technique of dilution and plating permits the simultaneous enumeration, identification, and even isolation of different living organisms of the sample. Accuracy of this procedure can be increased by the use of selective culture media, such as nitrogen free, for example, if the study refers to nitrogen-fixing cyanobacteria. But the main problem of this method is that it does not ensure all the species present in the soil will grow and develop on the plates; only those which can grow under those laboratory conditions and have akinetes or spores or some form of vegetative structure for fast reproduction will flourish. Even those species that grow on the plate can change their relative frequency, and it does not distinguish between active or inactive forms.

Direct counting and examination by light microscopy, although time-consuming, is the most satisfactory method. Nevertheless, it happens frequently that cyanobacterial filaments are more or less broken or separated in pieces and the accurate determination of species or even genera may be very difficult, if not impossible in some cases. This is because a homogenized soil suspension is used. The method thus requires some practice and expertise in systematics and morphology of cyanobacteria, and cannot be carried out by beginners. Another problem involved when counting soil samples is the great amount of dust or many other particles that can cover the organisms themselves or contribute to confusion. In these cases, the use of ultraviolet light is recommended, because chlorophylls produce a natural red-colored fluorescent emission when stimulated with ultraviolet light, which makes then very easily distinguishable from other particles in the suspension. Of course, a special microscope prepared with adequate illumination devices and several kinds of filters to protect the operator's eyes is necessary (Roger and Watanabe, 1982).

Some general sampling problems also affect all three enumerated techniques and are related to the distributional ecology of cyanobacteria in a rice field. The validity of organism enumeration (or, for the same reasons, nitrogenase activity measurements) in situ depends mainly on the density of sampling, which varies for a given degree of accuracy according to the distribution law of variables. Cyanobacterial populations and some other algae in soils have an uneven distribution that approximates a logarithmic normal pattern. Therefore, a high density of sampling is usually needed.

THE DIVERSITY OF CYANOBACTERIA

The main cyanobacterial species found in the area of Corrientes province, in Argentina, were the following, according to the classification of Bourrelly (1970) and following the identification keys by Anand (1990) and Desykachary (1959) (see Figure 18.1):

- Family Chroococcaceae: *Aphanocapsa banaresensis, A. elachista, Chroococcus limneticus, Gloeothece rupestre, Microcystis aeruginosa, M. elabens*
- Family Scytonemataceae: *Plectonema notatum*
- Family Microchaetaceae: *Microchaete tenera*
- Family Nostocaceae: *Anabaena laxa, A. doliolum, Calothrix marchica, Nostoc microscopicum*
- Family Oscillatoriaceae: *Lyngbya limnetica, L. kuetzingiana, Oscillatoria rubescens, Spirulina princeps, S. subsalsa*

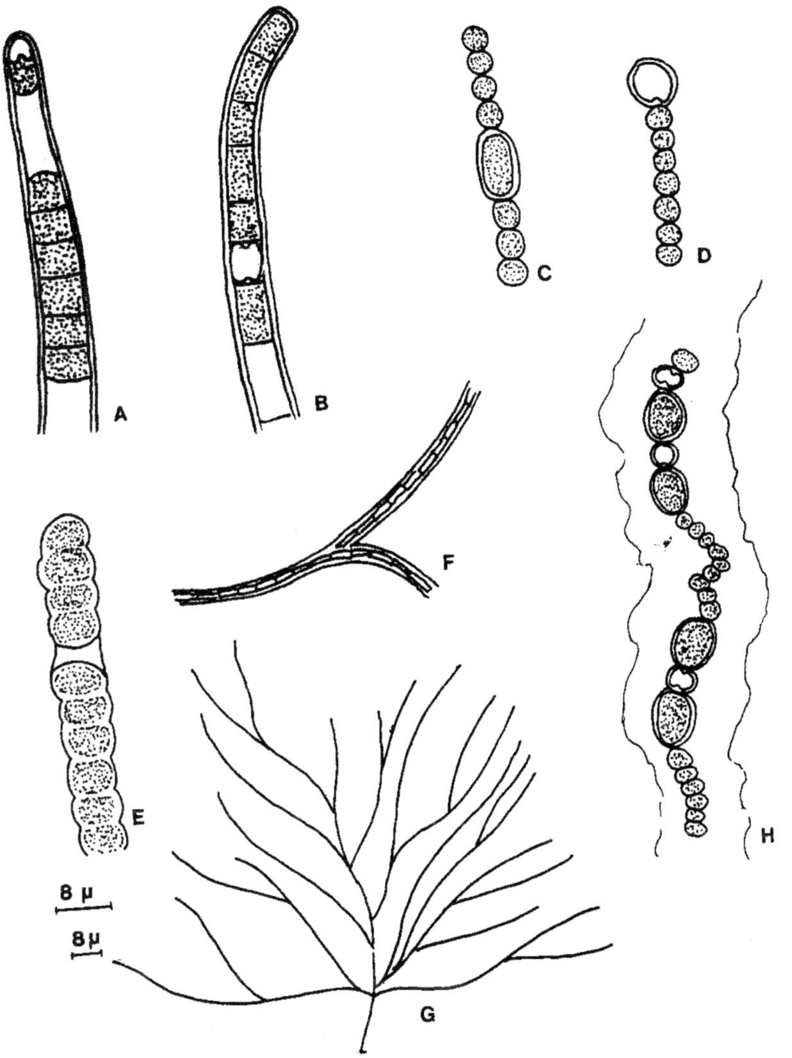

FIGURE 18.1. Cyanobacteria isolated from the rice fields of Corrientes. A-B: *Microchaete tenera;* C-D: *Anabaena laxa;* E: *Oscillatoria rubescens;* F-G: *Plectonema notatum;* H: *Nostoc microscopicum.* Bigger scale is for A-G; smaller one for H. All hand-drawings made with the help of a camera lucida.

In Valencia, Spain, the more conspicuous species found were the following, according to the same classification:

Family Nostocaceae: *Anabaena variabilis, Calothrix marchica, Nostoc punctiforme, Nodularia spumigena*

Among all of the previously mentioned species, those belonging to the family Nostocaceae are heterocystous forms and thus nitrogen-fixing organisms and potential biofertilizers. Several works (Prosperi et al., 1992; Prosperi and Fernandez Valiente, 1993) were carried out on the strains found in Valencia (Spain) that confirm their nitrogenase activity and how it is influenced by environmental factors in the rice fields (Prosperi et al., 1992).

Chroococcus limneticus, instead, is a unicellular nonheterocystous strain, that was very abundant in the soils of Corrientes (Argentina) but not in Spain (see Figure 18.2). A recent research work demonstrated that it is a

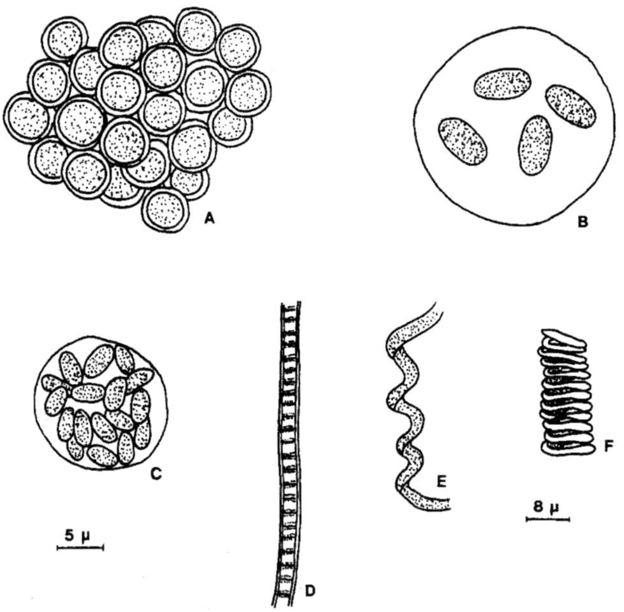

FIGURE 18.2. Cyanobacteria isolated from the rice fields of Corrientes. A: *Aphanocapsa banaresensis;* B: *Gloeothece rupestris;* C: *Microcystis elabens;* D: *Lyngbya kuetzingiana;* E: *Spirulina princeps;* F: *Spirulina subsalsa.* The 5 μ scale is for A, B, and C; the 8 μ scale is for D, E, and F. All hand-drawings made with the help of a camera lucida.

nitrogen fixer too (Rincón and Prosperi, unpublished); nitrogenase is protected from oxygen by means of the superoxide-dismutase enzyme and high respiration rate. Besides, in filamentous heterocystous forms, photosynthesis is conducted in vegetative cells, whereas nitrogenase activity takes place in the heterocysts. This provides a spatial separation between both activities; otherwise, the oxygen produced by photosynthetic activity would result in a serious inhibition of nitrogen fixation (Ernst et al., 1990; Prosperi et al., 1993). Being that *Chroococcus limneticus* is unicellular, as said, there is no possible spatial separation, but instead there is a temporal separation: photosynthesis occurs, obviously, during daylight, with very low nitrogenase activity, but in the dark periods the cells became very active in respiration and nitrogen fixation (Rincón and Prosperi, unpublished) (see Photos 18.1 to 18.3).

The input of nitrogen fertilizers to fields, when added as a supplement to cyanobacterial nitrogen fixation, reduces nitrogenase activity (Roger and Kulasooriya, 1980). At neutral to alkaline pH, ammonium inhibits nitrogenase synthesis and even heterocyst differentiation, giving rise to a progressive inhibition of nitrogenase activity as existing enzyme concentration diminishes (Guerrero and Lara, 1987).

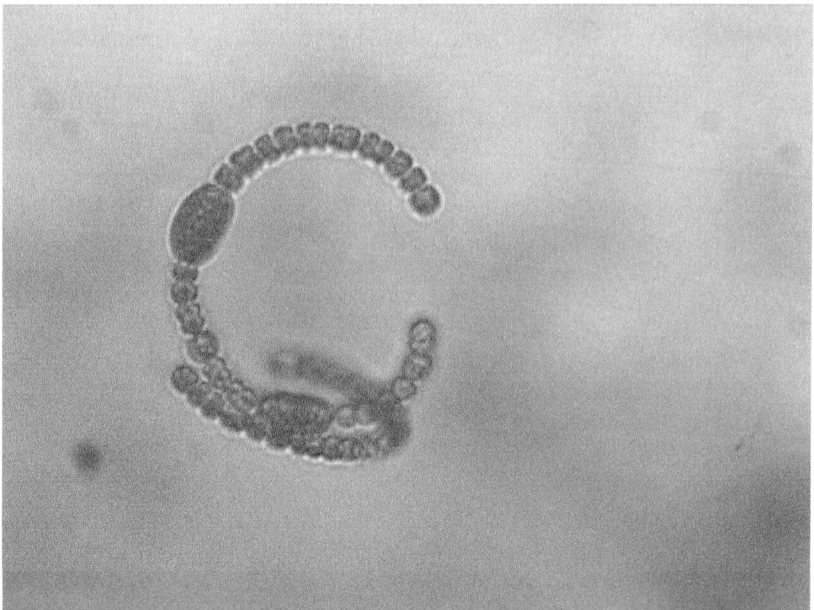

PHOTO 18.1. *Anabaena* sp. microphotograph.

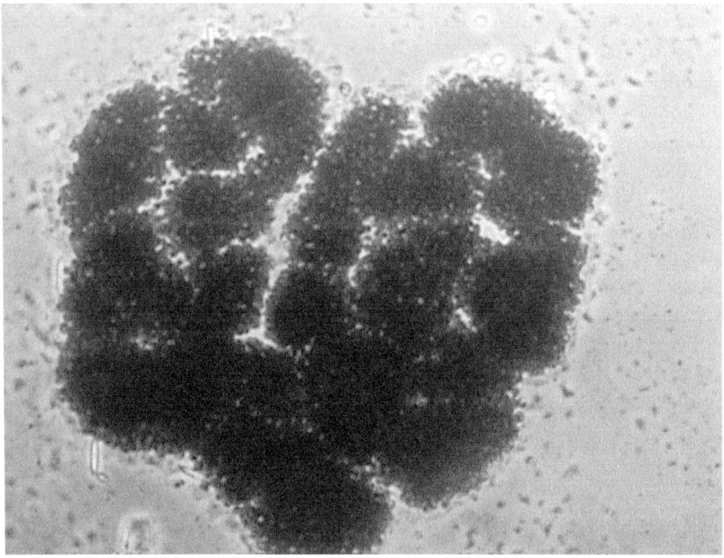

PHOTO 18.2. *Microcystis* sp. microphotograph.

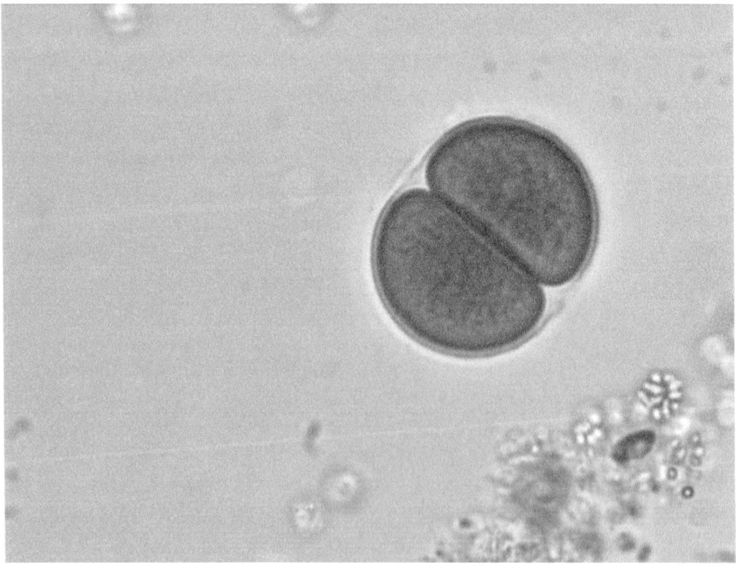

PHOTO 18.3. *Chroococcus* sp. microphotograph.

In addition to this mentioned long-term effect, it is known that ammonium produces a fast nitrogenase activity inactivation, due to a reversible modification of dinitrogenase reductase (Ernst et al., 1990).

The effect of pH on nitrogenase activity was measured using cyanobacteria isolated from the rice fields of Valencia (see Figure 18.3). The experiments conducted at pH 10 showed a more noticeable inhibition than those at pH 7 (Prosperi et al., 1993). This stronger inhibition at pH 10 could be the result of a change in membrane permeability of the microorganisms, thus allowing a massive difussion of nitrogen compounds in the cell. Of course, such high values of pH do not commonly happen under natural conditions, and the results shown here were obtained in laboratory conditions. Nevertheless, we found pH values near 9.5 both in the rice fields of Corrientes and those of Valencia exceptionally; that was on warm sunny

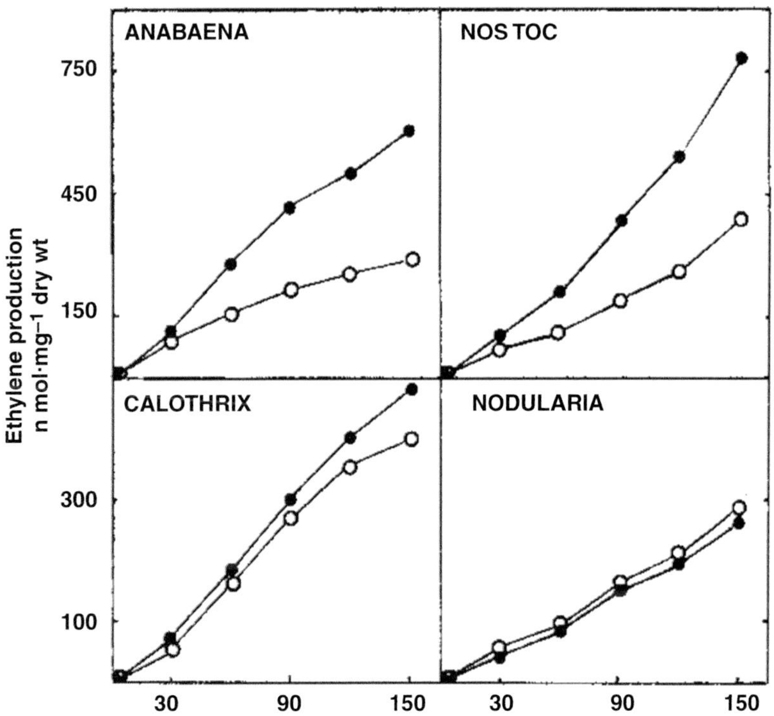

FIGURE 18.3. Effect of pH on nitrogenase activity of cyanobacteria isolated from the rice fields of Valencia. Organisms were cultured for 24 h at pH 7 (closed circles) or pH 10 (open circles). Values are the means of three different experiments, with a variability lower than 10 percent.

days, when the high levels of temperature and irradiance promoted active photosynthetic activity in cyanobacteria (data not shown).

There are evidences that light intensity influences nitrogen fixation: in general, it is possible to say that the higher the irradiance the higher the nitrogenase activity. This is true of course if cultures are irradiated with light intensities between physiological values.

During the light cultivation cycle in rice fields, underwater light intensity measured at full sunlight may range between 700 μmol photon·m^{-2}·s^{-1} at the onset of the cycle to only 5 μmol photon·m^{-2}·s^{-1} at the end of the cycle (Prosperi et al., 1992).

To check the response of some Spanish cyanobacterial strains, nitrogenase activity was assayed at limiting and saturating light intensities, with cells previously grown at 60 μmol photon·m^{-2}·s^{-1} (see Figure 18.4).

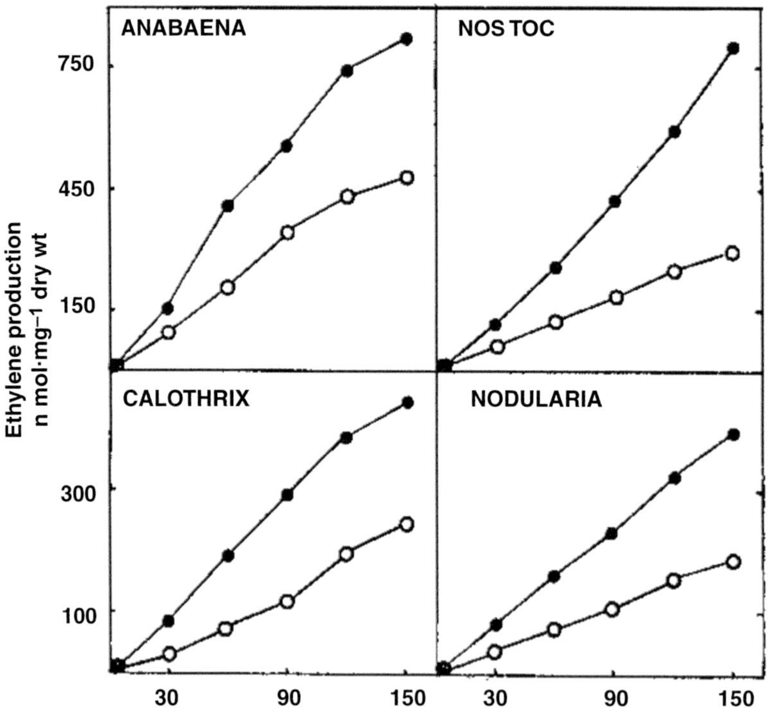

FIGURE 18.4. Short-term effect of irradiance on nitrogenase activity of cyanobacteria isolated from the rice fields of Valencia. Organisms were grown during 24 h at 60 μmol photon·m^{-2}·s^{-1}. Assays of nitrogenase activity were made at 200 (closed circles) or 5 (open circles) irradiation. Values are the means of three different experiments with a variability lower than 10 percent.

Respiration also plays an important role in the control of nitrogen fixation, as far as it can diminish the intracellular oxygen concentration. Oxygen inhibits nitrogenase activity at near the average atmospheric percentage (Fay, 1983).

Those organisms which are nitrogen fixers have developed several mechanisms to protect nitrogenase and, at the same time, grow under normal aerobic conditions (Prosperi, 1994b).

Comparing cells growing in a nitrogen-free medium and thus fixing nitrogen from the atmosphere, it is demonstrated that respiration and nitrogen fixation have almost opposite behaviors. Using a gradient of increasing oxygen concentrations (starting from 2 percent and up to 80 percent), nitrogenase activity is high at low oxygen, but decreases as oxygen increases. On the contrary, an initially low respiration rate becomes very high when oxygen increases. This suggests that oxygen consumption by means of respiration plays a protective role for nitrogenase activity, by lowering the intracellular oxygen concentration (Prosperi, 1994b).

FUTURE PERSPECTIVES

Calothrix marchica was the most promising strain tested in our previous works to be used as a biofertilizer for rice fields, as far as it could fix nitrogen at very high levels and could continue doing so even under severely restricted conditions, such as the presence of nitrates or ammonia in water and high pH values (Prosperi et al., 1993). Inhibition of nitrogenase happens in rice fields at the moment when chemical fertilizers are added, but after a certain time they are diluted by incoming irrigation water, thus restarting normal nitrogenase activity (Prosperi and Fernandez Valiente, 1993).

This particular strain of the genus *Calothrix* grows naturally, as an indigenous form, both in Valencia (Spain) and in Corrientes (Argentina). Many of its physiological properties are already known, and future research should be conducted to optimize its capabilities as a biofertilizer compatible with a moderate use of agrochemicals.

Another genus found growing naturally both in Argentina and in Spain is *Nostoc*. The nitrogen-fixation capabilities of several strains of *Nostoc* are well-known, either under laboratory conditions or in the field (Pantastico and Gonzalez, 1976; Martinez et al., 1978; Prosperi, 1986, 1989).

Future research on biofertilizers for South American and Mediterranean rice fields should give particular emphasis to the study of the various strains of *Calothrix* and *Nostoc*. Both genera bear the advantages of being indigenous in those areas and forming heterocystous filaments, with enlarged capabilities for nitrogen fixation and a strong ability to resist variable or unfavorable environmental conditions.

APPENDIX: IDENTIFICATION KEY FOR MOST COMMON CYANOBACTERIA IN RICE FIELDS OF ARGENTINA AND SPAIN

I. Cells forming trichomes
  A. With heterocysts
    1. Endings attenuated
      a. *Microchaete tenera*
    2. Endings not differentiated
    3. Forming benthic colonies
      a. *Nostoc microscopicum*
    4. Free-living, planktonic forms
    5. Cells spherical
      a. *Anabaena laxa*
    6. Cells shorter than wide
      a. *Anabaena doliolum*
  B. Without heterocysts
    1. Filaments with false branching
      a. *Plectonema notatum*
    2. Filaments without false branching
    3. Trichomes straight
    4. Trichomes 1 to 4 μm wide
    5. Trichomes 1 to 2 μm wide
      a. *Lyngbya limnetica*
    6. Trichomes 3 to 4 μm wide
      a. *Lyngbya kuetzingiana*
    7. Trichomes 6 to 8 μm wide
      a. *Oscillatoria rubescens*
    8. Trichomes coiled
    9. Trichomes 3 to 4 μm wide
      a. *Spirulina princeps*
    10. Trichomes 1.5 to 2 μm wide
      a. *Spirulina subsalsa*

II. Cells not forming trichomes
  A. Colonies up to eight elongated cells
    1. *Gloeothece rupestris*
  B. Colonies with more than eight spherical or elliptical cells
    1. Colonies microscopic, planktonic

a. Cells elliptical
  i. *Microcystis elabens*
b. Cells spherical
  i. *Microcystis aeruginosa*

2. Colonies macroscopic, nonplanktonic
  a. Cells up to 6 μ.m
    i. *Aphanocapsa banaresensis*
  b. Cells up to 2 μ.m
    i. *Aphanocapsa elachista*

REFERENCES

Agarwal A. (1979). Blue-green algae to fertilize Indian rice paddies. *Nature* 279: 181.

Anand A. (1990). *Handbook of blue-green algae* (of rice fields of south India). Bishen Singh Mahendra Pal Singh. 23 IA, Connaught Place, Dehra Dun 248001-1-80.

Bourrelly P. (1970). *Les algues d' eau douce*. N. Boubee & C.

De P. (1939). The role of blue-green algae in nitrogen fixation in rice fields. *Proc. R. Soc. London* 127 B: 121-139.

Dennis D. and Turpin D. (1990). *Plant physiology and molecular biology*. Longman. Singapore.

Desykachary T. (1959). *Cyanophyta*. ICAR.

Ernst A., Reich S., and Boger P. (1990). Modification of dinitrogenase reductase in the cyanobacterium *Anabaena variabilis* due to C starvation and ammonia. *J. Bacteriol.* 172: 748-755.

Fay P. (1983). *The blue-greens*. Southampton, England: Arnold Pub. Co.

Fernandez Valiente E., Quesada A., Nieva M., Leganés F., and Prosperi C. (1996) Contribution of biological nitrogen fixation to rice production in a Mediterranean rice field. In Rahmann et al. *Biological nitrogen fixation associated with rice production.* Dordrecht, the Netherlands: Kluwer Press.

Fernandez Valiente E., Quesada A., Prosperi C., Nieva M., and Leganés F. (1997). Short and long term effect of ammonium in wetland rice fields of Spain. *Biology and Fertility of Soils* 24: 353-357.

Guerrero M. and Lara C. (1987). Assimilation of inorganic nitrogen. In P. Fay and C. Van Baalen (eds.), *The cyanobacteria* (pp. 163-186). Amsterdam: Elsevier Science.

Gupta A. (1966). Algal flora and its importance in the economy of rice-fields. *Hydrobiologia* 28: 213-222.

Lee K. and Watanabe I. (1977). Problems of the acetylene reduction technique applied to water-saturated paddy soils. *Appl. Environ. Microbiol.* 34: 654-660.

Maggi M. and Prosperi C. (unpublished). Variación temporal de Cianobacterias en arrozales de Corrientes. Student level Thesis. National University of Córdoba, Argentina.

Martinez M., Evangelista C., and Pantastico J. (1978). *Nostoc commune* Vaucher as a potential fertilizer in rice-fish culture: A preliminary study. *The Philippine Journal of Crop Science* 2: 251-256.

Minchin F. (1986). Gaseous diffusion in cyanobacterial heterocysts and permeability to the gases nitrogen and oxygen. *Nature* (Lond.) 320: 483-484.

Pantastico J. and Gonzalez J. (1976). Culture and use of *Nostoc commune* as biofertilizer. *Kalikasan Philippines Journal of Biology* 5: 221-234.

Prosperi C. (1986). Effect of light on bud development in *Nostoc cordubensis* (Nostocaceae, Cyanophyta). *Phycologia* 25 (2): 262-265.

Prosperi C. (1989). Life cycle of *Nostoc cordubensis*. *Phycologia* 28: 501-503.

Prosperi C. (1994a). A cyanophyte capable of fixing nitrogen under high levels of oxygen. *J. Phycol.* 30: 22-224.

Prosperi C. (1994b). The role of respiration as a protection for nitrogenase activity in the cyanobacterium *Anabaena Flos-Aquae*. *Phyton* 55: 159-162.

Prosperi C. (2000). Cyanobacteria in human affaires. *Interciencia, Revista de Ciencia y Tecnología de América* 25 (6): 303-306.

Prosperi C. and Bellone E. (1994). Biological nitrogen fixation with non-legumes. *Microbial Releases* 2: 295-296.

Prosperi C., Boluda L., Luna C., and Fernandez Valiente E. (1992). Environmental factors affecting in vitro nitrogenase activity of Cyanobacteria isolated from rice fields. *J. of Applied Phycol.* 4: 197-204.

Prosperi C. and Fernandez Valiente E. (1993). Environmental factors regulating the effect of ammonium on nitrogenase activity in cyanobacteria from rice fields. In Hegazi et al. 1994: *Biological nitrogen fixation with non-legumes* (pp. 111-115). Cairo: The American University in Cairo Press.

Prosperi C., Luna C., and Fernandez Valiente E. (1993). Influence of pH, light intensity and oxygen on the short-term effect of ammonium on nitrogenase activity of Cyanobacteria from rice fields. *Environmental and Experimental Botany* 33 (4): 545-552.

Prosperi C., Pons S., and Maggi M. (1996) Ecological aspects of cyanobacteria from the rice fields of Corrientes (Argentina). In M. Rahmann, A. Podder, C. Hove, Z. Tahlmida Begum, T. Heulin, and A. Hartman (eds.), *Biological nitrogen fixation associated with rice production* (pp. 141-146). Dordrecht, the Netherlands: Kluwer Press.

Quesada A., Sanchez Maeso E., and Fernandez Valiente E. (1989). New incubation device for in situ measurement of acetylene-reduction activity in rice fields. *J. Applied Phycol.* 1: 195-200

Reddy P. and Roger P. (1988). Dynamics of algal populations and acetylene reducing activity in five soils inoculated with blue-green algae. *Biol. Fert. Soils* 6: 14-21.

Rincón A. and Prosperi C. (unpublished). Estudio taxonómico de Cianobacterias de los arrozales de Corrientes y actividad superóxido-dismutasa en Chroococcus limneticus. Student level thesis. National University of Cordoba (Argentina).

Roger P. (1988). Reconsidering the utilization of blue-green algae in wetland rice cultivation. In *International Simposium on "Biological N2 fixation associated with rice production and rice culture,"* Cuttack, India, 1-36.

Roger P. (1989a). Blue-green algae (cyanobacteria) in agriculture. In J. Dawson and P. Darts (eds.), *Microorganisms that promote plant productivity*. Nijhoff/w Junk pub.

Roger P. (1989b). Cyanobacteries et riziculture. *Bull. Soc. Bot. Fr.; Actual. Bot.*, 136: 67-81.

Roger P. and Kulasooriya S. (1980). *Blue-green algae and rice*. Manila, Philippines: IRRI.

Roger P. and Watanabe I. (1982). *Research on algae, blue-green algae and phototrophic* nitrogen fixation at the IRRI. Research paper series N° 78, 1-21, Manila, Philipinnes.

Sharma B. and Gupta R. (1983). Effect of algal application on rice yield in Jammu Division. *Phycos* 22: 176-179.

Shi D. and Hall D. (1988). *Azolla* and inmobilized cyanobacteria (blue-green algae): traditional agriculture to biotechnology. *Plants* 1: 5-11.

Singh P. and Singh A. (1983). Comparative studies on *Azolla* and blue-green algae bio fertilization to rice crop. In A. Shukla and S. Pandey (eds.), *Advances in applied phycology* (pp. 334-349). Kanpur. India.

Venkataraman G. (1981). Blue-green algae for rice production: A manual for its promotion. *FAO Soil Bull.* 46: 1-102.

Watanabe A. and Roger P. (1984). Nitrogen fixation in wetland rice field. In Subba Rao (ed.), *Current developements in biological nitrogen fixation* (pp. 237- 276). New Delhi: Oxford and IBH pub.

Whitton B. and Roger P. (1989). Use of blue-green algae and *Azolla* in rice culture. In R. Campbell and R. Mc Donald (eds.), *Microbial inoculation of crop plants* (pp. 89-100). Oxford: IRL Press at Oxford University Press.

Chapter 19

Piriformospora indica As a New and Emerging Mycofertilizer and Biotizer: Potentials and Prospects in Sustainable Agriculture

Sarika Shende
Kanchan Bhagwat
Prasad Wadegaonkar
Mahendra Rai
Ajit Varma

INTRODUCTION

The genus *Piriformospora indica* was discovered from the rhizospheric soil of two xerophytic plants, namely, *Prosopis juliflora* and *Zizyphus nummularia* (Verma et al., 1998). It was recovered from the desert of the state of Rajasthan in India. The genus belongs to Basidiomycota. During the screening of glomaceous fungi for their growth on agar medium, this fungus grew with scanty mycelium and later formed typically pear-shaped spores, and thus was named *P. indica* (Verma et al., 1998). Being a potential plant growth booster, the fungus has attracted researchers all over the world to harness its fullest potential in the field of sustainable agriculture. A review of the literature suggests that a significant contribution has been made on various aspects of *P. indica* with particular reference to its growth-promoting potential (Hindav et al., 1998; Sahay et al., 1998; Varma, 1998; Sahay, 1999; Sudha, 1998; Varma, Verma, et al., 1999; Varma, Sudha, et al., 1999; Bütehorn et al., 2000; Rai et al. 2000; Sahay and Varma, 2000; Rai et al., 2001; Singh, Singh, et al. 2001; Singh, Hampp, and Varma, 2001; Kumari, 2002; Malla et al., 2002; Rai and Varma, 2002; Singh et al., 2002; Singh, Singh, et al., 2003; Singh, Rexer, et al., 2003).

The first remarks on the symbiotic interactions between *P. indica* and terrestrial orchids was given by Blechert et al. (1999). Since then remarkable progress has been made to utilize biotechnological potentials of this culturable fungus in the field of horticulture and agriculture (Singh et al., 2002). The fungus has arbuscular mycorrhizal fungi (AMF)-like characteristics (Varma et al., 2001). Arbuscular mycorrhizal fungi are obligate endosymbionts and thus cannot be cultured in a laboratory. Moreover, the process of inoculum production is time-consuming and expensive. This problem can be solved by using *P. indica,* being culturable in natural and artificial media (Verma et al., 1998). As a matter of fact, the fungus *P. indica* has tremendous potential in the fields of industries dealing with tissue culture and also in agriculture.

The aim of this chapter is to discuss the discovery, systematics, mycorrhiza-like characteristics, role as a plant growth promoter and a biotizer, and future of *P. indica* in the field of sustainable agriculture.

THE MORPHOLOGY AND CULTURAL CHARACTERISTICS

The fungus grows on wide variety of media (Verma et al., 1998). Incubation under shaking retarded the growth in MM1 liquid culture, whereas no such negative effect is observed during cultivation on any other substrate. In fact, the fungal biomass is considerably enhanced on shaking cultures with *Aspergillus* medium (CM). On Moser b medium, the colonies appeared compact, wrinkled with furrows and constriction. The mycelium produced fine zonation and a great amount of white aerial hyphae.

The hyphae were highly interwoven and often adhered together with the appearance of simple interwined cord. Young mycelia are white, almost hyaline, thin-walled, septate, 0.7 to 3.5 μm in diameter. The hyphal walls show some external deposits at regular intervals, perhaps polysaccharides and/or some hydrophobic proteins, which stained deeply with toludine blue. Since, septation is irregular, the single compartments could contain more than one nucleus.

MYCELIAL PENETRATION IN HOST

Hyphae multiply within the cortical tissue and never traverse through the endodermis. Likewise they also do not invade the Arial portion of the plant (stem and leaves). A protein profile of the colonized root showed some qualitative and very little quantitative changes, but the protein profile of the leaf shows no qualitative and some quantitative changes.

ULTRASTRUCTURE OF HYPHAL WALL

Cell walls are very thin and show multilayered structure. The septa consist of dolipores with continuous parenthosomes. The dolipores are very prominent with a multilayered cross wall and median swelling, mainly consisting of electron transparent material. The parenthosomes are straight and have the same diameter as the corresponding dolipore. No pores of any kind could be detected, thereby showing that they are flat discs without perforation (Verma et al., 1998). The parenthosomes consist of an electron-dense outer layer and less dense inner layer which shows an inconspicuous dark line in the median region. These studies form the basis for the systematic position within the hymenomycetes.

Asexual State

Chlamydospores singly or in clusters, formed at the apex, are characteristically pear-shaped, (14-) 16-25 (-33) µm in length and X (9-) 10-17 (-20) µm in width. Very young spores have thin hyaline walls. At maturity these spores have walls up to 1.5 µm thick, which appeared two layered, smooth, and pale yellow. The cytoplasm of the chlamydospores is densely packed with granular material and usually with 8 to 25 nuclei.

The sexual state of *P. indica* is unknown.

MOLECULAR SYSTEMETICS

Molecular sequence data are useful for systematics, where morphological characteristics such as the taxonomically decisive sexual states are missing. A neighbor joining analysis on comparison of partial 18S rDNA sequences (525 nucleotide positions) placed *Piriformospora indica* within the Basidiomycota close to the *Rhizoctonia solani* of order Ceratobasidiales (Verma et al., 1998). A maximum likelihood analysis on an almost complete 18S rDNA sequence (1,550 nucleotide positions) confirmed this assignment. Similar results were obtained by distance and parsimony methods. A comprehensive phylogenetic analysis of Rhizoctonoids using sequences from mitochondrial and nuclear rDNA on more representatives should provide an insight into the evolution of this important group and clarify the evolutionary relationship of *P. indica* and *Rhizoctonia* within the Hymenonycetes. No existing genus covers all the characteristics of the new fungus. Other groups of Basidiomycota that may be related according to the molecular and ultrastructural data are not known to be able to produce chlamydospores or have a different way of life. Therefore, a new genus was established and the fungus was designated *Piriformospora indica* (Verma et al., 1998).

ROOT COLONIZATION POTENTIAL IN DIVERSE HOSTS

Piriformospora indica has the capacity to colonize a wide variety of plants. Some important plants that form novel host-symbiont combinations include *Zea mays, Nicotiana tabacum, Petroselinum crispum, Populas tremula, Setaria italica, Oryza sativa, Pisum sativum, Glycine max, Cicer arietinum, Solanum melongena, Artemisia annua, Bacopa monniera, Azadirachta indica, Dactylorhiza purpurella, D. maculata, Funaria hygrometrica, Spilanthes calva, Withania somnifera, Chlorophytum borivillianum, Dalbergia sissoo,* and *Terminalia arjuna*. The hyphae first colonized the surface and produced structures similar to the appressoria. Subsequently, they entered and traversed through the cortical cells and produced vesicles, and also differentiated into intracellular and coiled structures. At maturity, external and internal spores are formed like AM fungi.

IN VITRO RAPID PROLIFERATION OF ROOTS OF WITHANIA SOMNIFERA

Root organ culture of *Withania somnifera* is essential for the excessive production of secondary metabolites. The leaf explants from newly developed seedlings of *W. somnifera* were inoculated on half-strength Murashige and Skoog's medium with different concentrations of hormones. After root initiation within 8 to 10 days, when the roots were 0.2 cm in length, they were transferred to a hormone-free medium and *P. indica* was inoculated on the emerging roots of the explant. The cultures were incubated at 25 ± 2°C in the dark for 6 days. Encouragingly, the length of roots increased profusely as compared to control (see Photo 19.1).

RHIZOCTONIA-LIKE GROWTH PROMOTION OF ORCHIDS

The interaction of *P. indica* with orchid differs fundamentally from the described interaction with other higher plants. Blechert et al. (1999) inoculated two-year-old asymbiotically grown *Dactylorhiza maculata* with *P. indica*. They found that growth of the plants was significantly promoted. The germination rate of the seeds was not influenced by *P. indica,* but the growth rate of developing protocorms was increased by the fungus.

Varma et al. (2001) compared interaction of *P. indica* with *D. majalis* and *D. fuschii* and a natural mycorrhizal partner of *D. majalis,* identified as *Rhizoctonia* sp., to compare the interaction. The interaction of *P. indica*

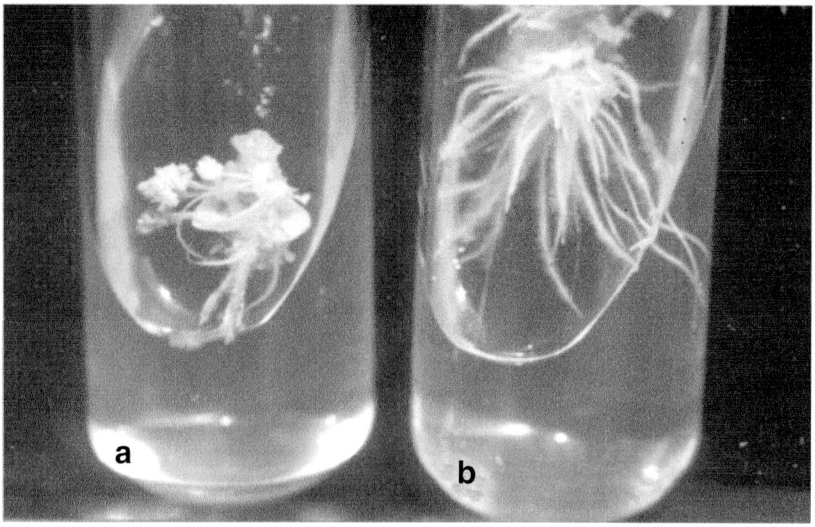

PHOTO 19.1. Rapid proliferation of roots of *Withania somnifera* by *P. indica:* a—control; b—with *P. indica.*

with the protocorms started with the colonization of the rhizoids. The hyphae forms small appressoria-like structures on the rhizoid surface and penetrates the cell with a narrow penetration neck. Within the cells, the hyphae expanded to a normal diameter and grow toward the protocorm. From the basal swellings of the rhizoids, inter- and intracellular hyphae appeared to be spreading. After the penetration of the cortical cells, especially in the the basal region of the protocorm, typical pelotons were formed. In the epidermal cells, hyphal coils never appeared. The first digested pelotomes were found 17 days after the inoculation.

The natural symbiont of *D. maculata, Rhizoctonia* sp. developed identical interaction structures and showed the same ontogeny of these structures as *P. indica*. Therefore, *P. indica* acts as a specific orchidaceous mycorrhizal fungus in all in vitro tested *Dactylorhiza* sp. The development of the interaction of *D. majalis* with *P. indica* is almost identical to the findings of Rasmussen (1990).

ENHANCEMENT OF SECONDARY METABOLITES

For the methanol extract (80 percent) of freshly harvested *P. indica* hyphae when used for analytical HPLC separation, the analysis showed seven peaks in the hyphae and one main peak in the culture filtrate (Varma et al., 2001). Preparative HPLC analysis of hyphal and culture filtrate showed a major peak identified as benzoic acid, and its analogs (benzoic acid, α-hydroxybenzoic acid, 3-4 dihydroxy benzoic acid, caffeic acid, ferulic acid, vanilic acid, cinnamic acid, and p-coumaric acid) showed no stimulation of the plant tested.

It is speculated that secondary metabolites might be involved in the regulation of mycorrhizal colonization (Fester et al., 1998). HPLC analysis of methanolic extract from six-week-old infected and noninfected plants showed quantitative but no qualitative changes in the root samples as a result of interaction of *P. indica* with barley, maize, and foxtail millet *(Setaria indica)* roots. The ultraviolet (UV) spectra obtained from the HPLC photodiode assay detector showed a cluster of peaks between 7.5 and 12.5 min of the HPLC chromatograph on interaction of maize, barley, rice and foxtail millet, with *P. indica* indicating the presence of indole derivatives, for example, tryptophane, tyrosine, and tyramine or their derivatives.

Root extracts of maize showed a significant presence of cyclic hydroxamine acids such as DIMBOA (2,4-dihydroxy-7-methoxy-1,4-benzoxazin-3-one) in wheat, but none in rice, barley, and setaria. The maize root colonization by *P. indica* showed enhancement of different benzoxazinone levels. HPLC analysis of methanol extracts of infected maize roots showed eight different peaks with a photodiode assay detector with typical UV spectra from benzoxazinone derivatives. None of these peaks is identical to DIMBOA or DIBOA. The chemical structure of these compounds was identified by spectrophotometric methods.

MECHANISMS INVOLVED DURING COLONIZATION

Cellular actions leading to reciprocal morphofunctional integration between symbionts during mycorrhizal establishment are complex. There is now evidence that the penetration of roots by endomycorrhizal fungi involves low and regulated production of a combination of cell-wall-degrading hydrolytic enzymes leading to an organized colonization of the plant root (Blilou et al., 1996). Investigations have demonstrated the production of pectinase, cellulase, xylanase, and chitinase (Garcia-Romera et al., 1990, 1991a,b, 1996; Varma and Bonfante, 1994; Perotto et al., 1995a,b, 1997) from the hyphae and mycorrhized roots. It seems that mycorrhizal fungi colonize the root tissues of their host plant by a combination of mechanical and enzymatic mechanisms (Bonfante and Perrotto, 1995; Gianinazzi-Pearson et al., 1996); very weak and localized production of enzymes might ensure that the viability of the host is maintained, defense responses are not triggered, and a high degree of compatibility is reached (Varma, Sudha, et al., 1999).

Hyphae of *P. indica* are able to intrude through the cells of the host root. Cell-wall-degrading enzymes are tested following the conventional assay methods. Carboxymethylcellulase, xylanase, and polygalactuonase are found in significant quantities in both the mycelial culture filtrate and the root samples, which are treated with *P. indica*. It is possible that cell wall penetration may be due to the dual action of hydrolytic enzymes and mechanical pressure, or either of these actions.

MYCORRHIZA-LIKE ENDOPHYTES

Piriformospora indica shows morphological, functional, and serological similarities with the established arbuscular mycorrhizal fungi, namely, *Glomus* and *Gigaspora*. But the data of 18S rDNA and presence of dolipore septum forms the basis to place *P. indica* in basidiomycetes instead of zygomycetes. *Piriformospora indica* possesses undulating hyphae with intermittent spores located at the terminal end and/or at the intercalary position. Spores are invariably pear shaped. Both the spores and hyphae are at least four to five times thinner and smaller in diameter than those of AMF. Like AMF fungi, *P. indica* have a wide host range, colonizing the roots of monodicots, including legumes. The basic and most important difference is that the new fungus can easily be cultured while the former are obligate endosymbionts. The comparison of arbuscular mycorrhizal fungi and *P. indica* has been discussed by Varma et al. (2001).

PROTOPLAST FUSION FOR BETTER UNDERSTANDING OF INTERACTION

Protoplasts have been isolated from all the taxonomic groups of fungi, including basidiomycetes (Peberdy, 1995). There are few reports on the isolation of protoplasts from the ectomycorrhizal basidiomycetes (Anunciacao et al., 1990; Barrett et al., 1989; Dias et al., 1996; Farquhar and Peterson, 1990; Hebraud and Fevre, 1988; Hampp et al., 1998; Kropp and Fortin, 1986), and there is only one report of protoplast isolation from ericoid mycorrhizal fungus (Bardi et al., 1999). However, there is no report for protoplast isolation from arbuscular mycorrhizal fungi. In 1999, all these experiments seemed a distant possibility for manipulating AMF due to the inability to culture them axenically. However, following the discovery of

P. indica, these obstacles no longer hindered research to understand the basis and outcome of the plant-fungus relationship. The protoplasts of *P. indica* have been successfully isolated and conditions have been optimized for their regeneration, in an effort to identify and develop systems with potential for transformation and other genetic study (Singh et al., 2001). This opens an important possibility to improve symbiosis by transgenic manipulation of the fungal component through the introduction of desirable genes.

Piriformospora indica is the most suitable candidate, based upon its wide host range, for the ease with which it colonizes and its hardiness in surviving adverse environmental conditions, to express foreign genes of importance to the host plants in a symbiosis-specific manner and confirm its benefit for biocontrol (Singh et al., 2000; Varma et al., 2001). A variety of fungal species have been transformed following incubation of protoplasts with DNA-encoding genes for selectable traits, such as antibiotic or pesticide resistance (Avalos et al., 1989; Punt et al., 1987; Van Engelenbert et al., 1989), required for amino acid synthesis (Yelton et al., 1984), or utilization of some novel nitrogen source, such as acetamide (Geisen and Leister 1989; Tilburn et al., 1983).

The development of techniques for the production, culture, and regeneration of *P. indica* protoplasts has opened up new possibilities for studies on physiological, biochemical, genetic, and molecular biological aspects of the organism concerned. Thus, it would be now possible to transform *P. indica* protoplasts with a reporter gene, and to apply the transformed strains to the rhizosphere to investigate colonization patterns, population dynamics, or dispersal under natural conditions (Singh et al., 2000; Varma et al., 2001). Other important experiments for transgenic manipulation of *P. indica* may involve genes for destroying pesticide residues in soil (McClung et al., 1989), for nitrate reduction (Daboussi et al., 1989), or for heavy metal resistance (Colpaert and van Assche, 1992). In such designs of genetically modifying the fungus, it will be important to avoid nontarget effects that could disrupt the rhizosphere ecology.

In recent years, green fluorescent proteins (GFP) have been developed as a reporter for expression, a tracer of cell lineage, and a label for host plants (Chalfie et al., 1994; Spelling et al., 1996). Thus, it would now be possible to transform *P. indica* protoplasts and to apply such strains to the rhizosphere to investigate population processes such as colonization patterns, dispersal of marked populations released to nature, microenvironmental relationships (Caldwell et al., 1992), and demographic factors (Andrews et al., 1987).

STRATEGIES FOR MASS PRODUCTION OF INOCULA

The growth of arbuscular mycorrhizal fungi is dependent on attachment of the hyphae to the parent spores, unless root infection has taken place. Azcon-Aguilar et al. (1986) suggested that over 90 percent germination can be obtained axenically on water agar in the absence of any mineral or organic nutrient supply.

There are many possible reasons why AMF fail to show extensive and continuous growth unless they are part of a symbiotic partnership with a host root. Hepper (1984) postulated that

- AMF may have simple nutritional requirements which, due to our lack of knowledge, have not yet been fulfilled;
- AMF may be necessary to supply some nutrient continuously at low concentrations in the host's metabolism; and
- as AMF have no saprophyte phase, under normal ecological conditions some part of the genome is in a suppressed condition and the host supplies the inducer to allow nucleic acid transcription. If this is true, then the postulated inducer must be almost universally present in plant roots and might be extractable.

The axenic culture of AMF continues to be one of the challenging goals in mycorrhizal research. However, following the discovery of the new isolate, *P. indica,* which mimics AMF in many respects and can be easily cultivated axenically on a wide range of synthetic media, for example, on minimal media (MM1) normally used for in vitro germination of AM fungi with 2 percent sugar or glucose as carbon source, on different media for *Aspergillus* sp. (CM1 and MM2), and on Moser b medium. Mass cultivation of the fungus can be easily achieved on simplified broth culture. The temperature range of the fungus growth is 20 to 35°C, the optimum temperature and pH being 30°C and 5.8, respectively. The axenic culture of *P. indica* represents a model system to understand the molecular basis of photo- and mycobiont interaction and would be helpful in unraveling the inherent mysteries of the nonculturable nature of arbuscular mycorrhizal fungi.

IMMOBILIZATION OF SPORES AND HYPHAE: A STEP FORWARD FOR INOCULUM PRODUCTION

The entrapment of enzymes (microbial cells inside the solid support, as stabilized reusable catalysts), commonly called immobilized enzymes, was proposed in the early 1900s by Nelson and Griffin (1916). Immobilization may be interpreted as the presence of organisms or individual animal or plant cells either on the surface of or within the particles. This phenomenon has been attributed to increased cell-to-cell contact, induced stress due to concentration gradients, and partial differentiation.

In the past, it was difficult to obtain large amounts of AMF inoculants for agronomic use as the spores of these obligate endosymbionts produce mycelium with limited independent growth (Hepper, 1984). Mycelium regenerated in the pure culture from intraradical vesicles have the ability to infect root cortices (Strullu and Romand, 1987). Ectomycorrhizal fungi, in contrast, can now be produced in a fermenter and by entrapping the mycelium in sodium/calcium alginate beads. Several researchers (Le Tacon et al., 1985; Mauperia et al., 1987) have found this to be an efficient tool for the production of inoculants of AMF. Immobilization can preserve the physiological properties of mycorrhizal roots, and promote the regeneration of mycelium and the formation of mycorrhizae (Strullu and Plenchette, 1990).

Varma et al. (2001) immobilized the hyphae and spores of *P. indica* in alginate beads and found that the former retain the germinating capacity even after two months when used as an inoculum in pot culture experiments using *Sorghum* species as the host. The colored alginate beads have also been prepared by incorporating low amounts of safranin (0.2 to 0.5 percent) and trypan blue (0.1 to 0.4 percent) without affecting the viability of the fungus. Now, this technology can be applied for the preparation of a dual culture loaded with *P. indica* and seeds of terrestrial orchids. The combination could make a breakthrough for the commercial micropropagation of terrestrial orchids, as the survival rate of these orchid seeds is normally extremely poor.

APPLICATION IN HORTICULTURE

Regeneration of plants by means of micropropagation is well-known. This has opened the door for the red data book plants. The micropropagation technique of plant biotechnology has a great impact on agriculture, viticulture, and arboriculture. There is a growing trend toward commercialization of plant tissue culture. The survival rate of micropropagated plantlets on field transfer is extremely low (Sahay and Varma, 2000). Micropropagation provides optimal growth conditions to plant, which differ from those available in field and pot environments, notably a weak vapor pressure deficit (VPD), reduced light and presence of sucrose, including such somaclonal variations as nonfunctional stomata and less developed cuticles that render the plants sensitive to water stress. Inoculation of micropropagated plantlets with active cultures of AMF appears to be critical for their survival and growth (Varma and Schuepp, 1995, 1996; Lovato et al., 1999; Rai, 2000). Although the mycorrhizal response of different species varies, this avoids transient transplantation shock and shunted growth on transfer to the field and promotes rooting in difficult plant strains, thereby increasing the rate of outplanting performance of micropropagated plants.

AMF are generally regarded as non-host-specific. However, the mycorrhizal dependency of the plant may vary with the fungus species or with the medium, which is regarded as a kind of functional host specificity (Clarke and Mosse, 1981; Fortuna et al., 1992). Guillemin et al. (1992) found that certain cultivar fungus associations are more efficient than others in enhancing plant growth.

Micropropagated plants inoculated in ex vitro experiments are highly mycorrhizae dependent. This fact has been documented in such high-value plants as grape vines (Ravolanirina et al., 1989; Schubert et al., 1990), *Vitis vinifera* (Schllenbaum et al., 1991), apple plum (Fortuna et al., 1992), pineapple (Lovato et al., 1992: Usoukainen and Vestberg, 1994), avocado (Azcon-Aguilar and Barea, 1992), strawberry, raspberry, hortensia (Varma and Schuepp, 1994a,b; Varma, 1995), woody legumes (Salamanca et al., 1992), and *Rhododendron* (Lemonie et al., 1992).

Transfer of tissue-culture-raised plantlets to field conditions after biological hardening with either established AMF or *P. indica* shows a good growth and higher survival rate. *Piriformaspora indica* colonize tissue-culture-raised plantlets of brahmi *(Bacopa monniera),* tobacco (*Nicotiana tabacum* cult *xanthi*), brinjal *(Solanum melongena),* and neem *(Azadirachta indica).* The positive data were also obtained from the micropropagated plants of *Bacopa monniera* and *Artemisia annua* (Sudha, 1998) and several other medicinal plantlets under field trials (Rai et al., 2001). In general, the rate of plantlet survival is much higher than for untreated control. Higher biomass is also recorded in the treated plants (Sahay and Varma, 1999, 2000).

Rai et al. (2000) reported the growth promotion of *Zea mays* and *Glycine max* by *P. indica*. Both the plants shows the maximum root/shoot length when inoculated with the mycelial disc of *P. indica* as compared to filtrate and control. The treated plants shows profuse rooting. The increase in growth of the plant is probably due to more uptake of water and minerals in general and phosphorus in particular. Plants showed dichotomy in lateral roots treated with the mycelial disc whereas in control main roots were divergent. The filtrate-treated roots were merged. This suggests the possibility of the presence of a growth-inhibitory metabolite in the filtrate of the fungus. Profuse production of red pigments in mycelia is also reported when treated with mycelial disc, which is more in mycelia than filtrate and control.

Commercially useful medicinal plants, namely *Spilanthus calva* and *Withania somnifera,* were inoculated with *P. indica* in nurseries and, subsequently, transferred to the field. A significant increase in growth and yield of both the plants was recorded as compared to uninoculated controls. There was a remarkable increase in shoot and root length as well as biomass of treated plants of *S. calva* and *W. somnifera*. The basal and leaf area of both plants was also relatively high in fungus-inoculated plants. The overall size, number of inflorescence, flower, and the seed production increased after treatment with fungus (see Table 19.1). Net primary productivity (NPP) was also superior as compared to control plants. This indicates the possibility of commercial exploitation of *P. indica* for large-scale cultivation of *S. calva* and *W. somnifera* by medicinal plant growers in association with root-colonizing friendly fungus (Rai et al., 2001).

Neem plants from pot cultures treated with mycorrhizal fungi such as *Glomus mosseae, Scutellospora gilmorei,* and *P. indica* exhibited a direct correlation between percent root infection to AM propagules. As compared to typical AM fungi, *Glomus mosseae* and *Scutellospora gilmorei,* the types of fungi recorded from the rhizosphere soil of neem, were relatively less when inoculated with *P. indica* (Singh, Singh, et al., 2003).

POTENTIAL FOR CONTROL OF PHYTOPATHOGENIC ROOTBORNE MICROBES

AMF are the major components of the rhizosphere. Literature suggests that they reduce soilborne diseases and/or the severity of diseases caused by phytopathogens (Dehne, 1982). When the *P. indica* is challenged with a virulent root and seed pathogen such as *Gaeumannomyces graminis,* it completely blocks the growth of pathogen. Culture filtrate of *P. indica* also completely stopped the growth of the pathogens (Varma et al., 2001). Similar inhibitory effects are also observed with interaction of *P. indica* and *Aspergillus sydowii.*

To control the soilborne pathogens, the use of fungicides is very common to manage the agriculture plants and to disinfect the soil. It has been found that a few fungicides exert a negative effect on mycorrhizal fungi, while certain fungicides show a positive influence, for example, *Bacopa moniera* when treated with fungicides Diathane M-45 (bisdithiocarbamate), Bavistin (carbendazim) and Baynate (thiophenate) at various concentrations and *P. indica* (Singh, Singh, et al., 2003). The plants treated with the fungus and any of the fungicides Diathane M-45 or Bavistin were healthy as compared to those which received only fungus. However, higher concentra-

TABLE 19.1. Influence of *Piriformospora indica* on shoot and root length and on percent root colonization of *S. calva* and *W. somnifera* in a field trial.

	Shoot length (cm)		**Root length (cm)**		**Percent colonization**	
Treatment	***S. calva***	***W. somnifera***	***S. calva***	***W. somnifera***	***S. calva***	***W. somnifera***
+ *P. indica*	28.76 (± 0.68)	67.3 (± 0.26)	14.46 (± 0.45)	16.53 (± 0.25)	62.33 (± 2.08)	73 (± 2.64)
Control	25.16 (± 0.20)	42.23 (± 0.2)	12.13 (± 0.15)	13.03 (± 0.15)	Nil	Nil

Note: The control plants were treated with an equal amount of autoclaved mycelium.

tions of Diathane M-45 showed a negative effect on plant survival. *Piriformospora indica* was resistant to Bavistin, higher concentration of mancozeb and Baynate was inhibitory to fungal growth, whereas Bavistin showed a positive and stimulatory effect on *P. indica* (Kumar, 2000).

CONCLUSIONS AND FUTURE DIRECTIONS

At the dawn of the new millennium, *Piriformospora indica* has generated a new enthusiasm among mycorrhizologists. The fungus has enormous potential in the field of sustainable agriculture for being easily culturable, a plant-growth enhancer, an inducer of secondary metabolites, and a biotizer. Research carried out on a wide variety of host plants has shown its root colonizing efficiency, which demonstrates that *P. indica* has less host specificity than arbuscular mycorrhizal fungi. However, much research is needed to screen and understand diverse groups of plants in order to reach to a final conclusion. The concentrated efforts are required to optimize fermentation techniques and to devise a simple and economically viable easy-to-use method for mass-scale inoculum production in agroforestry and horticulture. The use and purification of inoculum of arbuscular mycorrhizal fungi has been an inherent problem during application for acclamatization of micropropagated plantlets. Encouragingly, *P. indica* will be immensely useful for industries dealing with micropropagation due to its positive role in successfully overcoming the acclamatization phase. An interesting area that needs attention is to know more about the life cycle of *P. indica*. Its sexual phase is unknown and yet to be worked out. The discovery of a sexual phase will give new clues for developing a deeper understanding of the host-symbiont relationship.

The research should be focused on the molecular basis of interaction, signaling between the plant and *P. indica* and disease suppression. Efforts are needed to understand the role of this novel fungus in interaction against root-rot diseases caused by *Pythium, Phytophthora,* and *Aphanomyces*. The phenomenological changes in the host plant after inoculation with *P. indica* will also provide supporting evidence of what is happening inside the plant body. Furthermore, its possible role in alleviating stress under drought and heavy-metal-contaminated soil is to be investigated and is an urgent need. Finally, *P. indica* offers great opportunities for unlocking mysteries to know more about its strategic role in sustainable agriculture.

REFERENCES

Andrews JH, Kinkel LL, Berbee FM, and Nordheim EV (1987). Fungi, leaves and theory of island biogeography. *Microb Ecol.* 14: 277-290.

Anunciacao CE, Guimaraes WV, Araujo EF, Muchovej RMC, and Souza DRN (1990). Production and regeneration of protoplasts of *Pisolithus tinctorius. Biotechnol. Tech.* 4: 215-220.

Azcon-Aguilar C and Barea JM (1992). Interactions between mycorrhizal fungi and other rhizosphere microorganisms. In MF Allen (ed.), *Mycorrhizal functioning: An integrative plant-fungal process* (pp. 163-198). New York: Chapman and Hall.

Azcon-Aguilar C, Gianinazi-Pearson V, Fardaeu JC, and Gianinazi S (1986). Effect of VAM fungi and phosphate solubilizing bacteria on growth and nutrition of soyabean in a neutral-calcareous soil amended with p32-45ca tricalcium phosphate. *Plant and Soil.* 96: 3-15.

Avalos J, Geever RF, and Case ME (1989). Bialaphos resistance as a dominant selection marker in *Neurospora crassa. Curr Genet.* 16: 369-373.

Bardi L, Perotto S, and Bobfante P (1999). Isolation and regeneration of protoplasts from two strains of the ericoid mycorrhizal fungus *Oididendron maius:* Sensitivity to chemicals and heavy metals. *Microbiol Res.* 154: 105-111.

Barrett V, Lamke PA, and Dixon RK (1989). Protoplast formation from selected species of ectomycorrhizal fungi. *Appl. Microb. Biotechnol.* 30: 381-387.

Blechert O, Kost G, Hassle A, Rexer KH, and Varma A (1999). First remarks on the symbiotic interactions between *Piriformospora indica* and terrestrial orchids. In A Varma and B Hock (eds.), *Mycorrhizae: Structure, function, molecular biology and biotechnology,* Second edition (pp. 683-688). Berlin: Springer.

Blilou I, Martin J, and Ocampo JA (1996). Influence of cellulase on the susceptibility of non-host cabbage to colonization by Glomus intraradices. In C Azcon-Aguilar and JM Barea (eds.), *Mycorrhizas in integrated systems from genes to plant development* (pp. 215-217). Luxemburg: Official publications of the European Communities.

Bonfante P and Perrotto S (1995). Strategies of arbuscular mycorrhizal fungi when infecting host plants. *New Phytol.* 130: 3-21.

Bütehorn B, Rhody D, and Franken P (2000) Isolation and characterisation of Pitef1 encoding the translation elongation factor EF-1a of the root endophyte *Piriformospora indica. Plant Biology* 49: 687-692.

Caldwell DE, Krober DR, and Lawrence JR (1992). Confocal laser microscopy and digital image analysis in microbial ecology. *Adv Microb Ecol.* 12: 1-67.

Chalfie M, Tu Y, Euskirchen G, Ward WW, and Prasher DC (1994). Green fluorescent protein as a marker of gene expression. *Science.* 263: 802-805.

Clarke C and Mosse B (1981). Plant growth response of vesicular arbuscular mycorrhiza XII. Field inoculation responses of barley at two soil P levels. *New Phytol.* 87: 695-703.

Colpaert JV and van Assche JA (1992). The effects of cadmium and the cadmium-zinc interaction on the axenic growth of ectomycorrhizal fungi. *Plant Soil.* 145: 237-243.

Daboussi MJ, Djeballi A, Gerlinger C, Blaiseau PL, Bouvier I, Cassan M, and Lebrun MH (1989). Transformation of seven species of filamentous fungi using the nitrate reductase gene of *Aspergillus nidulans. Curr. Genet.* 15: 453-456.

Dehne HW (1982). Interaction between vesicular arbuscular mycorrhizal fungi and plant pathogens. *Phytopathology* 72: 1115-1119.

Dias ES, Araujo EF, Guimaraes WV, and Muchovej RMC (1996). Production and regeneration of protoplasts from the mycorrhizal fungus *Suillus granulatus. W. J. Microbiol. Biotechnol.* 12: 625-628.

Farquhar ML and Peterson RL (1990). Induction of protoplast formation in the ecto-mycorrhizal fungus *Paxillus involutus* by the root rot pathogen *Fusarium oxysporum. New Phytol.* 116: 107-113.

Fester T, Maier W, and Strack D (1998). Accumulation of secondary compounds in barley and wheat root in response to inoculation with an arbscular mycorrhizal fungi and co-inoculation with rhizosphere bacteria. *Mycorrhiza.* 8: 241-246.

Fortuna P, Citernisi S, Morini S, Giovanetti M, and Loreti F (1992). Infectivity and effectiveness of different species of arbuscular mycorrhizal fungi in micropropagated plants of plum root stock. *Agronomie.* 12: 825-830.

Garcia-Romera I, Garcia-Garrido JM, Martinez-Molina E, and Ocampo JA (1990). Possible influence of hydrolytic enzymes on vesicular arbuscular mycorrhizal infection in alfalfa. *Soil Biol Biochem.* 22: 149-152.

Garcia-Romera I, Garcia-Garrido JM, and Ocampo JA (1991a). Pectinase activity in vesicular-arbuscular mycorrhiza during colonization of lettuce. *Symbiosis.* 12: 189-198.

Garcia-Romera I, Garcia-Garrido JM, and Ocampo JA (1991b). Pectolytic enzymes in the vesicular-arbuscular mycorrhizal fungus *Glomus mosseae. FEMS Microbiol Lett.* 78: 343-346.

Garcia-Romera I, Garcia-Garrido JM, and Ocampo JA (1996). Hydrolytic enzymes in arbuscular mycorrhizae. In C Azcon-Aguilar and JM Barea (eds.), *Mycorrhizas in integral systems from genes to plant development* (pp. 234-237). Luxemberg: Official Publications from the European Communities.

Geisen R and Leister L (1989). Transformation of *Penicillium nalgiovense* with the amd sS gene of *Aspergillus nidulans. Curr Genet.* 15: 307-309.

Gianinazzi-Pearson V, Dumas-Gaudot E, Armelle G, Tahiri A, and Gianinazzi S (1996). Cellular and molecular defence-related root responses to invasion by arbuscular mycorrhizal fungi. *New Phytol.* 133: 45-68.

Guillemin JP, Gianinazzi S, and Trouvelots A (1992). Screening of arbuscular mycorrhiza fungi for establishment of micropropagated pineapple plants. *Agronomie.* 12: 831-836.

Hampp R, Stulten C, and Nehls U (1998). Isolation and regeneration of protoplast from ectomycorrhizal fungi. In A Varma (ed.), *Mycorrhiza manual* (pp. 115-126). Berlin: Springer.

Hebraud M and Fevre M (1988). Protoplast production and regeneration from mycorrhizal fungi and their use for isolation of mutants. *Can J Microbiol.* 34: 157-161.

Hepper CM (1984). Isolation and culture of vesicular arbuscular mycorrhizal (VAM) fungi. In CL Powell and DJ Bagyaraj (eds.), *VA mycorrhiza* (pp. 95-112). Boca Raton, FL: CRC Press.

Hindav R, Mondal N, Paul J, Ray M, Meera K, Sahay N, Sharma J, Singh A, Sudha, and Varma A (1998). One kilo tropical soil = 1 kilo gold: This is microbial science. In A Varma (ed.), *Microbes for health, wealth, and sustainable environment.* New Delhi, India: Malhotra Publishing House.

Kropp BR and Fortin JA (1986). Formation and regeneration of protoplasts from the ectomycorrhizal basidiomycete *Laccaria bicolor. Can J Bot.* 64: 1224-1226.

Kumar S. (2000). Characterization of *P. indica* a mycorrhiza-like-fungus. Master's thesis, School of Life Sciences, Jawaharlal Nehru University, New Delhi, India.

Lemonie MC, Gianinazzi S, and Gianinazzi-Pearson V (1992). Application of endomycorrhizae to commercial production of Rhododendron microplants. *Agronomie.* 12: 881-885.

Le Tacon F, Jung G, Mugnier J, Mitchelot P, and Mauperia C (1985). Efficiency of a forest nursery of an ectomycorrhizal inoculum produced in a fermentor and entrapped in polymeric gels. *Can J Bot.* 63: 1664-1668.

Lovato PE, Guillemin JP, and Gianinazzi S (1992). Application of commercial arbuscular endomycorrhizal fungal inoculants to the establishment of micropropagated grapevine rootstock and Pineapple plants. *Agronomie.* 12: 873-880.

Lovato PE, Schuepp H, Trouvelot A, and Gianinazzi A (1999). Application of arbuscular mycorrhizal fungi (AMF) in orchard and ornamental plants. In A Varma and B Hock (eds.), *Mycorrhizae: Structure, function, molecular biology and biotechnology,* Second edition (pp. 443-467). Berlin: Springer.

Maier W, Peipp H, Schmidt J, Wray V, and Strack D (1995). Levels of a terpenoid glycoside (blumenin) and cell wall-bound phenolics in cereal. Mycorrhizas. *Plant Physiol.* 109: 465-470.

Malla R, Singh A, Md Z, Yadav V, Suniti, Verma A, Rai M, and Varma A (2002). *Piriformospora indica* and plant growth promoting Rhozobacteria. In GP Rao, DJ Bhat, TN Lakhanpal, and C Manoharachari (eds.), *Frontiers of fungal diversity in India* (professor Kamal Commomoration Volume) (pp. 1-19).

Mauperia C, Mortier F, Garbaye J, le Tacon F, and Carr G (1987). Viability of an ectomycorrhizal inoculum produced in a liquid medium and entrapped in calcium alginate gels. *Can J Bot.* 65: 2326-2329.

McClung CR, Philips JD, Orbach MJ, and Dunlap JC (1989). New cloning vectors using benomyl resistance as a dominant marker for selection in *Neurospora crassa* and in other filamentous fungi. *Exp. Mycol.* 3: 299-302.

Nelson JM and Griffin EG (1916). Adsorption of invertase. *J Am Chem Soc* 38: 1109-1111.

Peberdy JF (1995). Fungal protoplasts. In U. Kueck (ed.), *The mycota II: Genetics and biotechnology* (pp. 49-60). Berlin: Springer-Verlag.

Perotto S, Bettini V, Favaron F, Alghisi P, and Bonfante P (1995a). Ericoid mycorrhizal fungi: Cellular and molecular basis of their interaction withhost plant. *Can J Bot.* 73: 557-568.

Perotto S, Bettini V, Favaron F, Alghisi P, and Bonfante P (1995b). Polygalacturonase activity and location in arbuscular mycorrhizal roots of *Allium porrum*. *Mycorrhiza.* 5: 157-165.

Perotto S, Coisson JB, Perugini I, Cometti V, and Bonfante P (1997). Production of protein degrading enzymes by ericoid mycorrhizal fungi. *New Phytol.* 135: 151-160.

Prasher DC (1995). Using GFP to see the light. *Trends Genet.* 11: 320-323.

Punt PJ, Oliver RP, Dingemanse MA, Powels PH, and van del Hondel CAMJJ (1987). Transformation of *Aspergillus* based on the hygromycin B resistance marker from *Escherichia coli. Gene.* 56: 117-124.

Rai MK. (2000). Current advances in mycorrhization in micropropagation. *In Vitro Cell. Biol.-Plant.* 37: 158-167.

Rai M, Acharya D, Singh A, and Varma A. (2001). Positive growth responses of the medicinal plants, *Spilanthus calva* and *Withania somnifera* to inoculation by *Piriformospora indica* in fields trial. *Mycorrhiza.* 11: 123-128.

Rai MK, Sahay N, and Varma A (2000). In vitro growth promotion of maize and soybean by *Piriformospora indica,* an emerging endophytes. *J Environ Biol.* 9(1):1-5.

Rai M and Varma A (2002). Field performance of *Withania somnifera* Dunal after inoculation with three species of *Glomus. J. Basic Appl. Mycol.* 1: 74-80.

Rasmussen H. (1990). Cell diffentiation and mycorrhizal infection in *Dactylorhiza majalis* (Rehb. F.) Hunt and Summeh. (Orchidaceae) during germination in vitro. *New Phytol.* 116: 137-147.

Ravolanirina F, Gianinazzi S, Trouvelot A, and Carre M (1989). Production of endomycorrhizal explants of micropropagated grspevine rootstocks. *Agriculture, Ecosystems and Environment.* 29: 323-327.

Sahay NS (1999). Interaction of *Piriformospora indica* with tissue culture raised plant. Doctoral thesis, Jawaharlal Nehru University, New Delhi.

Sahay NS, Sudha, Singh A, and Varma A (1998). Trends in endomycorrhizal research. *Indian J. Exp. Biol.* NISCOM. 36: 1067-1086.

Sahay NS and Varma A (1999). *Piriformospora indica:* A new biological hardening tool for micropropagated plants. *FEMS Microbiol Lett.* 181: 297-302.

Sahay NS and Varma A (2000). Biological approach towards increasing the survival rates of micropropagated plants. *Curr Sci.* 78: 126-129.

Salamanca CP, Herrera MA, and Barea JM (1992). Mycorrhizal inoculation of micropropagated woody legumes used in revegetation programmes for desertified Mediterranean ecosystems. *Agronomie.* 12: 869-872.

Schllenbaum L, Berta G, Ravolanirina F, Tisserant B, Ganinazzi S, and Fitter AH (1991). Influence of endomycorrhizal infection on root morphology in a micropropagated woody plant species *Vitis vinifera* L. *Annals of Botany* 68: 15-151.

Schubert A, Mazzitelli M, Ariusso O, and Eynaed I (1990). Effects of vesicular arbuscular mycorrhizal fungi on micropropagated grapevines: Influence of endophyte strain, P fertilization and growth medium. *Vitis* 29: 5-13.

Singh A, Rexer K-H, Kost G, Kaldorf M, Buscot F, Peskan T, Oelmüller R, and Varma A (2003). Surface microbiology of root colonising *Piriformospora indica*. In A Varma, L Abbott, D Werner, and R Hampp (eds.), *Plant surface microbiology*. Berlin: Springer Germany.

Singh A, Singh A, Kumari M, Kumar S, Rai MK, Sharma AP, and Varma A (2002). AMF-like-fungi: *Piriformospora indica*—A boon for plant industry. In BN Prasad (ed.), *Biotechnology in sustainable biodiversity and food security* (pp. 101-124). Enfield, NH: Science Publishers, Inc.

Singh A, Singh A, Kumari M, Rai MK, and Varma, A (2003). Biotechnological importance of *Piriformospora indica* Verma et al.-A novel symbiotic mycorrhiza-like fungus: An overview. *Indian J. of Biotechnology* 2: 65-75.

Singh A, Singh A, Rexer KH, Kost G, and Varma A (2001). Root endosymbiont: *Piriformospora indica*—A boon for Orchids. *The Journal of Orchid Society of India* (91-2) 15: 101-106.

Spelling T, Bottin A, and Kahmann R (1996). Green fluorescent protein (GFP) as new vital marker in the phytopathogenic fungus *Ustilago maydis. Mol. Gen Genet*. 252: 503-509.

Strullu DG and Plenchette C (1990).Encapsulation de la forme intraracinaire de *Glomus* dans alginate et utilization des capsules comme inoculum. *C R Acad Sci* Paris. 310: 447-452.

Strullu DG and Romand C (1987). Culture axenique de vesicules isolees a partir d`endomycorhizes et reassociation in vitro a des raciness de tomates. *C R Acad Sci Paris*. 305: 15-19.

Sudha (1998). In vitro study of endosymbionts associated with tissue culture-raised medicinal plants. Doctoral thesis. Hamdard University, New Delhi.

Tilburn J, Cazzocchio C, Taylor GG, Zabicky-Zissman JH, Mockington RA, and Davies RW (1983). Transformation by integration in *Aspergillus nidulans. Gene*. 26: 205-221.

Usoukainen M and Vestberg M (1994). Effect of inoculation with arbuscular mycorrhizas on rooting, weaning and subsequent growth of micropropagated malus (L) Moench. *Agricultural Science in Finland* 3: 269-279.

Van Engelenbert R, Smith R, Godsen T, van den Broek II, and Tudzynski P (1989). Transformation of *Claviceps purpurea* using a bleomycin resistance gene. *Appl. Microbiol. Biotechnol.* 30: 364-370.

Varma A. (1998). Mycorrhizae, the friendly fungi: What we know, what should we know and how do we know? In A Varma (ed.), *Modern laboratory manual—Mycorrhizae* (pp. 1-24). Berlin: Springer.

Varma A and Bonfante P (1994). Utilization of cell wall related carbohydrates by ericoid mycorrhizal endophytes. *Symbiosis*. 16: 301-313.

Varma A and Schuepp H (1994a). Infectivity and effectiveness of *Glomus intraradices* on micropropagated plants. *Mycorrhiza*. 5: 29-37.

Varma A and Schuepp H (1994b). Positive influence of arbuscular mycorrhizal fungi on in vitro raised hortensia plantlets. *Angew Bot*. 68: 108-115.

Varma A and Schuepp H (1995). Mycorrhizae: Their application in micropropagated plantlets. *Crit Rev Biotechnol*. 15: 313-328.

Varma A and Schuepp H (1996). Influence of mycorrhization on the growth of micropropagated plants. In KG Mukerji (ed.), *Concepts in mycorrhizal research* (pp. 113-132). Hand Book Vegetation Science Series. Dordrecht, the Netherlands: Kluwer.

Varma A, Singh A, Sudha, Sahay NS, Sharma J, Kumari M, Rana D, Thakeran S, Deka D, Bharti K, et al. (2001). *Piriformospora indica*: An axenically cultivable mycorrhiza-like endosymbiotic fungus. In *Mycota IX,* Springer Series, Germany, pp. 123-150.

Varma A, Sudha, Sahay NS, Singh A, Kumari M, Bharti K, Sarbhoy AK, Maier W, Walter MH, Strack D, and Franken P (1999). *Proceedings on the Symposium Pollution Abatement Through Biological Treatment of Industrial Effluents.* March 24, 1998. Central pollution Central Board (CPCB), Delhi.

Varma, A, Verma S, Sudha, Sahay N, Bütehorn B, and Franken P (1999). *Piriformospora indica,* a cultivable plant-growth-promoting root endophyte. *Appl Environ. Microbiol.* 65(6): 2741-2744.

Verma S, Varma A, Rexer KH, Hassel A, Kost G, Sarbhoy A, Bisen P, Buetehorn B, and Franken P (1998). *Piriformospora indica,* gen. Et sp. Nov., a new root-colonizing fungus. *Mycologia.* 90: 896-903.

Yelton MM, Hamer JE, and Timberlake (1984). Transformation of *Aspergillus nidulans* using a trp C plasmid. *Proc. Natl. Acad. Sci.* 81: 1470-1474.

Chapter 20

Matsutake: A Natural Biofertilizer?

Wang Yun
Ian Robert Hall

INTRODUCTION

Technically, most higher plants (ferns, soft wood trees, and flowering plants) do not have roots but mycorrhizae (literally "fungus-root"), a combined structure formed when a mycorrhizal fungus infects the roots of a higher plant (Harley and Smith, 1983). Although plants in a few families, such as the brassicas (Brassicaceae), nettles (Urticaceae), rushes (Juncaceae), and sedges (Cyperaceae), do not form any type of mycorrhizae, they are very much in the minority. Mycorrhizal relationships have been around for a very long time and are evident in 460-million-year-old Devonian fossils (Redecker et al., 2000). Over time the fungi have become very specialized so that most cannot survive unless they are in contact with a host plant. Many plants have also become equally dependent on mycorrhizal fungi. This association is generally a beneficial one, with the host plant providing the obligate mycorrhizal fungus with a place to live and organic nutrients. In return, the fungus provides benefits to the plant, for example, by stimulating the uptake of soil nutrients, especially phosphorus (P), and water (Harley and Smith, 1983). The mycorrhizal fungi are, therefore, the most natural of the biofertilisers although not all of them are mutually beneficial (Smith and Read, 1997).

Mycorrhizae are divided into two main groups—the ecto- and endomycorrhizae—with the second group further subdivided into several subgroups, including arbuscular, ericaceous, and orchid mycorrhizae. In the

---

We wish to thank our colleagues in Canada, China, England, Italy, Japan, South Korea, Sweden, the United States, and New Zealand for their support.

endomycorrhizae the fungus gets right inside the cells in the outer layers of the roots, whereas in ectomycorrhizae the fungus wraps itself around the outside of the host plant's fine roots, just like the fingers of a glove. Most groups of plants form only one type of mycorrhiza on their root system. For example, rhododendrons and azaleas form only the ericaceous type of mycorrhiza; pasture species form arbuscular mycorrhizae; and oaks form only ectomycorrhizae. Although members of a plant family usually form only one type of mycorrhizae, a few families have members that form either ectomycorrhizae or vesicular arbuscular mycorrhizae, for example, the Myrtaceae and Salicaceae.

Some of the Basidiomycetes and Ascomycetes that form ectomycorrhizae produce edible mushrooms that are highly sought after. Examples include truffles, porcini, and the Japanese delicacy, matsutake (Wang et al., 1997; Hall et al., 1998). While the dramatic decline of harvests of any edible mycorrhizal mushrooms has spurred interest in the development of methods for their cultivation, so far only a few species of truffles have been produced in commercial quantities (Chevalier and Frochot, 1997; Bencivenga, 1998; Chevalier, 1998; Olivier, 2000; Hall et al., 2001; Chevalier et al., 2002; Hall and Wang, 2002), although advances have been made toward the cultivation of some, including *Cantharellus cibarius, Lyophyllum shimeji,* and *Lactarius deliciosus* (Poitou et al., 1989; Fujita et al., 1990; Danell, 1994; Danell and Camacho, 1997; Wang and Hall, 2002). Despite this, many of the most expensive mycorrhizal mushrooms, including *Tricholoma matsutake* and *Tuber magnatum,* have defied cultivation (Hall and Wang, 2002). This has led some researchers to question whether matsutake and some of the other recalcitrant edible ectomycorrhizal mushrooms have other unrevealed aspects to their mode of existence (Wang et al., 1997; Gill et al., 1999, 2000) and that this lack of understanding has led those trying to cultivate them to use incorrect methodologies.

Tricholoma matsutake is one of the most highly prized edible ectomycorrhizal mushrooms and commands very high prices, particularly where Japanese cuisine is highly regarded. The world retail market for matsutake is difficult to estimate, but it would be close to US$500 million (Wang et al., 1997). Since World War II, matsutake production in Japan has declined from about 12,000 tonnes to less than 1,000 tonnes, but despite a century of research the cultivation of matsutake has remained elusive (Wang et al., 1997).

In this chapter we review past and recent research on matsutake in an attempt to consolidate the available information and investigate whether matsutake is a true mycorrhizal fungus and hence a truly natural biofertilizer.

MATSUTAKE IN THE COMMERCIAL WORLD

Matsutake (*matsu* = "pine"; *take* = "mushroom") have been collected in Japan for hundreds of years where, because of their unique flavor, they fill a similar niche in the Japanese gourmet market to that filled by black and white truffles in Europe (Wang et al., 1997). Matsutake contains a mixture of 1-octen-3-ol and methyl cinnamate, which gives the mushroom its special flavor. They are also rich in amino acids, vitamins, and minerals (Ohta, 1983; Terashita et al., 1991). In addition to their culinary value, matsutake have medicinal properties (Gong et al., 1999). They are also a symbol of autumn (Wang et al., 1997) and command great esteem as yokozuna, a position far above all other edible mushrooms (yokozuna is the highest title for Sumo). In a 1992 public opinion poll of the 100 best edible mushrooms in Japan, matsutake took first place (Wang, 1993). Unlike the common mushroom, at the "button" stage *T. matsutake* are elongated with a bulbous end (see Photo 20.1). Consequently, *matsutake* became the word for the male organ and was not used in polite company (Ogawa, 1978a).

In the 1940s, about 12,000 t of matsutake were collected in Japanese forests (Ogawa and Ito, 1989). Since then harvests have declined markedly

PHOTO 20.1. Grade 1 fruiting bodies of *Tricholoma matsutake.*

(see Figure 20.1), primarily because pests and diseases have affected *Pinus densiflora,* matsutake's primary host tree. *Pinus densiflora* has also been replaced with other forest species in Japan, and many modern forest management practices do not favor the growth and fruiting of matsutake (Ogawa, 1978a). Consequently, since the 1970s Japan has imported large quantities of matsutake, particularly from China, North Korea, and South Korea. Large quantities of the similar North American matsutake (pine mushroom or white matsutake), *Tricholoma magnivelare* (Peck) Redhead, have also been imported from Canada, the United States, and Mexico (Hosford et al., 1997; Berch and Wiensczyk, 2001) as well as *Tricholoma nauseosus* (Blytt) Kytöv. from Europe and North Africa (see Table 20.1). Currently, more than 3,000 metric tonnes of matsutake and related species are imported by Japan with a retail market value in excess of US$300 million (see Table 20.2). Recently *Tricholoma matsutake* has been found in Bhutan (Dhawa, personal communication); and in Sweden, *Tricholoma nauseosus* (Blytt) Kytöv. has been shown to be synonymous with *T. matsutake* (Danell, 2002), which suggests that both of these countries will become major suppliers to Japan.

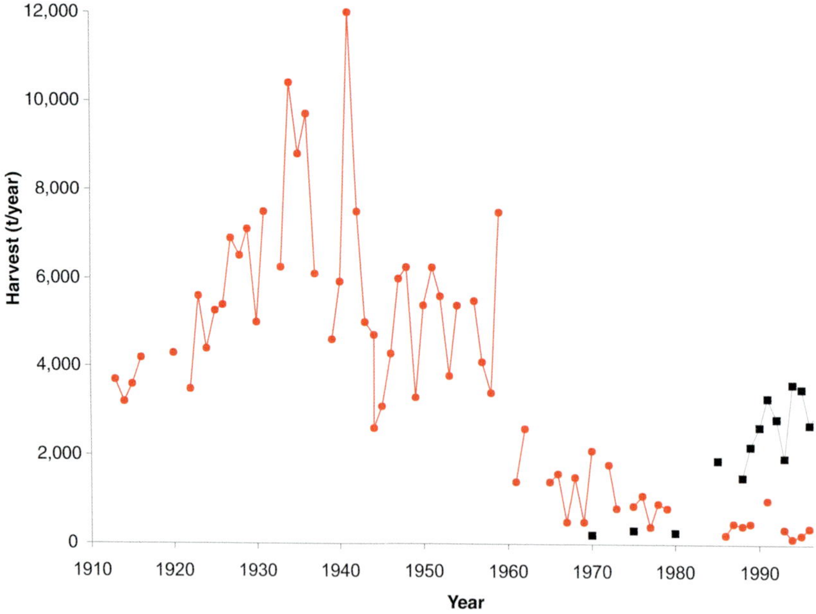

FIGURE 20.1. Combined *Tricholoma matsutake* production in Japan (1912-1998) represented by closed circles and combined imports of *T. matsutake* and related species (1970-1998) represented by closed squares.

TABLE 20.1. *Tricholoma matsutake,* fresh or chilled, imported by Japan between 1993 and 1997 (in tonnes).

Country	1993	1994	1995	1996	1997	Average
China	1,064	1,127	1,192	1,152	1,076	1,122
North Korea	383	1,760	1,141	541	615	888
South Korea	131	139	633	170	249	264
Bhutan	1.2	1.2	1.7	3.3	2.8	2
Morocco	19.5	72.2	1	85.8	124.6	61
Turkey	0	2.2	3.9	43.8	80	26
Canada	279	447	340	510	618	439
United States	51	47	164	172	284	144
Mexico	1.5	21.9	36	23.3	8.9	18
Total above	1,930	3,618	3,513	2,701	3,058	
Total all countries	1,943	3,622	3,515	2,703	3,059	2,968

TABLE 20.2. *Tricholoma matsutake,* fresh or chilled, imported by Japan between 1993 and 1997 (Yen: US$1 = ¥110).

	¥ millions					US$million Total 5 yrs (% all years)
	1993	1994	1995	1996	1997	
Bhutan	5	4	9	17	15.8	0.5
Canada	1,840	1,891	1,506	2,690	2,559	95.3
China	5,494	5,746	5,249	6,631	6,579	270.0
North Korea	2,291	6,928	4,074	1,060	2,794	155.9
South Korea	2,321	2,653	6,719	3,076	3,815	168.9
Mexico	78	100	206	156	73	5.6
Morocco	117	340	6	368	449	11.6
Turkey	0	4	12	140	256	3.7
United States	491	253	782	931	1,153	32.8
Total above	12,637	17,919	18,563	15,069	17,694	744.4
Total all countries	12,643	17,932	18,566	15,072	17,694	744.6
All countries (in US$million)	115	163	169	137	161	745

TAXONOMIC HISTORY OF MATSUTAKE

The taxonomic history of matsutake and other species that might be confused with it has been reviewed by Wang et al. (1997). Because of its past classification as a species of *Armillaria,* a genus containing a number of significant pathogens, it is worth noting the reasons for the current taxonomic position of matsutake. Because the various species of matsutake have an annulus on the stem similar to *Armillaria,* early in the 1900s they were called *Armillaria matsutake* S. Ito et Imai (Ito and Imai, 1925), *A. nauseosus* Blytt (1905), *A. caligata* Viviani (1834), *A. ponderosa* (Peck) Sacc. (Saccardo, 1887), and *A. arenicola* Murrill (1912). However, *Armillaria* has amyloid spores (Largent and Baroni, 1988), whereas matsutake species, like all *Tricholoma,* have inamyloid spores. As this feature is considered more important than the presence of an annulus, all species of matsutake are now placed in *Tricholoma*. Until 1925 *T. matsutake* from East Asia were classified as *T. caligatum* (Viv.) Rick. Ito and Imai (1925), then proposed a new species, *Armillaria matsutake* S. Ito and Imai. Singer (1949) accepted the new species and transferred it to *Tricholoma. Armillaria nauseosus* is now considered to be cospecific with *T. matsutake* (Bergius and Danell, 2000).

ECOLOGY

Vegetation Associated with **T. matsutake**

In the northeast of China *T. matsutake* is found in stands of *Pinus densiflora* with or without a few *Quercus mongolica* Fisch. The shrub and herb layers are not well developed and contain fewer than 20 species, which are considered indicator plants for matsutake-producing areas (Wang and Xie, 1982, 1983; Sun et al., 1989; Fu et al., 1999). These include *Rhododendron* spp., *Lespedeza bicolor* Turcz., *Atractylodes japonica* Koidz. ex Kitam., *Artemisia* spp., *Imperata cylindrica* (L.) Beauv., *Peucedanum terebinthaceum* (Fisch.) Fisch. ex DC., *Pedicularis* sp., *Platycodon grandiflorus* (Jacq.) A. DC., *Polygonatum odoratum* (Mill.) Druce, and *Carex callaitrichos* V. Krecz. In Korea, matsutake forests are very similar to those in northeast China (Lee, 1981, 1983) but contain a smaller component of broad-leaved tree species.

In southwest China, *T. matsutake* forests have canopy densities ranging from 0.6 to 0.7 and are either stands of pure pine, pine mixed with varying amounts of oak and other broad-leaved trees, or pure oak forests. The pine species are *P. densata* Mast., *P. yunnanensis* Franch., *P. massoniana* Lamb., *P. tabulaeformis* Carr., and *P. armandii* Franch. The oak species include *Q.*

semicarpifolia Smith, *Q. aquifolioides* Rehd. et Wils., *Q. senescens* Hand.-Mazz., *Q. pannosa* Hand.-Mazz., *Q. gilliana* Rehd. et Wils., *Q. spinnosa* David, *Castanopsis delavayi* Franch., *Lithocarpus sphaerocarpus* (Hick et A. Camus) A. Camus, and *L. dealbatus* Rehd. The other important broad-leaved trees are *Castanopsis orthacantha* Franch., *C. hystrix* A.DC., *Betula* spp., and *Populus* spp. As in northeast China, the shrub and herb layers are poorly developed.

Rhododendron, Carex, Festuca, and ferns are always found in matsutake-producing areas in Japan, China, and Korea, demonstrating the impoverished nature of the soils where it is found. Similarly, in Sweden *T. nauseosus* is found associated with pure 120 to 150 year old *Pinus sylvestris* forests where the shrub layer is composed of only *Vaccinum, Calluna,* and *Cladonia* lichens, which are also indicative of a low pH and poor soil nutrient status (Bergius and Danell, 2000).

More than 100 species of higher fungi have been found in *P. densiflora* forests and, according to Ogawa and Ito (1989), some of these, such as *Sarcodon scabrosus* (Fr.) Karst., *Ramaria botrytis* (Pers.) Ricken, *Lactarius piperatus* (Scop.: Fr.) S.F. Gray, and *Hydnum repandum* L.: Fr., compete with matsutake, although Korean mycologists consider *R. botrytis* to be beneficial rather than competitive (T.S. Lee, personal communication). Ogawa and Ito (1989) consider that such ectomycorrhizal fungi as *Suillus luteus* (L.: Fr.) S.F. Gray, *Lyophyllum shimeji* (Kawam.) Hongo, and *T. japonicum* Kawamura enhance matsutake production, although it may simply be that these fungi thrive in the areas that encourage the growth of matsutake. According to Lee et al. (1986), matsutake-producing forests have a greater diversity of fungal species than do nonproducing forests. Park et al. (1998) has suggested that an increase in the abundance of *Russula bella* Hongo, *Amanita pantherina* (DC.: Fr.) Secr., *Suillus bovinus* (L.: Fr.) O. Kuntze, *Boletus* sp., and *Laccaria amethystea* (Bull.: Gray) Murr. in *T. matsutake* forests was indicative of declining matsutake production. Thus, some fungal species may be conducive to matsutake production while others may have a less positive effect.

Shiro

Tricholoma matsutake and related species form distinctive white to pale gray fungal colonies just beneath the soil surface, which the Japanese refer to as *shiro* (white castle or place). The colonies are normally 50 to 150 mm thick, although they can be up to 350 mm thick in deep sandy soils, and consist of a compact mass of mycelia where everything in the shiro, including thick plant roots and fine rootlets, soil granules, and rocks, and gaps between soil granules, are colonized by the fungus (see Photo 20.2).

In Japan, China, and Korea, *T. matsutake* shiros usually occur on the middle to upper parts of hills and on sunny and partly shaded slopes (Wang et al., 1997; Gong et al., 1999). According to Lee (1983) and Ogawa (1978a) 60 percent of shiros are found within 25 m of ridges where there are generally 2,000 to 5,000 trees per hectare (ha), mainly *P. densiflora* about 4 to 5 m high; competing ectomycorrhizal fungi are not abundant; and shrub and herb layers are poorly developed. Matsutake production is greatest when the canopy provides only 50 to 80 percent cover and the shrub layer is relatively sparse. Typically, the soils have low fertility, are well drained and subhygrous to damp but not wet, rocky-sandy, or sandy-loam, with the litter layer about 30 to 50 mm deep. *Tricholoma magnivelare* shiros can be found on slopes, flat areas, and sand dunes in coastal areas of North America.

Shiros normally appear when *P. densiflora* are 10 to 20 years old and develop between the host trees rather than around them, unlike typical ectomycorrhizal fungi. The shiros then expand through the soil at a rate of 100 to 200 mm per year. Fruiting bodies first appear when the shiros are three to

PHOTO 20.2. A *Tricholoma matsutake* shiro in a *P. densiflora* forest in South Korea showing colonization of rocks, soil, and roots by *T. matsutake* mycelia.

four years old with production reaching a peak when the shiros become circular, arced, or irregularly shaped. Shiros remain active for 40 to 50 years and can still produce fruiting bodies for several years after the death of the host tree (Ogawa, 1975; Wang et al., 1997). When shiros touch, there is often a considerable rise in production for a few years before fruiting ceases completely.

Hur et al. (2001a) divided the shiro into four zones, whereas Ogawa (1978a) recognized seven:

- An outer zone at the "infection front" where only mycelia are found
- A zone containing newly infected roots
- A zone of maximum mycelial growth and root infections where the soil is extremely hydrophilic, and production of fruiting bodies depends on how well the zone is developed
- A zone where fruiting bodies are produced
- A powdery mycelial zone where the roots have begun to collapse and die
- An area where the soil is beginning to recover its normal state and structure
- The oldest zone, 200 to 300 mm from the infection front, where the soil has returned to normal (Ogawa and Ito, 1989)

The mycelia and mycorrhizae actively grow only in the first three zones.

The shiros of *T. magnivelare* are shallower and are only 100 mm deep under tan oak, *Lithocarpus densiflorus* (Hook: Arn.) Rehd., in Northern California (Wang et al., 1997). The shiros are normally in arcs, lines, and semicircles, while circles are rare. There are only five recognizable zones and fruiting bodies produced from the third zone (Hosford et al., 1997). The shiros produced by *T. bakamatsutake, T. caligatum, T. fulvocastaneum,* and other related species are very similar to those of *T. matsutake,* but may be shallower and the zones are not very clear (Hosford et al., 1997; Wang et al., 1997; Gong et al., 1999).

According to Japanese mycologists (Ohara and Hamada, 1967; Ogawa, 1977b) volatile antibiotics produced in the shiro eliminate most soil microorganisms. However, *Mortierella* spp. can be isolated from shiros (Tominaga and Komeyama, 1987), and ectomycorrhizae formed by *Cenococcum graniforme* (Sow.) Ferd., *Suillus* spp. and other basidiomycetes are common (Wang, 1995; Santillano, 1998).

Climatic Requirements

In Japan, *T. matsutake* and related species are found between latitudes 20° and 50°N, between 200 and 1,500 m elevation, generally on sunny slopes of up to 30°. In northeastern China and Korea, *T. matsutake* is found in similar situations, usually on the middle to upper parts of hills between 500 and 1000 m elevation where climatic conditions are generally similar to those in the matsutake growing areas of Honshu and Hokkaido (Wang and Xie, 1982; Lee, 1983). However, in southeastern China *T. matsutake* is found between 2,000 and 3,500 m on steep, sunny slopes of up to 75° where there is a humid, cool temperate climate (Wang et al., 1997; Gong et al., 1999). Additional climatic information for matsutake-producing regions can be found in publications by Lee (1983), Ogawa (1978b), Tominaga and Komeyama (1987), Park et al. (1995a,b), Hosford et al. (1997), Wang et al. (1997), Gong et al. (1999), Bergius and Danell (2000), and Berch and Wiensczyk (2001).

Soil Requirements

In Canada, the United States, North and South Korea, China, Japan, and Sweden, *T. matsutake, T. magnivelare,* and *T. nauseosus* are all found in well-drained sandy soils, and sandy loams developed on a variety of parent materials, including those of volcanic origin and granite. The litter layer varies in thickness from 5 to 100 mm and there is often a covering of mosses. Generally, the most productive soils are acidic, well drained, and infertile (Lee, 1983; Ogawa, 1978a; Tominaga and Komeyama, 1987; Hosford et al., 1997; Wang et al., 1997; Gong et al., 1999; Bergius and Danell, 2000; Berch and Wiensczyk, 2001).

Soils under productive stands in Korea have a pH between 4.5 and 5.5, 2.3 percent organic matter, 0.11 percent total nitrogen, 20 $\mu g \cdot mL^{-1}$ available P (unknown extraction procedure) and 12 $me \cdot 100\ g^{-1}$ cation exchange capacity (Lee et al., 1986). A similar pH is also found in Sweden (Bergius and Denall, 2000). Wang et al. (1997) analyzed soils from *T. matsutake* forests in Iwate and around Hiroshima, Japan, the Kuan Dian and Huidong district of China, and from *T. magnivelare* forests in British Columbia, Canada, using standard New Zealand Ministry of Agriculture and Forestry procedures (Cornforth and Sinclair, 1982). The soils had high C:N ratios, low pH, and low soil nitrogen and phosphorus levels, reflecting the relatively poor nutritional levels of their parent materials and the effect of a shallow litter. Surprisingly, the soils they sampled contained high sultur concentrations. This was also a feature of *T. matsutake*–infected soils collected by Wang (unpublished) in China, although further soil testing is required to determine if this is typical of all Chinese and Japanese matsutake soils. Inside the shiro the pH is typically in the range 4.5 to 5.5, while outside it is often a little less acidic, presumably because of the production of organic acids in the shiro by the dense fungal colony. Berch and Wiensczyk (2001) described in considerable detail the soil conditions of *T. magnivelare* forests in British Columbia.

Fruiting

In China, Japan, and Korea, matsutake fruits between July and early October, but yields are closely tied to the weather patterns, especially rainfall and temperature. Generally, *T. matsutake* yields are highest when rainfall is high in spring, summers are relatively hot and dry, and the autumn is moist and cool (Ogawa, 1978a; Lee, 1981; Tominaga and Komeyama, 1987). Primordia usually begin to form after mid-August when the 100 mm soil temperature is 16 to 19°C and there has been about 50 mm of gentle rain. From then on, 5 mm of rain every 3 days is ideal to ensure further growth of the fruiting bodies. However, if the soil temperature climbs higher than 19°C or drops below 15°C for days, primordia will abort. According to Sun et al. (1989) in Heilongjiang province, northeastern China, good harvests can be expected if the 100 mm soil temperature is between 16° and 19°C and 70 to 100 mm of rain falls in 10 showers between the last 10 days of July and the first 10 days of August. Alternatively, if the soil temperature is between 19° and 22°C good production will follow if there is a rainfall of no more than 100 mm of rain is spread over 8 showers.

Fruiting bodies are only found in shiros; primordia form just below the litter layer and are closely associated with matsutake-infected roots or a mixture of matsutake mycelia and soil material. Depending on soil temperature, moisture and other factors (Ogawa, 1978a; Tominaga and Komeyama, 1987; Sun et al., 1989; Hosford et al., 1997), the primordia take 10 to 20 days to develop into opened mushrooms with the optimum soil temperature between 18 and 21°C (Kinugawa, 1964a,b; Ogawa, 1978a; Tominaga and Komeyama, 1987). However, expansion of the primordia can occur at much lower soil temperatures, and in October 1989 the second author picked first-grade matsutake near Morioka when the night air temperatures were hovering around 7°C and the 100 mm soil temperatures about 13°C. As with *T. magnivelare,* the primordia normally stop growing when the ambient temperature falls below 5°C (Hosford et al., 1997) although expansion of matsutake primordia still occurs at 5°C (Ogawa, personal communication).

In a 1.5 ha experimental forest plot operated by the Kyoto Forest Experimental Station, Japan, production began in the twentieth year with one shiro producing 40 × 70 g matsutake, i.e., 2.8 kg per 1.5 ha. By the twenty-fifth year, there were 70 shiro producing a total of 20 kg per 1.5 ha and by the thirty-fifth year it is expected that there will be 100 to 200 shiros producing approximately 100 $kg \cdot ha^{-1}$. Production would then be expected to plateau until the stand reaches 80 years old. In contrast, in the United States 200-year-old *T. magnivelare* forests are still productive but this could be because these forests have multilayered canopies with trees of different ages. In Sweden fruiting begins in August, is not related to falls in temperature, and production ranges from 1 to 200 fruiting bodies per ha (Bergius and Danell, 2000).

In Canada and the United States, *T. magnivelare* fruits from late September to November although in southern Oregon, northern California, and very sheltered parts of British Columbia fruiting bodies can sometimes be found in December when the soil temperature is close to 0°C and the soil surface is covered with snow. Like *T. matsutake*, fruiting of *T. magnivelare* is closely tied to weather patterns, especially rainfall and temperature, but the interrelations between these variables are not well understood. Hosford et al. (1997), investigating fruiting of *T. magnivelare* in Washington, indicated that it occurred when cumulative rainfall reached 14 cm and continued until accumulation reached 35 cm or more. At the same time, average temperatures were mild and above 5°C. But in 1987 *T. magnivelare* did not produce the same mumbers of fruiting bodies at these sites because only 16 cm rain fell.

MORPHOLOGY OF MATSUTAKE INFECTIONS

Natural **T. matsutake** *Infections*

The roots of host plants become infected with *T. matsutake* as the shiro develops. At the same time almost every root, soil particle, and even rock within the shiro becomes colonized by matsutake hyphae (see Photo 20.3). Infections form on rootlets, short lateral roots, and also long lateral roots up to 2 mm in diameter. Rootlets and short lateral roots become swollen and clavate, or sticklike, with a short, brown, thin, uninfected neck where they are attached to the main root. New roots emerging from old infected roots also become infected, which gives the root mass a nodulose or witch's broom appearance (Photo 20.3). Rhizomorphs are absent. Apices are not normally infected, dichotomously branched roots are very rare, and no clusters of multibranched infected roots are found as would normally be pro-

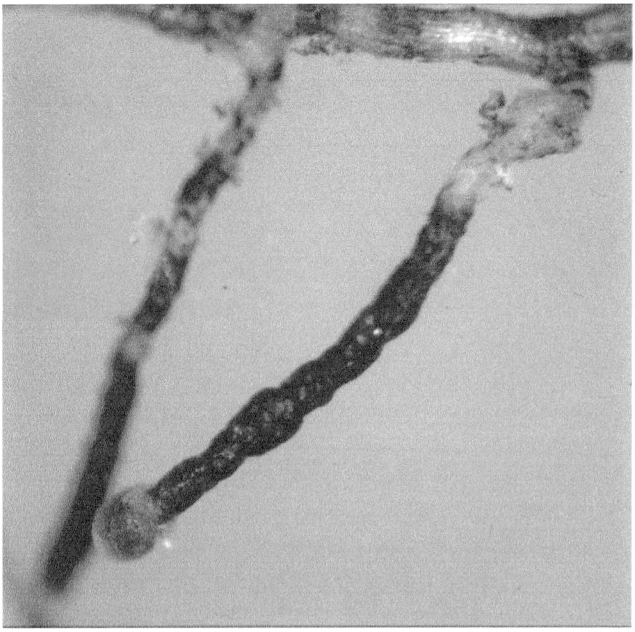

PHOTO 20.3. *Pinus densiflora* roots naturally infected by *Tricholoma matsutake* and a *Suillus*-like fungus.

duced by typical ectomycorrhizal fungi on pine roots. Infections formed by *T. magnivelare* and related species are similar to those produced by *T. matsutake* (Ogawa, 1977c, 1978b, 1979, 1981b; Ogawa and Ohara, 1978; Ohara and Ogawa, 1982; Wang, 1995). The species of host plant has little impact on infection morphology and anatomy (Ogawa, 1976, 1977a,b; Wang, 1995).

In the early stages of infection the light brown young roots become covered with white hyphae, but these were found to be rare in either autumn or spring root samples from Korea and China. The infected root surface blackens. Although the layer of white hyphae still covers the root surface at this stage, the sheathing hyphae are less abundant than in the earlier phase of infection (see Photo 20.4). Eventually the infected roots, which can be up to 5 mm in diameter, become almost completely black and the sheathing hyphae disappear. Finally infected roots wither, collapse, and become moribund before dying. The mycelia then become powdery and the soils containing the powdery mycelia become strongly hydrophobic.

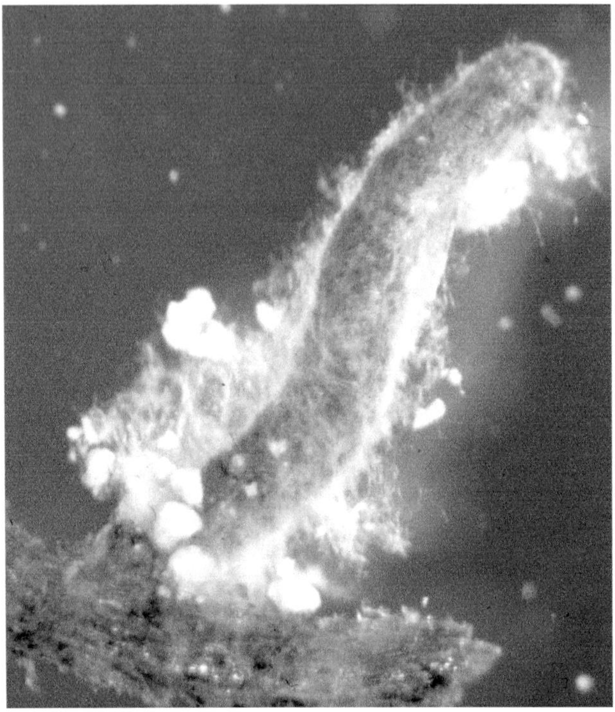

PHOTO 20.4. *Pinus densiflora* rootlet naturally infected with *Tricholoma matsutake* showing the sheathing hyphae and blackened root surface.

Mycorrhizal roots formed by other mycorrhizal fungi such as *Rhizopogon, Suillus,* and *Cenococcum geophilum* are often found adjacent to matsutake-infected roots and may even grow on an adjacent section of a single rootlet, although when this happens the two infections are clearly demarcated, both morphologically (see Photo 20.3) and anatomically (see Photo 20.5). Sometimes matsutake infected roots can be enveloped by a mycorrhizal fungus in what seems to be a hyperparasitic relationship. Terashima (1993) also reported multiple infections in *T. bakamatsutake* shiros.

Cross sections of *T. matsutake*–infected roots show no true mantle (see Photo 20.6), although as infection proceeds the roots become covered with a thick layer of disorganized, loosely arranged hyphae, sometimes mixed with soil materials. Moribund infected roots have no sheathing hyphae.

Tricholoma matsutake hyphae invade intercellular spaces of epidermal cells first and then the outer 2 to 3 cells of the cortex as the roots become in-

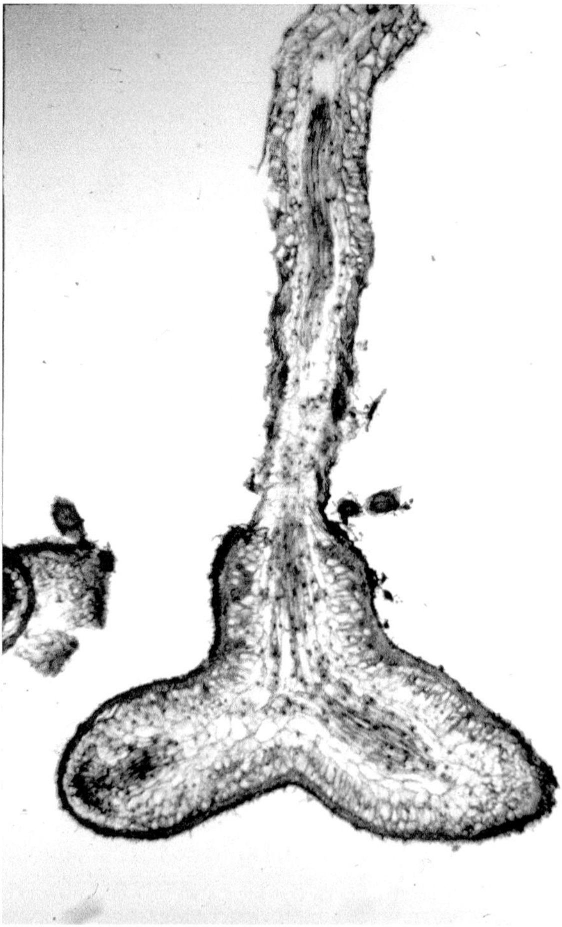

PHOTO 20.5. A cross section of a *P. densiflora* root with the lower part infected by *Tricholoma matsutake* and another typical ectomycorrhizal fungus.

fected (see Photo 20.6). Discrete brown droplets are normally produced on the interior walls of the cortical cells. The cortical cells do not become obliquely elongated, unlike in typical ectomycorrhizae (Harley and Smith, 1983). Often large hyphal masses form in the intercellular spaces of the cortical and epidermal cells and occasionally labyrinthine hyphal systems are formed between the cortical cells, similar to those formed by typical ectomycorrhizal fungi such as *Tuber* and *Lactarius*. These intercellular hyphae remain alive as the infected roots age and the cortical cell walls be-

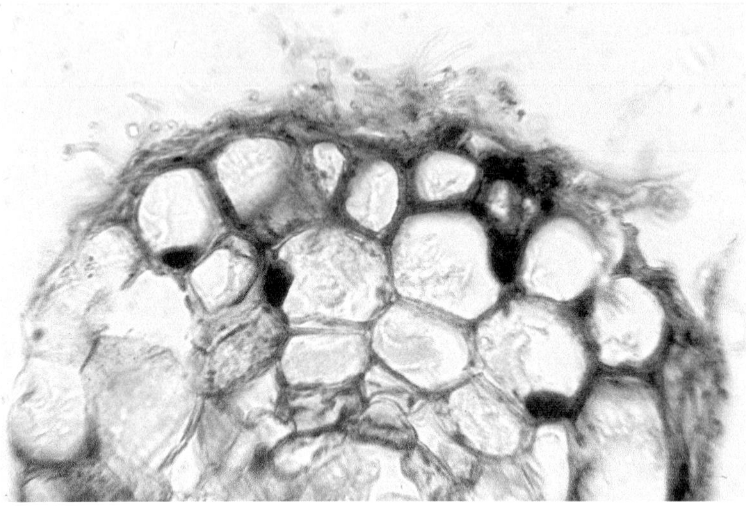

PHOTO 20.6. A cross section of a *P. densiflora* root collected from a forest infected with *Tricholoma matsutake* without a mantle and the matsutake hyphae invading the outer two to three cells of cortex.

come shriveled. Occasionally the intracellular hyphae penetrate the epidermal and cortical cells and produce haustoria-like structures, but these are rare (see Photo 20.7). We have yet to observe *T. matsutake* hyphae invading the stele as reported by Tominaga and Komeyama (1987). Thick-walled (2.5 to 3.5 μm) globose to subglobose chlamydospores 20 to 24 μm in diameter are often found in *T. matsutake* infections and at the interface between *T. matsutake* infections and an adjacent ectomycorrhizal infection. We have also seen them inside cortical root cells of *Pseudotsuga menziesii* infected by *T. magnivelare* collected from an old shiro at a five-year-old clear-cut site (see Photo 20.8) (Wang, unpublished data).

Brown-colored tanninlike substances, which appear to be a defensive strategy in response to infection, accumulate in epidermal and cortical cells of rootlets and short or long lateral roots (see Photos 20.3 and 20.6), giving old roots their characteristic blackened appearance. However, the blackening does not always indicate that infection is present; it can occur on uninfected roots adjacent to infected ones.

When infected roots regrow, the new roots usually become infected as soon as they emerge. Longitudinal sections show that these infections originate from external hyphae rather than internal hyphae, and typically there are zones at the junction of old and new roots that are free of intercellular

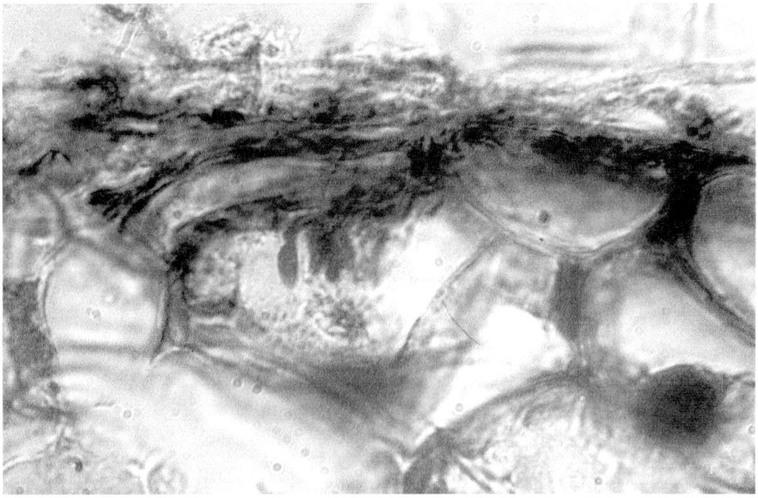

PHOTO 20.7. A cross section of a *P. densiflora* root collected from a forest infected with *Tricholoma matsutake* showing the haustoria-like hyphae that have invaded a cortical cell.

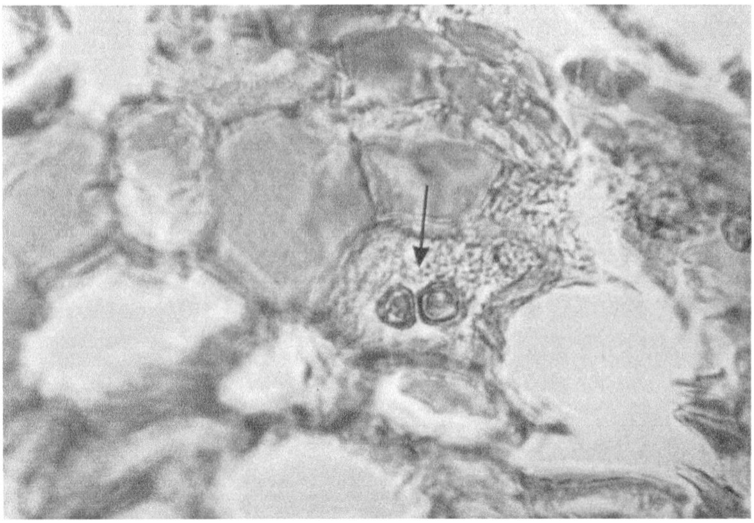

PHOTO 20.8. A cross section of a *Pseudotsuga menzezii* root collected from a forest infected with *Tricholoma magnivelare* showing chlamydospores inside the cortical cells.

hyphae. When infected roots branch, forming new roots, they become infected as soon as they emerge, but then infections originate from both the external and internal hyphae.

Synthesized T. matsutake *Infections*

In our experiments, when pine roots grow through a zone where there is active *T. matsutake* hyphal growth the roots usually become infected (see Photo 20.9). Early on, the root surfaces are covered with whitish mycelia from which the mycelia extend out into the media. After about two weeks

PHOTO 20.9. *Pinus densiflora* roots became infected with *Tricholoma matsutake* when they grew through the active mycelial zone.

the sheathing hyphae become very dense and at this point the roots appear very hairy (see Photo 20.10). As the infected roots age they become pale brown, then brown, but never black.

Infection is initiated by branched appressoria-like hyphae that attach to the root surface. From these, hyphae penetrate the epidermal cells or occasionally root hairs (see Photo 20.11) and then rapidly spread through the outer layers of the cortex (see Photo 20.12). At this stage the root surface has a sparse to dense covering of white downy hyphae and a layer of thin sheathing hyphae over the infected surfaces, except for the root tips of actively growing lateral roots, which remain uninfected. Sometimes, chlamydospores are produced by the sheathing hyphae. Eventually, the epidermal and cortical cells lose their integrity and appear crushed and detached (see Photo 20.13). At this stage the invading hyphae become densely interwoven with some hyphae becoming bulbous and swollen and producing

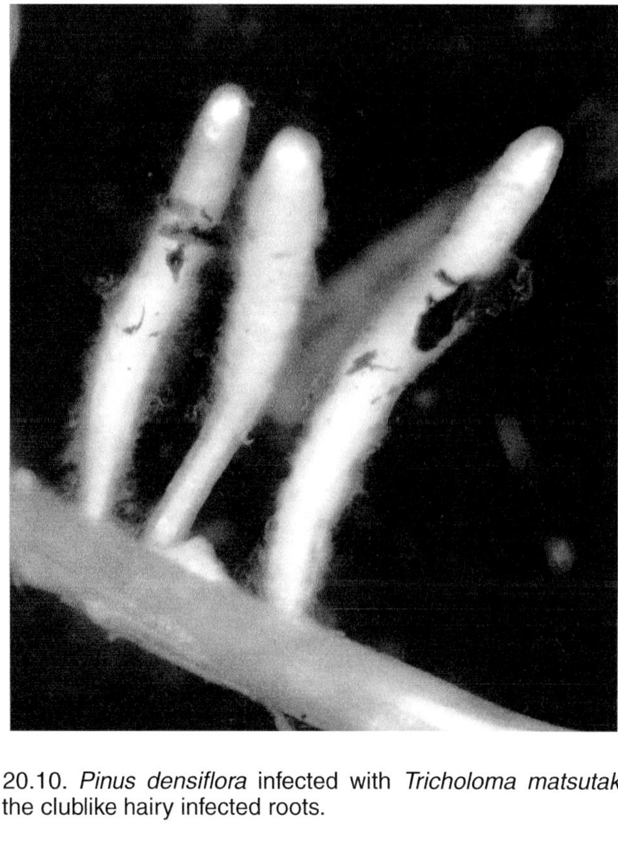

PHOTO 20.10. *Pinus densiflora* infected with *Tricholoma matsutake* in vitro showing the clublike hairy infected roots.

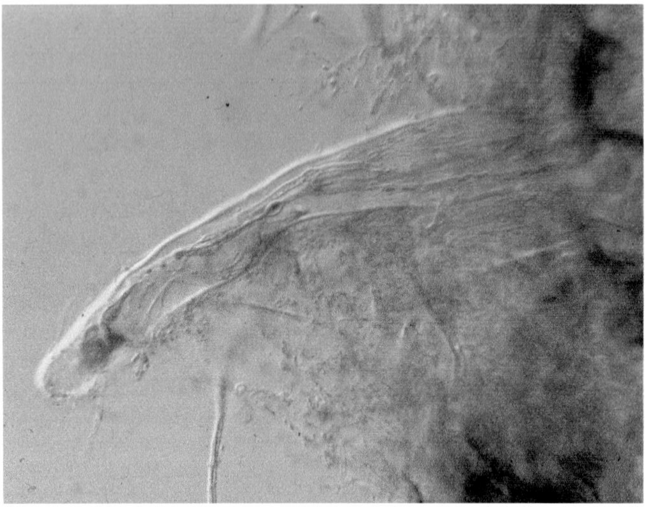

PHOTO 20.11. A cross section of a *Pinus radiata* root infected with *Tricholoma matsutake* in vitro showing the hyphae infecting via a root hair.

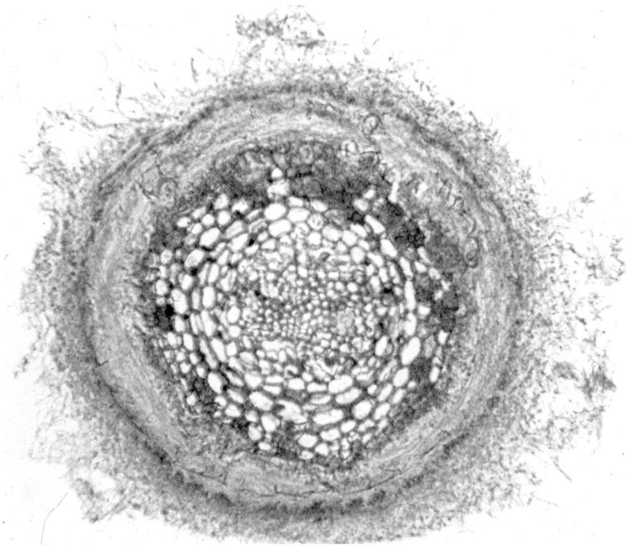

PHOTO 20.12. A cross section of infected *Pinus radiata* root infected with *Tricholoma matsutake* in vitro showing the hyphae rapidly spreading through the outer layers of the cortex.

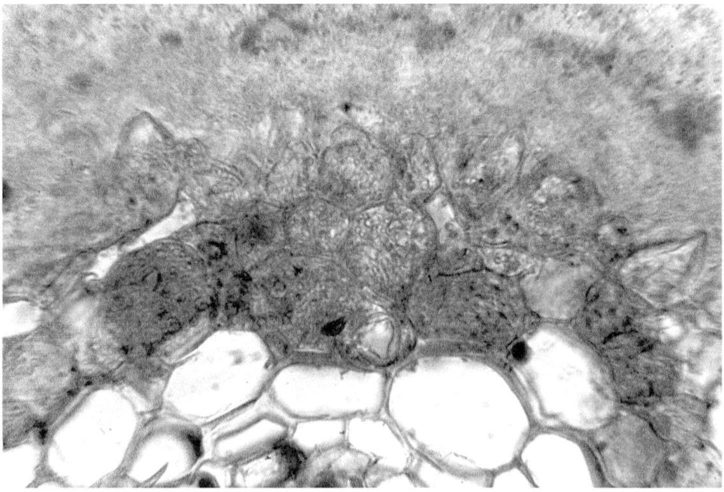

PHOTO 20.13. A cross section of infected *Pinus radiata* root infected with *Tricholoma matsutake* in vitro with chlamydospores formed inside the cortical cells and the general destruction of the cortex.

chlamydospores (see Photo 20.14). The infected root surfaces become covered by a thick layer of white hyphae, although this is not structured as in typical ectomycorrhizas and there is no Hartig net. Primodia or primodia-like hyphal aggregations occasionally develop on the surface of infected roots, severely disrupting the cortex (see Photo 20.15).

There is no obvious digestion of the hyphae by the host cells. However, the sheathing hyphae are not structured well. Occasionally intercellular infections are formed and large interwoven hyphal masses develop between the epidermal and cortical cells. As in nature, hyphae are never found inside the endodermis, but unlike old *T. matsutake* field infections no tanninlike substances are produced in infected roots. The macro- and microscopic features of synthesized infections by *T. magnivelare, T. bakamatsutake,* and *T. caligatum* on *Pinus radiata* and *P. densiflora* are identical to those produced by *T. matsutake* (Wang, unpublished data).

On *P. radiata, T. matsutake* infections develop mostly on the main roots whereas they form on both lateral roots and main roots of *P. densiflora*. Infected rootlets or short lateral roots are swollen, clavate and monopodial (see Photo 20.10), and dichotomous roots are rare. No clusters of multi-branched infected roots are produced. Often the root systems are poorly developed and the root caps became infected, but the root systems do not appear healthy and eventually growth ceases.

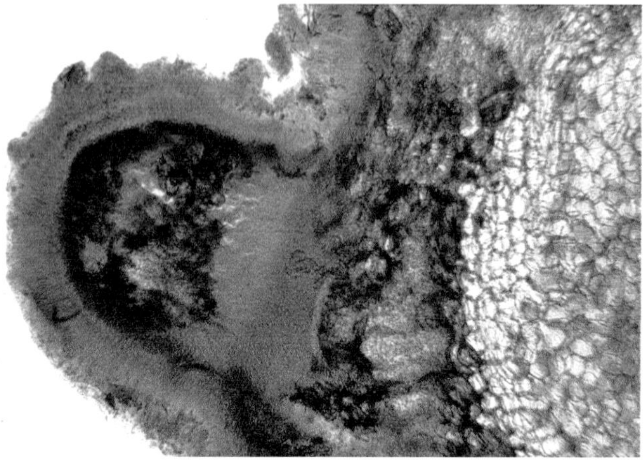

PHOTO 20.14. A cross section of a *Pinus radiata* root infected with *Tricholoma matsutake* in vitro showing a fruiting body primodium growing on the side of the root and the crushed and detached cortex.

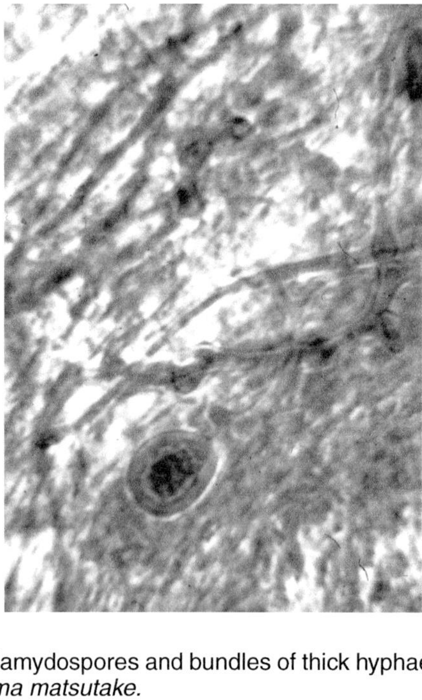

PHOTO 20.15. Chlamydospores and bundles of thick hyphae produced on agar media by *Tricholoma matsutake.*

CULTIVATION OF TRICHOLOMA MATSUTAKE

Methods have been developed for cultivating both *Tuber melanosporum* and *T. uncinatum* in the field (Hall et al., 2001; Chevalier et al., 2002). Like the routine methods for producing ectomycorrhizal plants used in the forestry industry, clean seedlings are inoculated with mycorrhizal fungi in greenhouses and then transplanted to the field once adequate infections have developed (Marx, 1980; Castallano and Molina, 1990; Hall et al., 2001; Chevalier et al., 2002). However, early attempts to cultivate *T. matsutake* using similar techniques failed both in Japan (Ogawa and Ito, 1989) and Korea (Lee, 1988b). Kawai and Ogawa (1981) and Kawai (1989) concluded that fruiting body formation in artificial plantations was "hopelessly difficult."

Improving Matsutake Yields Through the Management of Existing Forests

Following the disappointments of Japanese research on *T. matsutake* in the 1970s and early 1980s Tominaga and Komeyama (1987) and Ogawa and Ito (1989) turned their attention in developing techniques for maximum production of *T. matsutake* in existing forest. These methods have subsequently been adopted by others (Lee, 1988b; Park et al., 1995a,b; Koo and Milek, 1998; Gong et al., 1999). This was achieved by modifying the environment within the forests using the following techniques:

- Raking the litter layer to reduce its thickness to about 30 mm, which had the effect of increasing soil temperature and reducing soil moisture. Sometimes this was followed by the application of fertilizers (Park et al.,1995a,b; Wang et al., 1997; Koo and Milek, 1998; Gong et al., 1999).
- Reducing the canopy density by removing dead and dying trees, shrubs less than 1 m tall, and the judicious removal of mature trees (Ogawa and Ito, 1989; Wang et al., 1997).
- Removing ectomycorrhizal fungal fruiting bodies of species that might compete with *T. matsutake,* for example, *Sarcodon scabrosus* (Fr.) Karst., *Ramaria botrytis* (Pers.) Ricken, *Lactarius piperatus* (Scop.: Fr.) S.F. Gray, and *Hydnum repandum* L. ex Fr. (Ogawa and Ito, 1989).
- Inoculating the soil by spraying on *T. matsutake* spores or retaining some of the mature fruiting bodies on the forest floor. Stimulating fresh pine rootlet growth by digging ditches, removing the old roots,

and then filling the ditches with fresh soil (Ogawa and Ito, 1989)—a technique since used in Spain to stimulate the activity of *Tuber melanosporum* (Reyna et al., 2002).

- Using chemical and physical methods to protect *T. matsutake* forests from diseases, insects, birds, and other animals that can damage up to 40 percent of the fruiting bodies (Zhao, 1986; Ogawa and Ito, 1989; Gong et al., 1999).
- Modifying soil humidity and temperature by erecting irrigated tunnels over shiros (the Hiroshima method; Tominaga and Komeyama, 1987) or small plastic hoods or caps to cover shiros or individual fruiting bodies (Lee, 1988a).
- Protecting and managing *T. matsutake* forests using education and legislation, methods now widely used to control picking in American and Korean forests (Hosford et al., 1997; Pilz and Molina, 2002).

Establishing New Plantations Using Matsutake-Infected Plants

The application of the techniques just described have failed to halt the catastrophic decline in Japanese *T. matsutake* harvests over the past 60 years, from a peak in 1942 of around 12,000 tonnes to less than 1,000 tonnes now (see Figure 20.1; Hall and Wang, 2002). It became abundantly clear in the 1980s that to reverse the downward trend in production additional *T. matsutake* plantations would have to be established.

Attempts have been made to produce *T. matsutake* plants by planting uninfected pine seedlings in front of an advancing shiro (Ogawa and Ito, 1989). After infection occurs the young plants are then transplanted into nonmatsutake-producing areas (Lee, 1984)—a technique similar to Talon's technique for producing *Tuber melanosporum*–infected plants (Hall et al., 2001). However, to date, such techniques have failed to produce fruiting bodies. This has prompted new attempts to produce *T. matsutake*–infected plants using in vitro techniques (Eto, 1990; Hu, 1994; Wang, 1995; Gong et al., 1999; Yamada et al., 1999a,b; Guerin-Laguette et al., 2000).

Preparing of **T. matsutake** *Cultures*

The universal starting point with in vitro techniques has been the preparation of isolates from spores or mycorrhizae or more easily from fruiting bodies, an aim made all the more difficult for *T. matsutake* because isolates typically grow at only 1 cm per month (Ohta, 1990; Wang, 1995). It is therefore critical to choose the best agar, liquid, and nonagar solid media for cultivating the isolates. The first attempts to cultivate matsutake were made by Mimura (1915) and Masui (1927). However, it was not until the 1950s that a reliable technique was developed by Hamada (1950). Since then hundreds of isolates of *T. matsutake* and related species have been obtained in pure culture. Modified Melin-Norkrans' agar medium (MMN) has been the most commonly used for culturing ectomycorrihzal fungi (Molina and Palmer, 1982; Heinonen and Holopainen, 1991) although in Japan Hamada and revised Hamada media have been frequently used (Ogawa, 1978a) while Wang and other mycologists have used potato dextrose and malt extract agar (Gong et al., 1999; Wang, unpublished data).

Fresh matsutake fruiting bodies are collected from the forests and small triangular pieces of tissue are taken from the fresh fruiting bodies and placed in 30 × 200 mm test tubes or petri dishes containing revised Hamada or other suitable nutrient agar. Generally, it is better to attempt to prepare isolates as soon as possible after collecting the fresh mushroom, although sometimes storing the fruiting bodies for one to two days at 5°C can be successful, but not longer as this increases the likelihood of contamination (Wang, 1995). The isolates are then incubated at about 22°C. Generally, white hyphae develop from the inoculum 10 days after isolation, with fast-growing isolates covering a 100 mm dish in 60 days. Wang (1995) found that after storing fresh mushrooms of *T. magnivelare* for two days at 5°C, hyphae grew from about 50 percent of the inocula taken from stems but from only 25 percent of the inocula taken from caps. About 10 percent of the isolates were contaminated, with most of these originating from the edges of caps.

After storage of fresh *T. magnivelare* mushrooms for 21 days at 5°C, none of inocula from the mushroom caps produced hyphae and about half were contaminated. In contrast, almost 50 percent of the isolates from the stems had hyphal extension and about 33 percent were contaminated. MMN agar was far superior to revised Hamada agar (Wang, 1995). Isolates taken from the mushroom stem produced better results than from the cap and gill tissue, especially when the mushrooms were kept for a period after collecting from field. This contradicts the results from Japanese research (Hiromoto, 1960, 1961, 1975; Ogawa, 1978a).

Oyama et al. (1974) showed that in the vegetative phase the nutrient requirements for ectomycorrhizal fungi, such as *T. matsutake* and *Boletus edulis,* were identical with those for saprobic fungi, such as *Lentinus edodes,* and formulated an "ideal" medium. In 1990 Ohta developed a new medium for mycelial growth of mycorrhizal fungi, including matsutake. Ogawa and Kawai (1976) found that Hamada agar plus a filtrate of needles or roots of *Pinus densiflora* supported better growth of *T. matsutake* than Hamada agar, whereas Wang (1995) found that the reverse was true, perhaps because of the different pine species used, the extraction procedure, and the presence of tannin and polyphenol substances that may have inhibited growth (Ogawa and Kawai, 1976).

Identification of Matsutake Isolates

Faster-growing contaminants commonly contaminate very slow-growing fungal species such as *T. matsutake,* in particular, *Mortierella* spp. (Tominaga and Komeyama, 1987; Wang, 1995). It is, therefore, very important to verify the identity of the isolates before proceeding. This can be achieved by comparing the morphology of the isolates on agar, although molecular techniques are a more reliable means of identification (Kikuchi et al., 2000; Guerin-Laguette et al., 2002).

On revised Hamada agar Wang (1995) found that matsutake isolates produced round, pure white, thick, raised colonies, with very fine bundles of fungal threads and sometimes 5 to 6, and occasionally up to 15, grooves radiating from the colony center. Colonies were silken with very even edges and in the center of the colony there were usually abundant aerial hyphae forming a papilla from which fungal threads projected. When Wang (1995) grew isolates on media rich in sugars, very large papillae and hyphal threads were produced; the size of the papillae apparently was determined by the amount of glucose in the agar. On MMN agar, isolates produced irregularly shaped, very hairy, whitish colonies without a silken sheen, with sinuous edges, grooves; papillae were indistinct; and more hyphae penetrated the agar. *Tricholoma matsutake* isolates that were isolated from different host plants produced morphologically similar colonies on the same agar media. However, occasionally a single *T. matsutake* isolate produced colonies that differed morphologically.

In Wang's (1995) experiments *T. bakamatsutake* colonies were very similar to those produced by *T. matsutake,* except that abundant, brown, thick-walled, obovoid chlamydospores 8.5 to 12.0 (to 15.0) × 5.5 to 6.5 (to 9) μm in diameter were produced and the agar became discolored brown. *Tricholoma fulvocastaneum* isolates have flocculent colonies with two zones, the exterior one being more hairy. The colonies produced by *T. magnivelare* and *T. caligatum* isolates were similar to *T. matsutake* colonies. *Tricholoma magnivelare* isolates occasionally produced brown pigmentation on revised Hamada agar—as with *T. bakamatsutake,* a feature never observed in *T. matsutake* isolates. *Tricholoma fulvocastaneum* colonies on revised Hamada had two clear zones, an interior smooth zone with grooves and an exterior zone, hairy, without grooves, while *T. matsutake* colonies on revised Hamada agar were uniformly smooth (Wang, 1995).

Hyphae of *T. matsutake* are made up of long hyaline, hyphal cells, normally 2 to 3 μm in diameter without clamp connections. They show a tendency to form bundles with thick hyphae (up to 6 μm), inflated round hyphae, and thick- or thin-walled inflated cells. Very short branching hyphae occur on short sections of a normal hypha, and chlamydospores are produced, in particular from older cultures (see Photo 20.16). Dense, spring-like, apical, hyphal coils (10.5 to 18.9 μm long × 12.6 to 27.3 μm in diameter) are sometimes produced and project from the surface of *T. matsutake* agar. This unique hyphal structure has also been found in cultures of *T. magnivelare* and *T. caligatum* on revised Hamada agar (Wang, 1995).

Wang (1995) found that the older isolates of a species grew faster than younger ones on agar, in liquid media, and on nonagar solid substrates, suggesting that older isolates have become adapted to the artificial media following repeated subculturing. Older isolates were also no less able to infect

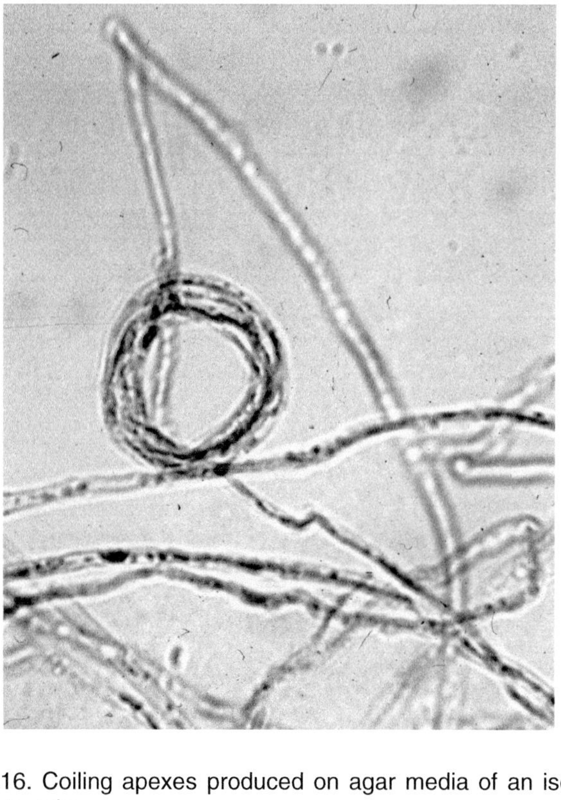

PHOTO 20.16. Coiling apexes produced on agar media of an isolate of *Tricholoma matsutake.*

host plants in vitro than younger isolates. This is the opposite of what normally happens with ectomycorrhizal fungi where subculturing generally reduces a culture's ability to form mycorrhizae (Molina and Palmer, 1982). Wang (1995) found that *Mortierella* isolates isolated from *T. matsutake* fruiting bodies grew much faster than the matsutake species and, at one month old, become pink due to abundant sporangispores produced from the colony. These features were sufficient to clearly distinguish them from matsutake isolates.

Preparation of Liquid Inocula—Growth of Matsutake Isolates in Liquid Media

Tricholoma matsutake cultures are often raised in MMN, revised Hamada, MNC, revised PDA, and Oyama nutrient solutions (Ogawa, 1978a; Wang, 1995; Yamada et al., 1999a) by placing two 5 mm plugs of inocula taken from actively growing margins of cultures growing on nutrient agar into 28 mL McCartney vials containing a small quantity of broken glass and 5 to 10 mL of MMN or other nutrient solutions and then incubated at 22°C. Normally it takes 50 days for the mycelia to reach the surface of the solution.

In nature, matsutake grows in acid soils with a pH range of 5.0 to 6.0, but on artificial media matsutake hyphae can grow from pH 3.0 to 6.5 (Ogawa, 1978a). Oyama et al. (1974), Kawai and Ogawa (1976), and Ohta (1990) reported that the optimum pH of nutrient solution was around 5.0 while Kim (1984) suggested an optimum pH of 4.5 for matsutake spore germination. Wang (1995) found that hyphae grew best in liquid media between pH 4.5 and 6.0, but failed to grow at pH 3.0. This tallied with growth on solid media.

Growth of Matsutake Isolates on Nonagar Solid Media

Vermiculite or mixtures of vermiculite/peat moss (9/1, v/v) are commonly used for cultivating ectomycorrhizae (Molina and Palmer, 1982) and have been used to cultivate *T. matsutake* (Ogawa, 1978a; Kawai and Ogawa, 1981; Tominaga and Komeyama, 1987; Hu, 1994; Wang, 1995; Gong et al., 1999; Guerin-Laguette et al., 2002; Vaario et al., 2002). Sand and other soil materials have also been used (Ogawa, 1978a; Tominaga and Komeyama, 1987). However, Wang (1995) found that substrates containing perlite such as peat/perlite/ vermiculite (1/1/1, v/v) or *Sphagnum* moss/perlite/vermiculite (1/1/1, v/v) and peat/perlite (1/3, v/v) with a pH of 4 to 6.5 and an incubation temperature of 22°C produced better results. Recently, Guerin-

Laguette et al. (2002) found that the incorporation of surfactants or vegetable oils into soils or pine bark can also enhance growth.

Preparation of Sterile Seedlings

The method we have used (Wang, 1995) to sterilize seeds is a modification of one used by Molina and Palmer (1982) which we adopted after experiments showed that germination on agar was poor. The sterilized seeds are first placed in sterile petri plates containing filter paper. When the radical is about 1.5 cm long the germinated seeds are transferred into vials or test tubes with nutrient agar and checked regularly for contamination. Seedlings that remain sterile can be used when they are two weeks old.

The containers we use (Wang, 1955) for growing sterilized seedlings are 40 × 300 mm glass tubes covered with inverted autoclavable plastic beakers. Each tube is filled with 120 mL of substrate moistened with nutrient solution. In contrast, Yamada et al. (1999a) used widemouth 300 or 600 mL plastic bottles with 150 or 300 mL of substrate moistened with MNC or Ohta solution. Sterility was achieved by covering the beakers and the upper part of the tubes with three layers of polyvinylchloride film containing two vents, about 6 mm in diameter, covered with a fluorocarbon membrane filter. Guerin-Laguette et al. (2000) used culture plates (222 × 90 × 10 mm) containing 150 mL of sterilized soil and a perforated plate lid covered with sterilized adhesive filter membranes.

Infecting Plants Under Sterile Conditions

Normally liquid cultures are used to inoculate seedlings (Wang, 1995; Yamada et al., 1999a; Guerin-Laguette et al., 2000). In our work, mycelia in 10 mL of a liquid culture were first broken by shaking and then simply poured into the inoculation tubes (Wang, 1995). Liquid culture inocula have also been used in Japan (Eto, 1990; Yamada et al., 1999a; Guerin-Laguette et al., 2000). They first homogenized the mycelia with a blender (AM-3 Ace Homogenizer, Nihonseiki Ltd), incubated the mycelial mass for three days, rinsed it three times in sterilized modified SH solution on a nylon mesh filter with a pore size of 24 × 30 μm, and then resuspended the hyphae in Ohta liquid solution. Sterile seedlings were then aseptically immersed in the suspension for an hour before transplanting into containers filled with sterilized soil. The remaining hyphal suspension was then applied to the seedlings with a sterilized syringe.

Agar plugs, 3 × 5 mm, cut from the edge of *T. matsutake* fungal colonies on agar plates, have also been used to inoculate seedlings (Wang, 1995;

Yamada et al., 1999a). In our work (Wang, 1995) these were placed on the surface of nutrient agar adjacent to sterile seedlings. However, generally liquid inocula produced superior infections (Eto, 1990; Wang, 1995; Yamada et al., 1999a; Guerin-Laguette et al., 2000). Typically, inoculated seedlings have been incubated in growth chambers at 20 to 25°C under 3,000 to 6,000 lux light with a 16-hour day or 24 hours continuous light.

Potting Up Infected Seedlings

Once seedlings become infected they need to be transplanted into larger containers. We used pots and similar containers filled with potting mixture and transplanted seedlings covered with transparent polyethylene bags and left them in the laboratory at room temperature (20 to 22°C) or placed them in mist chambers for one to two weeks. However, some researchers reported that once the bags were removed and the pots transferred to the greenhouse bench *T. matsutake* infections were deteriorated and eventually disappeared (Eto, 1990; Wang, 1995; Guerin-Laguette et al., 2000).

Factors Affecting Matsutake Infection

Wang (1995) found that container, substrate, nutrient solution, isolate, species, and age of host plant affected growth of *T. matsutake* and seedlings, and subsequent level of infection. Complex interactions occurred between the isolates, substrates, nutrient solutions, and host plants. Three-month-old seedlings were more suitable for mycorrhizal synthesis. In nature, *T. matsutake* has the ability to stimulate the rooting, branching, and elongation of pine roots (Ogawa, 1978a), but on artificial media *T. matsutake* infection depressed root growth, especially of weak or heavily infected seedlings. Wang (1995) attributed this to the limited space in test tubes and found much lower mortalities and healthier plants in nonagar solid media. He also observed that 40-year-old cultures were just as likely to infect roots as younger ones, which is contrary to the common finding that cultures of ectomycorrhizal fungi quickly lose their ability to form mycorrhizae (Marx and Daniel, 1976).

RELATIONSHIP BETWEEN MATSUTAKE AND ITS HOST PLANTS

In 1927 Masui suggested that both *T. matsutake* and *T. caligatum* species were "parasitic mycorrhizal fungi," while Hiromoto (1963a,b) suggested that *T. matsutake* had saprobic tendencies. Hamada (Ogawa, 1982) finally concluded that the relationship between *T. matsutake* and its host plants fell somewhere between the two poles that he described as "plant parasitic on fungus" (*Gastrodia* on *Armillaria*) and "fungus parasitic on plant" (matsutake on *Pinus*). Subsequent work in Japan by researchers, such as Ogawa (1978a), Tominaga (Tominaga and Komeyama, 1987), Eto (1990) and others, has supported the view that *T. matsutake* is mycorrhizal but has some pathogenic traits.

Garrett (1960) stated that there was a continuous gradation of behavior rather than sharp boundary lines between pathogen, saprobe, and mycorrhizal trophic states, and that environment could affect the nature of an association, a view reiterated by Lewis (1973) and Harley and Smith (1983). In the mid-1990s Wang (1995) and Wang et al. (1997) resurrected this view and suggested that fungi such as *T. matsutake* occupied zones in a triangle bounded by the three terms and that during its life cycle its trophic state might drift toward one pole or another. Wang (1995) found that non-mycorrhizal fungi such as *Mortierella* can produce pseudomycorrhizae or form symbiotic associations with trees under sterile conditions. In contrast, isolates of *Boletus edulis* can lose their ability to form mycorrhizae after growing on agar media (Wang, unpublished data) while Ohta (1990) reported that highly selected isolates of the ectomycorrhizal fungus *Lyophyllum shimeji* produced mature fruiting bodies in pure culture. Since then, Japanese-based researchers (Gill et al., 1999; Yamada et al., 1999a,b; Gill et al., 2000; Guerin-Laguette et al., 2000; Vaario et al., 2002) have concluded that although *T. matsutake* is mycorrhizal with *Pinus densiflora,* the fungus does have saprobic and parasitic traits (Gill et al., 2000; Vaario et al., 2002).

Although *T. matsutake* is found in forests and is associated with the roots of trees, many features suggest that it is nothing but a typical mycorrhizal fungus. These features are highlighted below.

Parasitic Features

- *Tricholoma matsutake* infections are often sheathed by a mass of hyphae. However, these sheathing hyphae are unstructured, often mixed with soil materials, and eventually become detached from the root surface. Often the sheathing hyphae are completely absent (Wang, 1995; Yamada et al., 1999a; Gill et al., 1999, 2000). These sheathing hyphae, therefore, are quite dissimilar from true ectomycorrhizal mantles (Wang, 1995).
- The Hartig nets in natural and synthesised *T. matsutake* infections are quite different from the Hartig nets produced by typical ectomycor-

rhizal fungi such as *Rhizopogon rubescens* and *Amanita muscaria* (Ogawa, 1978a; Wang, 1995; Yamada et al., 1999; Gill et al., 1999, 2000)—the cortical cells are not obliquely elongated as normally occurs in typical ectomycorrhizae (Harley and Smith, 1983) and the intercellular hyphae develop unevenly and in masses in the intercellular spaces of the cortex and epidermis.

- Matsutake does not produce rhizomorphs, which are characteristic of such typical ectomycorrhizal fungi as *Rhizopogon rubescens, Lactarius deliciosus,* and *Amanita muscaria* (Ogawa, 1981a; Read, 1992; Wang, unpublished data). Ogawa (1981a) regarded those mycorrhizal fungi without rhizomorphs and mantles, as parasitic and those with well-developed rhizomorphs, mantles and Hartig net as symbiotic. Formation of rhizomorphs is very important, not only for transportation of nutrients to the host but also for infecting new roots (Wang and Hall, 2002).
- The way in which *T. matsutake* infects its host plant is very different from typical ectomycorrhizal fungi, such as *Pisolithus tinctorius, Hebeloma,* and *Laccaria*.
- Intracellular infections are found in both natural and synthesized infections. These cause severe damage to the cortex, particularly in synthesized infection (Ogawa, 1978a; Wang, 1995). This does not occur in typical ectomycorrhizae.
- Infection by *T. matsutake* does not stimulate dichotomous root branching, as is normal with ectomycorrhizae. Instead, *T. matsutake*–infected roots take on a witch's-broom-like appearance.
- Tannins are produced in the *T. matsutake*–infected roots, which indicates a host defense response to infection. This does not usually occur with mycorrhizal infections.
- *Tricholoma matsutake* is most likely to be found in stands of trees that appear to be in poor condition, where branches leave the trunks at all angles, and foliage is chlorotic (Ogawa and Ito, 1989). It is not known whether trees in this condition favor matsutake growth or this is a consequence of infection by matsutake (Ogawa, personal communication). In contrast, normal ectomycorrhizae enhance the uptake of nutrients by host plants.

Saprobic Features

- Hiromoto (1963a,b) suggested that matsutake could be saprobic. Further saprobic features of matsutake were revealed in our research (Wang, 1995; Wang et al., 1997).

- The development of a colony in soil, as represented by the shiro, is more like those produced by saprobic fungi, such as *Tricholoma mongolicum* Imai (Wang, unpublished data; Read, personal communication) and *Lepista* spp. (Ogawa, 1981a,b).
- In the shiro the matsutake mycelia colonized not only host plant's fine roots but everything, including rocks, dead roots, and soils (Wang 1995), a typical feature of saprobic fungi (Ogawa, 1981a,b; Wang, 1995).
- The fruiting bodies of typical ectomycorrhizal fungi usually form in circles or part circles with the trunk of the tree at the center. Similarly, the brûlé of *Tuber melanosporum* (Périgord black truffle, Hall et al., 2001), which indicates the position of active mycorrhizal infections, usually surrounds the host. However, in *T. matsutake* the shiros form between trees and then gradually get larger, often growing toward them. In this regard *T. matsutake* resemble forest-dwelling saprobic fungal species.
- Our investigations show that chlamydospores or chlamydospore-like cells are commonly formed by *T. matsutake* on agar, in liquid cultures, and on natural and synthesized infected roots (see Photo 20.16). Similarly, Shimazono (1979) found that *T. bakamatsutake* also produced chlamydospores, while Wang (Wang and Hall, 1996, 1998) found chlamydospores in the root cortices of *Pseudotsuga menziesii* infected with *T. magnivelare* in Washington State, United States, in an area that had been clear felled five years previously (Photo 20.9). The mycelia looked as fresh as new ones. The invading hyphae disappeared from the infected roots but formed again inside cortical cells. This certainly supports the suggestion that matsutake can be saprobic at some stage in its life cycle (Wang, 1995; Wang and Hall, 1998).
- Chlamydospores are often formed in cultures by lower fungi. They are rarely formed by Basidiomycetes and then only by saprobic and parasitic species.
- Hyphal coil apexes were produced by matsutake isolates on agar culture (Wang, 1995; Wang and Hall, 1996, 1998; Wang et al., 1997). Erland (personal communication, 1994) suggested these structures were very similar to the trap rings of predacious fungi, none of which are known to produce mycorrhizae. Predacious fungi such as *Pleurotus, Hohenbuehelia, Hyphoderma,* and *Resupinatus,* which live in N-deficient environments, are primarily saprobes (Thorn and Barron, 1983; Tzean and Liou, 1993; Thorn and Tsunda, 1993; Dijksterhuis et al., 1994).

- When invading cortical cells, the matsutake hyphae pass through the cell walls and finally digest them. It has been suggested that matsutake may have saprobic ability, producing extracellular enzymes (Wang, 1995). Terashita and Matashi (1987) and Hur et al. (2001a) have found extracellular enzymes (amylases and proteinases) showing that matsutake can metabolize polymers.

Other Atypical Mycorrhizal Features

- Techniques normally used to produce mycorrhizal-infected plants (Molina and Palmer, 1982; Heinonen and Holopainen, 1991) have failed.
- Neither Wang (1995; Wang et al., 1997) nor Yamada et al. (1999a) were able to maintain matsutake infections when infected seedlings were transplanted into nonsterile media.
- *Tricholoma matsutake* and related species are separated from the genus *Armillaria* on the basis of whether they produce amyloid or inamyloid spores. Consequently, significant pathogens now reside in the genus *Armillaria,* for example, *A. mellea* and *A. novae-zelandiae.* Furthermore, Ma et al. (1988) have determined that taxonomically *Tricholoma populinum* Lge., a species believed to be ectomycorrhizal (Trappe 1962), is in fact a potent pathogen of *Populus* spp.
- *Armillaria mellea* can form an association with the achlorophyllous orchid Gastrodia at the same time as being pathogenic on another host plant (Campbell, 1962; Hood, 1992). A similar association is formed by *Tricholoma magnivelare* with *Allotropa virgata,* another achlorophyllous orchid (Castellano and Lauma, personal communication; Wang, 1995; Hosford et al., 1997).

Clearly *T. matsutake* is not a typical mycorrhizal fungus. Instead, the morphological and anatomical characteristics of *T. matsutake* infections suggest that this fungus better fits the definition of "pseudomycorrhiza" (Robertson, 1954; Mikola, 1965; Marks and Foster, 1973), "superficial ectomycorrhiza" (Malajczuk et al., 1987), or "primitive symbiont" (Wang, 1995). The term *pseudomycorrhiza* was coined by Melin in 1917 to describe black mycorrhizae with or without a poorly formed mantle and/or Hartig net, intracellular root infections, poor dichotomy of the infected roots, and a physiologically more parasitic than mutually symbiotic tendency (Mikola, 1965). The term *superficial ectomycorrhiza* was used by Malajczuk et al. (1987) to describe the mycorrhizae formed by some *Hysterangium* and *Cortinarius* species with *Eucalyptus* in Australia. These form colonies very similar to *T. matsutake* shiros. The roots become enveloped with loose, uneven hyphae that radiate into the surrounding soil and litter, binding soil particles and organic matter, as well as thin sheathing hyphae. Inter- and intracellular infections occur in roots, tannins accumulate in the infected roots, and the epidermal cells of host roots are not elongated by the infection process. While Warcup (1980) demonstrated growth stimulation of hosts by superficial ectomycorrhizae, a closer comparison of these and *T. matsutake* infections would be desirable. The term *primitive symbiont* was proposed by Wang (1995) to describe matsutake species that have three phases or respects in their life cycles—symbiotic, parasitic, and saprobic—which could suggest matsutake species might evolve from a saprobic fungus and become a primitive symbiont.

FUTURE CHALLENGES

If *T. matsutake* is a mild facultative parasite that produces structures we believe to be mycorrhizal, then clearly the methods that have been used in attempts to introduce it to new areas are quite likely to have been inappropriate. A relatively small amount of fungal tissue on an infected root may simply not have the inoculum potential to withstand the onslaught of browsing insects and competition from other fungi in an unsterile environment. Similarly, the environmental conditions in a culture tube held at 22°C will be quite different from those in an open pot on a greenhouse bench where there are likely to be wild swings in soil moisture and temperature.

On the other hand, if *T. matsutake* forms mycorrhizae (or pseudomycorrhizae or superficial ectomycorrhizae) on young roots and then becomes a mild parasite as the roots age, while at the same time exhibiting considerable saprobic abilities within the soil, we need to approach the cultivation of *T. matsutake* from another angle. Attempts to introduce *T. matsutake* by transplanting sections of shiro have been as unsuccessful, as have efforts to introduce the fungus to infected plants (Ogawa and Ito, 1989) in Korea (Lee, 1988a), China (Fu et al., 1999), and New Zealand (Wang, unpublished data). We believe that the key to solving the problem lies in a thorough understanding of the shiro—the conditions that exist within it; how it is sustained; the role of roots, true mycorrhizae, and companion organisms within the shiro; and, above all, whether the fungus is dependent on the host tree for sustenance and/or simply because the host tree maintains environmental conditions within the soil that are conducive to fungal growth. For example, Danell (2002) and Högberg et al. (1999) have suggested that C-13 measurements or C-14 labeling could ascertain whether carbon in fruiting bodies originates from soil materials or from photosynthesis. Or it might be that there is a need for a simultaneous supply of nutrients from both the host plant and the soil (Guerin-Laguette et al., 2002; Vaario et al., 2002).

If *T. matsutake* has significant pathogenic tendencies, particularly late in the infection process, then inoculating sterile seedlings only with *T. matsutake* is perhaps doomed to failure. A possible way around this problem might be to establish normal ectomycorrhizal fungi on sterile host seedlings and, when well established, introduce *T. matsutake* into the system. This might be achieved, for example, in culture chambers similar to those used by Danell (1994) to establish *C. cibarius* on host plants.

Another challenge is maintaining stocks of *T. matsutake* given the ever-increasing pressure by pickers. Matsutake forests need to be protected and well managed (Ogawa, 1978a; Koo and Milek, 1998; Gong et al., 1999; Berch and Wiensczyk, 2001; Pilz and Molina, 2002). In China and in East Asian countries, plunder harvesting has taken its toll—shiros have been damaged and harvests from once-productive forests have declined, often dramatically. Recently, in Jilin province, China, *T. matsutake* became a protected species by being placed on the Regulation Regarding Protection of Jilin Wildlife and Plants list. In Liaoning, Jilin, and Hilongjiang provinces, *T. matsutake* natural reserves have been established for research (Wang, unpublished data) while in Canada and the United States, where matsutake forests are generally well managed, their sustainability is being monitored (Koo and Milek, 1998; Berch and Wiensczyk, 2001; Pilz and Molina, 2002). Education regarding the harvesting and management of forests is also being provided on both sides of the Pacific. However, in China the financial temptation of making more money in one day from picking *T. matsutake* than could otherwise be earned in a month is simply too much and even severe penalties do not seem to be a sufficient deterrent.

CONCLUSIONS

Tricholoma matsutake is one of the most highly prized edible ectomycorrhizal mushrooms. It commands very high prices, particularly where Japanese cuisine is highly regarded. Since World War II matsutake production in Japan has declined from more than about 12,000 tonnes to less than 1,000 tonnes. Despite a century of research, the cultivation of matsutake has remained elusive (Wang et al., 1997) due to a lack of understanding about its biotrophic status. Traditionally, matsutake is considered a typical ectomycorrhizal mushroom (Lee, 1988a; Yamada et al., 1999a,b; Gong et al.,

1999), even though its parasitic and saprobic habits have been indicated (Masui, 1927; Hiromoto, 1963a,b; Ogawa, 1978a, 1982). Recently, more evidence has been found that strongly suggests matsutake is not a typical ectomycorrhizal but a pseudomycorrhizal, superficial, or primitive mycorrhizal fungus (Wang, 1995).

Research on the biotrophic status of matsutake and new ways of cultivating matsutake are definitely needed. Managing and protecting the existing matsutake forests is also a top priority of matsutake research.

REFERENCES

Bencivenga M. (1998). Ecology and cultivation of *Tuber magnatum* Pico. In *Proceedings of the First International Meeting on Ecology, Physiology and Cultivation of Edible Mycorrhizal Mushrooms.* Uppsala, Sweden, July. Available online at <http://www.mykopat.slu.se/mycorrhiza/edible/proceed/bencivenga.html>.

Berch M.S. and Wiensczyk M. (2001). *Ecological descriptions and classification of some pine mushroom* (Tricholoma magnivelare) *habitat in British Columbia.* Ministry of Forests Research Program, Southern Interior Forest Extension and Research Partnership, British Columbia.

Bergius N. and Danell E. (2000). The Swedish matsutake (*Tricholoma nauseusus* sun. *T. matsutake*): Distribution, abundance and ecology. *Scand. J. For. Res.* 15: 318-325.

Blytt A. (1905). Norges hymenomyceter, Videnskabs-Selsk. Skrifter!. Math.-nat.Kl. 1904 no. 6: 1-164.

Campbell E.O. (1962). The mycorrhiza of *Gastrodia cunningahmii* Hook. F. *Transactions of Royal Society of New Zealand* 1(24): 289-296.

Castellano M.A. and Molina R. (1990). Mycorrhizae in the biological components: Nursery pests and mycorrhizae. In R.G. Nisley (ed.), *The container tree nursery manual,* Volume 5 (pp. 101-171). Washington, DC: Public Affairs Office.

Chevalier G. (1998). The truffle cultivation in France: Assessment of the situation after 25 years of intensive use of mycorrhizal seedlings. *Proceedings of the First International Meeting on Ecology, Physiology and Cultivation of Edible Mycorrhizal Mushrooms.* Uppsala, Sweden, July. Available online at <http://www.mykopat.slu.se/mycorrhiza/edible/proceed/chevalier.html>.

Chevalier G. and Frochot H. (1997). *La Truffe de Bourgogne,* Tuber uncinatum *Chatin.* Eds. Pétrarque, Levallois-Perret.

Chevalier G., Gregori G., Frochot H., and Zambonelli A. (2002). The cultivation of the Burgundy truffle. In I.R. Hall, Y. Wang, E. Danell, A. Zambonelli (eds.), *Edible mycorrhizal mushrooms and their cultivation: Proceedings of the Second International Conference on Edible Mycorrhizal Mushrooms,* CD-ROM. New Zealand Institute for Crop & Food Research Limited, Christchurch.

Cornforth I.S. and Sinclair A.G. (1982). *Fertilizers and lime recommendations for pastures and crops in New Zealand.* Wellington: New Zealand Ministry of Agriculture and Fisheries.

Danell E. (1994). *Cantharellus cibarius:* Mycorrhiza formation and ecology. Comprehensive summaries of Uppsala dissertations from the faculty of science and technology 35. Acta Universitatis Upsalensis, Uppsala, Sweden. pp. 1-56.

Danell E. (2002). Current research on matsutake in Sweden. In I.R. Hall, Y. Wang, E. Danell, A. Zambonelli (eds.), *Edible mycorrhizal mushrooms and their cultivation: Proceedings of the Second International Conference on Edible Mycorrhizal Mushrooms*. New Zealand Institute for Crop and Food Research Limited, Christchurch.

Danell E. and Camacho F.J. (1997). Successful cultivation of the golden chanterelle. *Nature* 385: 303.

Dijksterhuis J., Sjellena J., Veenhuis K.A., and Hatder W. (1994). Competitive interactions between two nematophageous fungi during infection and digection of the nematodes *Panagrellus redivivus. Journal of Mycol. Res.* 98 (12): 1458-1462.

Eto S. (1990). Cultivation of the pine seedlings infected with *Tricholoma matsutake* by use of in vitro mycorrhizal synthesis. *Bulletin of the Hiroshima Prefectural Forestry Experiment Station* 24:1-6 (in Japanese).

Fu W.J., Xu G.B., Chen Y.Q., Wu T.Z., Liang Y.J., and Liu W. (1999). Matsutake's semi-artificial and artificial cultivation. *Edible Fungi of China* 18(6): 10-12.

Fujita H., Fujita T., and Ito T. (1990). Study on cultivation technique of *Lyophyllum shimeji* using infected tree seedling. *Annual report of the forestry experimental station, Kyoto prefecture* (in Japanese.)

Garrett S.D. (1960). Parasitic growth habit: Mycorrhizal fungi. In *Biology of root-infecting fungi* (pp. 92-106). London: Cambridge University Press.

Gill W.M., Guerin-Laguette A., Lapeyrie F., and Suzuki K. (2000). Matsutake—Morphological evidence of ectomycorrhiza formation between *Tricholoma matsutake* and host roots in pure *Pinus densiflora* forest stand. *New Phytologist* 147(2): 381-388.

Gill W.M., Lapeyrie F., Gomi T., and Suzuki K. (1999). *Tricholoma matsutake*—An assessment of in situ and in vitro infection by observing cleared and stained whole roots. *Journal of Mycorrhiza* 9(4): 227-231.

Gong M.Q., Chen Y., Wang F.Z., and Chen Y.L. (1999). *Song Rong.* Kunming: Yunna Science and Technology Press.

Guerin-Laguette A., Vaario L., Gill W.M., Lapeyrie F., Matsushita N., and Suzuki A. (2000). Rapid in vitro ectomycorrhizal infection on *Pinus densiflora* roots by *Tricholoma matsutake. Mycoscience* 41: 389-393.

Guerin-Laguette A., Vaario L., Gill W.M., Matsushima N., Suzuki A., and Lapeyrie F. (2002). Growth stimulation of a shiro-like mycelium of *Tricholoma matsutake* on artificial substrate following addition of non-ionic surfactants or vegetable oils. In I.R. Hall, Y. Wang, E. Danell, A. Zambonelli (eds.), *Edible mycorrhizal mushrooms and their cultivation: Proceedings of the Second International Conference on Edible Mycorrhizal Mushrooms*. Christchurch, New Zealand, July 3-6. New Zealand Institute for Crop and Food Research Limited, Christchurch.

Hall, I.R., Brown, G., and Byars, J. (2001). *The black truffle: Its history, uses and cultivation,* Second edition. CD ROM. New Zealand Institute for Crop and Food Research Limited, Christchurch, New Zealand.

Hall I.R., Lyon A.J.E., Wang Y., and Sinclair L. (1998) Ectomycorrhizal fungi with edible fruiting bodies 2. *Boletus edulis. Economic Botany* 52(1): 44-56.

Hall I.R. and Wang Y. (2002). Truffles and other edible mycorrgizal mushrooms-Some new crops for the Southern Hemisphere. In I.R. Hall, Y. Wang, E. Danell, and A. Zambonelli (eds.), *Edible mycorrhizal mushrooms and their cultivation: Proceedings of the Second International Conference on Edible Mycorrhizal Mushrooms.* Christchurch, New Zealand, July 3-6. New Zealand Institute for Crop and Food Research Limited, Christchurch.

Hamada M. (1950). A study on the physiology and ecology of Japanese matsutake (pre-report). *The Botanic Magazine* 63: 40-41.

Harley J.L. and Smith S.E. (1983). *Mycorrhizal symbiosis.* London: Academic Press.

Heinonen T. and Holopainen T. (1991). Maintenance of ectomycorrhizal fungi. In J.R. Horris, D.J. Read, and A.K. Varma (eds.), *Methods in microbiology,* Volume 23 (pp. 413-422). London: Academic Press.

Hiromoto K. (1960). Isolation and pure culture of the mycelia of *Armillaria matsutake* S. Ito et Imai, the important edible mushroom in Japan. *Journal of the Botanic Magazine* 73: 326-333 (in Japanese).

Hiromoto K. (1961). Culture of mycelia from the fruiting body Hymenomycetes by "the Hymenium method." *Journal of the Botanic Magazine* 74: 154-159 (in Japanese).

Hiromoto K. (1963a). Life-relation between *Pinus densiflora* and *Armillaria matsutake.* I. Components in the pine needle decoction concerned with the growth of *Armillaria matsutake. Journal of the Botanic Magazine* 76: 264-272 (in Japanese).

Hiromoto K. (1963b). Life-relation between *Pinus densiflora* and *Armillaria matsutake.* II. Mycorrhiza of *Pinus densiflora* with *Armillaria matsutake. Journal of the Botanic Magazine* 76: 292-298 (in Japanese).

Hiromoto K. (1975). Memorandom of the pure culture of mycelia of higher fungi by "the Hymenium isolation method." *Journal of Transactions of the Mycological Society of Japan* 16: 198-203 (in Japanese).

Högberg P., Plamboeck A.H., Taylor A.F.S., and Fransson P.M.A. (1999). Natural 13C abundance reveals trophic status of fungi and host-origin of carbon in mycorrhizal fungi in mixed forests. *Proceedings of the national Academy for Sciences, USA* 96: 8534-8539.

Hood I.A. (1992). *An illustrational guide to fungi on wood in New Zealand.* Auckland, New Zealand: Auckland Press.

Hosford D., Pilz D., Molina R., and Amanranthus R. (1997). *Ecology and management of the commercially harvested american matsutake mushroom.* Portland, OR: USDA, Forest Service, PNW Research Station. PNW-GTR-412.

Hu H.T. (1994). Study on the relationship between *Pinus taiwanensis* and *Tricholoma matsutake* (I) Semi-aseptic mycorrhizal synthesis of *Tricholoma matsutake* and *Pinus taiwanensis. Quart. Journ. Exp. For. Taiwan Univ.* 8(3): 47-54.

Hur T.C., Ka K.H., Joo S.H., and Terashita T. (2001a). Characteristic of the amylase and its related enzymes produced by ectomycorrhizal fungus *Tricholoma matsutake. Journal of Mycobiology* 29(4): 183-189.

Hur T.C., Park H., and Joo S.H. (2001b). Dynamics of root systems in the fairy rings of *Tricholoma matsutake. Journal of Korean For. Soc.* 88(4): 451-461.

Ito S. and Imai S. (1925). On the taxonomy of shiitake and matsutake. *Journal of the Botanical Magazine* 39: 219-328.

Kawai M. (1989). Future and feasibility of cultivation of mycorrhizal mushrooms. In *Abstracts of Second Symposium of Mushroom Technology and First Seminar of Fungal Culture Preservation Research Association,* Kyoto, Japan, pp. 48-53 (in Japanese).

Kawai M. and Ogawa M. (1976). Studies on the artificial production of *Tricholoma matsutake* (Ito et Imai) Sing.: IV: Studies on seed culture and trial for the cultivation on solid media. *Journal of Transactions of the Mycological Society of Japan* 17: 499-505 (in Japanese).

Kawai M. and Ogawa M. (1981). Some approaches to the cultivation of a mycorrhizal fungus, *Tricholoma matsutake* (Ito et Imai) Sing. In N.G. Hair (ed.), *Mushroom science XI. Proceedings of the Seventh International Scientific Congress on the Cultivation of Edible Fungi, Australia* (pp. 869-883). Sydney, Australia: Congress Organization Committee.

Kikuchi K., Matsushita N., Guerin-Laguette, A., Ohta A., and Suzuki K. (2000). Detection of *Tricholoma matsutake* by specific ITS primers. *Journal of Mycol. Res.* 104(12): 1427-1430.

Kim C.H. (1984). A study of multiplication response of *"Tricholoma matsutake"* (Pine mushroom) conidio to cultural media environment. *Journal of the Korean Forestry Society* 64: 33-41 (in Korean).

Kinugawa K. (1964a). Prediction of the time of development and the crop-abundance in the wild edible mushrooms, *Tricholoma matsutake* (S. Ito et Imai) Sing. *Transactions of the Mycological Society of Japan* 4: 165-166 (in Japanese).

Kinugawa K. (1964b). Some information on the environments of the habitat of *Tricholoma matsutake. Transactions of the Mycological Society of Japan* 5: 16-21 (in Japanese).

Koo C.D. and Milek E.M. (1998). Finacial analysis of vegetation control for sustainable production of songyi *(Tricholoma matsutake)* in Korea. *Journal of Korean Forestry Society* 87(4): 519-527.

Largent D.L. and Baroni T.J. (1988). *How to identify mushrooms to genus IV: Modern genera.* Mad Rever Press, Ereka.

Lee K.J., Kim Y.S., Lee T.S., and Kim K.S. (1986). A comparative study on the mushroom productions between matsutake-producing and non-producing *Pinus densiflora* stands. *Journal of Korean Forestry Society* 72: 27-31 (in Korean).

Lee T.S. (1981). Ecological environments and yield production of matsutake in Korea. In *Report of Symposium on matsutake cultivation method 7 October 1987. Journal of Forestry Experimental Farm, Korea Forest Administration, Seoul, Korea:* 36-44 (in Korean).

Lee T.S. (1983). Survey of the environmental conditions of the habitat of *Tricholoma matsutake* in Korea. *Journal of Wood Science and Technology* 11(6): 34-44.

Lee T.S. (1984). Technique on production of pine musthroom in Korea. *Proceedings of Research and Technique on Production of Pine Mushroom* 22: 15-25.

Lee T.S. (1988a). Increasing yields and promoting quality of pine mushroom by cap-covering or soil covering. In A. Mahadevan, N. Raman, and K. Natarajan (eds.), *Mycorrhizae for green Asia: Proceedings of the First Asian Congress on Mycorrhizae, Madras, India* (pp. 224-230). Madras, India: Alamu Printing Works.

Lee T.S. (1988b). Status of mycorrhizal research in Korea. In A. Mahadevan, N. Raman, and K. Natarajan (eds.), *Mycorrhizae for green Asia: Proceedings of the First Asian Congress on Mycorrhizae, Madras, India* (p. 16). Madras, India: Alamu Printing Works.

Lewis D.H. (1973). Concepts in fungal nutrition and origin of biotrophy. *Journal of Biological Review* 48: 261-278.

Ma R.W., Zhao Y.L., and Liu, J.S. (1988). Observation on poplar root rot caused by *Tricholoma populinum. Journal of Forest Pest and Disease.* 1: 15-16 (in Chinese).

Malajczuk N., Dell B., and Bougher N.L. (1987). Ectomycorrhiza formation on *Eucalypus* III. Superficial ectomycorrhizas initiated by *Hysteragium* and *Cortinarius* species. *Journal of New Phytologist* 105: 421-428.

Marks G.C. and Forster R.C. (1973). Structure morphogenesis and ultrastructure of ectomycorrhizae. In G.C. Marks and T.T. Koslowski (eds.), *Ectomycorrhizae* (pp. 1-41). New York: Academic Press.

Marx D.H. (1980). Ectomycorrhizal fungus inoculations: A tool for improving practices. In P. Mikila (ed.), *Tropical mycorrhiza research* (pp. 13-71). London: Oxford Univ. Press.

Marx D.H. and Daniel W.J. (1976). Maintaining cultures of ectomycorrhizal and plant pathogenic fungi in sterile water cold storage. *Canadian Journal of Microbiology* 22: 238-341.

Masui K. (1927). A study of ectotrophic mycorrhizas of woody plants. *Memoirs of the College of Science, Kyoto Imperial University, Series B* 3: 156-214.

Mikola P. (1965). Studies on the ectendotrophic mycorrhizal of pine. *Journal of Acta Ferestalia Fennica* 79: 1-56.

Mimura S. (1915). Research the culture of matsutake. *Journal of Bull. Forest Exp. Stat. Meguro, Tokyo* 115.

Molina R. and Palmer J.G. (1982). Isolation, maintenance and pure culture manipulation of ectomycorrhizal fungi. In N.C. Schenck (ed.), *Methods and principles of mycorrhizal research* (pp. 115-129). St. Paul, MN: The American Phytopathological Society.

Murrill, W.A. (1912). The Agaricaceae of the Pacific Coast—I. *Journal of Mycologia* 4: 205-217.

Ogawa M. (1975). Microbial ecology of mycorrhizal fungus, "Shiro" in *Tricholoma matsutake* (Ito et Imai) Sing. in pine forest I fungal colony (shiro) of *Tricholoma matsutake. Journal of Bull. Gov. For Exp. Sta.* 272: 79-121 (in Japanese).

Ogawa M. (1976). Microbial ecology of "Shiro" in *Tricholoma matsutake* (Ito et Imai) Sing. and its allied species. III *Tricholoma matsutake* in *Picea plehnii* forests. *Transactions of the Mycological Society of Japan* 17: 188-198.
Ogawa M. (1977a). Microbial ecology of "Shiro" in *Tricholoma matsutake* (Ito et Imai) Sing. and its allied species. IV *Tricholoma matsutake* in *Tsuga diversifolia* forests. *Transactions of the Mycological Society of Japan* 18: 20-33.
Ogawa M. (1977b). Microbial ecology of "Shiro" in *Tricholoma matsutake* (Ito et Imai) Sing. and its allied species. V *Tricholoma matsutake* in *Tsuga sielboldii* forests. *Transactions of the Mycological Society of Japan* 18: 34-46.
Ogawa M. (1977c). Microbial ecology of "Shiro" in *Tricholoma matsutake* (Ito et Imai) Sing. and its allied species. VI *Tricholoma fulvocastaneum* in *Quercus serrata-Quercus acutissimai* forests. *Transactions of the Mycological Society of Japan* 18: 286-297.
Ogawa M. (1978a). *Biology of matsutake mushroom.* Tokyo: Tsukiji Shokan (in Japanese).
Ogawa M. (1978b). Microbial ecology of "Shiro" in *Tricholoma matsutake* (Ito et Imai) Sing. and its allied species. VII *Tricholoma fulvocastaneum* Hongo in *Castanopsis cuspidata* forest. *Transactions of the Mycological Society of Japan* 19: 37-46 (in Japanese).
Ogawa M. (1979). Microbial ecology of "Shiro" in *Tricholoma matsutake* (Ito et Imai) Sing. and its allied species. IX *Tricholoma ponderosum* in *Pseudotsuga menziesii-Tsuga heterophylla* and *Pinus contorta* forests. *Transactions of the Mycological Society of Japan* 20: 370-382 (in Japanese).
Ogawa M. (1981a). Ecological characters of ectomycorrhizal fingi and their mycorrhizae. An introduction to the ecology of higher fungi. *International Union of Forest Research Organizations World Congress* XVII: 89-95.
Ogawa M. (1981b). Microbial ecology of "Shiro" in *Tricholoma matsutake* (Ito et Imai) Sing. and its allied species. X *Tricholoma robustum* in *Pinus densiflora* and *P. pumila* forests and *T. zellerii* in *P. contorta* forests. *Transactions of the Mycological Society of Japan* 22: 231-245 (in Japanese).
Ogawa M. (1982). In memorium Dr Minoru Hamada, Matsutake Sensei. *Journal of Transactions of the Mycological Society of Japan* 23: 191-93.
Ogawa M. and Ito I. (1989). *Is it possible to cultivate matsutake?* Tokyo: Sou Shin Press (in Japanese).
Ogawa M. and Kawai M. (1976). Studies on the artificial reproduction of *Tricholoma matsutake* (Ito et Imai) Sing. III. Effects of growth promoting natural products on the vegetative growth of *T. matsutake. Transactions of the Mycological Society of Japan* 17: 492-498 (in Japanese).
Ogawa M. and Ohara H. (1978). Microbial ecology of "Shiro" in *Tricholoma matsutake* (Ito et Imai) Sing. and its allied species. VIII *Tricholoma bakamatsutake* in *Quercus mongolica* var. *grosseserrata* forest and *Quercus serrata* forests. *Transactions of the Mycological Society of Japan* 19: 391-405 (in Japanese).
Ohara A. and Hamada M. (1967). Disappearance of bacteria from the zone of active mycorrhizas in *Tricholoma matsutake* (S. Ito et Imai) Singer. *Journal of Nature* 217: 528-529.

Ohara A. and Ogawa M. (1982). Microbial ecology of "Shiro" in *Tricholoma matsutake* (S. Ito et Imai) and its allied species. XI *Tricholoma caligatum* in *Cedrus libanotica* forests. *Transactions of the Mycological Society of Japan* 23: 365-377 (in Japanese).

Ohta A. (1983). Quantitative analysis of odorous compounds in the fruit bodies of *Tricholoma matsutake. Transactions of the Mycological Society of Japan* 24 (24): 185-190.

Ohta A. (1990). A new medium for mycelial growth of mycorrhizal fungi. *Transactions of the Mycological Society of Japan* 31: 1323-1324 (in Japanese).

Olivier, J.M. (2000) Progress in the cultivation of truffles. In L.J.L.D. Van Griensven (ed.), *Mushroom science* XV: *Science and cultivation of edible fungi,* Volume 2 (pp. 937-942). Rotterdam, the Netherlands: Balkema.

Oyama Y., Yoshida T., and Taguchi H. (1974). The artificial cultivation of mycorrhizal-forming basidiomycetes. In *Mushroom Sciences* IX (Part I), *Proceedings of the Ninth International Scientific Congress on the Cultivation of Edible Fungi, Tokyo:* 719-731.

Park H., Ka K.H., Ryoo C.I., Kim K.S., and Kim H.J. (1995a). Ectomycorrhizal mushroom occurrentce around the fairy ring of *Tricholoma matsutake* at a pine-mushroom forests. *Korean Journal of Mycology* 26(3): 306-313.

Park H., Kim K.S., and Koo C.D. (1995b). Effects of climatic conditions in September on pine-mushrooms yield and methods for overcoming the limiting factors in Korea. *Journal of Korean Forestry Society* 86: 399-404.

Pilz D. and Molina R. (2002). Commercial harvests of edible mushrooms from the forests of the Pacific Northwest United States: Issue, management, and monitoring for sustainability. *Forest Ecology and Management* 155: 3-16.

Poitou N., Mamoun M., Ducamp M., Guinberteau J., and Olivier J.M. (1989) Mycorrhization controlee et culture experimentale du champ de *Boletus (= Suillus) granulatus* et *Lactarius deliciosus. Proceedings of the Twelfth International Congress on the Science and Cultivation of Edible Fungi* (pp. 551-563). Braunschweig, Germany.

Read D.J. (1992). The mycorrhizal mycelia. In M.F. Allen (ed.), *Mycorrhizal functioning* (pp. 102-133). London: Chapman and Hall.

Redecker D., Kodner R., and Graham L.E. (2000). Glomalean fingi from the Ordovician. *Journal of Science* 289:1920-1921.

Reyna S., Rodriguez Barreal J.A., Folch L., Pérez-Badía R., Domínguez A., Saiz-De-Omecaña J.A., and Zazo. (2002). Techniques for inoculating mature trees with *Tuber melanosporum* Vitt. In I.R. Hall, Y. Wang, E. Danell, and A. Zambonelli (eds.), *Edible mycorrhizal mushrooms and their cultivation. Proceedings of the Second International Conference on Edible Mycorrhizal Mushrooms,* CD-ROM. New Zealand Institute for Crop and Food Research Limited, Christchurch.

Robertson N.F. (1954). Studies on the mycorrhizal of *Pinus sylvestris* I. The pattern of development of mycorrhizal roots and significance for experimental studies. *New Phytologist* 53: 253-283.

Saccardo P.A. (1887). Sylloge Hymenomycetum. 1. Agaricianae, Padua.

Santillano D.D. (1998). A study of the *Tricholoma magnivelare* shiro using a taxon-specific oligonucleotide in the Sierra Juarez of Oaxaca Mexico. Available online at <http://www-mcnair.berkeley.edu/98journal/dsantillano/>.

Shimazono H. (1979). Comparative studies on morphological characteristics of colonies of *Tricholoma matsutake*, *T. fulvocastaneum* and *T. bakamatsutake* on agar media. *Transactions of the Mycological Society of Japan*. 20: 176-184.

Singer R. (1949). *Agaricales in modern taxonomy*. Republica Argentina, Tucuman. pp 5-832.

Smith S. and Read D. (1997). *Mycorrhizal symbiosis* (Second edition). London: Academic Press.

Sun C.B., Nan L.S., and Bei D.Z. (1989). Preliminary report of study of matsutake. *Journal of Edible Fungi* 5: 7-9 (in Chinese).

Terashima T. (1993). Distribution and external morphology of mycorrhizal roots at shiros of *Tricholoma bakamatsutake* in a mixed forest of *Pasania edulis* and *Castanopsis cuspidate* var. *sieboldii. Trans. Mycol. Soc. Japan* 34: 495-505.

Terashita T. and Matashi K. (1987). Production and some properties of carboxyl proteinases from *Tricholoma matsutake. Trans. Mycol. Soc. Japan* 28: 245-256.

Terashita T., Takeuchi T., Wakayama S., Sekino Y., Terui S., and Yoshikawa K. (1991). Effect of low-temperature on the production of l-octen-3-ol by the vegetative mycelia of *Tricholoma matsutake. Trans. Mycol. Soc. Japan* 320: 477-484.

Thorn R.G. and Barron G.L. (1983). Carnivorous mushrooms. *Journal of Science* 224: 76-78.

Thorn R.G. and Tsunda A. (1993). Interactions between *Pleurotus* species, nematodes, and bacteria on agar and in wood. *Trans. Mycol. Soc. Japan.* 34: 49-464.

Tominaga Y. and Komeyama, S. (1987). *Practice of matsutake cultivation.* Tokyo: Youken Press.

Trappe J.M. (1962). Fungus associates of ectotrophic mycorrhizae. *Journal of Botanic Review* 28: 538-605.

Tzean S.S. and Liou J.Y. (1993). Nematophageous *Resupinatus* Basidiomycetous fungi. *Journal of Phytopathology* 83(10): 1015-1020.

Vaario L.M., Guerin-Laguette A., Matsushima N., Suzuki K., and Lapeyrie F. (2002). Saprobic potential of *Tricholoma matsutake:* Growth over pine bark treated with surfactants. *Journal of Mycorrhiza* 12(1): 1-5.

Viviani D. (1834). I fungi d'Italia e principalmente le loro specie Mangereccie, velenose, o sospette descritte ed illustrate con tavole desegnate, e colorite dal vero. Genova.

Wang Y. (1993). One hundred of best edible mushrooms in Japan. *Foreign Edible Fungi* 1: 7-9 (in Chinese).

Wang Y. (1995). *Tricholoma matsutake.* Doctoral thesis, University of Otago, Dunedin, New Zealand, 3 volumes.

Wang Y. and Hall I.R. (1996). Morphology and anatonomy of matsutake mycorrhizas. In T.M. Szaro and T.D. Bruns (eds.), *First International Conference on Mycorrhizae, August 4-9, 1996, University of California at Berkeley USA, Program and Abstracts* (p. 131). Berkeley: University of California.

Wang Y. and Hall I.R. (1998). Matsutake—A prized edible mushroom in Japan. In U. Ahonen-Jonnarth, E. Danell, P. Fransson, O. Karen, B. Lindahl, I. Rangel, and R. Finlay (eds.), *Second International Conference on Mycorrhizae,* July 5-10, Uppsala, Sweden, Program and Abstracts (p. 182).

Wang Y. and Hall I.R. (2002). The cultivation of *Lactarius deliciosus* and *Rhizopogon rubescens* (shoro) in New Zealand. In I.R. Hall, Y. Wang, E. Danell, and A. Zambonelli (eds.), *Proceedings of the Second International Conference on Edible Mycorrhizal Mushrooms,* Christchurch, New Zealand, July 3-6. New Zealand Institute for Crop and Food Research Limited, Christchurch.

Wang Y., Hall I.R., and Evans L. (1997). Ectomycorrhizal fungi with edible fruiting bodies 1. *Tricholoma matsutake* and related fungi. *Journal of Economic Botany* 51(3): 311-327.

Wang Y. and Xie Z.X. (1982). Study on *Tricholoma matsutake. Edible Fungi* 2: 7-8.

Wang Y. and Xie Z.X. (1983). A preliminary survey of ectomycorrhizal fungi of some forest trees in China. *Acta Mycologica Sinica* 2: 59-61.

Warcup J.H. (1980). Ectomycorrhizal association of Australian indigenous plants. *New Phytologist* 81: 531-535.

Yamada A., Meada K., and Ohmasa M. (1999a). Ectomycorrizal formation of *Tricholoma matsutake* isolates on *Pinus densiflora* in vitro. *Journal of Mycoscience* 40: 455-463.

Yamada A., Meada K., and Ohmasa M. (1999b). Ectomycorrizal formation of *Tricholoma matsutake* on *Pinus densiflora. Journal of Mycoscience* 40: 193-198.

Zhao S.F. (1986). *Matsutake—Song-Rong.* Changchun, China: Jilin Science and Technology Press.

Index

Page numbers followed by the letter "f" indicate figures; those followed by the letter "t" indicate tables.

Abies, 248
Abiotic stresses, 273, 289, 428
Abscisic acid (ABA), 62, 277, 294
Acacia
  distribution of, 206, 207
  mycorrhizal fungi interactions with, 119, 120t, 126
  rhizobia interactions with, 119, 120t, 205, 208, 209, 210, 211
Acacia ampliceps, 210
Acacia cyclops, 210
Acacia holosericea, 120t, 121, 121t, 122, 122f, 123, 125, 127
Acacia nilotica, 210
Acacia raddiana, 119, 120t
Acacia redolens, 210
Acacia senegal, 205
Acaricides, 381
Acaulospora, 405
Acaulospora morrowiae, 336
ACC deaminase, 149, 150, 154
Acer, 248
Acetamide, 484
Acetobacter diazotrophicus, 7-8
Acetylene, 462-463
Acetylene reduction activity (ARA), 423, 462, 463
Achlorophyllous orchids, 530
Achromobacter, 9
Acidic soils, 423
Actinomycetes. *See also Frankia*
  in cultures, 37
  functions of, 31, 33
  in phosphorus solubilization, 107
Actinomycetes *(continued)*
  population, influences on, 109
  sulphur oxidation, role in, 146
  symbiotic relationships of, 289
Actinorhizal plants, 5, 6, 203
Adenosine triphosphate (ATP), 252, 437
Aeration, 377, 446
Aeromonas, 17
Aeroponics, 306-307, 364, 405
Afforestation, 265
Africa, agriculture in, 169, 421, 441
Agar culture. *See also* Hamada agar; Modified Melin-Norkrans' agar medium (MMN)
  for arbuscular mycorrhizal fungi, 406, 485
  for glomaceous fungi, 477
  for matsutake, 518f, 520, 523, 523f, 524, 525, 526, 529
  for mycorrhizal fungi, 527
  preparation of, 36-37
Aglaophyton, 291
Agriculture, intensive, 185
Agriculture, mycorrhizae in, 224-226, 291, 342, 395-396. *See also* Arbuscular mycorrhizal fungi (AMF), in agriculture
Agriculture, productivity of, 191-192
Agriculture, subsidized, 196
Agriculture, sustainable. *See also term* agriculture in *after country name, e.g.* India, agriculture in
  benefits of, 193, 195, 196, 398-399

Agriculture, sustainable *(continued)*
defined, 137-138f, 224, 289
evolution of, 399-400
fungi in, 224-226, 341, 342, 395-396, 407-408, 477, 478, 490
goal of, 400
in poor nations, 165-166
requirements of, 396
shift toward, 396, 400
in tropics, 398
Agriculture management practices, 193-194, 196, 223-224, 226-230, 232, 233, 341
Agrobacterium, 206, 212
Agrobacterium radiobacter, 139t
Agrobacterium rhizogenes, 363
Agrobacterium tumefaciens, 205
Agronomics, 189
Akinetes, 421, 423, 463
Alanine, 438
Albizzia, 207
Alcaligenes, 7
Alcohols, 127
Alder, 150, 244, 395
Alfalfa, 8, 17, 69, 70, 153, 166, 191
Algae, 423, 434, 442, 459, 463-464. *See also* Cyanobacteria (blue-green algae, BGA); Hamada agar
Algae culture. *See* Agar culture; Modified Melin-Norkrans' agar medium (MMN)
Algal mutants, 427
Algalization. *See* Cyanobacterial inoculation and inoculants (algalization)
Alkaline phosphatase, 61
Allied crops, 193, 195, 196
Allium cepa, 61, 404
Allorhizobium, 3, 185
Allotropa virgata, 530
Alnus, 141, 248
Alternative farming systems, 397-398, 399
Aluminum, 68, 192
Amanita, 248
Amanita muscaria, 68, 528
Amanita pantherina, 503
Amanita phalloides, 14
AMF. *See* Arbuscular mycorrhizal fungi (AMF)
Amines, 127
Amino acids
auxins diminished by, 255
constitution of, 183
in cyanobacteria, 438
formation of, 32-33, 34, 62
from host plant to fungus, 292
level, fungi impact on, 59, 60
micorrhizal organism dependence on, 257
micorrhizal plant impact on, 354
salt tolerance, role in, 93
sulphur-containing, 254
synthesis of, 484
transfer, 403
and water stress, 279, 280
Ammonia (NH_3+)
nitrates derived from, 34
nitrogen reduced to, 51, 184
in organic compound, 377-378
production of, 95, 183, 428, 435, 437
releasing, 427, 438
in soil, 32
volatilization, 420, 424
Ammonifiers, 41
Ammonium
assimilation of, 253
compounds, 93
effects of, 444, 461
nitrification of, 278
Ammonium nitrogen (N), 100
Ammonium sulfate, 419
Amplified ribosomal DNA restriction analysis (ARDRA), 30
Amylase, 104, 530
Anabaena, 94, 420, 421, 435, 436f, 438, 440, 446, 467f, 469f, 470f
Anabaena azollae, 6, 90, 91f, 92, 139t, 427
Anabaena cylindrica, 419
Anabaena doliolum, 94, 372
Anabaena laxa, 465f, 472
Anabaena variabilis, 427

Anaerobic conditions, 381
Anaerobic sulphur mineralizing, 42
Andropogon grass *(Andropogon gayanus),* 365-366
Angiosperms, 243
Angular leaf spot, 161
Anhydrous ammonia, 398
Animal manures, 141, 223, 378, 379, 397
Anthracnose *(Colletotrichum obiculare),* 161, 163
Antibacterial compounds, 256
Antibiosis (biopesticide mode of action), 158, 159-160
Antibiotic production, 160, 161, 162, 256
Antibiotic resistance, 484
Antifungal compounds, 256
Antioxidant enzymes, 280
Antioxidant mechanisms, 273
Antiviral compounds, 256
Apatite, 252
Aphanocapsa banaresensis, 466f, 473
Aphanocapsa elachista, 473
Aphanomyces, 490
Aphanothece, 92
Apple, 299, 301, 302
Apple plum, 487
Aquaporins, 272, 283
Arabidopsis thaliana, 213
Arachis hypogaea, 60
Arboriculture, 486
Arbuscular mycorrhizal (AM) inoculation and inoculants
  amount of, 374-375
  application of, 11-12, 226-230, 231-232, 300, 307-308, 371, 373-375, 373t, 377
  benefits of, 402-403
  cost of, 373, 375-376t, 383, 407, 412
  of cuttings, 301
  evaluation of, 374-375
  in horticultural systems, 298, 299-300, 310
  of legumes, 53, 383
Arbuscular mycorrhizal inoculation and inoculants *(continued)*
  for micropropagated plants, 486-487
  need for, determining, 12-13, 355-357
  in nurseries, 301-302, 304, 305
  production of, 11, 231, 306-307, 341, 404-406, 412, 486, 490
  of seedlings, 225, 263, 339, 341, 355-362, 368-376t, 380, 383-384, 525-526, 532
  successful, factors of, 13, 128-129, 226-230, 300, 302, 355-362
  types and selection of, 306, 369-370, 372t, 407, 409t-411t
Arbuscular mycorrhizal fungi (AMF)
  in agriculture, 224-230, 341, 363, 365, 406-407, 408, 408t, 412
  bacteria, interaction with, 16, 334-336, 335t
  benefits of, 311, 325-326, 341, 370, 402
  as biofertilizers, 11-13, 141, 226, 297-298, 303
  biotechnology, 232, 310, 384, 406
  changes produced by, 327t
  characteristics of, 290-291
  classification of, 325, 395, 403-404
  composts, integration with, 230-231
  disease control, role in. *See under* Disease control
  experiments and field studies, 341
  formation of, 51-52, 303
  functions of, 52, 225
  growth of, 485
  identification of, 310
  lipid metabolism of, 60
  metal tolerance of, 65
  morphology and nutrition, 403
  nematodes, interaction with, 336-340t
  nitrate reductase activity in, 278
  nitrogen effect on, 379
  nitrogen uptake and transfer, role in, 273, 293
  and organic amendments, 229-230
  pathogenic fungi, interaction with, 329-334

Arbuscular mycorrhizal fungi (*continued*)
pesticide impact on, 229, 305, 380-381
phosphate absorption by, 292
Piriformospora indica resemblance to, 477-478, 480, 483, 485, 490
plant uptake of metals, effect on, 68-69
production of, 361, 362-368
and protoplasts, 483
replacing lost, 299, 353, 380
and *Rhizobium* interactions, 52-58, 295
root colonization by, 62, 233
selection and evaluation of, 230, 305-306, 309, 340, 360-362, 370, 383
soil microbiota, interaction with, 295-296
synergistic effect of, 340
water uptake and transfer, role in, 274-276f
Arbuscular mycorrhizal plants
in agriculture, 291
in drought, 277, 280-283, 281f, 282f, 283, 284, 294
health of, 292-293
nutrient concentration in, 53-54
osmotic stress alleviation in, 272
stress, tolerance of, 294, 406
Arbuscular mycorrhizal symbioses. *See also* Vesicular arbuscular mycorrhizae (VAM)
AMF growth dependent on, 485
benefits of, 353-355, 406
causes and effects of, 293, 328f
drought stress and, 280, 281f, 282f, 283, 284
formation of, 325, 326, 497, 498
fungicide effect on, 305
horticultural production, impact on, 299-300
in horticultural systems, 298-300
initial events of, 71
interactions during, 72
management of, 233, 302-309
nitrate reductase activity in, 278
in nurseries, 302, 303-305
Arbuscular mycorrhizal symbioses (*continued*)
as nutrient uptake and transfer system, 273
and osmotic stress alleviation, 274
and *Rhizobium* symbioses, 69-70
stimulation of, 153
study of, 342
successful, factors of, 307-308
synthesis of, 301-302
verification of, 308-309, 360, 361-362, 362f
versus chemical fertilizers and pesticides, 226, 228
and water potential, 277
Arbuscules, 246, 290, 292, 326, 403
Arbutus, 247
ARDRA (amplified ribosomal DNA restriction analysis), 30
Argentina, agriculture in, 185, 419, 461-462, 466, 469-470, 471, 472-473
Arginine, 59
Arid soils, 6
Armillaria, 502, 527, 530
Armillaria mellea, 263, 530
Armillaria nauseosus, 502
Armillaria novae-zelandiae, 530
Artemisia, 502
Artemisia annua, 480, 487
Arthrobacter, 17
Ascomycetes, 51, 243, 247, 248, 498
Ascorbate peroxidase, 280, 284
Ascorbate-glutathione, 283
Asexual state, 479, 490
Ash, 244, 258
Asia, agriculture in, 418, 428, 459-460
Aspartic acid, 438
Aspen, 243
Aspergillus, 478, 485
Aspergillus awamori, 107
Aspergillus sydowii, 488
Associative bacteria, 203
ATP (adenosine triphosphate), 252, 437
Atractylodes japonica, 502
Aulosira, 92, 420, 435, 436f, 440, 446
Aulosira fertilissima, 422
Autotrophic growth, 291
Autotrophic mycorrhizal systems, 242
Autotrophic organisms, 242
Autotrophic sulphur oxidizers, 42-43

Auxins, 56, 62, 63, 70, 166, 254, 255, 301
Avena sativa (oat), 68, 185, 194, 297
Avocado, 299, 302, 487
Axenic cultures, 306, 483, 485
Azadirachta indica (neem), 423, 480, 487, 488
Azadiractin, 100
Azaleas, 498
Azoarcus, 7, 145
Azolla (water fern), 90-92, 101, 103, 104, 109, 459
Azolla microphylla, 90f, 101, 104
Azolla-Anabaena, 6, 90, 91f
Azorhizobia, 104
Azorhizobium, 3, 109, 142-143
Azorhizobium caulinodans, 99, 104, 105, 105f, 106, 106f, 185
Azospirillum, 7, 10, 17, 107, 109, 153, 154, 166, 170, 184, 257
Azospirillum brasilense, 139t, 144, 148-149, 151-152
Azospirillum lipoferum, 107, 139t
Azotobacter, 8, 17, 109, 144, 166, 184
Azotobacter chroococcum, 145
Azotobacter nigricans, 18
Azotococcus, 184

B vitamins, 153, 256
Bacillus, 7, 9, 10, 145, 153, 160, 166, 213, 257, 295
Bacillus circulans, 145, 160
Bacillus polymyxa (Paenibacillus polymyxa), 10, 107
Bacillus subtilis, 139t, 159, 166
Bacillus thuringiensis (Bt), 139t, 155, 159, 169
Bacopa monniera (brahmi), 480, 487, 488
Bacteria. *See also bacteria type*
- associative, 203
- biofertilizer influence on, 109
- culture media for, 36, 43-44, 95-96f, 97f
- experiments and field studies, 18, 19
- fungi, interaction with, 16-17, 123-127, 123t, 124f, 126t, 151

Bacteria *(continued)*
- inoculation with, 18, 19, 139t, 155, 196, 213-214. *See also* Cyanobacterial inoculation and inoculants (algalization)
- and mycorrhizae, 122, 123, 129
- number and diversity of, 137
- paranodulation, use in, 109
- in rhizoplane, 142f
- sulfate-reducing, 43-44
- sulphur oxidation, role in, 146
- symbiotic, 204, 214, 289

Bacterial wilt, 335-336
Bahia grass *(Paspalum notatum),* 364, 365, 404
Banana, 299, 302
BAP (benzylamino purine), 104
Bark, 405
Barley *(Hordeum vulgare)*
- and arbuscular mycorrhizal fungi, 297
- in developing countries, 185
- field studies of, 59, 195
- growth stimulation in, 8
- and *Piriformospora indica,* 482
- roots, 159
- yield of, 208

Basidiomycetes, 51, 243, 248, 483, 498, 505, 529
Basidiomycota, 477, 479
Bavistin (carbendazim), 95, 488, 490
Bean. *See also* Common bean *(Phaseolus vulgaris);* Faba bean *(Vicia faba);* Field beans *(Phaseolus);* Soybean
- disease control in, 161
- inoculation of, 166, 193, 299
- maize intercropped with, 187
- mineral uptake by, 69
- mycorrhizal mutants, 70
- nematode protection, 336
- in South America, 188
- symbioses, 152, 192-193

Beech, 123, 243, 251, 252
Beijerinckia, 8, 145, 184
Beijerinckia mobilis, 8

Benomyl, 228, 305, 380
Benzimidazole fungicides, 305
Benzylamino purine (BAP), 104
Betula, 68, 248, 503
Betula pendula, 8
Betulaceae family, 247
Bhendi, 107
Biliproteins, 435
Biochemical pesticides. *See* Biopesticides, bacterial
Biocides. *See* Pesticides
Biocontrol agents (BCAs), 159, 160, 294, 304, 340
Biocontrolling activities, 147, 154, 155, 164, 167, 484
Biodiversity, 46, 204-206, 232, 383, 404, 406
Biodynamic farming, 397
Biofertilizers. *See also* Cyanobacterial biofertilizers; Plant-growth-promoting rhizobacteria (PGPR); Rice; *term* biofertilizers *under* Fungi; Wild-legume rhizobia
  agriculture, sustainable, role in, 90, 137, 138f, 139-140, 169, 170, 232, 399
  bacteria, influence on, 109
  biopesticides, dual role as, 157
  defined, 140-141, 297, 417, 433, 445
  future of, 429
  legume products as, 207, 211
  for legumes, 90, 108
  mycorrhizal fungi as, 10-13, 15, 226, 290, 297-298, 303, 341
  nitrogen fixation, role in, 3, 140, 203, 212
  nutrients supplied by, 109, 140, 143-145
  phosphobacterial, 107-108
  promotion of, 167
  in rice cultivation, 428, 461
  use breakdown by crop, 139t
Biofungicides, 155, 157
Biogeochemical cycles, 29, 34, 47. *See also specific cycle type, e.g.* Nutrient cycling
Bioherbicides, 157, 162
Bioinsecticides, 157
Biological farming, 397, 398, 399
Biological nitrogen fixation (BNF)
  Azolla-Anabaena association, 90-91
  bacteria capable of, 166
  benefits of, 89-90, 187, 195, 196, 203, 212
  cost-benefit analysis of, 426, 427
  cyanobacteria role in, 417, 418, 422, 434, 438, 439, 459
  with fertilizer, 402
  improvements needed in, 108-109
  interest in, 185-186
  by leguminous crops, 96, 399
  maximizing, 188
  photodependent, 423, 440, 443
  process of, 89-90
  in rice production, 108
  in tropics, 419
Biological pest control, 397
Biopesticides
  agriculture, sustainable, role in, 137, 138f
  bacterial, 155-158, 156t-157t, 166, 170
  definition and categories of, 154-155, 157
  modes of action, 158-164
  use, increase in, 168-169, 223-224
  use breakdown by crop, 139t
Bioprotectors, arbuscular mycorrhizal fungi as, 226
Bioregulators, 341
Biosorption, 67
Biotechnology
  arbuscular mycorrhizal fungi (AMF), applications of, 232, 310, 384, 406
  biopesticide development, 169
  and microorganisms, 400-401
  micropropagation technique, 486-487

Biotechnology *(continued)*
mycorrhizal, 299, 396
in nitrogen fixation, 212-214
plant-growth-promoting rhizobacteria (PGPR), applications of, 170
and soil productivity, 398
Biotic conditions, 260-261
Biotic stresses, 273, 289, 428
Biotin, 255
Biotrophic status, 291, 532, 533
Birch, 243, 395
Black root rot of tobacco, 159, 163
Black walnut, 244
Blackleg, soilborne, 169
Blue-green algae (BGA). *See* Cyanobacteria (blue-green algae, BGA)
Boletus, 503
Boletus edulis, 521, 527
Bonemeal, 108
Brachiaria grasses *(Brachiaria decumbens),* 365, 366
Bradyrhizobium, 3, 4, 53, 54, 63, 152, 204, 205, 206
Bradyrhizobium elkanii, 205, 361
Bradyrhizobium japonicum, 17, 64, 139t, 145, 153, 205, 282f, 361
Bradyrhizobium yuanmingense, 205
Brahmi *(Bacopa monniera),* 480, 487, 488
Brassica rapa, 61
Brassicas (Brassicaceae), 497
Brassins, 150
Brazil
agriculture in, 185, 186, 187, 192, 193, 194, 361, 384
fungi production in, 361
Brazilian wood species, 356, 357-359f
Brinjal *(Solanum melongena),* 480, 487
Broad bean, 166
Broadcasting, surface, 423
Bryophytes, 291
Bt *(Bacillus thuringiensis),* 139t, 155, 159, 166
Burkholderia, 206, 214
Burkholderia cepacia, 150, 151, 157, 159
Burk's culture medium, 40-42
Burn sites, 258, 260
Bynate (thiophenate), 488, 490

Cacti, 291
Cadmium (Cd), 65, 66, 68, 293
Cajanus cajan, 194
Calamagrostis epigejos, 68
Calcium (Ca), 57, 63-64, 192, 244, 255, 258, 262
Calcium carbonate, 398
Calcium phosphate, 443
Calcium phytate, 252
Calluna lichens, 503
Calothrix, 92, 435, 436f, 440, 469f, 470f, 471-472
Calothrix marchica, 471
Canavalia ensiformis, 194
Canola, 146, 148f, 227
Cantharellus cibarius, 17, 498, 532
Captan, 380
Carbofuran, 104
Carbohydrates, 31, 93, 253, 255, 259, 354
Carbon
allocation and transfer of, 259, 293, 412
availability of, 183
competition for, 158
compounds, 225, 242, 290, 292
cycles, 29, 32, 37, 140
as growth requirement, 443
in mycorrhizal plants, 118
for N_2-fixing bacteria, 52
organic, 438, 440
sequestration of, 194
sources of, 242
Carbon dioxide (CO_2), 32, 101, 127, 196, 258, 354, 428, 439
Carboxymethylcellulase, 483
Carex, 503
Carex callaitrichos, 502

Carnation, 158, 161
Carpinus, 248
Carrier materials, 447-448, 449
Carrot, 333-334
Cassava, 187
Castanopsis delavayi, 503
Castanopsis hystrix, 503
Castanopsis orthacantha, 503
Casuarina, 203
Casuarina cunninghamiana, 6
Catalase, 280, 284
Cd (cadmium), 65, 66, 68, 293
Cell division in plants, 63
Cellulase, 104, 334, 482
Cellulolytic culture, 38
Cellulolytic enzymes, 257
Cellulolytic microorganisms, 140
Cellulose, 31, 32, 254, 259
Cenchrus ciliaris, 404
Cenococcum, 248, 251
Cenococcum geophilum, 510
Cenococcum graniforme, 505
Centrosema *(Centrosema pubescens),* 366
Ceratobasidiales, 479
Cereal crops, 105, 106, 187, 207, 208
CH_4, 101
Charcoal, 367
Chemical fertilizers. *See* Fertilizers
Chemical pesticides. *See* Pesticides
Chenopodiaceae family, 291
Chenopodium quinona, 381
Chitin, 67, 160
Chitinase, 334, 482
Chitooligosaccharides, 71
Chlamydospores, 515, 517, 518f, 523, 529
Chloride, 293
Chloridium paucisporum, 247
Chloris gayana, 404
Chlorophycae (green algae), 459
Chlorophyll, 444, 464
Chlorophyll degradation products, 463
Chlorophytes, 441
Chlorophytum borivillianum, 480
Chroococcaceae family, 464
Chroococcidiopsis, 435
Chroococcus, 468f
Chroococcus limneticus, 466-467
Chroococcus minor, 94
Cicer, 185, 211
Cicer arietinum, 480
Citric acid, 66, 127
Citrus, 193, 295, 298, 299, 339, 355, 371, 372t, 381, 402
Citrus limon, 60
Cladosporium herbarum, 145
Clavibacter, 139t
Clay, 11, 366, 367, 368, 374, 377, 405, 448
Clear-cutting, 262
Climatic conditions, 260
Cloning, 369
Clostridium, 8, 166
Clover, 65, 69, 153, 166, 191, 193, 232, 280
Coconut coir, 448
Coffee, 193, 299, 355, 370, 371, 373, 375, 379, 382
Coleus, 404
Colletotrichum obiculare (anthracnose), 161, 163
Colombia, agriculture in, 188
Colony forming units (CFU)
  determining, 35, 36, 37, 39, 40, 41, 46-47
  and inocula production, 420
  numbers, altering, 31
  nutrient cycle, role in assessing, 29
Common bean *(Phaseolus vulgaris),* 61, 63, 187, 192, 195, 204, 207
Competition (anti-disease/biopesticide mode of action), 158-159, 162, 256, 257, 296, 329, 331
Competition among fungi, 250-251, 329, 331, 370
Composts and composting, 140, 230-231, 398
Conservation cropping, 194
Conservation tillage, 399
Consociation, 193

Consultative Group on International Agricultural Research, 224
Contamination, control for, 365
Copper (Cu)
  absorption of, 402
  fertilizer, 298
  in mycorrhizal plants, 53-54, 57, 58, 68
  transfer of, 66, 69
  uptake of, 292
Copper sulfide, 67
Coprosma robusta, 404
Cork oak, 14
Corn *(Zea mays),* 159, 208, 364, 365, 404, 480, 487. *See also* Maize
Cortinarius, 530
Corylus, 248
Cotton, 107, 166, 299, 336
Cottonseed, 158
Counting, direct of soil specimens, 463, 464
Cover crops, 193, 194
Cow manure, 378, 379
Cowpea *(Vigna unguiculata),* 53, 57, 58, 194, 339
Crops
  as arbuscular mycorrhizal plants, 291
  growth cycle, 440
  and livestock integration, 399
  production and yield, 296, 398, 401, 406, 421-424, 433
  protection, 406
  residues, 207, 211, 223, 232, 397
  rotation, 193, 195, 196, 223, 226-227, 228, 229, 232, 397, 399
Crotalaria, 206
Cruciferae family, 291
Cu (copper). *See* Copper (Cu)
Cucumber, 161, 299
  growth stimulation in, 8
Cucumber beetle, 161
Cucumber mosaic cucumovirus (CMC), 161
Cucurbit wilt disease, 161
Cultivars, 154
Cultures and culture media
  actinomycetes in, 37
  aeroponic and hydroponic, 405
  agar. *See* Agar culture; Hamada agar; Modified Melin-Norkrans' agar medium (MMN)
  axenic, 306, 483, 485
  for bacteria, 36, 43-44, 95-96f, 97f
  Burk's, 40-42
  cellulolytic, 38
  cyanobacterial (blue-green algae, BGA), 440, 446, 463
  evaluating, 170
  for fungi, 36-37
  inoculation with, 263
  Lapage, 43-44
  liquid, 520-521, 524, 525, 529
  matsutake, 520-522, 524
  monoxenic, 363-364
  nonagar, 520-521, 526
  pot, 306, 365, 401, 404-405, 486
  proteolytic, 39
  root organ, 405-406
  selective, 463
  soilless, 310, 405
  spores, 401
  Starkey's, 42-43
Cupressus, 247, 248
Cuttings, 300, 301
Cyanide production (biopesticide mode of action), 158, 163
Cyanobacteria (blue-green algae, BGA)
  acid-tolerant, 94
  benefits of, 424-425
  classifying and identifying, 434, 460, 463-464, 466-467, 472-473
  crop yield, impact on, 421-424
  cultivation of, 420-421, 428
  culture, 440, 446, 463
  distribution of, 434
  diversity of, 464, 466-467, 469-471
  genetic modifications in, 428
  growth, 440, 441-443, 445, 459

Cyanobacteria *(continued)*
heterocystous, 144, 425, 435, 436f, 437, 441, 445, 466, 472
indigenous versus foreign, 422, 425-426, 460
metabolism of, 443
morphology of, 434-435, 464
nitrogen processes of, 420, 421, 422, 426-427, 438, 443-444. *See also* N_2-fixing cyanobacteria
nitrogenase inhibition, overcoming, 437
nutrients utilized by, 443-444
production and application of, 6-7
in rice cultivation, 6, 90, 103, 417, 418-419, 422, 423, 424, 425-426, 427, 438, 440-443, 459-462, 464, 465f, 466f, 469, 469f, 470f, 472-473
salinity, tolerance to, 93-94, 444-445
survival of, 422-423, 424
Cyanobacterial biofertilizers
benefits of, 7, 417, 418, 439-440, 445, 446, 461
candidates for, 424
cost and profitability of, 427, 450
distribution of, 449
effectiveness of, 460
market potential of, 450
for nonflooded cultures, 419
production of, 446-448, 449, 450
promotion and education concerning, 449-450
for rice cultivation, 434, 439, 449, 450, 459, 471
storage and shelf life of, 447, 448
study of, 434
technology of, 450
Cyanobacterial inoculation and inoculants (algalization)
effectiveness of, 421-424, 425-426
immobilization of, 95-96f, 97f, 427
nitrogen, impact on, 420
production of, 97f, 420-421
for rice, 92-93, 418-419, 439, 445-446, 459
Cyanobacterial inoculation and inoculants *(continued)*
sources of, 450
storage of, 421
techniques, 460
technology, 446-448
Cyanocobalamin, 438
Cylindrospermum, 92, 435, 436f, 440
Cyperaceae family, 291, 497
Cysteine, 33, 254
Cystine, 33, 254
Cytokinins, 10, 56, 62, 63, 70, 149, 150, 166, 195, 254, 255, 256, 294, 301
Cytosolic osmoregulation, 272

Dactylorhiza maculata, 480
Dactylorhiza majalis, 480, 481
Decaying wood, 258
Dehydrogenase, 104, 440
Deoxyribonucleic acid (DNA), 183, 252, 273, 438
Dermocarpa, 435
Derxia, 184
Desmodium, 206
Desertification, 55, 56
DGGE (denaturing gradient gel electrophoresis), 30
Diathane M-45 (bisdithiocarbamate), 488, 490
Diazotrophic bacteria, 54, 106, 144, 145, 168, 183-184
Diazotrophic rhizobia, 53
Dicyandiamide (DCD), 101
Dinitrogenase reductase, 444
Disease control
arbuscular mycorrhizal fungi role in, 231, 294, 296, 309, 326, 328f, 354, 369, 370, 383, 408, 488
with biopesticides, 155, 157, 158-159, 160-161, 162-163
fungal and fungicide role in, 159, 169, 259, 369, 380, 488, 490
inoculations, multiple in, 19

Disease control *(continued)*
through legume use, 195
microbial, 401
mycorrhizae role in, 117, 256, 326-328f
pesticides, conventional, resistance to, 170
plant hormone role in, 294-295
plant-growth-promoting rhizobacteria role in, 69, 129, 138, 168
root diseases. *See* Root diseases, control of
and tolerance, 326, 329
in trees, 244
DNA (deoxyribonucleic acid), 183, 252, 273, 438
Douglas fir, 121, 124, 127, 129, 130t, 256, 258
Drought
adapting to, 272
arbuscular mycorrhizal plants during. *See* Arbuscular mycorrhizal plants, in drought
free radicals, increase due to, 272-273
mycorrhizal versus nonmycorrhizal plants during, 277
nitrate reductase decrease due to, 278, 279t
and oxidative stress, 280-283
plant growth and yield, impact on, 271
proline increase due to, 58, 280
stomatal behavior during, 277
tolerance of, 11, 209, 211, 244, 254, 259, 273, 274, 282, 283, 354, 406, 423, 490. *See also under specific plant type*

Earthworm humus, 378
Ecological specificity, 250, 265. *See also* Host specificity
Ectendomycorrhizae, 247-248, 290
Ectomycorrhizae
characteristics of, 290, 497, 498, 511
forest association with, 243-244, 245f-246f, 247, 248, 261-262, 519
formation of, 255, 505
in nurseries, 262
phosphorus mobilization by, 252
reduction, factors in, 258-261, 262, 526
superficial (defined), 530-531
on trees, 243-244, 395
Ectomycorrhizal (EM) fungi
bacteria interaction with, 17, 124, 126, 129
as biofertilizers, 13-14
competition among, 250-251, 504
cultivation of, 524
cultures, 521, 524
development of, 119f
disease control by, 256
environmental conditions, response to, 250
inoculations with, 18, 119, 121, 256, 259, 261, 262
matsutake, effect on, 503
metal tolerance of, 68
nitrogen metabolism, role in, 253
production of, 486
symbioses, 153
trees, impact on, 241-242, 254, 509
typical versus matsutake, 527-528, 529, 530, 532-533
vitamin production by, 255
Effective microorganisms, 400-401
Egypt, agriculture in, 207, 209, 211, 418, 421
Elaphomyces, 248
Elm, 244
Emergence-promoting rhizobacteria (EPR), 69. *See also* Mycorrhiza helper bacteria (MHB); Rhizobacteria
Endogonaceae, 246

Endomycorrhizae, 244, 290, 291-292, 300-302, 395, 497-498
Endomycorrhizal (EM) fungi, 54, 246, 305, 482
Endomycorrhizines, 334
Endophytes (endophytic plant-growth-promoting rhizobacteria), 141-142, 195
Enterobacter, 7, 17, 150, 153, 184
Enterobacter agglomerans, 145
Enterobacter cloacae, 158, 164
Entrophospora colombiana, 370
Environment impact
  of desertification, 55
  of fertilizers, 1, 165, 186, 195, 196, 296, 417, 433, 461
  of pesticides, 154, 165, 169, 230, 399
  studies, 47
Environmental protection, 400
Environmental Protection Agency (EPA), 155, 157
Enzyme activity
  antioxidant, 280
  increasing, 354
  in matsutake, 530
  in mycorrhizal plants, 60-62, 482-483
  in nitrate assimilation, 278
  nitrogen role in, 183
  oxygen antithetical to, 435
  plant growth-promotion and PGPR in production, 150, 160, 166, 168, 401
  regulation, 434
  in rice fields, 92, 434
  in soil, 104, 109, 257
  in symbioses, 58
Enzymes, immobilized, 485
Ergosterols, 260
Ericaceae family, 260
Ericaceous mycorrhizal fungi, 483, 497, 498
Erosion control, 193, 203, 399
Erwinia carotovora, 166
Escherichia, 184
Escherichia coli, 164, 279
Esfenvalerate, 161
Ethanol, 158
Ethylene, 127, 150, 256, 462-463
Eucalyptus, 243, 395, 530
Eucalyptus, 14, 247, 248
Eucalyptus diversicolor, 125
Eukaryotic algae, 442, 459
Exudates, 118

Faba bean *(Vicia faba),* 70, 193, 207, 212
Fabaceae, 383
Fagaceae family, 247, 260
Fagus, 248
Fallow systems, 226-227
Family farms, decline of, 399
FAO (Food and Agriculture Organization), 185
Farming practices. *See* Agriculture management practices; Alternative farming systems
Farmyard manure (FYM), 419, 433
Fatty acids, low-molecular-weight, 127
Fe (iron). *See* Iron (Fe)
FeEDTA solution, 45
Ferns, 503
Ferric and ferrous iron, 147
Fertilizers. *See also* Biofertilizers; Nitrogen (N), fertilizer
  arbuscular mycorrhizal fungi levels, effect on, 228-229
  biological nitrogen fixation (BNF) through, 402
  mineral, 224, 379
  necessity of, 398
  organic, 141, 419
  for seedlings, 380
  slow-release, 405
  in substrates, 377
  in tilled versus untilled soil, 228
  world consumption of, 296, 396-397

Fertilizers, chemical
alternatives to, 226, 232, 289, 296-298, 299, 300, 310, 341, 397, 408, 417
demand for, 433
excessive, dangers of, 357
negative consequences of, 1, 195, 196, 223, 232, 296, 417, 426, 433, 461. *See also under* Environment impact
problems of, eliminating, 401
selective use and reduction of, 398, 399, 400, 406, 426
sulfate, 146
Festuca, 503
Festuca rubra, 68
Field beans *(Phaseolus),* 54, 185, 206
Firs, 395
Flavonoids, 105-106, 151-152, 168, 333
Flooding
of rice fields, 227
tolerance for, 96, 105
Floricultural crops, 299
Fluorescent pseudomonads. *See Pseudomonas fluorescens* (fluorescent pseudomonads)
Folic acid, 438
Food
supply and demand, 186, 196, 296, 398, 433
technology, 400
Food and Agriculture Organization (FAO), 185
Forests and forestry
disease control in, 138
ectendomycorrhizal associations of, 247-248
ectomycorrhizal associations of, 241, 242, 243-244, 250-251, 258-262, 519
endomycorrhizal associations of, 244, 246
fertilizers and fertilization in, 259, 519
fungal inoculation in, 13, 14
Forests and forestry *(continued)*
mushrooms in, 499-500, 502-503, 508, 519-520, 532, 533
mycorrhizal associations of, 243-249, 252-258, 265
mycorrhizal helper bacteria role in, 130
trees, 243-248, 265, 519
vegetation, 368
Fosetyl-Al, 380
Fossil fuels, 427
Foxtail millet *(Setaria indica),* 482
Frankia, 5-6, 8, 17, 109, 139t, 141, 203
Free radicals, 66, 272-273
Fruit crops, 299
Fruit production, 300
Fruit trees, 244
Functional microorganism groups
actinomycetes and heterotrophic bacteria, 31, 37
assessing, 34-35
carbon cycle, 32, 37
classification of, 47
filamentous fungi, 31
nitrogen cycle, 32-33, 39
overview, 29, 46
phosphorus cycle, 33-34, 44-45
sulphur cycle, 33
Functionality versus biodiversity, 46
Fungal sterols, 260
Fungi. *See also* Arbuscular mycorrhizal fungi (AMF); Ectomycorrhizal (EM) fungi; Endomycorrhizal (EM) fungi; *mushroom type, e.g.:* Matsutake; Mushrooms; Mycorrhizal fungi; Pathogenic fungi
bacteria interaction with, 16-17, 123-127, 123t, 124t, 126t, 151
biofertilizer influence on, 109
as biofertilizers, 10-15, 141
chemical processes of, 33
competition among, 250-251, 329, 331, 370
culture media for, 36-37

Fungi *(continued)*
  diversity of, 13, 15
  experiments and field studies, 18, 19
  filamentous, 31, 67
  indigenous, arbuscular mycorrhizal (AM) as supplement for, 231
  inoculation and inoculants, 15, 18, 19, 302. *See also* Arbuscular mycorrhizal (AM) inoculation and inoculants; Ectomycorrhizal (EM) fungi, inoculations with; Mycorrhizal inoculation and inoculants; *Piriformospora indica,* inoculation with
  mantle (sheathing), 256, 290, 527, 528
  matsutake, effect on, 503
  metal tolerance of, 65, 66, 67-68
  nonmycorrhizal in tree roots, 248
  in phosphate solubilization, 107
  removal, selective, 519
  root colonization by, 14-15, 62, 233
  selection of, 396
  stress, tolerance of, 259
  sulphur oxidation, role in, 146
  in symbioses, 72, 150, 153-154, 289
Fungicides, 95, 169, 189, 228, 259-260, 304-305, 380-381, 488, 490. *See also* Biofungicides
Fungistasis, 256
Furadon granule (Carbofuran, 3 percent), 446
Fusarium, 146, 157, 256, 294
Fusarium culmorum, 160
Fusarium moniliforme, 332
Fusarium oxysporum (fusarium wilt), 158, 161, 162, 256, 334
Fusarium solani, 332

Gaeumannomyces graminis, 488
Gaeumannomyces graminis var. *tritici,* 159
Gall root nematode, 336
Gallium (Ga), 67
Gamma irradiation, 101
Gas exchange, 276-277
Gastrodia, 527, 530
Genetic diversity, 170
Genetic engineering, 212-213, 400
Gibberellic acid (GA), 149, 150, 438
Gibberellins, 9, 56, 166, 254, 256
Gigaspora, 304, 308, 405, 483
Gigaspora margarita, 58, 370, 373, 375
Global warming, 296
Glocobacter, 435
Gloeocapsa, 92, 435
Gloeocapsa polydermatica, 94
Gloeothece, 435
Gloeothece rupestris, 466f, 472
Glomaceae family, 403
Glomales order, 291, 403
Glomus, 55, 56, 61, 120t, 153, 229, 329, 364, 403, 405, 483
Glomus aggregatum, 368
Glomus caledonium, 57
Glomus clarum, 58, 153, 370, 373
Glomus deserticola, 275, 277, 279t
Glomus etunicatum, 60, 61, 305, 331, 365, 368, 370
Glomus fasciculatum, 54, 60, 62, 68, 275, 277, 279t, 304-305, 331, 332, 404, 405
Glomus fistulosum, 334
Glomus geosporum, 61
Glomus hoi, 334
Glomus intraradices, 119, 120t, 123, 125-126, 277, 281, 281f, 282f, 298, 305, 363, 366, 370
Glomus macrocarpum, 334, 339, 375
Glomus manihotis (Manihotina), 366-367
Glomus monosporum, 59, 61, 62
Glomus mosseae, 58, 60, 61, 62, 69, 123, 153, 229, 279t, 280, 281, 281f, 282f, 331, 334, 370, 488
Glomus versiforme, 298
Gluconacetobacter diazotrophicus, 142, 144, 168

Glutamate, 279, 437
Glutamate dehydrogenase (GDH), 253
Glutamate synthase (GOGAT), 253
Glutamic acid, 438
Glutamine, 437
Glutamine oxoglutarate amino transferase (GOGAT), 437
Glutamine synthetase (GS), 253, 427, 437
Glutathione reductase (GR), 273, 280, 282-283, 284
Glycine, 194
Glycine betaine, 279
Glycine max, 480, 487
Glycoprotein, 295
Glyricidia, 207
Grafting, 381
Grain legumes, 207
Grain yield, 421
Graminaceous crops (grass), 185, 186, 193, 196, 208, 228, 291, 366
Grapes, 339, 402, 486
Grapevines, 299, 302, 487
Grass-clover system, 225
Gravel, porous, 421
Grazing, 423, 427, 460
Green algae (Chlorophycae), 459
Green fluorescent proteins (GFP), 484
Green manures, 91, 96, 100, 101, 103, 104, 109, 141, 193, 223, 397, 425, 433
Greenhouse effect, 194, 196
Greenhouses, 341, 365, 402, 404, 519, 526
Groundnut, 57
Groundwater, 169
Groundwater contamination, 166, 399
Growth chambers, 404
Growth regulators, 223, 397
GSA dehydrogenase, 279
Guinea colonial grass *(Panicum maximum),* 366, 404
Gum, 244
Gunnera, 141
H_2S_4, 262
Haber-Weiss reaction, 273
Halo blight, 161
Hamada agar, 521-522, 523, 524
Hardwood trees, 246
Hartig net, 243, 247, 251, 253, 290, 517, 527-528, 530
Hazel trees, 121
HCN (hydrogen cyanide), 159, 161, 163
Health, human and animal, 400
Hebeloma, 528
Hebeloma crustuliniforme, 123, 128f
Hedysarum corronarium, 207
Helper bacteria, 150-151, 153. *See also* Mycorrhiza helper bacteria (MHB)
Hemlock, 243
Herbaspirillum, 7, 184
Herbicides, 189, 304, 354, 381, 397, 399, 424. *See also* Bioherbicides
Herbs, 206, 207
Heterocysts, 435, 437-438, 444
Heterotrophic bacteria, 31, 257
Heterotrophic fungus, 290
Heterotrophic organisms, 33, 34, 203, 242, 257
Hickory, 243
Hippophaë rhamnoides, 5
Hippophaë tibetana, 17
Hohenbuehelia, 529
Hormocysts, 421
Hormones, 294-295, 401, 480. *See also* Phytohormones; Plant growth, hormones
Hortensia, 487
Horticultural systems, mycorrhizal symbioses in, 298-300
Host specificity, 250, 265, 291, 360, 402, 487, 490
Host-pathogen association, 249
HPLC separation and analysis, 481-482
Humic acid, 464
Humus, 242, 252, 258, 378
Humus-formation process, 32

Hydnangium carneum, 14
Hydnum repandum, 503, 519
Hydrangea, 302
Hydrocarbons, 305
Hydrogen, 183, 443
Hydrogen cyanide (HCN), 159, 161, 163
Hydrogen peroxide, 273
Hydrolitic enzymes, 482, 483
Hydroponics, 307
Hydroxyl radicals, 273
Hymenonycetes, 479
Hyphoderma, 529
Hypochytridiomycetes, 67
Hysterangium, 530

IAA (indole-3-acetic acid), 149, 151, 154, 166, 195, 255
IBA (indolebutyric acid), 295
Ice plants, 272
Illumination, 380, 381, 382
Imperata cylindrica, 208, 502
India, agriculture in, 90, 94, 96, 418, 420, 433, 441, 446, 450
Indigofera, 206
Indole acetic acid, 438
Indole compounds, 255
Indolebutyric acid (IBA), 295
Indole-3-acetic acid (IAA), 149, 151, 154, 166, 195, 255
Indonesia, agriculture in, 187-188
Induced systemic resistance (biopesticide mode of action), 158, 160-161, 163, 195
Inga (*Inga edulis* Mart.), 383
Inoculation and inoculants
  bacteria, 18, 19, 139t, 155, 196, 213-214
  for beans, 166, 193, 299, 361
  with cultures, 263
  cyanobacterial (BGA). *See* Cyanobacterial inoculation and inoculants (algalization)
  dual, 53, 150-151, 153-154, 295, 333, 383

Inoculation and inoculants *(continued)*
  effectiveness of, 401
  evaluation of, 166, 170, 190, 310, 361
  experiments, 18, 19
  with fungi (*see* Arbuscular mycorrhizal (AM) inoculation and inoculants; Ectomycorrhizal (EM) fungi, inoculations with; Fungi, inoculation and inoculants; Mycorrhizal inoculation and inoculants; *Piriformospora indica,* inoculation with
  immobilized, 95-96f, 97f, 427, 486
  for legumes. *See under* Legumes
  limitations and challenges, 190-191
  for maize, 104, 105, 107, 123
  for matsutake, 524, 525-526, 532
  microbial, 137, 140-141. *See also* Biofertilizers
  of micropropagated plants, 486-487
  modes of action of, 170
  multiple, 15-18, 19, 69, 150, 360, 383
  need for, establishing, 188
  nursery, 6
  production of, 362-368, 485, 486, 490
  regulations and standards, 19
  research and field experiments, 412
  rhizobia, 4-5, 53, 152t
  for seedlings, 263
  soil, 263
  with spores, 263-264
  techniques, 190-191
  technological innovations in, 401
  use increase, projected, 168
Inorganic acids, 34
Inorganic fertilizers, 303-304
Inorganic ions, 279
Inorganic substrates, 367
Insect-flea beetles, 169
Insecticides
  alternatives to, 408
  arbuscular mycorrhizal fungi, impact on, 229, 305
  bioinsecticides, 157
  nursery use of, 304

Insecticides *(continued)*
  plant-growth-promoting rhizobacteria treatment versus, 161
  reducing use of, 398, 399
  as seed treatment, 169
  for seedlings, 381
  tolerance to, 189
Integrated pest management, 399
Intensive agriculture, 185
Intercropping, 187, 196, 207, 208
International Network on Soil Fertility and Fertilizer Evaluation for Rice (INSFFER), 103
Ions, mobility of, 225
Ipomea batatas (sweet potato), 105, 364
Iron (Fe). *See also* Siderophore production (biopesticide mode of action)
  algae effect on, 440
  carrriers, 162
  chelation of, 145, 147
  in cyanide production, 163
  forms of, 444
  importance of, 67
  ions, Haber-Weiss reaction in, 273
  metalloenzymes containing, 66
  protein, 444
  reduced, 192
  regulation of, 424
  sources of, 109, 298
  uptake of, 57
Iron phosphate, 252
Irradiance, 470
Irrigation, 380, 381-382
Isoflavonoids, 333
Isopentenyladenine (2-iP), 63
Isopentenyladenosine (IPA), 63
Isozymes, 54, 55

Japan
  agriculture in, 418, 421, 498
  forestry in, 499-500, 508, 519-520
Jasmonic acid, 294
Java (Indonesia), agriculture in, 187-188
Juniperus, 247, 248

K (potassium). *See* Potassium (K)
Kallar grass, 7
Kinetin, 104
Kiwifruit, 299, 301, 302
Klebsiella, 7, 17
Klebsiella pneumoniae, 145
Kudzu *(Pueraria phaseoloides)*, 366
Kummerowia stipulacea, 206

Lablab purpureus, 207, 208, 209
Laccaria, 528
Laccaria amethystea, 503
Laccaria bicolor, 121, 127, 129, 130t
Laccaria fraterna, 125
Laccaria laccata, 124, 125, 126t, 256
Lactarius, 248, 511
Lactarius deliciosus, 498, 528
Lactarius piperatus, 503, 519
Laminae, 58, 59
LaMotte-Morgan soil test, 398
Lapage culture, 43-44
Larch *(Larix)*, 121, 243, 248, 249, 258
L-asparaginase activity, 104
Late embryogenesis abundant (LEA) proteins, 272, 283-284
Lathyrus sativus, 194
LCO (lipochitin oligosaccharides), 72
Leaching, 258
Lead (Pb), 65
Leaf blast, 426-427
Leaf pruning, 207
Leghemoglobin, 209, 212
Legume-rhizobia symbiosis
  benefits of, 140, 185-186, 195
  challenges of, 186-187
  chemical signals, 58, 70
  diazotrophic microorganisms, role in, 184
  international practices and trends, 187-188

Legume-rhizobia symbiosis *(continued)*
and nitrogen, 52-53, 195, 203, 211
plant-growth-promoting rhizobacteria stimulating, 141, 143, 150, 151-153, 152t, 166, 170
study and model of, 72, 109
Legumes
in adverse conditions, 213, 282
in alternative farming systems, 397
bacteria infecting, 3, 204
as biofertilizers, 207, 211
biofertilizers for, 90, 108
crop, 56-58
grain, 207
growth, improving in, 54
herbs, 206, 207
in Indonesia, 188
inoculation of, 4, 5, 53, 190, 192, 207-208, 213-214, 295, 379, 383
mutations, 70
mycorrhizal fungi, interactions with, 119
nitrogen processes of, 51, 52, 143, 195, 209, 295, 383
as nitrogen source, 203, 399
products, 207, 211
root colonization of, 483
salt tolerance in, 213
shrubs, 206
spores, 366
trees, 204-205, 206-207, 208
utilization of, 193-194, 195, 196
wild, rhizobia from (*see* Wild-legume rhizobia)
woody, 55-56, 487
Leguminosae family, 184, 185
Lentil *(Lens culinaris)*, 58, 187, 208, 213
Lentinus edodes, 521
Lepista, 529
Lespedeza, 205
Lespedeza bicolor, 502
Lettuce, 68, 195, 274-275, 276f, 280, 281f, 299
Leucaena, 193, 207, 208, 209
Leucaena leucocephala, 54, 205, 210, 211, 379
Leucerne *(Medicago sativa)*, 57, 65, 207, 209, 211
Lichens *(Cladonia)*, 503
Light, 441
Light intensity, 470
Light microscopy, 464
Lignin, 32, 61, 254
Linseed, 227
Lipid metabolism, 60
Lipid peroxidation, 273
Lipids, 260, 282
Lipochitin oligosaccharides (LCO), 72
Lipochitooligosaccharides, 70, 71
Lipopolysaccharides (LPS), 210
Liquid cultures, 520-521, 524, 525, 529
Lithocarpus dealbatus, 503
Lithocarpus densiflorus (tan oak), 505
Lithocarpus sphaerocarpus, 503
Livestock feed additives, 223, 397
Logged sites, 260
Lolium perenne, 58, 60
Lotus corniculatus, 211, 212
Lotus japonicus, 212
Lotus maroccanus, 209
Low-input farming, 399
Lupinus, 4
Lupinus angustifolius, 194
Lyngbya, 435
Lyngbya kuetzingiana, 466f, 472
Lyngbya limnetica, 472
Lyophyllum shimeji, 498, 503, 527
Lysates, 118

Macronutrients, classification of, 443
Macroptilium, 206
Macroptilium atropurpureum (siratro), 208
Magnesium (Mg), 258, 443
Maize
arbuscular mycorrhizal colonization in, 227
bacteria associated with, 7, 145

Maize *(continued)*
bean intercropped with, 187
biofertilizers, growth-stimulating, 157
cultivation, effects of continuous, 228
in developing nations, 185
field trials with, 195
inoculation of, 104, 105, 107, 123
metals, exposure to, 68
and *Piriformospora indica,* 482
population increase pressures, 186
rotation, 229
in South America, 188
yield, increasing, 194, 196
Major intrinsic protein (MIP) family, 272, 283
Malathion, 446
Malic acid, 127
Malus, 247
Mancozeb, 490
Manganese (Mn), 57, 58, 66, 68, 192, 298, 424, 440
Manihotina *(Glomus manihotis),* 366-367
Mannitol, 279
Manures. *See* Animal manures; Farmyard manure (FYM); Green manures; Organic manures
Maple, 244
Matsutake
classification of, 502
climate and soil requirements for, 506-507
colonies, 503-505, 504f, 507, 508, 520, 522, 524, 529, 531, 532
conservation measures, 520, 532, 533
cultures, 520-522, 524
fruiting of, 499f, 500, 505, 507-508, 519, 521, 524, 527, 532
infections, morphology of, 508-518
inocula with, 524, 525-526, 532
isolates, 520-521, 522-525, 523f, 526
Matsutake *(continued)*
Japanese import of, 500-501t
life cycle of, 527, 531
overview of, 498
pathogenic tendencies of, 527, 532
production and harvest of, 499-500, 500f, 504, 508, 519-526, 531-532, 533
varieties of, 500
vegetation associated with, 500, 502-503, 504-505, 508-518f, 526-531. *See also Pinus densiflora*
MB (mycorrhizal benefit), 361, 362f
MBC (minimum bacterial concentration), 160
MD (mycorrhizal dependence), 355-359f, 379, 380, 383-384, 402, 487
Mechanical cultivation, 397
Medicago truncatula, 70, 207, 212
Medical technology, 400
Medicinal plants, 487, 488, 499
Meloidogyne, 294, 336
Meloidogyne incognita, 339
Mesorhizobium, 3, 204, 205, 206
Metabolism and metabolizing, 60, 253, 400-401, 443, 444
Metabolites. *See also* Siderophore production (biopesticide mode of action)
growth-inhibitory, 487
production of, 159, 161-162, 163, 254-256, 257
secondary, 333, 354, 403, 480, 481-482, 490
Metalaxyl, 380
Metals
adverse effects of, 1-2, 64-65
in fungicides, 259
ions, 273
removal of, 65
tolerance to, 11, 65, 66, 67-68, 229-230, 354, 490
uptake of, 66, 67, 68-69
Methanol, 185

Methionine, 254
Methyl bromide, 294, 366
Methyl cellulose, 371
Methylobacterium, 206, 214
Methylobacterium nodulans, 185
Mexico, agriculture in, 192
Mg (magnesium), 258, 443
MHB (mycorrhiza helper bacteria). *See* Mycorrhiza helper bacteria (MHB)
MIC (minimum inhibitory concentration), 160
Microbial activity, conditions affecting, 30
Microbial inoculants, 137, 140-141. *See also* Biofertilizers
Microbial pesticides. *See* Biopesticides, bacterial
Microbiological Resources Center (MIRCEN), 188
Microchaetaceae family, 464
Microchaete tenera, 465f, 472
Micrococcus, 17
Microcystis, 94, 468f
Microcystis aeruginosa, 473
Microcystis elabens, 466f, 473
Microflora, 118, 124-125, 257, 261, 263, 427, 433
Micronutrients, 45, 424, 443
Microorganisms, classifying, 401
Micropropagation and micropropagated plants, 232, 300, 301-302, 374, 381, 486-487, 490
Millet (*Panicum miliaceum* L.), 365, 482
Mimosa affinis, 205
Mine spoils, 65
Mineral substrates, 367
Mineral-bearing rocks, 397
Minimum bacterial concentration (MBC), 160
MIP (major intrinsic protein) family, 272, 283
Mn (manganese). *See* Manganese (Mn)
MNC solution, 525
Modified Melin-Norkrans' agar medium (MMN), 521, 522, 524
Mokichi Okada Association (MOA), 397
Molecular biology, 30, 46, 72, 212
Molybdenum, 109, 187
Monocultures, 227
Monoxenic cultures, 363-364
Montmorillonite clay, 448
Montmorilonite, calcinated, 367
Mortierella, 505, 522, 524, 527
Moss, 367
Mucilages, 118, 144
Mucuna pruriens, 208
Multanimetti (clay), 448
Mushrooms. *See also under mushroom type, e.g.* Matsutake
  colonies, 531
  cultivation of, 498, 519-526, 532, 533
  wild, 499-500, 502-503, 519-520, 532, 533
Mutants, 70, 72, 427
Mutualism range, 361, 362f
Mycorrhiza helper bacteria (MHB), 18, 118, 123t, 124-130, 124f, 126t, 128f, 130t, 257
Mycorrhizae (mycorrhizal symbioses). *See also* Arbuscular mycorrhizal symbioses; Ectomycorrhizae; Endomycorrhizae; Vesicular arbuscular mycorrhizae (VAM)
  in agriculture. *See* Agriculture, mycorrhizae in; Arbuscular mycorrhizal fungi (AMF), in agriculture
  arbuscular mycorrhizal fungi growth, requirement for, 371, 373
  and bacteria, 122, 123, 129
  as biocontrol agents, 294
  defined, 289-290, 395, 397
  disease control, role in, 117

Mycorrhizae *(continued)*
drought tolerance enhancement by, 282, 283
ecological significance of, 241-242
forest association with, 243-249, 265
formation of, 248-250, 255, 262, 263, 265, 309, 531
infected root staining, 45-46
legume dependence on, 55
managing, 232, 233
in nurseries, 261, 263
nutrient cycling, role in, 230-231
overview of, 241
plant growth, effect on, 252-258, 293-294, 300, 412
plant physiology, effect on, 292-295
practical aspects of, 262-264
rhizobial infection process, impact on, 120t
root colonization in, 482-483
signal function in, 71
survival without, 260
types of, 290-292, 497-498
water absorption, role in, 117
Mycorrhizal benefit (MB), 361, 362f
Mycorrhizal dependence (MD), 355-359f, 379, 380, 383-384, 402, 487
Mycorrhizal fungi. *See also* Arbuscular mycorrhizal fungi (AMF); Ectomycorrhizal (EM) fungi; Endomycorrhizal (EM) fungi
bacteria, interaction with, 31, 118, 123-127, 123t, 124f, 126t, 142
benefits of, 117-118, 128-129, 225, 273
as biofertilizers. *See* Biofertilizers, mycorrhizal fungi as
changes produced by, 326-329
chemical processes of, 71
competition among, 251
diversity of, 56, 264, 265
ericaceous, 483, 497, 498
in forests, 248-249, 261-262
fungicide effect on, 305, 488
host range of, 249

Mycorrhizal fungi *(continued)*
legumes, interactions with, 119
loss of, 301
matsutake as, 527
metal tolerance of, 67-68
native, 335-336, 341
nitrogen uptake and transfer, role in, 293
in nurseries, 310-311
nutrients, obtaining, 33, 264, 273
optimal conditions for, 250
pesticides, impact on, 304-305
plant nutrition, role in. *See* Nutrients, mycorrhizal fungi contribution to
population shifts in, 258
rhizosphere of, 16
selection of, 262
stress, tolerance of, 260, 262, 264
transgenic, 310
typical versus matsutake, 530-531
Mycorrhizal inoculation and inoculants, 263-264, 265, 303, 519. *See also* Arbuscular mycorrhizal (AM) inoculation and inoculants; Ectomycorrhizal (EM) fungi, inoculations with; *Piriformospora indica,* inoculation with
Mycorrhizal mutants, 70, 72
Mycorrhizal plants
absorption surface of, 292
copper concentrations of, 53-54, 57, 58, 68
drought tolerance of, 277, 278, 279t, 280-283, 284
enzyme activity in, 60-62, 482-483
gas exchange of, 276-277
metal uptake by, 68-69
physiology of, 354
production of, 530
roots of, 118
versus nonmycorrhizal plants, 53-54, 59-60, 225, 229, 252, 277, 361, 362f, 378f, 527
water uptake by, 274-276f, 283

Mycorrhizal seedlings, 377-382, 378f. *See also* Arbuscular mycorrhizal (AM) inoculation and inoculants, of seedlings
Mycorrhizal soil, 277-278
Mycorrhizosphere
  composition of, 293
  defined, 231, 290, 325
  effect, 118-123
  ethylene in, 256
  formation of, 295, 325, 367
  overview of, 257-258
Myrtaceae, 498
Myxosarcina, 435

N_2-fixing bacteria
  beneficial organisms, 401
  carbon and energy supply for, 51
  chemical processes of, 32, 33
  diazotrophic, 183, 184
  evolving abilities of, 196
  free-living, 8, 39-40, 105, 109, 166, 184, 212, 257
  global, 196
  heterotrophic bacteria, interaction with, 31
  inoculation with, 191, 379
  as microbial biofertilizers, 3
  overview and examples of, 7-8
  paranodulation, use in, 109
  plant-associative, 184
  plant-growth-promoting rhizobacteria (PGPR), 17, 143-145, 144t, 154, 168
  selection of, 295, 361
  in stem nodules, 185
  stimulation of, 100
  symbiotic, 3-6, 8, 203, 211, 212
  use breakdown by crop, 139t
N_2-fixing cyanobacteria, 6, 90, 92-94, 103, 109, 417, 418, 419, 420, 422, 423, 424, 425, 426, 427-428, 434, 435, 437-438, 439, 440, 442, 445, 449, 459, 460, 461, 462, 471-472
NaCl (sodium chloride), 94, 209, 210, 444, 445
National Academy of Sciences, 399
National Research Council, 399, 400
Nature farming, 397, 400, 401
Neem *(Azadirachta indica),* 423, 480, 487, 488
Neem cake, 98f, 100-101, 103, 104
Nematicides, 229, 339, 381
Nematodes, 336-340t, 337t-338t, 369, 380
Nettles (Urticaceae), 497
NFT (nutrient-film technique), 405
Niacin, 438
Nickel, 293
Nicotiana tabacum (tobacco). *See* Tobacco *(Nicotiana tabacum)*
Nif gene, 435, 437
Nimbidin, 100
Nimbin, 100
Nimbosterin, 100
Nitrate leaching, 186
Nitrate reductase (NR), 62, 278, 279t
Nitrate reduction, 444, 445, 484
Nitrates, 32, 33, 34, 62, 444, 461
Nitric acid, 34
Nitrobacter, 100
Nitrogen (N)
  arbuscular mycorrhizal fungi, effect on, 379
  compounds, 461
  cycles, 29, 33, 39, 140
  deficiency in, 203
  demand for, 396
  demineralization processes, 32-33
  fertilizer, 1, 2, 100, 101, 103, 184, 186, 195, 196, 212, 229, 259, 297, 303, 379, 399, 402, 417, 423, 427, 428, 433, 467
  growth limiting nutrient, 2, 51, 425
  IAA production suppressed by, 255
  in ionic form, 443
  metabolism, 253, 444
  mineral, 194, 203
  mycorrhizae and, 244, 253, 262, 264
  N_2 molecule, formation of, 183, 437

Nitrogen *(continued)*
organic, availability of, 254, 264
overview of, 443-444
sources of, 62, 194, 195, 203, 241, 278, 379, 419, 423, 427, 428-429, 484
storage of, 438
uptake and transfer of, 273, 293, 420
use efficiency (NUE), 95, 98f, 100-101, 103, 303-304, 427
volatilization, 258
Nitrogen (N_2) fixation. *See also* Biological nitrogen fixation (BNF)
asymbiotic, 69, 143-145
bacteria associated with. *See* N_2-fixing bacteria; N_2-fixing cyanobacteria
biofertilizer role in, 3, 140, 203, 212
capacity, increasing (thru genetics), 212-213
control and regulation of, 437, 471
effects of, 18
by free-living organisms, 8, 39-40, 51
green manures, 100, 101, 103, 104
growth-promoting substances as cause of, 153, 166
heterocysts and, 435, 437-438
iron role in, 444
in legumes, 51, 52, 143, 195, 209, 383
levels, determining, 94, 462-463
limiting factors in, 53
nitrogenase activity inhibited in, 30
process of, 203
rate, 105
requirements for, 56, 64, 143, 186-187, 192, 213-214, 438
in rice, 104-106, 419-420, 423
salinity impact on, 445
stimulation of, 17
symbiotic, 3-6, 8, 51, 192, 206-208, 214
in trees and shrubs, 119
Nitrogenase activity, 30, 95, 105, 143, 183, 209, 424, 435, 437, 438, 462, 463, 466, 467, 469, 469f, 470, 470f, 471
Nitrogenase inhibition, overcoming, 437
Nitrogenase synthesis, 437, 444
Nitrogen-free malate (Nfb), 39-40
Nitrosomonas, 100
Nitrous oxide (N_2O), 223
Nodularia, 469f, 470f
Nodules
formation of, 166, 186, 214
morphogenesis, 212
protein of, 209
in roots, 63, 70, 104-106, 105f, 204, 210, 212
on stems, 185
Nonagar cultures, 520-521, 526
Nonmycorrhizal plants versus mycorrhizal plants. *See under* Mycorrhizal plants
Northern blot, 281
Norway spruce, 68
Nostoc, 92, 94, 141, 418, 419, 420, 423, 435, 436f, 440, 442, 446, 469f, 470f, 471-472
Nostoc commune, 419
Nostoc microscopicum, 465f, 472
Nostoc piscinale, 94
Nostocaceae family, 464, 466
No-tillage systems, 196
Nucleic acid fragments, sequencing of, 30
Nurseries
fungal inocula at, 13, 14, 365, 366, 367, 373, 382, 487
herbicides in, 381
plant production at, 290, 300-309, 310-311
Nut trees, 244
Nutrient cycling, 29, 46, 47, 140, 230-231, 232, 258, 273, 289, 355, 401, 408, 433, 438, 445
Nutrient-film technique (NFT), 405

Nutrients. *See also* Phosphorus (P); Sulphur; *specific nutrient type, e.g.* Iron
absorption of, 355, 360
arbuscular mycorrhizal fungi contribution to, 225-226, 230, 231, 233, 383, 402
availability of, 377, 429, 445
biofertilizers, 109, 140, 143-145
classification of, 443
competition for, 158-159, 162, 360
consumer demand for, 396
deficiency in, 257
growth limiting, 2, 12, 51, 57, 145, 425
imbalance in, 192, 264
microorganisms as source of, 2
mineral, 296, 297, 487
mycorrhizal fungi contribution to, 34, 56-57, 241, 242, 244, 252, 273
in organic farming, 224
plant-growth-promoting rhizobacteria as source of, 145, 153, 168
ratios, 398
recycling of, 383
replacing lost, 296
reserves of, 355
transfer of, 242, 244, 253, 273, 289, 292, 293, 353-354, 360, 395, 403, 528
uptake of, 241, 252, 273, 292-293, 297, 406, 497
use efficiency, 297
Nutritional solution, 363, 364-365, 405, 524, 526

Oak, 14, 243, 395, 498, 502-503, 505
Oat *(Avena sativa),* 68, 185, 194, 297
Obligate biotrophs, 11, 341, 362
Obligate endosymbionts, 483
Obligate mycorrhizae, 247-248, 291
Ocimum basilicum, 280
Off-farm wastes, 397
Ohta solution, 525
Oil palm, 299, 302
Oligisaccharins, 71
Olive, 301, 302
On-farm wastes, 399
Onion, 59, 60, 61, 299
Onobrychis ptolemaica, 211
Oomycetes, 67
Open-reading frames (ORFs), 437
Orchid mycorrhizae, 497
Orchids, 477, 480-481, 486, 530
Organic acids, 34, 127, 257
Organic amendments, 229-230, 232
Organic compounds, 366, 377-378
Organic farming, 223-226, 232, 397, 400, 401, 426
Organic fertilizers, 141, 419
Organic food products, 427
Organic manures, 224, 231, 232, 233. *See also* Animal manures; Green manures
Organic matter, 377-379, 378t, 446
Organic residues, 401
Organic substrates, 367
Organic wastes, 223, 397, 401
Oscillatoria, 92, 94, 435
Oscillatoria rubescens, 465f, 472
Oscillatoria salina, 94
Oscillatoriaceae family, 464
Osmoregulation (osmotic adjustment), 271, 278-279, 283
Osmotic stress
alleviation, 272, 274
and arbuscular mycorrhizal symbiosis, 284
defined, 271
responses to, 271, 272
tolerance of, 272
Oxalic acid, 66
Oxidative stress, 272-273, 280-283, 282f
Oxygen (O_2)
for arbuscular mycorrhizal fungi, 381
availability of, 183

Oxygen *(continued)*
consumption of, 143-144
cyanobacteria, input by, 420
diffusion of, 212
exclusion of, 435
growth, requirement for, 443
for microorganisms, 401
nitrogenase activity inhibited by, 471
photoevolution, 424
for seedlings, 377
Oyama nutrient solution, 524
Ozone layer, 223, 424

Paddy, 108
Paenibacillus, 16, 18
Paenibacillus polymyxa (Bacillus polymyxa), 10, 107
Panicum maximum (guinea colonial grass), 366, 404
Panicum miliaceum L. (millet), 365, 482
Panthotenic acid, 255, 438
Pantoea agglomerans, 145
Papaya, 379
Paraglomus occultum, 365
Paranodulation, bacteria use in, 109
Paraquat, 381
Parasitic fungi, 309, 526, 527-528, 529, 531, 533
Parasitism (biopesticide mode of action), 158, 163-164
Parasponia, 141
Paspalum notatum (bahia grass), 364, 365, 404
Passion fruit, 378, 378f, 379
Pasture species, 498
Pastures, 185, 191, 193, 203, 208
Pathogen-host association, 249
Pathogenic bacteria, 334-336
Pathogenic fungi, 329-334, 527
Pathogens, control and protection against. *See* Disease control
Pathogens, risk factors for, 259, 263, 401
Paxillus involutus, 68, 127, 128f, 259
Pb (lead), 65
PB (phosphorus benefit), 361, 362f
PCNB (Pentachloronitrobenzene), 380
PCR (polymerase chain reaction), 30
PCR-DGGE (polymerase chain reaction-denaturing gradient gel electrophoresis), 122-123
PCR-RFLP (polymerase chain reactions/restriction fragment length polymorphism), 119, 121, 121t, 122f, 205
PDA nutrient solution, revised, 524
PDM (phosphate-dissolving microorganisms), 107
Pea
in Brazil, 187
in developing nations, 185
disease control in, 158
inoculation of, 166
root nodules of, 63, 70
symbioses, 152
Peach, 299, 302
Peanut, 166, 333
Pear, 299, 302
Pearl millet, 107, 208
Peat, 366, 367, 368, 405, 524
Peat moss, 405, 524
Pecan, 243
Pectinase, 166, 334, 482
Pectolytic enzymes, 257
Pedicularis, 502
Penicillium digitatum, 107
Pentachloronitrobenzene (PCNB), 380
Pepper, 299, 304
Peptides, 32
Pericopsis mooniana, 207-208
Perlite, 11, 366, 377, 405, 524
Peroxidase activity, 60, 61
Peru, agriculture in, 188
Pest attack, protection against, 425
Pest management, integrated, 399
Pesticides. *See also* Biological pest control; Biopesticides
agricultural production, role in, 426

Pesticides *(continued)*
conventional, alternatives to, 223, 226, 298, 310, 397, 408
cyanobacteria, effect on, 424
degradation of, 428
demand for, 427
environmental hazards of, 154, 165, 169, 230, 399, 426
health hazards of, 399, 426
mycorrhizal fungi, impact on, 226, 228, 229, 304-305, 380-381
mycorrhization, impact on, 357
nursery dependence on, 304
plant, 155
problems of, eliminating, 401, 427-428
reducing use of, 399-400, 406
residues, destroying, 484
resistance to, 170, 484
for seedlings, 380
tolerance to, 92
Peucedanum terebinthaceum, 502
Phaseolus (field beans), 54, 185, 206
Phaseolus vulgaris. See Common bean *(Phaseolus vulgaris)*
Phenolic compounds, 256
Phenols, 333, 334
Pheromones, 155
Phialophora finlandia, 247
Phormidium, 94, 435
Phosphatase activity, 104, 297, 298, 440
Phosphate solubilizing
bacteria role in, 145, 195, 225, 439
biofertilizer role in, 140
inorganic, 33-34, 44-45
microorganisms (PSM), 33-34, 295
Phosphates
absorbing, 230, 292
compounds, 297
dissolving microorganisms (PDM), 107
fertilizer (P_2O_5), 228, 229, 311, 396
inorganic, 61
mobilization of, 424, 446
organic, 295
Phosphates *(continued)*
plant assimilation and processing of, 230, 252-253, 292
reducing, 299
rock, 146-147, 295
transport of, 354
uptake of, 443
Phosphobacterial biofertilizer, 107-108
Phosphorus (P)
absorption of, 354, 355, 380, 402
arbuscular mycorrhizal fungi, impact on, 228, 303, 304, 309, 361, 362, 366, 377, 378, 379-380
arbuscular mycorrhizal fungi as source of, 225, 230, 233, 273, 292, 293, 297-298, 311, 402
availability of, 297-298, 440
benefit (PB), 361, 362f
and cyanobacteria, 422, 460
cycles, 29, 33-34, 44-45
cytokinin influence on, 255
deficiency, overcoming, 53, 54, 109
fertilizer, 1, 2, 275, 276f, 279t, 280, 281f, 282, 297, 304
fixing capacity, 402
growth limiting nutrient, 2, 12, 57, 145
improved nutrition, 274
increasing, 57, 62, 153, 258
in ionic form, 443
mycorrhizae and, 244, 262, 264, 329
and mycorrhizal dependence, 356-357, 358f-359f
overview of, 443
processes, isozyme role in, 54
reduced, 192
requirements for, 64
scarcity in desertified ecosystems, 55
solubilizing bacteria (PSB), 9, 17, 56, 69, 107, 153, 154
sources of, 11, 33, 295
storage of, 438

Phosphorus *(continued)*
tillage effect on levels of, 227, 228
uptake and transfer of, 253, 257-258, 297, 298, 311, 402, 487, 497
Phosphorylation, 279
Photosynthates, 273, 360, 382, 439
Photosynthesis, 32, 51, 54, 153, 242, 290, 354, 437, 440, 445
iron role in, 444
Phycomycetes, 246
Phyllobacterium, 17
Phytoalexins, 333
Phytohormones, 143, 147-150, 149t, 151, 153, 154, 166, 168, 257. *See also* Plant growth, hormones
Phytopathogenic organisms, 380
Phytophthora, 294, 334, 490
Phytophthora cinnamomi, 263
Phytophthora fragariae, 331
Phytophthora nicotiana var. *parasitica,* 331
Phytosanitary control, 193
Phytosiderophores, 147
Phytostimulators, 137
Phytotoxin production, 162
Picea, 248
Picea abies, 125
Pigeon pea, 57, 58
Pigment analysis, 464
Pinaceae family, 243, 247, 260
Pine
bark, 525
forests, 263, 502, 503
growth stimulation in, 150
mycorrhizae on, 243, 243f-244f, 245f, 246, 247, 248, 250
roots, 255, 509, 509f, 510f, 511f, 512f, 513f, 514-517, 514f, 515f-517f, 518f, 526
seedlings, 259, 520
Pineapple, 299, 302, 487
Pinitol, 279
Pinus, 527
Pinus armandii, 502
Pinus densata, 502
Pinus densiflora, 500, 502, 503, 504, 504f, 509f, 510f, 511f, 512f, 513f, 514f, 515f, 517, 521, 527
Pinus massoniana, 502
Pinus nigra, 125
Pinus radiata, 124, 263, 516f-517f, 517, 518f
Pinus sylvestris, 125, 252, 256, 503
Pinus tabulaeformis, 502
Pinus yunnanensis, 502
Piriformospora indica
agriculture, sustainable, role in, 477, 478, 490
classification of, 479
cultivation of, 485
disease control, role in, 488, 490
experiments and studies of, 481-482, 484
genetic modification of, 484
horticulture applications of, 486-488, 490
host plants and symbioses, 478, 480-483, 481f, 484, 490
inoculation with, 480, 486, 488, 490
life cycle and sexual phase of, 479, 490
morphology and characteristics of, 478, 479
plant growth promotion by, 487, 488, 489t, 490
protoplasts of, 483-484
Pisolithus, 122f, 248
Pisolithus alba, 120t, 121, 122, 122f, 123, 125, 127
Pisolithus tinctorius, 13, 14, 528
Pisum arvense, 194
Pisum sativum, 60, 207, 480
Plant growth
arbuscular mycorrhizal fungi role in, 230, 232, 290, 294, 298, 311, 360, 402-403, 406
hormones, 8, 9-10, 62-63, 69, 255
index, 355

Plant growth *(continued)*
limiting nutrients for, 2, 12, 51, 57, 145, 425
mycorrhizae role in, 252-258, 293-294, 300
promotion, 52, 438, 439, 480-481. *See also Piriformospora indica,* plant growth promotion by
regulating substances, 56
Plant-growth-promoting rhizobacteria (PGPR)
arbuscular mycorrhizal fungi, interaction with, 295, 383
benefits of, 165-166, 168, 169, 195, 257
as biofertilizers, 143-154, 168, 170
as biopesticides, 155, 158, 161, 170
definition and overview of, 8-9, 138
in disease control, 129
emergence-promoting rhizobacteria (EPR), 69
endophytic, 141-142
hosts, relationships with, 141-143
limitations of, 165, 167, 169-170
with multiple modes of action, 154, 168, 170
N_2-fixation, role in, 17, 143-145, 144t, 154, 168
regulating, 167
rhizospheric, 141, 142-143
root growth, role in, 147-149t, 148f
soybean nodulation, 17
success factors, 164-165
sulphur-oxidizing, 146-147
testing and evaluation of, 164, 167, 170
viability and shelf life of, 164-165
Plants
cell division in, 63
defense mechanisms, 333-334
genetic manipulation of, 106
materials, decomposition of, 32, 33
metal uptake of, 66, 67, 68-69
nutrients. *See* Nutrients
Plants *(continued)*
physiology, mycorrhizae effects on, 292-293
processes, adverse effects of, 168
productivity of, 225, 290
regeneration of, 486-487
from seeds, 300
stress, tolerance of, 271, 273, 309, 311, 325-326, 339, 341, 354, 369. *See also under plant species name*
water potential of, 278
Plasma membrane, 272
Plating on soil suspension, 463
Platycodon grandiflorus, 502
Plectonema, 92, 420, 435, 446
Plectonema notatum, 465f, 472
Pleurocapsa, 435
Pleurotus, 529
Plowing, moldboard, 227
Plum, 299, 302
Pollution, 294, 406, 407
Poly alkene bags, 448
Polybags, 447
Polygalactuonase, 483
Polygonatum odoratum, 502
Polyhouses, 449
Polymerase chain reaction (PCR), 30
Polymerase chain reaction-denaturing gradient gel electrophoresis (PCR-DGGE), 122-123
Polymerase chain reactions/restriction fragment length polymorphism (PCR-RFLP), 119, 121, 121t, 122t, 205
Polyphosphatase activity, 62
Polyphosphate, 252
Polyphosphate hydrolase, 54
Polysaccharides, 67, 210, 439
Polythene lined tanks, 447, 449
Poplar, 243
Populus, 503, 530
Porcini, 498
Pot cultures, 306, 401, 404-405, 486
Potash (K_2O), 396

Potassium (K), 57, 62, 109, 192, 244, 255, 258, 262
  in ionic form, 443
Potassium chloride, 398
Potato
  and arbuscular mycorrhizal fungi, 226, 299
  cyanide effect on, 163
  disease control in, 166
  plant growth, 162
  roots, 195
  in South America, 188
Potato dextrose agar, 36-37
Potting mixes, artificial, 303, 402, 526
Pratylenchus, 294
Predacious fungi, 529
Primitive symbiont (defined), 530, 531, 533
Prokaryotes, 183. *See also* N_2-fixing bacteria
Proline, 58-59, 279-280
Propagules, 261, 374, 375
Propanil, 424
Propiconazole, 259
Prosopis, 206, 207, 208, 209, 211
Prosopis chilensis, 205
Prosopis juliflora, 62, 477
Proteinase, 530
Proteins
  constitution of, 183
  denaturation of, 273
  green fluorescent (GFP), 484
  iron (Fe), 444
  late embryogenesis abundant (LEA), 272, 283-284
  lipid, 282
  major intrinsic, 272, 283
  in mycorrhizal versus nonmycorrhizal roots, 59-60
  of nodules, 209
  pathogenesis related, 334
  sources of, 62
  soy, 59, 185
  synthesis, 254, 445
  water channel, 272, 283
Proteolytic culture, 39
Protoplasts, 483-484
Protozoa, 33, 41-42, 257
Pruning, 207
Prunus, 302
PSB (phosphorus-solubilizing bacteria). *See* Phosphorus (P), solubilizing bacteria
Pseudobactin, 166
Pseudomonas, 7, 9, 17, 139t, 153-154, 159, 161, 162, 166, 184, 294, 295
Pseudomonas aeruginosa, 158
Pseudomonas aureofaciens, 159
Pseudomonas chlororaphis, 145, 159-160
Pseudomonas fluorescens (fluorescent pseudomonads), 17, 118, 121, 122t, 124, 125, 126t, 127, 129, 139t, 150, 151-152, 153, 158, 159, 161, 162, 163, 257
Pseudomonas monteilii, 125-126
Pseudomonas putida, 145, 158, 162, 164
Pseudomonas striata, 107, 108
Pseudomonas syringae, 139t, 153
Pseudomycorrhizae, 527, 530-531, 533
Pseudotsuga, 250
Pseudotsuga menziesii, 125, 512, 513f, 529
PSM (phosphate-solubilizing microorganisms), 33-34, 295
Pteridophytes, 291
Pueraria phaseoloides (kudzu), 366
Pumice stone, 421
Pyrenochaeta lycopersici, 162
Pyricularia oryzae, 426
Pyridoxine, 438
Pyrus, 247, 248
Pythium, 294, 490
Pythium splendans, 158
Pythium ultimum, 158, 159, 162, 164, 340

Quartz sand, 368
Quercus, 248

Quercus aquifolioides, 503
Quercus gilliana, 503
Quercus mongolica, 502
Quercus pannosa, 503
Quercus robur, 125
Quercus semicarpifolia, 502-503
Quercus senescens, 503
Quercus spinnosa, 503
Quinclorac, 424
Quintozene, 380

Radish, 145, 154, 161
Rainforest trees, 291
Ralstonia, 206, 214
Ralstonia solanacearum, 332, 335-336
Ramaria botrytis, 503, 519
Rape
 growth stimulation of, 154
 nitrogen and protein content, increasing, 18
 oilseed, 9
 seed, 145, 228
 treating, 17-18
Raspberry, 299, 302, 487
Reams fertility system, 398
Red clover, 153
Reforestation, 130, 203, 261, 265, 382-383
Renewable resources, managing, 310
Respiration, 471
Resupinatus, 529
Revegetation, arbuscular mycorrhizae and *Rhizobium* use in, 55
Rhizobacteria, 56, 69, 118, 257. *See also* Mycorrhiza helper bacteria (MHB); Plant-growth-promoting rhizobacteria (PGPR)
Rhizobia. *See also under rhizobia type, e.g.* Azorhizobia
 and arbuscular mycorrhizal fungi, 383
 classification of, 184-185
 infection process of, 120t
Rhizobia *(continued)*
 mycorrhizosphere effect of, 118-119
 N_2 fixation by, 3-5
 root systems, isolation from, 121t
 strains, selection of, 188-190
 studies of, 196
 survival of, 191, 192
 symbioses of, 3-5, 141, 192-193
 wild-legume. *See* Wild-legume rhizobia
Rhizobium
 aluminum-tolerant, 192-193
 and arbuscular mycorrhizal fungi (AMF) interactions, 52-58, 295
 calcium requirements for, 63-64
 classification of, 204, 205, 206
 free-living, 60, 109
 genetic modification of, 212-213
 inoculation, effects of, 17, 62
 legume symbiosis. *See* Legume-rhizobia symbiosis
 lipid metabolism of, 60
 nitrogen fixation, role in, 3, 51, 52
 in phosphate solubiliization, 145
 phosphorus requirements for, 64
 plants associated with, 7
 salt-tolerant, 209, 213
 strain selection, factors for, 3-4, 5
 symbioses, 69, 70-71, 151-153, 166
Rhizobium etli, 195, 205, 208
Rhizobium galegae, 205
Rhizobium japonicum, 60, 63
Rhizobium leguminosarum, 58, 63, 145, 192, 195
Rhizobium leguminosarum bv. *phaseoli,* 211
Rhizobium meliloti, 64, 139t, 209, 210, 212
Rhizobium phaseoli, 63
Rhizobium tropici, 192
Rhizoctonia, 480, 481
Rhizoctonia solani, 160, 162, 164, 479
Rhizomorphs, 528
Rhizopogon, 248, 250, 510
Rhizopogon luteolus, 124

Rhizopogon rubescens, 528
Rhizosphere
  composition of, 354, 498
  defined, 118, 289, 290
Rhizospheric fungi. *See* Arbuscular mycorrhizal fungi (AMF)
Rhizospheric plant-growth-promoting rhizobacteria (PGPR), 141, 142-143
Rhodes grass, 208
Rhodococcus, 17
Rhododendrons, 487, 498, 502, 503
Ribonucleic acid (RNA), 183, 252, 281f
Riboside zeatin, 255
Rice
  Azolla biofertilizer for, 90-92, 101, 103, 109
  Azospirillum biofertilizer for, 107
  bacteria associated with, 6, 7, 108, 109, 195. *See also* Cyanobacteria (blue-green algae, BGA), in rice cultivation
  fertilization of, 298, 423
  field soils, 445
  flooding of, 227
  inoculation of. *See under* Cyanobacterial inoculation and inoculants (algalization)
  natural bioresources for, 102f, 108-109
  nitrogen sources for, 104-106, 419-420, 423, 428-429, 438, 461, 462
  nitrogen use efficiency (NUE) of, 95, 98f, 101, 103, 109
  nutritional requirements for, 419-420
  organic, 427
  and *Piriformospora indica,* 482
  production and yield, 417, 421, 428, 433-434, 439, 440
  roots, nodule-like structures in, 104-106, 105f
  Sesbania biofertilizer for, 96, 99-100
Rice *(continued)*
  straw, 104, 448
  temperature, 442
  tropical, 418-419
  xylem of, 106, 106f
Ringer's solution, 35-36
Rivularia, 440
RNA (ribonucleic acid), 183, 252, 281f
Rock phosphate, 146-147, 295, 398
Root diseases, control of
  by bacteria, 16
  by biological control agents, 294
  by fungi, 15, 244, 326, 329-336, 330t-331t, 332t, 335t, 339, 341-342, 406, 490
  fungi-caused, 335t
  fungi-compost integration, 230
  inocula, 18
  tobacco, 159, 163
Roots. *See also under plant type*
  of arbuscular mycorrhizal plants, 62, 233
  bacteria associated with, 142f
  characteristics, 355-356
  colonization of, 52, 326, 332, 356, 360, 367, 369, 371, 372t, 379, 480, 482-483, 528, 529. *See also* Fungi, root colonization by; Legumes, root colonization of; Mycorrhizae (mycorrhizal symbioses), root colonization in
  function of, 118
  growth, 147-149t, 148f, 150, 526
  measuring, 308-309
  mycorrhizal, 225, 251, 252, 497
  nodule bacteria. *See* Wild-legume rhizobia
  nodule development in. *See* Nodules, in roots
  nonmycorrhizal, 225, 252
  organ culture, 405-406
  rot pathogens, 59
  staining, 45-46, 308
Roses, 277, 280
Rubber, 193, 299

Rushes (Juncaceae), 497
Russia, agriculture in, 418
Russula bella, 503
Ryegrass, 208, 228

S. cerevisiae, 66-67
Sacarose, 279
Safranin, 486
Salicaceae family, 247, 498
Salix, 247, 248
Salt and salinity
 plant growth and yield, impact on, 271, 444
 tolerance of, 93-94, 209-211, 213, 294, 354, 406, 439, 444-445
Sand, 11, 366, 368, 377, 405, 421, 524
Saprobic fungi, 309, 521, 526, 527, 528-530, 531, 533
Saprophytism and saprophytes, 30, 242, 485
Sarcodon scabrosus, 503, 519
Sclerocystis, 403
Sclerotium, 294
Sclerotium rolfsii, 164, 340
Scorpiurus vermiculatus, 207
Scutellospora, 405
Scutellospora gilmorei, 488
Scutellospora heterogama, 336
Scytonema, 92, 420, 423, 435, 436f, 440
Scytonemataceae family, 464
Secretions, 118
Sedges (Cyperaceae), 291, 497
Seedbeds, 366
Seedlings
 handling of, 380-382
 inoculation of. *See* Arbuscular mycorrhizal (AM) inoculation and inoculants, of seedlings
 production of, 353, 355, 368-369, 377-380, 382, 383
 selection of, 382
 sterile, 525, 532
 stress, tolerance of, 383
 transplanting of, 526, 530
Seeds
 plants from, 300
 treatment of, 169
Semihydroponics system, 366
Sequatchie loam, 277
Serratia, 17, 152
Serratia marcescens, 161, 164
Sesbania, 96, 99-100, 184, 208, 209
Sesbania aculeata, 96, 99, 100
Sesbania rostrata, 96, 99f, 100, 101, 103, 104, 185, 212
Sesbania sesban, 207
Sesbania speciosa, 96, 104
Setaria, 482
Sewage sludge, 1-2, 229-230
Shale, 367
Shiros (matsutake colonies), 503-505, 504f, 507, 508, 520, 522, 524, 529, 531, 532
Shoots
 of arbuscular mycorrhizal plants, 62
 dry mass, 372t, 378f
 growth and yield, 57
 water content, 277
Shrubs, 119, 206, 260-261
Siderophore production (biopesticide mode of action), 147, 154, 157, 158, 161-163, 166, 192
Simazine, 381
Sinorhizobium, 3, 153, 204, 205, 206, 208
Sinorhizobium fredii, 361
Sinorhizobium meliloti, 4, 210
Siratro *(Macroptilium atropurpureum),* 208
Sitka spruce, 121
Slash burning, 258
SOD (superoxide dismutase), 59, 60-61, 280, 281-282, 284
Sodium chloride (NaCl), 94, 209, 210, 444, 445
Soil
 aggregation of, 354, 360, 439, 440, 446
 amendments, 398
 arid, 6

Soil *(continued)*
conditioner, 399
conditions, adverse, 264
conservation, 225, 395, 396
cover, 193
density, 258
erosion control. *See* Erosion control
fertility, 109, 140, 191-192, 193, 206, 209, 214, 398, 426-427, 429, 433, 445
fertilization, 261
forest, 248-249, 261-262
fumigation, 261
metal removal from, 65
microbiological equilibrium, 401
microbiota, interaction with arbuscular mycorrhizal fungi (AMF), 295-296
microorganisms, 295-296
mycorrhizal, 277-278
pH, 68-69, 142, 250, 264, 303, 365, 370, 401, 423, 442, 443, 460, 469, 469f, 503
productivity, 296
properties, 249-250, 273, 360, 424, 506-507
quality, 406
recuperation, 193
samples, collecting and analyzing, 34-35, 35-46, 463-464
specificity, 250
sterilization, 404
structure, 258, 277, 295
suspension, plating on, 463
sustainability, 406
water potential of, 278
water retention of, 277-278
Soilless culture, 310, 405
Soilrite, 11
Solanum melongena (brinjal), 480, 487
Sorghum, 193
Sorghum, 7, 104, 105, 107, 486
Sorghum *(Sorghum bicolor),* 365, 404
Sorghum/Sudan grass hybrid, 208, 364, 365
South America, agriculture in, 188, 419. *See also* Argentina, agriculture in; Brazil, agriculture in; Uruguay, agriculture in
Soybean
alkaline phosphatase activity in, 61
biofertilizer for, 108
in Brazil, 186, 187, 361
disease control in, 158
drought effect on, 282-283, 282f
growth of, 151, 154
mycorrhizal versus nonmycorrhizal, 229
nematode protection, 339
nitrogen fixation in, 166, 185, 192
nodulation, 17
phosphorus solubilizing, 108, 145
as protein source, 59, 185
restrictions on, 227
Rhizobium-AMF interactions and, 53, 54, 70
roots, 148-149, 339
seed, 158
symbioses, 152, 153
Spain, agriculture in, 419, 461, 462, 466, 469-470, 471, 472-473
Species, improving, 170
Sphagnum moss, 524
Spilanthus calva, 488, 489t
Spinach, 228
Spirulina princeps, 466f, 472
Spirulina subsalsa, 466f, 472
Sponge blocks, synthetic, 421
Spores
amyloid versus inamyloid, 530
in arbuscular mycorrhizal fungi inocula, 11, 369, 370, 371, 375, 377
counts, increasing, 404, 405
cultures, 401
fungi identification through, 403-404
germination of, 405, 406, 425
immobilized, 486
in mycorrhizal fungi inocula, 263-264, 306

Spores *(continued)*
of *Piriformospora indica,* 483, 486
production of, 361, 365, 366, 367, 381, 421, 463
Spruce, 243, 256
Starch, 31, 32
Starch casein agar, 37
Starch hydrolizers, 37-38
Starkey's culture medium, 42-43
Stigonema, 435, 436f
Stizolobium niveum, 194
Stomatal conductance, 277
Straw carrier, 448
Strawberries, 228, 299, 302, 331, 334, 487
Streptomyces, 146
Stylosanthes *(Stylosanthes guianenses),* 366
Subsidized agriculture, 196
Substrates
and arbuscular mycorrhizal fungi, 11, 303, 304, 363, 364, 365-368, 369-370, 405
fungi in, detecting, 404
inorganic, 367
for matsutake, 524, 525, 526
metabolizing, 400-401
mineral, 367
nonagar solid, 523
organic, 367
for seedlings, 377-380, 378f, 382, 384
Sucrose, 51, 486
Sudan grass *(Sorghum sudanense),* 208, 364, 365
Sugarcane, 7-8, 9, 142
Sugarcane mealybug, 142
Sugars, 164, 257, 279, 280, 292, 403
Suillus, 505, 510
Suillus bovinus, 503
Suillus grevillei, 249
Suillus luteus, 252, 503
Suillus tridentinus, 249
Sulfate, 293
Sulfate fertilizers, 146
Sulfate uptake, 254
Sulfate-reducing bacteria, 43-44
Sulfides, 66-67
Sulfur
compounds, 127
cycles, 29, 33, 34, 42
mineralizing, anaerobic, 42
oxidizers and oxidation of, 42-43, 145, 146-147, 166
plant nutrient, role as, 146, 254
sources of, 109
Sunflower, 107, 227
Superficial ectomycorrhiza (defined), 530-531, 533
Superoxide dismutase (SOD), 59, 60-61, 280, 281-282, 284
Superoxide radicals, 273
Superphosphate, 92, 420, 446
Surfactants, 525
Sustainable agriculture. *See* Agriculture, sustainable
Sweet potato *(Ipomea batatas),* 105, 364
Sweetgum, 244
Sycamore, 244
Symbioses. *See also* Arbuscular mycorrhizal symbioses; Ectomycorrhizae; Endomycorrhizae; Mycorrhizae (mycorrhizal symbioses); *term* symbioses *under type of plant*
bacterial, 184
formation of, 249
as life cycle phase, 527, 531
molecular aspects of, 342
parasitic and mutualistic, 242
plant-growth-promoting rhizobacteria, role in, 141, 143
processes, 72
and species classification, 403
Symbiotic bacteria, 204, 214, 289
Symbiotic effectiveness (SE), 360, 361-362, 362f
Synechococcus, 434
Synechocystis, 94, 434

Tamarind (*Tamarindus indica* L.), 383
Tan oak *(Lithocarpus densiflorus),* 505
Tannins, 248, 528, 531
Technology transfer, 310
Temperature in rice fields, 442
Terrestrial orchids, 477, 486
Thelephora terrestris, 261
Thiamine, 255
Thielaviopsis basicola, 59, 163
Thiobacillus, 146
Thiram, 380
Tilia, 247
Tillage practices, 227-228, 232
Tissue-culture-raised plants, 486, 487
Tobacco *(Nicotiana tabacum),* 59, 60, 159, 161, 163, 164, 334, 480, 487
Tobacco necrosis, 161
Tolypothrix, 92, 420, 435, 436f, 440, 446
Tolypothrix tenuis, 94, 418
Tomato, 145, 161, 162, 299, 331, 335-336, 339, 375
Tonoplast, 272
Toxins, accumulation, avoiding, 365
Toxins from harmful microorganisms, 401
Trace elements, 264
Trans-ribosylzeatin, 256
Trans-zeatin, 256
Trees. *See also name of individual species, e.g.* Beech
  bark, 367
  disease control in, 138
  mycorrhizae and mycorrhization of, 129-130, 241-242, 243-248, 252, 265
  and nitrogen, 119
  and plant-growth-promoting rhizobacteria, 138
  removal, selective, 519
  roots of, 242
  stress, tolerance of, 244
Triadimefon, 380
Trichoderma, 294, 340
Trichodesmium, 6
Tricholoma, 502
Tricholoma bakamatsutake, 505, 510, 517, 522, 529
Tricholoma caligatum, 505, 517, 522, 523, 526
Tricholoma fulvocastaneum, 505, 522
Tricholoma japonicum, 503
Tricholoma magnivelare, 500, 504, 505, 506, 507, 508, 509, 512, 513f, 517, 521, 522, 523, 529, 530
Tricholoma matsutake. See Matsutake
Tricholoma mongolicum, 529
Tricholoma nauseosus, 500, 503, 506
Tricholoma populinum, 530
Trichomes, 435, 472
Trifluraline, 381
Trifolium fragiferum, 209, 211
Trifolium pratense, 60
Trifolium repens, 60
Trifolium squarrosum, 207
Trifolium subterraneum, 404
Trigonella, 62
Triptic soy broth (TSB), 160
Triticum aestivum (wheat). *See* Wheat *(Triticum aestivum)*
Trophic versus saprobic and pathogenic states, 527
Tropical plantation crops, 299
Tropics, agriculture in, 398, 418-419
Truffles, 121, 243, 498, 499, 529
Trypan blue, 486
Tryptic soy agar (TSA), 36
Tryptophan, 255
Tuber, 248, 511
Tuber magnatum, 498
Tuber melanosporum, 519, 520, 529
Tuber uncinatum, 519

Ultraviolet (UV) spectra, 482
Ultraviolet light, 464
Ultraviolet-B (UVB) radiation, 424
United Nations Educational, Scientific, and Cultural Organization (UNESCO), 186

United Nations Environment Programme (UNEP), 186
United States, agriculture in, 419
Urea, 33, 100, 461
Urea super granules (USG), 103
Urease, 104, 440
Urticaceae, 497
Uruguay, agriculture in, 420-421, 422, 424
USSR, former, agriculture in, 107

Vaccinum lichens, 503
Vegetable crops, 299
Vegetable oils, 525
Vegetational cover, 203
Velvet bean, 194
Vermiculite, 11, 364, 366, 368, 377, 405, 524
Verticillium, 294
Verticillium dahliae, 333
Vesicular arbuscular mycorrhizae (VAM), 246, 253, 290, 395, 498
Vetch *(Vicia sativa),* 194
Vicia faba (faba bean), 70, 193, 207, 212
Vicia villosa, 194
Vigna sinensis, 207
Vigna unguiculata (cowpea), 53, 57, 58, 194, 339
Vitamin B12, 438
Vitamins, 153, 255, 256, 257, 292, 401
Viticulture, 486
Vitis vinifera, 487

Wastes
  decomposing, 401, 433, 445
  off-farm, 397
  on-farm, 399
  organic, 223, 397, 401
  recycling, 399
  residues, 433, 445
Wastes *(continued)*
  and sewage as plant nutrition source, 1-2
  treatment of, 400
Water
  absorption, 117
  binding, 272
  channel proteins, 272, 283
  deficit (*see* Drought)
  holding capacity of soil, 439, 446
  logging, 445, 446
  pollution, 407
  potential, 277, 278
  relations, 274, 283, 293
  retention of soil, 277-278
  uptake and transfer of, 253-254, 271, 274-276f, 283, 293, 403, 487, 497
  use efficiency, 354
Weedicides, 92, 157
Weeds, 170, 300, 366, 368, 380, 381, 397
Westiellopsis, 94, 435, 436f, 440
Wheat *(Triticum aestivum)*
  and arbuscular mycorrhizal fungi, 226, 228, 297
  as carrier, 448
  and cyanobacteria, 419
  in developing nations, 185
  disease control in, 159-160, 163
  field trials with, 195
  increasing growth of, 7, 108
  leaves, 273, 277
  phosphorus solubilization and, 145
  population increase pressures, 186
  roots, 61
  rotation, 228
  seed, 193
  water potential, 277
  yield, increasing, 196
Wilcoxina mikolae, 247
Wildfires, 258
Wild-legume rhizobia
  in adverse conditions, 209-211, 213
  as biofertilizers, 204, 214
  nitrogen-fixation in, 206-208
  taxonomy and biodiversity of, 204-206

Willow, 243
Withania somnifera, 480, 481f, 488, 489t
Woody plants, 5, 55-56, 153, 302, 487
Woody shrubs, 260-261

Xanthomonas, 145
Xylanase, 482, 483

Yellow passion fruit, 378, 378f, 379

Zea mays. See Corn *(Zea mays);* Maize
Zeatin, 63, 255
Zinc (Zn)
- algae effect on, 440
- arbuscular mycorrhizal interaction with, 57, 58, 65, 402
- experiments, 68, 69
- fertilizer, 298
- plant uptake of, 53, 66, 228, 293, 402
- regulation of, 424
- sources of, 109, 298
- tolerance, 68

Zinc (Zn) sulfate, 68
Zizyphus nummularia, 477
Zygomycetes, 51, 291, 403, 483